AF882885

Jacek Koronacki, Zbigniew W. Raś, Sławomir T. Wierzchoń, and Janusz Kacprzyk (Eds.)

Advances in Machine Learning I

Studies in Computational Intelligence, Volume 262

Editor-in-Chief
Prof. Janusz Kacprzyk
Systems Research Institute
Polish Academy of Sciences
ul. Newelska 6
01-447 Warsaw
Poland
E-mail: kacprzyk@ibspan.waw.pl

Further volumes of this series can be found on our homepage: springer.com

Vol. 242. Carlos Artemio Coello Coello, Satchidananda Dehuri, and Susmita Ghosh (Eds.)
Swarm Intelligence for Multi-objective Problems in Data Mining, 2009
ISBN 978-3-642-03624-8

Vol. 243. Imre J. Rudas, János Fodor, and Janusz Kacprzyk (Eds.)
Towards Intelligent Engineering and Information Technology, 2009
ISBN 978-3-642-03736-8

Vol. 244. Ngoc Thanh Nguyen, Rados law Piotr Katarzyniak, and Adam Janiak (Eds.)
New Challenges in Computational Collective Intelligence, 2009
ISBN 978-3-642-03957-7

Vol. 245. Oleg Okun and Giorgio Valentini (Eds.)
Applications of Supervised and Unsupervised Ensemble Methods, 2009
ISBN 978-3-642-03998-0

Vol. 246. Thanasis Daradoumis, Santi Caballé, Joan Manuel Marquès, and Fatos Xhafa (Eds.)
Intelligent Collaborative e-Learning Systems and Applications, 2009
ISBN 978-3-642-04000-9

Vol. 247. Monica Bianchini, Marco Maggini, Franco Scarselli, and Lakhmi C. Jain (Eds.)
Innovations in Neural Information Paradigms and Applications, 2009
ISBN 978-3-642-04002-3

Vol. 248. Chee Peng Lim, Lakhmi C. Jain, and Satchidananda Dehuri (Eds.)
Innovations in Swarm Intelligence, 2009
ISBN 978-3-642-04224-9

Vol. 249. Wesam Ashour Barbakh, Ying Wu, and Colin Fyfe
Non-Standard Parameter Adaptation for Exploratory Data Analysis, 2009
ISBN 978-3-642-04004-7

Vol. 250. Raymond Chiong and Sandeep Dhakal (Eds.)
Natural Intelligence for Scheduling, Planning and Packing Problems, 2009
ISBN 978-3-642-04038-2

Vol. 251. Zbigniew W. Ras and William Ribarsky (Eds.)
Advances in Information and Intelligent Systems, 2009
ISBN 978-3-642-04140-2

Vol. 252. Ngoc Thanh Nguyen and Edward Szczerbicki (Eds.)
Intelligent Systems for Knowledge Management, 2009
ISBN 978-3-642-04169-3

Vol. 253. Roger Lee and Naohiro Ishii (Eds.)
Software Engineering Research, Management and Applications 2009, 2009
ISBN 978-3-642-05440-2

Vol. 254. Kyandoghere Kyamakya, Wolfgang A. Halang, Herwig Unger, Jean Chamberlain Chedjou, Nikolai F. Rulkov, and Zhong Li (Eds.)
Recent Advances in Nonlinear Dynamics and Synchronization, 2009
ISBN 978-3-642-04226-3

Vol. 255. Catarina Silva and Bernardete Ribeiro
Inductive Inference for Large Scale Text Classification, 2009
ISBN 978-3-642-04532-5

Vol. 256. Patricia Melin, Janusz Kacprzyk, and Witold Pedrycz (Eds.)
Bio-inspired Hybrid Intelligent Systems for Image Analysis and Pattern Recognition, 2009
ISBN 978-3-642-04515-8

Vol. 257. Oscar Castillo, Witold Pedrycz, and Janusz Kacprzyk (Eds.)
Evolutionary Design of Intelligent Systems in Modeling, Simulation and Control, 2009
ISBN 978-3-642-04513-4

Vol. 258. Leonardo Franco, David A. Elizondo, and José M. Jerez (Eds.)
Constructive Neural Networks, 2009
ISBN 978-3-642-04511-0

Vol. 259. Kasthurirangan Gopalakrishnan, Halil Ceylan, and Nii O. Attoh-Okine (Eds.)
Intelligent and Soft Computing in Infrastructure Systems Engineering, 2009
ISBN 978-3-642-04585-1

Vol. 260. Edward Szczerbicki and Ngoc Thanh Nguyen (Eds.)
Smart Information and Knowledge Management, 2009
ISBN 978-3-642-04583-7

Vol. 261. Nadia Nedjah, Leandro dos Santos Coelho, and Luiza de Macedo de Mourelle (Eds.)
Multi-Objective Swarm Intelligent Systems, 2009
ISBN 978-3-642-05164-7

Vol. 262. Jacek Koronacki, Zbigniew W. Raś, Sławomir T. Wierzchoń, and Janusz Kacprzyk (Eds.)
Advances in Machine Learning I, 2010
ISBN 978-3-642-05176-0

Foreword

Professor Richard S. Michalski passed away on September 20, 2007. Once we learned about his untimely death we immediately realized that we would no longer have with us a truly exceptional scholar and researcher who for several decades had been influencing the work of numerous scientists all over the world - not only in his area of expertise, notably machine learning, but also in the broadly understood areas of data analysis, data mining, knowledge discovery and many others. In fact, his influence was even much broader due to his creative vision, integrity, scientific excellence and exceptionally wide intellectual horizons which extended to history, political science and arts.

Professor Michalski's death was a particularly deep loss to the whole Polish scientific community and the Polish Academy of Sciences in particular. After graduation, he began his research career at the Institute of Automatic Control, Polish Academy of Science in Warsaw. In 1970 he left his native country and hold various prestigious positions at top US universities. His research gained impetus and he soon established himself as a world authority in his areas of interest – notably, he was widely considered a father of machine learning.

His contacts with the Polish scientific community were very close over all the years; in the last couple of years he was an affiliate scientist at the Institute of Computer Science, Polish Academy of Sciences, Warsaw. This relation culminated some years ago with his election to the rank of Foreign Member of the Polish Academy of Sciences, a distinction granted to only a small number of world-wide best scientists, including numerous Nobel Prize and other prestigious awards winners.

Professor Michalski was one of those active members of the Polish Academy of Sciences who were always interested in solving whatever problems we had, always ready to help us in shaping the research policy of the Academy and discuss with us all difficult issues that are these days unavoidable in any large and prestigious research organization with so many strong links to science worldwide. He was always ready to offer us his deep understanding and scholarly vision of the future of the human scientific endeavor. As President of the Polish Academy of Sciences I sense very personally an enormous loss coming from no longer being able to ask for his opinion and advice.

I wish to congratulate the editors of these scholarly volumes, Professors Jacek Koronacki, Zbigniew Raś, Sławomir T. Wierzchoń and Janusz Kacprzyk, for their initiative to pay the tribute to the memory of Professor Michalski. Having known him for many years they realized that the best way to honor his life achievements would be to prepare a collection of high quality papers on topics broadly perceived as Professor Michalski's main interest and to present *in memoriam* volumes of the contributions written by those who had the luck to be his friends or, at least, to meet him on various occasions. I am really impressed that so many prominent authors have accepted the invitation and I thank all of them most deeply.

I believe the memory of Professor Richard S. Michalski should remain with us for ever. Hopefully, these volumes will contribute to reaching this objective in the most appropriate and substantial way.

Professor Michał Kleiber
President
Polish Academy of Sciences

Preface

This is the first volume of a large two-volume editorial project we wish to dedicate to the memory of the late Professor Ryszard S. Michalski who passed away in 2007. He was one of the fathers of machine learning, an exciting and relevant, both from the practical and theoretical points of view, area in modern computer science and information technology. His research career started in the mid-1960s in Poland, in the Institute of Automation, Polish Academy of Sciences in Warsaw, Poland. He left for the USA in 1970, and since then had worked there at various universities, notably, at the University of Illinois at Urbana – Champaign and finally, until his untimely death, at George Mason University. We, the editors, had been lucky to be able to meet and collaborate with Ryszard for years, indeed some of us knew him when he was still in Poland. After he started working in the USA, he was a frequent visitor to Poland, taking part at many conferences until his death. We had also witnessed with a great personal pleasure honors and awards he had received over the years, notably when some years ago he was elected Foreign Member of the Polish Academy of Sciences among some top scientists and scholars from all over the world, including Nobel prize winners.

Professor Michalski's research results influenced very strongly the development of machine learning, data mining, and related areas. Also, he inspired many established and younger scholars and scientists all over the world.

We feel very happy that so many top scientists from all over the world agreed to pay the last tribute to Professor Michalski by writing papers in their areas of research. These papers will constitute the most appropriate tribute to Professor Michalski, a devoted scholar and researcher. Moreover, we believe that they will inspire many newcomers and younger researchers in the area of broadly perceived machine learning, data analysis and data mining.

The papers included in the two volumes, Machine Learning I and Machine Learning II, cover diverse topics, and various aspects of the fields involved. For convenience of the potential readers, we will now briefly summarize the contents of the particular chapters.

Part I, "Introductory Chapters", contains – first of all a more general chapter which presents the most notable research interests and accomplishments of Professor Richard S. Michalski, and his inspiration and impact of the field of machine learning, data mining, and related areas. The other chapters may be viewed as being directly influenced by Professor Michalski's ideas, and employ tools and techniques that may be viewed as straightforward extensions and generalizations of his works.

- **Janusz Wojtusiak and Kenneth A. Kaufman** ("Ryszard S. Michalski: The Vision and Evolution of Machine Learning") discuss some basic elements and aspects of the vision and contributions of Ryszard S. Michalski who has pioneered so many areas and methods of machine learning. The authors offer a brief summary of what they believe are the most important aspects of Professor Michalski's research, and present the vision of machine learning that he communicated to them personally on multiple occasions. The most important topics mentioned in this chapter are: natural induction, knowledge mining, AQ learning, conceptual clustering, VL1 and attributional calculus, constructive induction, the learnable evolution model, inductive databases, methods of plausible reasoning, and the inferential theory of learning.

- **Marcus A. Maloof** ("The AQ Methods for Concept Drift") deals with the concept drift which occurs when the target concept that a learner must acquire changes over time. It is present in applications involving user preferences (e.g., calendar scheduling) and adversaries (e.g., spam detection). The author first mentions his earlier works, based on Michalski's AQ algorithm, and more recent ones based on ensemble methods, as well as some implementations of several methods that other researchers have proposed. In the chapter, the author provides a survey of his results obtained since the mid-1990s using the Stagger concepts and learning methods for concept drift. The author's methods based on the AQ algorithm and the ensemble methods, and the methods of other researchers are also examined. It is shown that, for the Stagger concepts, the dynamic weighted majority with an incremental algorithm for producing decision trees as the base learner and the systems based on the AQ11 algorithm (notably an AQ11 system with partial instance memory and Widmer and Kubat's window adjustment heuristic) achieve the best performance.
- **Krzysztof J. Cios and Łukasz A. Kurgan** ("Machine Learning Algorithms Inspired by the Work of Richard Spencer Michalski") first define the field of inductive machine learning and then describe Michalski's basic AQ algorithm. Next, the authors' two machine learning algorithms, the CLIP4: a hybrid of rule and decision tree algorithms, and the DataSqeezer: a rule algorithm, are discussed. The development of the latter two algorithms was inspired to a large degree by Michalski's seminal paper on inductive machine learning from1969 which was one of first attempts devise inductive machine learning algorithms that generate rules.
- **Janusz Kacprzyk and Grażyna Szkatuła** ("Inductive Learning: A Combinatorial Optimization Approach") propose an improved inductive learning method to derive classification rules correctly describing (at least) most of the positive examples and do not correctly describe (at least) most of the negative examples. First, a pre-analysis of data is performed to assign higher weights to those values of attributes which occur more often in the positive than in the negative examples. The inductive learning problem is represented as a modification of the set covering problem which is solved by an integer programming based algorithm using elements of a greedy algorithm or a genetic algorithm, for efficiency. The results are very encouraging and are illustrated on a thyroid cancer data set.

Part II, "General Issues", contains contributions which deal with more general issues related to machine learning, data mining, knowledge discovery, and other topics relevant to the volumes and to be considered in the constituting contributions.

- **Pinar Donmez and Jaime G. Carbonell** ("From Active to Proactive Learning Methods") consider the situation, which is common in machine learning, that unlabeled data abounds, but expert-generated labels are scarce. Therefore, the question is how to select the most informative instances to label, as some labeled examples can be much more useful than others in training classifiers to minimize errors. Active learning is the process of selecting the best examples to label in order to improve classifier performance, where "best" is typically defined as minimizing the loss function of the classifier or otherwise directly minimizing its estimated classification error. Alternatives to active learning attempt to use unlabeled data directly in helping to train classifiers, without obtaining additional labels. These efforts range from transductive support vector machines to co-training and other semi-supervised methods, but they cannot in general rival active learning. Combining ideas from both semi-supervised and active learning remains a largely unexplored area for future research.
- **Nada Lavrač, Johannes Fürnkranz, and Dragan Gamberger** ("Explicit Feature Construction and Manipulation for Covering Rule Learning Algorithms") deal with features as the main rule building blocks of rule learning algorithms. In contrast to a common practice in classification rule learning, it is argued in the chapter that the separation of feature construction and rule construction processes has a theoretical and practical justification. An explicit usage of features enables a unifying framework of both propositional and relational rule learning and procedures for feature construction in both types of domains are presented and analyzed. It is demonstrated that the presented procedure for constructing a set of simple features has the property that the resulting feature set enables the construction of complete and consistent rules whenever possible, and that the set does not include obviously irrelevant features. It is also shown that feature relevancy may improve the effectiveness of rule learning. The concept of relevancy in the coverage space is illustrated, and it is shown that the transformation from the attribute to the feature space enables a novel, theoretically justified way of handling unknown attribute values. The same approach makes it possible that the estimated imprecision of continuous attributes can be taken into account, resulting in the construction of features that are robust to attribute imprecision.
- **Lisa Torrey, Jude Shavlik, Trevor Walker, and Richard Maclin** ("Transfer Learning via Advice Taking") describe a transfer method in which a reinforcement learner analyzes its experience in the source task and learns rules to use as an advice in the target task. The rules, which are learned via inductive logic programming, describe the conditions under which an action is successful in the source task. The advice taking algorithm used in the target task allows a reinforcement learner to benefit from rules even if they are imperfect. A mapping, which is provided by the human being, describes the alignment between the source and target tasks, and may also include advice about the differences between them. Using three tasks in the RoboCup simulated soccer domain, the authors demonstrate that this transfer method can speed up reinforcement learning substantially.

Part III, "Classification and Beyond", deals with many aspects, methods, tools and techniques related to broadly perceived classification which is a key issue in many areas, notably those related to the topics of the present volume.

- **Pavel Brazdil and Rui Leite** ("Determining the Best Classification Algorithm with Recourse to Sampling and Metalearning") discuss the determination of the best classification algorithm which is the best one for a given task. A method is described which relies on relatively fast pairwise comparisons involving two algorithms. This method is based on a previous work and exploits sampling landmarks, that is, information about learning curves besides classical data characteristics. One key feature of this method is an iterative procedure for extending the series of experiments used to gather new information in the form of sampling landmarks. Metalearning plays also a vital role. The comparisons between various pairs of algorithms are repeated and the result is represented in the form of a partially ordered ranking. Evaluation of the approach is done by comparing the partial order of algorithms that has been predicted to the partial order representing the supposedly correct result. The results of the evaluation show that the method has good performance and could be of help in practical applications.
- **Michelangelo Ceci, Annalisa Appice, and Donato Malerba** ("Transductive Learning for Spatial Data Classification") are concerned with learning classifiers to classify spatial data. Accordingly, they address such issues as: heterogeneity of spatial objects, an implicit definition of spatial relationships among objects, spatial autocorrelation and the abundance of unlabelled data which potentially convey a large amount of information. The first three issues are due to the inherent structure of spatial units of analysis which can be easily accommodated if a (multi-)relational data mining approach is considered. The fourth issue demands the adoption of a transductive setting which aims at making predictions for a given set of unlabelled data. Transduction is also motivated by the contiguity of the concept of positive autocorrelation which typically affects spatial phenomena, with the smoothness assumption which characterizes the transductive setting. The authors investigate a relational approach to spatial classification in a trans ductive setting. Computational solutions to the main difficulties met in this approach are presented. In particular, a relational upgrade of the naïve Bayes classifier is proposed as a discriminative model, an iterative algorithm is designed for the transductive classification of unlabelled data, and a distance measure between relational descriptions of spatial objects is defined in order to determine the k-nearest neighbors of each example in the dataset. Computational solutions have been tested on two real-world spatial datasets. The transformation of spatial data into a multi-relational representation, and experimental results, are reported and commented.
- **Krzysztof Dembczyński, Wojciech Kotłowski, and Roman Słowiński** ("Beyond Sequential Covering – Boosted Decision Rules") consider boosting, or forward stagewise additive modeling, a general induction procedure which appears to be particularly efficient in binary classification and regression tasks. When boosting is applied to induction of decision rules, it can be treated as generalization of sequential covering because it approximates the solution of the prediction task by sequentially adding new rules to the ensemble without

adjusting those that have already entered the ensemble. Each rule is fitted by concentrating on examples which were the hardest to classify correctly by the rules already present in the ensemble. The authors we present a general scheme for learning an ensemble of decision rules in boosting framework, using different loss functions and minimization techniques. This scheme, called ENDER, is covered by such algorithms as SLIPPER, LRI and MLRules. These algorithms are compared on benchmark data.

- **Matti Saarela, Tapio Elomaa, and Keijo Ruohonen** ("An Analysis of Relevance Vector Machine Regression") consider the relevance vector machine, a Bayesian framework for learning sparse regression models and classifiers, and present two kinds of analytic results for the regression setting. First, a full characterization of the behavior of the relevance vector machine is analytically determined when the columns of the regression matrix are orthogonal, and, second, some results concerning the scale and rotation invariance of the relevance vector machine are given. Practical implications of the results obtained are considered and a scenario is presented in which the results obtained can be used to detect potential weakness in the relevance vector machine framework.
- **Zbigniew W. Raś, Agnieszka Dardzińska, and Wenxin Jiang** ("Cascade Classifiers for Hierarchical Decision Systems") consider hierarchical classifiers in which the classification occurs first on a low-level with highly specific pieces of input data, and then the individual pieces of data are combined systematically and classified on a higher level iteratively until one output is produced, the overall classification of the data. The authors follow a controlled devise type of an approach. The initial group of classifiers is trained using all objects in an information system partitioned by values of the decision attribute at all of its granularity levels. Only values of the highest granularity level are used to split the information system into information sub-systems each one of which is built by selecting objects of the same decision value. These sub-systems are used for training new classifiers at all granularity levels of its decision attribute. Next, the authors split each sub-system further by sub-values of its decision value. The obtained tree structure with groups of classifiers assigned to each of its nodes is then called a cascade classifier. Given an incomplete information system with a hierarchical decision attribute, the problem of training classifiers describing the values of the decision attribute at its lowest granularity level is considered. The MIRAI database of music is used in computational examples.
- **Gisele L. Pappa and Alex A. Freitas** ("Creating Rule Ensembles from Automatically-Evolved Rule Induction Algorithms") propose a new evolutionary algorithm based method for creating an ensemble of rule sets consisting of two stages. First, an evolutionary algorithm (more precisely, a genetic programming algorithm) is used to automatically create complete rule induction algorithms. Secondly, the automatically evolved rule induction algorithms are used to produce rule sets that are then combined into an ensemble. In this second stage, they investigate the effectiveness of two different approaches for combining the votes of all rule sets in the ensemble and two different approaches for selecting which subset of evolved rule induction algorithms should be used to produce the rule sets that will be combined into an ensemble.

- **Ugo Galassi, Marco Botta, and Lorenza Saitta** ("Structured Hidden Markov Model versus String Kernel Machines for Symbolic Sequence Classification") focus in their paper on the specific task of classifying and tagging symbolic sequences, by introducing a characterization approach, based on Structured Hidden Markov Models, and compare its performance with a widely used discriminative approach, i.e. kernel machines. This task is particularly relevant to several applications in many fields, such as molecular biology, web log analysis, network traffic analysis, user profiling, etc. In order to assess the validity of the proposed approach, the authors design an artificial benchmark such that the regularities to be discovered are well known and allow for a controlled evaluation of the real capabilities the learning algorithms investigated. The obtained results allow to point out the major advantages and weaknesses of the investigated approaches in the specific classification task addressed.

Part IV, "Soft Computing", deals with various problems, issues and aspects related to broadly perceived machine learning, data analysis, data mining and the like for which some tools and techniques belonging to the area of soft computing have been proved effective and efficient. Notably, the papers in this part utilize various tools from fuzzy sets and rough sets theory to devise measures that can help quantify and monitor various tasks and processes related to the topics cover by the volumes, and to devise effective and efficient tools and techniques.

- **Ronald R. Yager** ("Partition Measures for Data Mining") investigates a number of measures associated with partitions. The first of these is congruence measures, which are used to calculate the similarity between two partitions. He provides a number of examples of this type of measure. Another class of measures investigated are prognostication measures. These measures, closely related to a concept of containment between partitions, are useful in indicating how well knowledge of an objects class in one partition predicts its class in a second partitioning. Finally, the author introduces a measure of the non-specificity of a partition which measures a feature of a partition related to the generality of the constituent classes of the partition. A common task in machine learning is to develop rules that allow us to predict the class of an object based upon the value of some features of the object. The more narrowly we categorize the features in the rulet, the better we can predict an objects classification. However, counterbalancing this is the fact that to many narrow feature categories are difficult for human experts to cognitively manage, and this introduces a fundamental issue in data mining. The author shows how the combined use of his measures of prognostication and non-specificity allow us to navigate this issue.
- **Jens Christian Hühn and Eyke Hüllermeier** ("An Analysis of the FURIA Algorithm for Fuzzy Rule Induction") elaborate on a novel fuzzy rule-based classification method, called FURIA, which stands for "Fuzzy Unordered Rule Induction Algorithm", that has recently been developed as an extension of the well-known RIPPER algorithm. It learns fuzzy rules instead of conventional rules and unordered rule sets instead of rule lists, and also - to deal with uncovered examples - it makes use of an efficient rule stretching method. First experimental results have shown that FURIA significantly outperforms the original RIPPER

in terms of classification accuracy. Elaborating on the advantages of a fuzzy approach, the influence of rule fuzzification on the performance of the algorithm is analyzed. Moreover, going beyond the conventional classification problem, the authors investigate the performance of FURIA in the context of bipartite ranking, in which a fuzzy approach appears to be even more appealing.

- **Jerzy W. Grzymała-Busse and Witold J. Grzymała-Busse** ("Increasing Incompleteness of Data Sets – A Strategy for Inducing Better Rule Sets") are concerned with the verification of the following hypothesis: for some complete (i.e., without missing attribute vales) data sets it is possible to induce better rule sets (in terms of an error rate) by increasing incompleteness (i.e., removing some existing attribute values) of the original data sets. Detailed results of experiments are presented showing that rule sets induced from some incomplete data sets are significantly better than the rule set induced from the original (complete) data set, with the significance level of the 5% two-tailed test. Additionally, criteria for inducing better rules by increasing incompleteness are discussed.
- **Yaile Caballero, Rafael Bello, Leticia Arco, Maria García, and Enislay Ramentol** ("Knowledge Discovery Using Rough Set Theory") consider the use of Rough Set Theory (RST) for broadly perceived data analysis, to be more specific show the possibility of using this theory to generate a priori knowledge about a dataset. A proposal is developed for previous characterization of training sets, using RST estimation measurements. This characterization offers an assessment of the quality of data in order to use them as a training set in machine learning techniques. The proposal is experimentally studied using well known international databases and some known classifiers such as MLP, C4.5 and k-NN. The use of the approach proposed yields satisfactory results.
- **A.E. Hassanien, H. Al-Qaheri, V. Snášel, and J.F. Peters** ("Machine Learning Techniques for Prostate Ultrasound Image Diagnosis") consider the estimation of prostate location and volume which is essential in determining a dose plan for ultrasound-guided brachytherapy, a common prostate cancer treatment. Since the manual segmentation is difficult, time consuming and prone to variability, the authors present a machine learning scheme, employing a combination of fuzzy sets, wavelets and rough sets, for analyzing prostrate ultrasound images in order diagnose prostate cancer. To address the image noise problem they first utilize an algorithm based on type-II fuzzy sets to enhance the contrast of the ultrasound image. This is followed by performing a modified fuzzy c-mean clustering algorithm in order to detect the boundary of the prostate pattern. Then, a wavelet features are extracted and normalized, followed by application of a rough set analysis for discrimination of different regions of interest to determine whether they represent cancer or not. The experimental results obtained, show that the overall classification accuracy offered by the employed rough set approach is high compared with other machine learning techniques including decision trees, discriminant analysis, rough neural networks, and neural networks.
- **Marek Kowal and Józef Korbicz** ("Segmentation of breast cancer fine needle biopsy cytological images using fuzzy clustering") presents some research results concerning an automatic diagnosis system of breast cancer based on cytological images of FNB (Fine Needle Biopsy). The work concentrates on the

image segmentation phase, which is employed to find nucleus in cytological images. The accuracy and correctness of the image segmentation algorithm is a critical factor for successful diagnosis due to the fact that case classification is based on morphometrical features extracted form segmented nucleus. The presented approach to image nucleus segmentation is based on the FCMS (Fuzzy C-means with shape function) clustering algorithm. The novelty of the presented approach is that the clustering process is conducted in color space but the searched objects must have an arbitrarily defined shape. Simulations and experimental results are included to illustrate the effectiveness of the proposed approach.

Part V, "Machine Learning for Robotics", deals with various aspects of learning, object recognition and related issues focused on robotics.

- **Reinaldo A.C. Bianchi, Arnau Ramisa, and Ramón López de Mántaras** ("Automatic Selection of Object Recognition Methods using Reinforcement Learning") present a method that uses reinforcement learning to decide which algorithm should be used to recognize objects seen by a mobile robot in an indoor environment, based on simple attributes extracted on-line from the images, such as mean intensity and intensity deviation. Usually, selecting which algorithms should be used by a mobile robot computer vision system is a decision made a priori by the system developer, based on past experience and intuition, without taking into account information that can be found in the images and in the visual process. According to the authors, two state-of-the-art object recognition algorithms can be selected to perform the task they posed: the constellation method proposed by Lowe together with its interest point detector and descriptor, the scale-invariant feature transform, and the vocabulary tree approach of Nistér and Stewénius. A set of empirical evaluations is conducted using a image database acquired with a mobile robot in an indoor environment, and results obtained show that the approach adopted is very promising.
- **Ivan Bratko** ("Comparison of Machine Learning for Autonomous Robot Discovery") considers autonomous robot discovery through experimentation in the robot's environment. The applicability of machine learning methods with respect to various levels of robot discovery tasks, from extracting simple laws among the observed variables, to discovering completely new notions that were never explicitly mentioned in the data directly, are discussed. First, some illustrative experiments in robot learning within the XPERO European project are shown. Then, criteria for a comparison of learning methods and a systematic list of types of learning or discovery tasks are dealt with, and the suitability of chosen machine learning methods for these tasks is considered.
- **Claude Sammut and Tak Fai Yik** ("Multistrategy Learning for Robot Behaviours") try to overcome the problem that the pure reinforcement learning does not scale well to domains with many degrees of freedom and particularly to continuous domains. They introduce a hybrid method in which a symbolic planner constructs an approximate solution to a control problem, and then a numerical optimisation algorithm is used to refine the qualitative plan into an operational policy. The authors use the method developed to solve the problem of learning a stable walking gait for a bipedal robot, and .use their approach to illustrate the benefits of a multistrategy approach to robot learning.

Part VI, "Neural Networks and Other Nature Inspired Approaches", deals with the development and applications of various nature inspired paradigms, approaches and techniques. Notably, diverse aspects related to neural networks, evolutionary computation, artificial immune systems, etc. are considered, showing their potentials and applicability.

- **Joachim Diederich, Alan B. Tickle, and Shlomo Geva** ("Quo Vadis? Reliable and Practical Rule Extraction from Neural Networks") consider rule extraction from neural networks, first thoroughly reviewing the state-of-the-art and pointing to some of the obstacles, namely a lack of evaluation techniques in experiments and larger benchmark data sets. Then, the authors argue for the view that rule extraction from neural networks should be an interactive process which actively involves the user (domain expert). They conclude stating that the kind and magnitude of problems that can be solved with neural networks have recently changed significantly and, therefore, that the evaluation of rule extraction from neural networks should now be based on large scale problems. Of interest are now problems which involve millions of pattern vectors and domains where there are many underlying rules to be discovered.
- **Vladimir G. Red'ko and Danil V. Prokhorov** ("Learning and Evolution of Autonomous Adaptive Agents") consider a model of evolving populations of self-learning agents and analyze the interaction between learning and evolution. They consider agent-brokers that predict stock price changes and use these predictions for selecting actions. Each agent is equipped with a neural network adaptive critic design for behavioral adaptation. We discuss three cases in which either learning, or evolution, or both, are active in our model. The authors show that the Baldwin effect can be observed in the model, viz., an originally acquired adaptive policy of best agent-brokers becomes inherited over the course of the evolution. Additionally, they analyze the influence of neural network structure of adaptive critic design on learning processes.
- **A. Frolov, D. Húsek, I. P. Muraviev, and P. Y. Polyakov** ("Learning and Unlearning in Hopfield-like Neural Network Performing Boolean Factor Analysis") consider some issues related to factor analysis based on the assumption that signals are random combinations of latent factors. The goal is to find factor representation in the signal space (factor loadings) and the contributions of factors into the original signals (factor scores). Boolean factor analysis, introduced by the authors and based on the Hopfield-like neural network, is dealt with. Due to the Hebbian learning rule, neurons of a factor become connected more tightly than other neurons and hence factors can be revealed as attractors of the network dynamics by random search. It is shown that two global spurious attractors may appear and become dominant, thus preventing the search for successful factors. To eliminate these attractors there is proposed a special unlearning procedure, which is shown to be equivalent to the effect of a special inhibitory neuron included into the common Hopfield network. The efficiency of the proposed learning and unlearning rules in Boolean factor analysis are investigated analytically and by computer simulations.

We are happy that we initiated and now are able to finalize this exceptional two-volume editorial project the scope and size of which is extraordinary. First, and most importantly, we have gathered astonishingly many eminent researchers and scholars from all parts of the world who have been actively working on a broadly perceived area of machine learning, data analysis, data mining, knowledge discovery, etc. They have contributed with great papers in which a synergistic and inspiring combination of a comprehensive state of the art material, deep technical analysis, novel applications, and much vision and new look on the past, present and future of the fields have been included. Second, the sheer size of this editorial project is exceptional, but the readers will obtain a rare view of what is the best, most original and promising in the areas. This all is certainly the best tribute that our research community can pay to Professor Ryszard Michalski. He had been inspiring all of us for so many years, and many of us owe him so much because his works, and contacts with him, have shaped our careers and maybe even life.

In an email to us, sent upon submitting her chapter to our editorial project, Professor Katharina Morik summarized it all in an unsurpassed way, albeit based on just one personal recollection: "*Ryszard would not want to have stories told and memories shared but just to see scientific work. Hence, I abstained from expressing my feelings [in the chapter submitted]. In 1986, I was just an unknown scientist in Berlin when inviting him to a workshop on knowledge representation and machine learning, admitting that I could not pay any royalty. To him, only the subject mattered. He came, and I learned a lot from him. He was kind of a godfather to machine learning those days, but he was open to the inductive logic programming approach which we were developing then. This gave me confidence that in science you don't need to have connections, be famous, or offer a lot of money: just work hard, discuss, and share the enthusiasm, that's it. Ryszard has never lost this attitude*"..

We wish to thank all the authors for their excellent contributions and an efficient collaboration in this huge and demanding editorial project. Special thanks are also due to Dr. Tom Ditzinger, Senior Editor, and Ms. Heather King, both from Engineering/Applied Sciences of Springer, who have provided much encouragement and support.

Warsaw
August 2009

Jacek Koronacki
Zbigniew W. Raś
Sławomir T. Wierzchoń
Janusz Kacprzyk

Table of Contents

Part I: Introductory Chapters

Part II: General Issues

Part III: Classification and Beyond

Part IV: Soft Computing

Part V: Machine Learning for Robotics

Part VI: Neural Networks and Other Nature Inspired Approaches

Part I
Introductory Chapters

Ryszard S. Michalski: The Vision and Evolution of Machine Learning

Janusz Wojtusiak and Kenneth A. Kaufman

Machine Learning and Inference Laboratory
George Mason University
Fairfax, VA 22030, USA
jwojt@mli.gmu.edu, ken.kaufman@gmail.com

Abstract. Ryszard S. Michalski was an outstanding scientist who dedicated his life to research and discovery. He pioneered so many areas and methods of machine learning that it is not possible to describe them properly in one chapter. Thus, we present a brief summary of what we believe are the most important aspects of his research, and present the vision of machine learning that he communicated to us on multiple occasions. The most important topics mentioned in this chapter are: natural induction, knowledge mining, AQ learning, conceptual clustering, VL1 and attributional calculus, constructive induction, the learnable evolution model, inductive databases, methods of plausible reasoning, and the inferential theory of learning.

Keywords: Machine Learning, Natural Induction, Concept Learning, Conceptual Clustering, Knowledge Mining.

This chapter is dedicated to Dr. Ryszard S. Michalski, our mentor and friend, who opened for us the wonderful world of machine learning.

1 Introduction

The very fast growth of machine learning and related disciplines is attributed to its great importance and applicability in almost all aspects of modern society. Learning is considered one of the most important aspects of intelligent systems, thus work in this area has been attempted since the very beginning of the field of artificial intelligence over half a century ago. Dr. Ryszard S. Michalski was among the most significant contributors to machine learning. His involvement in the formation of the field, including organization of the first workshops and conferences on machine learning, the first books in the area, including [29][30], is only a small part of the overall contributions.

In this chapter we overview selected research areas originated by Dr. Michalski, and present the vision and directions of the field presented to us over many years of work with him. We understand that it is not possible to summarize four decades of intense research and over 350 publications in one chapter; thus, we emphasize the newest aspects of his research, and those we believe are the most significant.

J. Koronacki et al. (Eds.): Advances in Machine Learning I, SCI 262, pp. 3–22.
springerlink.com © Springer-Verlag Berlin Heidelberg 2010

We start this chapter with an overview of natural induction, an approach to inductive learning that puts a human recipient of learned knowledge in a central position. A key part of natural induction concerning representation of knowledge is described in Section 3. We also discuss learning methods and paradigms originated by Dr. Michalski, including the learnable evolution model, and inductive databases. We conclude this chapter by describing his selected contributions to theory of machine learning, and describe software built to teach and entertain.

2 Natural Induction

Before describing selected methods and algorithmic contributions of Ryszard Michalski to the field of machine learning, let us start by describing *natural induction*, which can be viewed as a central aspect of his research over the past decades. The actual name "natural induction" did not appear until the 1990s, but its principles can be found in different forms back at least to the early 1970s. The name natural induction consisting of two parts suggests that induction is done in a natural way. The induction, or more precisely inductive reasoning or learning, is a falsity preserving reasoning process in which generated knowledge is generated from observations/examples and background knowledge. The word *natural* relates to the knowledge resulting from learning that is in a form easy to understand by people. Thus, natural induction refers to an inductive learning process whose results are designed to be natural to people.

There are several forms of knowledge that are natural to people, including natural language descriptions [43], easy to interpret rules, graphical representations, relatively small decision trees, Bayesian networks, etc. These forms of knowledge are considered to be "transparent box", in contrast to "black box" representations that may provide very good predictions, but be hard to understand. The latter can be exemplified by neural networks, random forests, support vector machines, and many other models often used in machine learning.

In order to present results of reasoning in forms easily understandable, one can either transform the learned knowledge, or learn directly in a language natural to people. While the former can be exemplified by all kinds of visualization methods, the latter requires reasoning directly in some language that corresponds directly to natural language statements. This and other issues regarding representation of knowledge are described in next section.

One important representation of knowledge that directly corresponds to natural language is an attributional rule, described in section 3, and exemplified in Figure 1. The rule can be paraphrased: "*The plan to do is to run experiments, if it is a weekend, the weather is rainy and cold, the available workstation has a clock speed at least 2GHz, and the available laboratory is lab1 or lab2, except for when there is a server malfunction.*" The value "weekend" is a higher-level value of a structured (hierarchical) attribute "Day", the attribute "Weather" is a *compound attribut*e that takes a conjunction of values. The pairs of numbers within conditions represent numbers of positive and negative examples satisfying these conditions. The entire rule is satisfied by 27 positive and 1 negative example, as indicated by "p=27, n=1" in the rule's annotation, and its quality, defined by (6), is 0.9.

```
[Plan to do = run experiments]
    [Day = weekend: 38,131] &
    [Weather = rainy and cold: 22, 5] &
    [Available workstation clock speed >= 2GHz: 50,5] &
    [Available lab = lab1 or lab3: 43,57]
 √   [Server malfunction: 0,3] : p=27, n=1, Q=0.9
```

Fig. 1. An example of an attributional rule

Two graphical representations of knowledge developed by Dr. Michalski are *general logic diagrams* (GLD), and *concept association graphs* (CAG). In GLDs, each cell represents a combination of input attributes, and examples are marked within cells by their class [19]. Complexes (see section 3) are represented by rectangles. An example of a GLD with 19 examples and 2 complexes is presented in Figure 2. A CAG represents rules as a labeled graph with varying thicknesses of the links indicating the strength of the knowledge component represented by the link [11]. An example is presented in Figure 3.

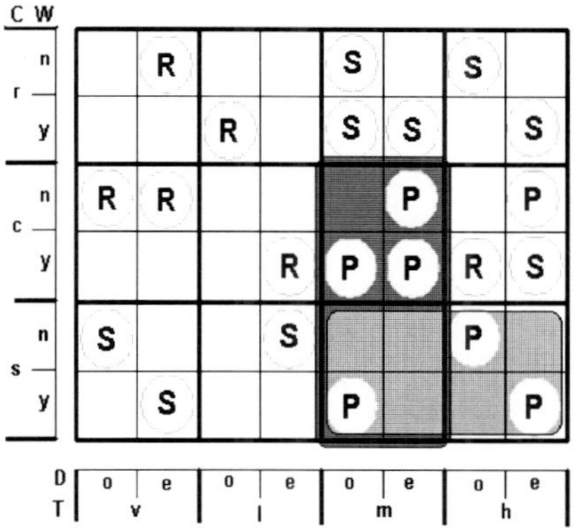

C – Condition, W – Wind, T – Temperature, D – Daytype, r – rain, c – cloudy, s – sunny, n – no, y – yes, v – very low, l – low, m – medium, h – high, o – workday, e – weekend, P – Play, R – Read, S - Shop

Fig. 2. A general logic diagram (reproduced from [34])

Natural induction is an important part of *knowledge mining*, an approach to data analysis that is used to derive high level concepts from data and background knowledge [13]. In contrast to traditional data mining, this approach is not limited to analyzing large amounts of data. To the contrary, knowledge mining can analyze both large and small amounts of data and, more importantly, use problem-oriented background knowledge. It can be viewed as a next step following in the evolution of data analysis and knowledge creation methods.

Natural induction requires human-oriented representation of knowledge and learning algorithms that can utilize power of that representation. In the next two sections we describe the creation of attributional calculus, a representation language of natural induction, and learning methods that can utilize this language.

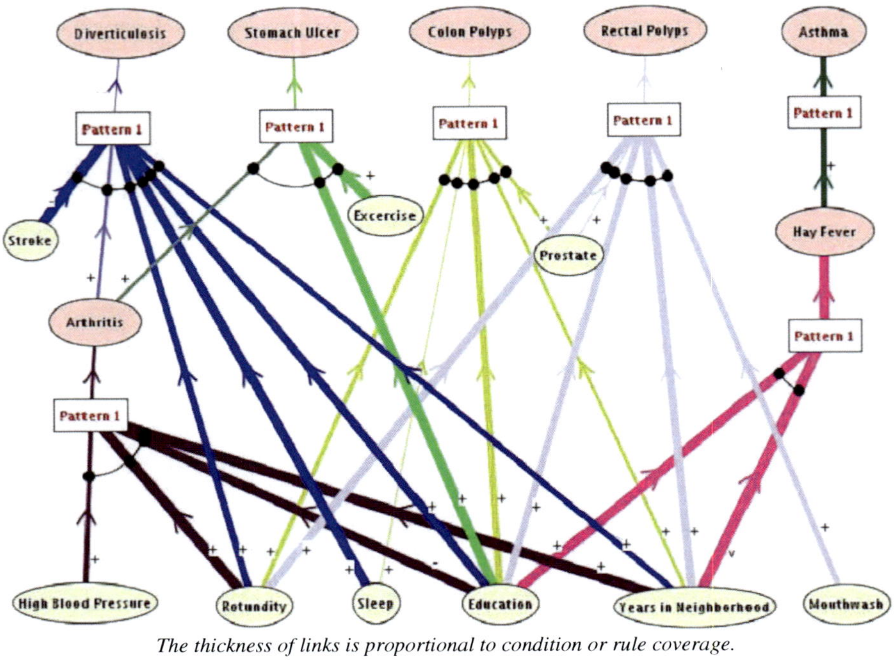

Fig. 3. A concept association graph representing seven patterns in a medical database (reproduced from [34])

3 Representation

Appropriate representation of knowledge and data is critical for successful application of machine learning methods. To go along with his nascent A^q algorithm, Michalski [20] developed the VL1 knowledge representation language, a variable-valued logic with an expressive power somewhere between those of propositional and predicate calculus. The availability of such internal operators as range and internal disjunction made it suitable for representing compact, easily understood rulesets. Contrast that with, for example, the traditional-style decision trees that still persist to this day, in which several levels of nodes may be needed to subdivide a single continuous attribute into the ranges needed for the learned classifier.

But to take advantage of the power of such a representation, it was necessary to have available all data types that mirrored our natural cognitive experience – nominal, discrete-ordered, hierarchically structured, cyclic, continuous, set-valued, etc. – and be able to operate upon and reason with them as needed. Thus, as part of the AQ packages developed and distributed over the years (see section 4.1), the operators to handle the calculus of (most of) these data types at all stages – input, extension against, selection, trimming, and output – was built in.

This would be enough for many applications, but many was not enough. The representation of knowledge soon included annotations, so that it would be easy to access and work with and from the metaknowledge as well as the basic discovered facts.

Now it would be easy to access the support and confidence levels of different knowledge components, and use these to guide your data exploration process.

The VL1 language was subsequently extended to the VL2 representation language [16] that added the existential quantifier. Now the natural induction schema extended to structured descriptions, in which components of the whole and relations among them could be understandably expressed; such descriptions would be utilized in the INDUCE programs for learning structured descriptions (see section 4.2), the CLUSTER conceptual clustering programs (see section 4.5), and the SPARC program for qualitative sequence analysis, discovery and prediction (see section 4.3).

In 2004, Michalski presented a further extension of these languages, *Attributional Calculus* [26]. Attributional calculus further simplified representation of concepts we use naturally by introducing constructs for counting attributes (e.g., at least 3 of these 4 conditions must be present) and relationships among attributes (e.g., height < length + width). Another important idea formalized in attributional calculus concerns using unknown, not-applicable, and irrelevant meta-values. Although all three values appeared in different forms in the early work of Dr. Michalski, it is in attributional calculus where their semantics are first described together. Unknown meta-values, are the most popular, and represent situations in which a regular value exists, but it is not recorded for some reason. Not-applicable meta-values are used when the regular values do not exist. For example the value of an attribute describing pregnancy in male patients does not exist, thus it is not applicable. The irrelevant meta-value is used when a regular value exists, but it is considered not to be relevant to the current learning task. The two latter types of meta-values constitute problem background knowledge.

In attributional calculus, learned knowledge is represented in the form of *attributional rules,* which consist of *attributional conditions*. An attributional condition takes the form (1):

$$[L\ rel\ R{:}\ A] \tag{1}$$

where L is an attribute, an internal conjunction or disjunction of attributes, a compound attribute, or an expression; *rel* is one of =, >, <, ≤, ≥, :, or ≠; R is an attribute value, an internal disjunction of attribute values, an attribute, an internal conjunction of values of attributes that are constituents of a compound attribute, or an expression, and A is an optional annotation that may list |p| and |n| values for the condition, defined as the numbers of positive and negative examples, respectively, that satisfy the condition, the condition's consistency defined as |p|/(|p|+|n|), and other measures of conditions' quality. There are several different forms of attributional rules allowed by attributional calculus. Three important forms of attributional rules are (2) - (4).

$$\text{CONSEQUENT} \Leftarrow \text{PREMISE} \tag{2}$$

$$\text{CONSEQUENT} \Leftarrow \text{PREMISE} \llcorner \text{EXCEPTION} \tag{3}$$

$$\text{CONSEQUENT} \Leftarrow \text{PREMISE} \ulcorner \text{PRECONDITION} \tag{4}$$

where PREMISE, CONSEQUENT, EXCEPTION, and PRECONDITION are *complexes*, that is, conjunctions of attributional conditions. An EXCEPTION can also be an explicit list of examples that constitute exceptions to the rule. The rules (2) are interpreted that the CONSEQUENT is true whenever the PREMISE is true. The rules (3) are interpreted that the CONSEQUENT is true whenever the PREMISE is true, except for when the EXCEPTION is true. The rules (4) are interpreted that the CONSEQUENT is true whenever the PREMISE is true, provided that the PRECONDITION is true. Each rule may be optionally annotated with several parameters such as numbers of covered examples (positive and negative), the rule complexity etc.

Static representation used by *selective induction* methods (most inductive learning methods including rule learners, decision tree learners, Bayesian networks, etc.) did not seem sufficient for many applications. This sparked the development of *constructive induction*, in which a representation space is changed to better fit the machine learning problem being considered [22]. This change of representation space may include removing irrelevant attributes, adjusting discretization of existing attributes, including operations on hierarchies, and most importantly the construction of new attributes. The latter can be in the form of arithmetic or logic expressions, counting attributes (generalization of quantifiers), or some other special forms. A general classification of constructive induction methods include data-driven constrictive induction (DCI) in which modifications of the representation space are based on the analysis of input data, hypothesis-driven constructive induction (HCI) in which modifications of the representation space are based on the analysis of learned hypotheses, knowledge-driven constructive induction (KCI) in which modifications to the representation space are based on background knowledge provided to the system, and multistrategy constructive induction (MCI) that dynamically selects appropriate DCI, HCI, and KCI methods based on a learning problem's characteristics. Some recent work on constructive induction includes the investigation of different specificities of advices provided to the system and the use of constructive induction in intelligent evolutionary optimization [50].

4 Learning

This section gives a brief overview of machine learning methods developed or co-developed by Ryszard Michalski over the past four decades. These methods were implemented in numerous computer programs. Most of them follow a general separate and conquer approach to inductive learning. They take data and background knowledge and generate new knowledge, often in the form of attributional rules.

4.1 AQ: Learning Attributional Rules from Examples

The well-known family of AQ programs originated with the A^q algorithm for solving the general covering problem [14][18]. In fact, the AQ programs perform a simplified version of the original algorithm. AQ programs pioneered the separate and conquer approach to rule learning [6]. Numerous implementations and extensions of the method were developed using different programming languages and on different platforms over the years. Among the best known AQ implementations are AQ7 [26], AQ11 [38], AQ15c [47], AQ17 [1], AQ19 [32], and most recently AQ21[53].

The general problem to which AQ programs are applied is learning from examples. Given a set of examples e_1, e_2, .. e_n, belonging to classes c_1, …, c_k, find a set of rules (5), where $C_{i,j}$ are complexes, that describe the input data. Note that this is the form

(2) of rules. Some AQ implementations are also able to create rules in the forms (3) and (4) that include exceptions and preconditions.

$$\begin{array}{c} c_1 \Leftarrow C_{1,1},\, c_1 \Leftarrow C_{1,2},\, \ldots,\, c_1 \Leftarrow C_{1,t1} \\ c_2 \Leftarrow C_{2,1},\, c_2 \Leftarrow C_{2,2},\, \ldots,\, c_2 \Leftarrow C_{2,t2} \\ \ldots \\ c_k \Leftarrow C_{k,1},\, c_k \Leftarrow C_{k,2},\, \ldots,\, c_k \Leftarrow C_{k,tk} \end{array} \tag{5}$$

A basic version of the AQ learning algorithm is presented in Figure 4. It takes as input a set of positive examples of a concept, P, and set of negative examples, N, belonging to all other classes (examples not belonging to the learned concept), and a quality measure. It returns a complete and consistent hypothesis in the form of an *attributional ruleset*, optimized according to the given *lexicographical evaluation functional* (LEF). In the case when input data consists of multiple classes, learning is repeated for each class against all other classes.

```
Hypothesis = null
While P is not empty
  Select a seed example e from P
  Generate star G(e, N)
  Select the best rule R from G(e, N) according to LEF, and
    include it in Hypothesis
  Remove from P all examples covered by the selected rule
Return learned Hypothesis
```

Fig. 4. Basic AQ algorithm

The central part of the algorithm is the generation of star *G(e, N)* given a seed *p* and set of negative examples *N*. The star is a set of maximally general rules covering the seed *e*, but not covering any negative example from *N*. A star is constructed by intersecting partial stars which are generated using the *extension-against* operator [22]. The extension-against that takes two data points and creates a set of single-condition rules. The rules are maximal generalizations of one data point (the seed) that does not cover the second data point (a selected negative example). The set of rules created by extension-against is called a *local star*. An intersection of local stars for all negative examples is the star *G(e, N)*. To narrow down a possibly very large number of intermediate generalizations, AQ uses a beam search that at each step of star generation keeps no more than a parameter-defined number of best rules, as determined by the given pattern quality measure, as defined by the LEF. LEF evaluates rules through a sequence of criteria with defined thresholds. Different AQ programs have implemented different LEF criteria, but most of them select by default rules that cover the most positive examples and are the simplest. Many other criteria have been introduced that can be tuned to a specific problem by the user. A complete list of LEF criteria available in the AQ21 system is presented in [48].

To select a rule from a star, the algorithm also uses a LEF, but possibly with different criteria than for the beam search. One important extension of the method is to learn approximate theories from noisy data, and discover strong regularities, patterns, in the data. To do so, AQ at each step of star generation is able to optimize rules according to the $Q(w)$ rule quality measure. The measure is given by (6) and has been defined by Michalski and Kaufman [33].

$$Q(w) = cov^{w} * consig^{1-w} \tag{6}$$

where

$$cov=|p| / |P| \tag{7}$$

and

$$consig=((|p| / (|p| + |n|)) - (|P| /(|P| + |N|))) * (|P| + |N|) / |N| \tag{8}$$

Here, $|p|$ and $|n|$ are the numbers of positive and negative examples covered by the rule, and $|P|$ and $|N|$ are the numbers of positive and negative examples in the training dataset, respectively. The $Q(w)$ definition (6) assures in practice that $consig \geq 0$ which means that a rule's prediction is better than a blind guess of the target class. The actual implemented formula handles also the case when $consig < 0$, but such rules are always dropped as not useful.

Optimization of rules consists of several possible operations which may lead to improvement of rules' quality. The operations are: abstraction of conditions, specialization of conditions, removal of conditions, and removal of entire rules.

Numerous other modifications and extensions of the AQ learning algorithm have been made over years. For example, some AQ versions use several seeds (to protect the method against noise), employ different concept representations (attributional or relational), generate rules with different interrelationships (independent, disjoint or sequentially ordered covers), use different methods for handling data inconsistency (minimum, maximum, free and statistic-based generalization), learn rules in a batch of incremental mode, seek rules that represent the best trade-off between their consistency, coverage and simplicity, use different criteria of rule optimality, involve operators for deriving more relevant attributes (data-driven, hypothesis-driven or multistrategy constructive induction), apply prior knowledge (a- and l-rules, knowledge-driven constructive induction), post-optimize learned descriptions (TRUNC/s, TRUNC/sg and TRUNC/nl), generate single or alternative descriptions, learn rules with exceptions or preconditions, handle unknown, not-applicable and irrelevant, meta-values, learn in standard multiclass mode or without a contrast set, etc.

The evolution of AQ-based learning systems was not linear, and no one system had encompassed all of these methodological features. In recent years, Dr. Michalski along with his group of collaborators initiated an attempt to implement most of the AQ learning methodology developed over many years into one system, AQ21. Its goal is to become a laboratory for natural induction, that is, to allow users to learn attributional rules of different types, present results in easy to interpret forms, such as natural language, and allow experimentation exploiting large numbers of parameters to specify users' preferences. The program, which is being constantly extended, is the most advanced AQ implementation to date, and this effort continues at the GMU Machine Learning and Inference Laboratory.

4.2 INDUCE: Learning Structures

Learning structural descriptions in the form of easy to understand rules is realized by a class of programs called INDUCE. In contrast to AQ programs, INDUCE does not take input in the form of labeled examples, but rather in the form of sets of rules describing considered objects. Its goal is to arrive at a smaller set of rules consistent with original ones. The INDUCE programs are multipurpose and applicable to a wide class of problems, but they are particularly suitable for learning structural descriptions of classes of objects in which the number of relations between objects is not known in advance. The method takes as an input a set of rules (9) describing objects in classes $c_1, c_2, \ldots, c_n$.

$$\begin{array}{c} c_1 \Leftarrow C_{1,1},\ c_1 \Leftarrow C_{1,2},\ \ldots,\ c_1 \Leftarrow C_{1,r1} \\ c_2 \Leftarrow C_{2,1},\ c_2 \Leftarrow C_{2,2},\ \ldots,\ c_2 \Leftarrow C_{2,r2} \\ \ldots \\ c_k \Leftarrow C_{k,1},\ c_k \Leftarrow C_{k,2},\ \ldots,\ c_k \Leftarrow C_{k,rk} \end{array} \tag{9}$$

The goal is to find a set of rules (10) where $ri \geq ti$, $i=1..k$. The obtained set of rules is not only more compact (fewer rules), but it is also a generalization of input rules.

$$\begin{array}{c} c_1 \Leftarrow C_{1,1},\ c_1 \Leftarrow C_{1,2},\ \ldots,\ c_1 \Leftarrow C_{1,t1} \\ c_2 \Leftarrow C_{2,1},\ c_2 \Leftarrow C_{2,2},\ \ldots,\ c_2 \Leftarrow C_{2,t2} \\ \ldots \\ c_k \Leftarrow C_{k,1},\ c_k \Leftarrow C_{k,2},\ \ldots,\ c_k \Leftarrow C_{k,tk} \end{array} \tag{10}$$

The INDUCE method follows the star generation algorithm similar to one used in AQ, but in addition to extension-against, it uses other generalization operators [22]. It is more general than the AQ example learner, but is also less efficient. Because most real word learning problems can be represented in the form of labeled examples, development of AQ systems has progressed much further than that of INDUCE.

4.3 SPARC: Learning Sequences

Another type of learning is done through *part-to-whole generalization* in which given observations of a part of an observed set of objects, the goal is to hypothesize the entire set of objects. One specific case of this type of learning concerns learning and predicting sequences (ordered sets) of objects. Among the many real world applications of predicting sequences are predicting stock market behavior, predicting outcomes of medical treatments, predicting gene sequences, predicting computer users' behavior, and so forth.

The problem considered here is to predict events (objects) that follow a given sequence of objects. Given the sequence of events in the form (11), the goal is to find events e_{n+1}, e_{n+2}, and so on.

$$e_1, e_2, e_3, e_n \tag{11}$$

Each of the objects e_i may be described in terms of one or more attributes (12), where $v_{i1}, v_{i2}, .. v_{ik}$ are values of attributes $X_1, \ldots, X_k$ in the event e_i. These attributes may be of different types, e.g. nominal, structured, ordinal, cyclic, ratio.

$$e_i = (v_{i1}, v_{i2}, \ldots, v_{ik}) \tag{12}$$

In some specific situations it may not be necessary to predict entire events e_{n+1}, e_{n+2}, …, but rather values of selected attributes in these events. It may also be appropriate to predict not one, but several plausible events that fit into the sequence in a particular place. In this *non-deterministic prediction problem*, the goal is to discover some properties of events in the sequence, not necessarily precisely predict them [3].

Computer programs called SPARC represent a multistrategy approach to discovering sequences, by combining the *periodic conjunctive model*, *lookback decomposition model*, and *disjunctive normal form* models. Two programs were created, SPARC/E, a specialized system designed for playing the eleusis card game [3][35], and the SPARC/G general purpose sequence discovery tool [36]. The programs work in three steps. In the first step, constructive induction methods are applied to derive attributes that may better characterize sequences of events. In the second step, rules characterizing sequences are derived by applying AQ learning to data prepared for each type of model. In the last step, final rules are selected based on their consistency and simplicity. These rules can be later used to predict plausible continuationa of the input sequence.

4.4 ABACUS: Learning Equations

Quantitative discovery deals with discovering equations from data. The ABACUS approach to quantitative discovery is able to incorporate symbolic attributes in learning equations, cope with irrelevant attributes, and work with noisy data. Its main advantage is that it is able do discover multiple equations, each characterizing a subset of data, and provide preconditions under which the equations should be used [5].

The ABACUS method works in two steps. First, in the *equation discovery module*, it analyzes input data and creates equations. If more than one equation is needed, the data is split into subsets. Because of the enormous size of the possible search space of possible equations ABACUS uses a number of heuristics. One such heuristic involves constructing and searching proportionality graphs of attributes that are *qualitatively proportional* (or *inversely quantitatively proportional*). For such attributes, a given percentage of values of one attribute grows (or decreases) with values of other attribute for other attributes constant. Other heuristics include checking for compatibility of units, detection of redundancy in formulas, and tautology detection. In the second step, a *precondition generation module* generates logic preconditions characterizing the subsets of data for which equations were created in the first step. To do so, ABACUS employs the AQ learning method.

The system was applied to multiple datasets, and was able to correctly discover, for example, Stoke's law and the law of conservation of kinetic energy.

4.5 CLUSTER: Conceptual Clustering

Learning from observation is, after concept learning from examples, the second largest class of problems in machine learning. The goal is to construct meaningful classes from observed objects. Traditional clustering methods group objects based on their similarity, without regard to what global concepts they form. The answer to this problem was proposed by Michalski and is called *conceptual clustering* [21][41] in which objects are grouped together based on their *conceptual cohesiveness* (13). This measure is based not only on the distance between objects, but more importantly on the set of concepts, C, that are available for describing the objects, and other objects, E, in the dataset.

$$\textit{conceptual cohesiveness}(O_1, O_2) = \mathrm{f}(O_1, O_2, C, E) \tag{13}$$

An example of a problem in which standard distance-based clustering methods may incorrectly form clusters is presented in Figure 5. The two marked points are the closest objects to one another in the figure, so they would quickly and invariably be assigned to he same cluster by many methods. In contrast, a conceptual clustering method that has been provided with the concepts of letters and their shapes would be able to correctly create two clusters representing the X and the Y.

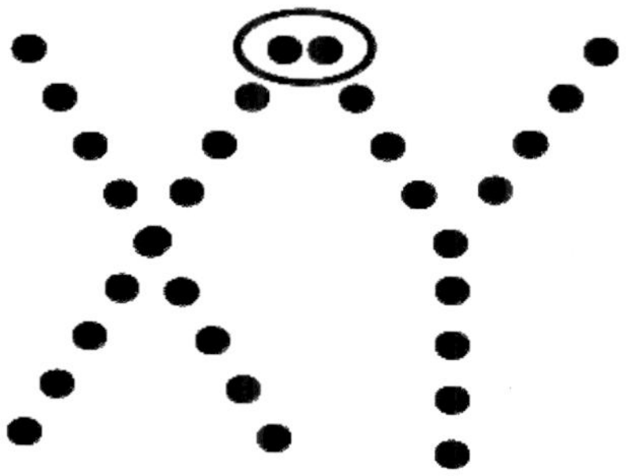

Fig. 5. An example of a conceptual clustering problem

The basic algorithm used in CLUSTER implementations of conceptual clustering takes as an input a set of objects, *E*, the desired number of clusters, *k*, and an LEF expressing criteria of clusters' quality. It starts by randomly selecting *k* seed examples, $E_s=\{e_1, ..., e_k\}$. For each seed example o_i a star $G(o_i, E_s\backslash\{o_i\})$ is generated. Then a clustering is created by selecting disjoint rules from the stars according to the LEF. If termination criteria are not met, a new set of seeds is selected from examples covered by rules in the clustering, and the procedure is repeated [40].

Recent research on conceptual clustering concerns building goal-oriented clusters from the perspective of different *viewpoints* [45]. A viewpoint is defined as a subset of attributes in the representation space that can be meaningfully combined. For example, one may cluster students in a database from the viewpoint of their intellectual abilities, while other may want to cluster students based on their demographics. These two viewpoints require clearly different sets of attributes, which are drawn from the set of all attributes available in database.

5 Learnable Evolution Model

The learnable evolution model (LEM) is an evolutionary optimization method that employs machine learning to direct the evolutionary process [24][25]. Specifically, LEM creates general hypotheses indicating regions in the search space that likely contain optimal solutions and then instantiates these hypotheses to generate new candidate solutions. In order to apply concept learning, LEM creates two groups of individuals that are respectively high- and low-performing according to the fitness function being optimized. These individuals can be selected from the current population or a combination of current and past populations of individuals. The group of high-performing individuals is called H-Group and the group of low-performing individuals is called L-Group. Once the groups are selected, LEM applies concept learning to create a general hypothesis describing the H-Group in contrast to the L-Group. The hypotheses are then instantiated to create new candidate solutions. In the final step, a new population is assembled from old and new individuals, and the process is repeated until stopping criteria are met.

Very successful initial implementations of the learnable evolution model sparked development of the third generation of LEM software, called LEM3. It extends many ideas found in the original LEM methodologies, some of which are unique in the field of evolutionary computation. The general flow diagram of LEM3's algorithm is presented in Figure 6. In addition to components found in standard evolutionary computation methods, such as generation of an initial population, evaluation of individuals, and selection of individuals, LEM3 includes several novel components. It dynamically selects one or more *innovation* methods to create new individuals. These methods are: *learn & instantiate*, the aforementioned main mechanism for creating new individuals in LEM3; *probe*, to apply traditional operators such as mutation and crossover, *search locally*, to apply a user-defined local search method; and *randomize*, to add to the current population a number of randomly created individuals, or restart the evolutionary process. One of the major novelties of LEM3 is the ability to automatically adjust representation space through constructive induction [49][50].

Theoretical and experimental work indicates that LEM is particularly suitable for optimization problems in which the fitness evaluation is costly. This is because of the trade-off between significantly shorter evolution length [25][52], and more complex learning and instantiation when compared to simple operators used in evolutionary

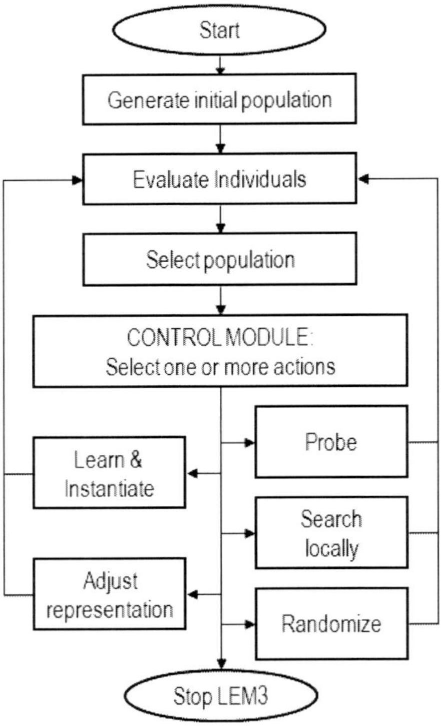

Fig. 6. Diagram of the LEM3 algorithm (reproduced from [49])

computation. Moreover, the use of machine learning to guide evolutionary computation extends the applicability of LEM. For example, because of the use of AQ21 as a learning module in LEM3, it is able to handle optimization problems naturally described using different types of attributes (nominal, structured, ordinal, cyclic, interval, ratio, and compound) and background knowledge provided to the learning program [51].

A class of LEM-based systems for heat exchanger optimization has been developed. These include the ISHED system for optimizing evaporators and ISCOD system for optimizing condensers [4][10]. These specialized systems combine LEM's learning and instantiation operators with specialized probing operators that are specifically designed to work with heat exchangers. Based on the ISHED and ISCOD systems, Michalski and Kaufman proposed a general LEMd methodology for optimizing complex systems [31].

6 Inductive Databases and Knowledge Scouts

Imagine a database system that can answer queries about items not present in the database. By applying inductive learning to data and background knowledge stored in the database an *inductive database* is able to answer such queries. In fact, an inductive database system should integrate many inductive and deductive reasoning methods on data and knowledge stored in the database. Results of reasoning are added to the existing knowledge in the database and can be reused when answering further queries. The latter feature distinguishes the concept of *inductive databases* proposed by Michalski and his collaborators, from those often found in literature. The key idea behind inductive database is one of *knowledge system* that combines database and knowledge base. The term knowledge system signifies integration of database and a

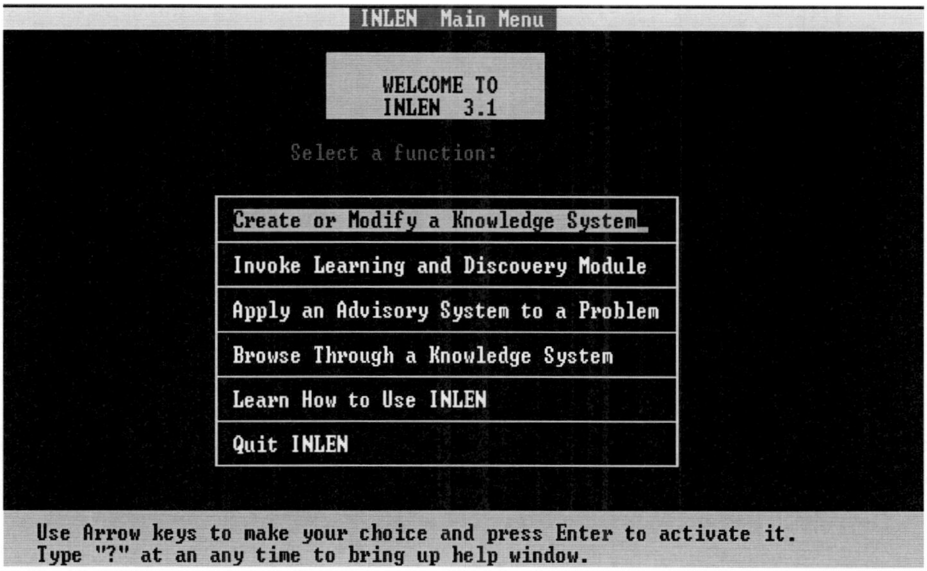

Fig. 7. The INLEN system main menu

relevant knowledge base to support knowledge mining and knowledge application. Knowledge bases contain knowledge created by experts and knowledge generated by inductive learning operators applied to data in the database and to knowledge in the knowledge base.

The idea of inductive database originated in 1980s through development of computer systems QUIN [28][46], ADVISE [27], and AURORA [7]. It was later extended in the inductive database system INLEN (INference and LEarNing) [7][14], and most recently in its successor, VINLEN [12][15]. Both INLEN and VINLEN provide interfaces that allow users to access knowledge system and apply knowledge generation and application operators, as depicted in Figures 7 and 8.

A knowledge scout is a script that allows intelligent search for answers in an inductive database [11]. It usually involves multiple steps of inductive reasoning and knowledge application in which one step depends on results of previous steps. One script language that can create knowledge scouts is *knowledge generation language* (KGL) developed within the INLEN system [12]. The second generation of such a language is *knowledge query language* (KQL) which is being implemented in the VINLEN system. One way of interpreting knowledge scouts is that they are intelligent agents operating inside the inductive database, whose purpose is to seek requested relevant knowledge.

Through knowledge scouts or a graphical interface, the users of VINLEN have access to numerous operators for data and knowledge manipulation, knowledge creation,

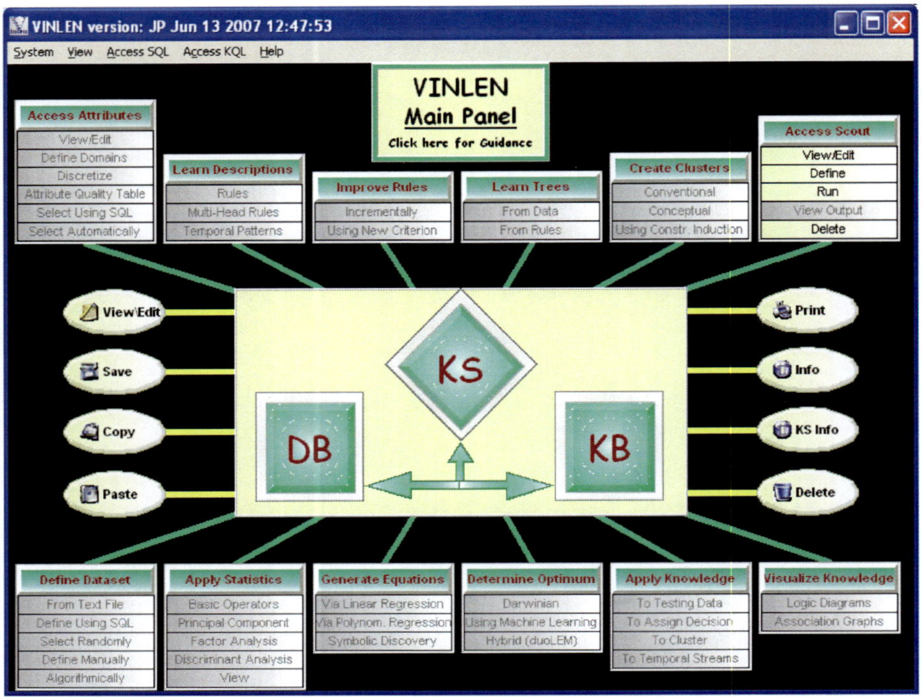

Fig. 8. The VINLEN system graphical interface

and knowledge application. These operators are grouped in categories such as *learn description*, *improve rules*, *learn trees*, *create clusters*, *apply statistics*, *generate equations*, and *determine optima*. Each of the categories consists of multiple operators, as shown in Figure 8. Because of the large variety of different operators and forms in which they generate knowledge, VINLEN is oriented so that the results of one operator are understood by other, applicable, operators. This enables the creation of knowledge scouts in which operators are applied to the results of applying other operators.

At the time that this chapter is being written, the VINLEN system is still under development, and new operators and functionalities are being added. This work follows the original design outlined by Ryszard Michalski.

7 Theoretical Aspects of Machine Learning

With his Theory and Methodology of Inductive Learning [22], Michalski defined several axes on which to categorize the inductive process, including input type (classified examples or unclassified observations), data types, whether or not counterexamples were presented, the types of covers learned and their relation to each other, and the use of original or constructed attributes. In this framework, he utilized VL1-style languages to express learned knowledge with high comprehensibility, which could be employed in concept learning, conceptual clustering, qualitative prediction, integrated qualitative and quantitative discovery, etc.

But more than that, he was fascinated by how we learn, for only through such an understanding could machines be made to truly emulate human learning process. He collaborated with numerous cognitive scientists, and among the fruits of this collaboration were the logic and theory of *plausible reasoning* [2], and *the inferential theory of learning* [23].

To utilize both of these ideas, Michalski noticed that we store many concepts in hierarchies. If an entity about which you are trying to extrapolate knowledge is a close relative (e.g., they share a common parent or grandparent in the hierarchy) of an entity whose characteristics you know, it stands to reason *plausibly* that the unknown entity may share characteristics similar to those of the known entity. Minnesota is near Iowa. They grow lots of corn in Iowa. Maybe they grow lots of corn in Minnesota too. Conversely, Arizona is far from Iowa and has a hotter, drier climate. They probably don't grow much corn there.

Thus, similization and dissimilization operators allow the induction of relationships between objects and concepts through the traversal of concept hierarchies. The inferential theory extended this idea by viewing all learning operators as implementers of *knowledge transmutations*, which would perform various tasks upon the network of hierarchies. Generalization would climb a hierarchy, while specialization would descend it. Abstraction would select only certain available information about an entity, while concretion would restore such information. Other transmutations would alter the hierarchies themselves and their interrelationships.

On these theoretical foundations, it was possible to explore the integration of diverse knowledge transmutations and their associated learning strategies into multistrategy systems. It is no coincidence that Ryszard Michalski was a prime impetus behind the first workshops, conferences and books on multistrategy learning.

8 Education

Given his belief in understandable knowledge representations, it is natural that Dr. Michalski also sought to educate all comers in the wider field of machine learning. Behind his impetus, the Machine Learning and Inference Laboratory produced a series of educational tools that could also serve as a springboard for research in the field.

In 1986, he entered into an agreement with representatives of the Boston Museum of Science to produce an introduction to and demonstration of inductive learning. The result was ILLIAN, a system that presented five different learning programs: AQ, INDUCE, CLUSTER, SPARC and ABACUS (see section 4). For each of these, the user was able to accept challenges on the types of problems these programs were capable of solving, and then create their own problems in the same toy domains and witness the programs' performance on them. They could also see brief descriptions of the learning algorithms used.

ILLIAN toured eight cities as part of the exhibition "Robots and Beyond: The Age of Intelligent Machines" before permanently taking root in Boston. Counters had been installed to track usage, and a report after the tour had concluded indicated that the program had been run over half a million times.

ILLIAN was oriented toward the general public, who may have had no AI exposure. In order to allow users go somewhat further into depth, an expanded version of the system, called EMERALD [9] doubled as a research tool. More complex problems were displayed and available to the user to create for the learning programs.

Fig. 9. A screen from the iAQ program

EMERALD was displayed at classes in several countries, and distributed to hundreds of interested institutions. A recent reimplementation of one of EMERALD's modules, AQ, with a modern graphical interface is offered in the iAQ program [39][44], one of whose screens is presented in Figure 9.

9 Conclusion

In his research on machine learning, Ryszard S. Michalski contributed through new theories, algorithms, and organizational skills. He looked at problems differently than most researchers, which provided the field with his unique perspective, allowing him to originate several directions of research. Many of these directions, such as conceptual clustering, constructive induction, or natural induction, continue to evolve all over the world. His vision of the field has inspired many.

Perhaps the most important aspect of his research concerns the understandability of knowledge created by machine learning systems. While a majority of researchers in the field agree with this principle, very little has been actually done in this direction. Researchers construct more and more complicated models that slowly expand the barrier of predictive accuracy, and at the same time are often not understandable even by other researchers in the area. Dr. Michalski's ideas of natural induction or knowledge mining deserve further research in order to win back comprehensibility along with predictive accuracy.

This chapter briefly describes what we believe are some of the most important contributions of Dr. Michalski to machine learning research. It is, however, not possible to give enough attention to many important aspects of work that appeared in over 350 publications. In the references section of this chapter, readers can find many important works, some of which are already considered classic in the field. Among the "conceptual offspring" worth mentioning, but not included in this chapter are: variable-precision logic, two-tiered concept representation, dynamic recognition, attributional ruletrees, dynamic interlaced hierarchies, and many others.

A comprehensive review of applications of natural induction and conceptual clustering is in a recent technical report [34]. Descriptions and lists of publications of the mentioned works and the many works of Dr. Michalski not mentioned here are maintained at the website of the GMU Machine Learning and Inference Laboratory, at http://www.mli.gmu.edu.

References

[1] Bloedorn, E., Wnek, J., Michalski, R.S., Kaufman, K.: AQ17 A Multistrategy Learning System The Method and Users Guide. Reports of the Machine Learning and Inference Laboratory, MLI 93-12, School of Information Technology and Engineering, George Mason University, Fairfax, VA (November 1993)

[2] Collins, A., Michalski, R.S.: The Logic of Plausible Reasoning: A Core Theory. Cognitive Science 13, 1–49 (1989)

[3] Dietterich, T.G., Michalski, R.S.: Discovering Patterns in Sequence of Events. Artificial Intelligence Journal 25(2), 187–232 (1985)

[4] Domanski, P.A., Yashar, D., Kaufman, K., Michalski, R.S.: An Optimized Design of Finned-Tube Evaporators Using the Learnable Evolution Model. International Journal of Heating, Ventilating, Air-Conditioning and Refrigerating Research 10, 201–211 (2004)

[5] Falkenhainer, B., Michalski, R.S.: Integrating Quantitative and Qualitative Discovery in the ABACUS System. In: Kodratoff, Y., Michalski, R.S. (eds.) Machine Learning: An Artificial Intelligence Approach, San Mateo, CA, vol. III, pp. 153–190. Morgan Kaufmann Publishers, San Francisco (1990)
[6] Fürnkranz, J.: Separate-and-Conquer Rule Learning. Artificial Intelligence Review 13(1), 3–54 (1999)
[7] International Intelligent Systems, Inc.: User's Guide to AURORA 2.0: A Discovery System. International Intelligent Systems, Inc. (1988)
[8] Kaufman, K.: INLEN: A Methodology and Integrated System for Knowledge Discovery in Databases. Ph.D. Dissertation, School of Information Technology and Engineering, Reports of the Machine Learning and Inference Laboratory, MLI 97-15, George Mason University, Fairfax, VA (November 1997)
[9] Kaufman, K., Michalski, R.S.: EMERALD 2: An Integrated System of Machine Learning and Discovery Programs to Support Education and Experimental Research. Reports of the Machine Learning and Inference Laboratory, MLI 93-10, School of Information Technology and Engineering, George Mason University, Fairfax, VA (September 1993)
[10] Kaufman, K., Michalski, R.S.: ISHED1: Applying the LEM Methodology to Heat Exchanger Design. Reports of the Machine Learning and Inference Laboratory, MLI 00-2, George Mason University, Fairfax, VA (2000)
[11] Kaufman, K., Michalski, R.S.: A Knowledge Scout for Discovering Medical Patterns: Methodology and System SCAMP. In: Proceedings of the Fourth International Conference on Flexible Query Answering Systems, FQAS 2000, Poland, October 25-28, pp. 485–496 (2000)
[12] Kaufman, K., Michalski, R.S.: The Development of the Inductive Database System VINLEN: A Review of Current Research. In: International Intelligent Information Processing and Web Mining Conference, Zakopane, Poland (2003)
[13] Kaufman, K., Michalski, R.S.: From Data Mining to Knowledge Mining. In: Rao, C.R., Solka, J.L., Wegman, E.J. (eds.) Handbook in Statistics. Data Mining and Data Visualization, vol. 24, pp. 47–75. Elsevier/North Holland (2005)
[14] Kaufman, K., Michalski, R.S., Kerschberg, L.: Mining for Knowledge in Databases: Goals and General Description of the INLEN System. In: Piatetski-Shapiro, G., Frawley, W.J. (eds.) Knowledge Discovery in Databases. AAAI Press/The MIT Press, Menlo Park (1991)
[15] Kaufman, K., Michalski, R.S., Pietrzykowski, J., Wojtusiak, J.: An Integrated Multi-task Inductive Database VINLEN: Initial Implementation and Early Results. In: Džeroski, S., Struyf, J. (eds.) KDID 2006. LNCS, vol. 4747, pp. 116–133. Springer, Heidelberg (2007)
[16] Larson, J., Michalski, R.S.: Inductive Inference of VL Decision Rules. In: Invited paper for the Workshop in Pattern-Directed Inference Systems, Hawaii, and published in SIGART Newsletter, May 23-27, vol. 63, pp. 38–44. ACM, New York (1977)
[17] Michalski, R.S.: On the Quasi-Minimal Solution of the General Covering Problem. In: Proceedings of the V International Symposium on Information Processing (FCIP 1969), Yugoslavia, Bled, October 8-11. Switching Circuits, vol. A3, pp. 125–128 (1969)
[18] Michalski, R.S.: Synteza wyrazen minimalnych i rozpoznawanie symetrii funkcji logicznych. Prace Instytutu Automatyki PAN, Zeszyt 1992, Warszawa, Instytut Automatyki Polskiej Akademii Nauk (1971)
[19] Michalski, R.S.: A Geometrical Model for the Synthesis of Interval Covers. Report No. 461, Department of Computer Science, University of Illinois, Urbana, June 24 (1971)
[20] Michalski, R.S.: A Variable-Valued Logic System as Applied to Picture Description and Recognition. In: Nake, F., Rosenfeld, A. (eds.) Graphic Languages. North-Holland Publishing Co., Amsterdam (1972)

[21] Michalski, R.S.: Knowledge Acquisition Through Conceptual Clustering: A Theoretical Framework and an Algorithm for Partitioning Data into Conjunctive Concepts. Journal of Policy Analysis and Information Systems 4(3), 219–244 (1980)
[22] Michalski, R.S.: A Theory and Methodology of Inductive Learning. In: Michalski, R.S., Carbonell, T.J., Mitchell, T.M. (eds.) Machine Learning: An Artificial Intelligence Approach, pp. 83–134. TIOGA Publishing Co., Palo Alto (1983)
[23] Michalski, R.S.: Learning = Inferencing + Memorizing: Basic Concepts of Inferential Theory of Learning and Their Use for Classifying Learning Processes. In: Foundations of Knowledge Acquisition. Machine Learning, vol. 2, pp. 1–41 (1993)
[24] Michalski, R.S.: Learnable Evolution: Combining Symbolic and Evolutionary Learning. In: Proceedings of the Fourth International Workshop on Multistrategy Learning (MSL 1998), Desenzano del Garda, Italy, June 11-13, pp. 14–20 (1998)
[25] Michalski, R.S.: Learnable Evolution Model Evolutionary Processes Guided by Machine Learning. Machine Learning 38, 9–40 (2000)
[26] Michalski, R.S.: Attributional Calculus: A Logic and Representation Language for Natural Induction. Reports of the Machine Learning and Inference Laboratory, MLI 04-2, George Mason University, Fairfax, VA (April 2004)
[27] Michalski, R.S., Baskin, A.B.: Integrating Multiple Knowledge Representations and Learning Capabilities in an Expert System: The ADVISE System. In: Proceedings of the 8th International Joint Conference on Artificial Intelligence, Karlsruhe, West Germany, pp. 256–258 (1983)
[28] Michalski, R.S., Baskin, A.B., Spackman, K.A.: A Logic-based Approach to Conceptual Database Analysis. In: Sixth Annual Symposium on Computer Applications in Medical Care (SCAMC-6), George Washington University Medical Center, Washington, DC, pp. 792–796 (1982)
[29] Michalski, R.S., Carbonell, T.J., Mitchell, T.M. (eds.): Machine Learning: An Artificial Intelligence Approach. TIOGA Publishing Co., Palo Alto (1983)
[30] Michalski, R.S., Carbonell, T.J., Mitchell, T.M. (eds.): Machine Learning: An Artificial Intelligence Approach, vol. II. Morgan Kaufmann Publishers, Inc., Los Altos (1986)
[31] Michalski, R.S., Kaufman, K.: Intelligent Evolutionary Design: A New Approach to Optimizing Complex Engineering Systems and its Application to Designing Heat Exchangers. International Journal of Intelligent Systems 21, 12 (2006)
[32] Michalski, R.S., Kaufman, K.: The AQ19 System for Machine Learning and Pattern Discovery: A General Description and User's Guide. Reports of the Machine Learning and Inference Laboratory, MLI 01-2, George Mason University, Fairfax, VA (2001)
[33] Michalski, R.S., Kaufman, K.: Learning Patterns in Noisy Data: The AQ Approach. In: Paliouras, G., Karkaletsis, V., Spyropoulos, C.D. (eds.) ACAI 1999. LNCS (LNAI), vol. 2049, pp. 22–38. Springer, Heidelberg (2001)
[34] Michalski, R.S., Kaufman, K., Pietrzykowski, J., Wojtusiak, J., Mitchell, S., Seeman, W.D.: Natural Induction and Conceptual Clustering: A Review of Applications. Reports of the Machine Learning and Inference Laboratory, MLI 06-3, George Mason University, Fairfax, VA (June 2006)
[35] Michalski, R.S., Ko, H., Chen, K.: SPARC/E(V.2), An Eleusis Rule Generator and Game Player. Reports of the Intelligent Systems Group, ISG 85-11, UIUCDCS-F-85-941, Department of Computer Science, University of Illinois (February 1985)
[36] Michalski, R.S., Ko, H., Chen, K.: Qualitative Prediction: The SPARC/G Methodology for Inductively Describing and Predicting Discrete Processes. In: Expert Systems. Academic Press Inc., London (1986)
[37] Michalski, R.S., Larson, J.: AQVAL/1 (AQ7) User's Guide and Program Description. Report No. 731, Department of Computer Science, University of Illinois, Urbana (June 1975)

[38] Michalski, R.S., Larson, J.: Incremental Generation of VL1 Hypotheses: The Underlying Methodology and the Description of Program AQ11. Reports of the Intelligent Systems Group, ISG 83-5, UIUCDCS-F-83-905, Department of Computer Science, University of Illinois, Urbana (January 1983)

[39] Michalski, R.S., Pietrzykowski, J.: iAQ: A program that discovers rules. AAAI-07 AI Video Competition. In: Twenty-Second Conference on Artificial Intelligence (AAAI 2007), Vancouver, British Columbia, July 22-26 (2007)

[40] Michalski, R.S., Stepp, R.: Learning from Observation: Conceptual Clustering. In: Michalski, R.S., Carbonell, T.J., Mitchell, T.M. (eds.) Machine Learning: An Artificial Intelligence Approach, pp. 331–363. TIOGA Publishing Co., Palo Alto (1983)

[41] Michalski, R.S., Stepp, R., Diday, E.: A Recent Advance in Data Analysis: Clustering Objects into Classes Characterized by Conjunctive Concepts. In: Kanal, L., Rosenfeld, A. (eds.) Progress in Pattern Recognition, vol. 1, pp. 33–55. North-Holland, Amsterdam (1981)

[42] Michalski, R.S., Wojtusiak, J.: Reasoning with Meta-values in AQ Learning. Reports of the Machine Learning and Inference Laboratory, MLI 05-1, George Mason University, Fairfax, VA (June 2005)

[43] Michalski, R.S., Wojtusiak, J.: Generalizing Data in Natural Language. In: Kryszkiewicz, M., Peters, J.F., Rybiński, H., Skowron, A. (eds.) RSEISP 2007. LNCS (LNAI), vol. 4585, pp. 29–39. Springer, Heidelberg (2007)

[44] Pietrzykowski, J.: Demonstration and Application of Rule Discovery Methods Using iAQ. In: Workshop on Building Computational Intelligence and Machine Learning Virtual Organizations, George Mason University, Fairfax, VA, October 24, pp. 39–44 (2008)

[45] Seeman, W.D., Michalski, R.S.: The CLUSTER3 System for Goal-oriented Conceptual Clustering: Method and Preliminary Results. In: Proceedings of The Data Mining and Information Engineering 2006 Conference, Prague, Czech Republic, July 11-13 (2006)

[46] Spackman, K.A.: QUIN: Integration of Inferential Operators within a Relational Database. Reports of the Intelligent Systems Group, ISG 83-13, UIUCDCS-F-83-917, M.S. Thesis, Department of Computer Science, University of Illinois, Urbana (1983)

[47] Wnek, J., Kaufman, K., Bloedorn, E., Michalski, R.S.: Inductive Learning System AQ15c: The Method and User's Guide. Reports of the Machine Learning and Inference Laboratory, MLI 95-4, George Mason University, Fairfax, VA (March 1995)

[48] Wojtusiak, J.: AQ21 User's Guide. Reports of the Machine Learning and Inference Laboratory, MLI 04-3, George Mason University, Fairfax, VA (September 2004)

[49] Wojtusiak, J.: Handling Constrained Optimization Problems and Using Constructive Induction to Improve Representation Spaces in Learnable Evolution Model. Ph.D. Dissertation, College of Science, Reports of the Machine Learning and Inference Laboratory, MLI 07-3, George Mason University, Fairfax, VA (November 2007)

[50] Wojtusiak, J.: Data-driven Constructive Induction in the Learnable Evolution Model. In: Proceedings of the 16th International Conference Intelligent Information Systems, Zakopane, Poland, June 16-18 (2008)

[51] Wojtusiak, J.: The LEM3 System for Multitype Evolutionary Optimization. Computing and Informatics (accepted for publication, 2009)

[52] Wojtusiak, J., Michalski, R.S.: The LEM3 Implementation of Learnable Evolution Model and Its Testing on Complex Function Optimization Problems. In: Proceedings of Genetic and Evolutionary Computation Conference, GECCO 2006, Seattle, WA, July 8-12 (2006)

[53] Wojtusiak, J., Michalski, R.S., Kaufman, K., Pietrzykowski, J.: The AQ21 Natural Induction Program for Pattern Discovery: Initial Version and its Novel Features. In: Proceedings of the 18th IEEE International Conference on Tools with Artificial Intelligence, Washington, D.C, November 13-15 (2006)

The AQ Methods for Concept Drift

Marcus A. Maloof

Department of Computer Science, Georgetown University, Washington, DC 20057-1232, USA
maloof@cs.georgetown.edu

Abstract. Since the mid-1990's, we have developed, implemented, and evaluated a number of learning methods that cope with concept drift. Drift occurs when the target concept that a learner must acquire changes over time. It is present in applications involving user preferences (e.g., calendar scheduling) and adversaries (e.g., spam detection). We based early efforts on Michalski's AQ algorithm, and our more recent work has investigated ensemble methods. We have also implemented several methods that other researchers have proposed. In this chapter, we survey results that we have obtained since the mid-1990's using the Stagger concepts and learning methods for concept drift. We examine our methods based on the AQ algorithm, our ensemble methods, and the methods of other researchers. Dynamic weighted majority with an incremental algorithm for producing decision trees as the base learner achieved the best overall performance on this problem with an area under the performance curve after the first drift point of .882. Systems based on the AQ11 algorithm, which incrementally induces rules, performed comparably, achieving areas of .875. Indeed, an AQ11 system with partial instance memory and Widmer and Kubat's window adjustment heuristic achieved the best performance with an overall area under the performance curve, with an area of .898.

Keywords: online learning, concept drift, AQ algorithm, ensemble methods.

1 Introduction

Concept drift is a phenomenon in which the concept that a learner is attempting to acquire changes over time. Such change occurs in applications such as intelligent user interfaces, in which user tasks and preferences change over time, spam detection, in which spammers change the characteristics of their messages, and intrusion detection, in which crackers change their tactics. Learning in the presence of concept drift is also known as learning a shifting target, learning from time-changing data streams, and learning in non-stationary environments.

Since the mid-1990's, we have developed, implemented, and evaluated a number of algorithms and systems for learning in the presence of concept drift [2, 11, 12, 13, 18, 20, 21]. We based early efforts on Michalski's AQ algorithm [18, 20, 21], a covering algorithm that has had a prominent role in the development of the field of machine learning since 1969 [22]. More recently, we have examined ensemble methods, which use multiple models for prediction, instead of a single model [2, 11, 12, 13].

This chapter is a retrospective of our work on learners that cope with concept drift. We review our methods based on the AQ algorithm, but we also do so with our more

J. Koronacki et al. (Eds.): Advances in Machine Learning I, SCI 262, pp. 23–47.
springerlink.com © Springer-Verlag Berlin Heidelberg 2010

recent ensemble methods and with the methods of other researchers. We focus on results for these methods on the Stagger concepts [30, 34]. It was the first problem introduced for evaluating learners on drifting concepts, and it has been a centerpiece in a number of evaluations (e.g., [2, 11, 12, 13, 18, 20, 21, 30, 34]).

The problem consists of three targets concepts. The first is conjunctive, the second is disjunctive, and the third is internally disjunctive. The feature space is small, consisting of three attributes, each with three values. However, its perceived simplicity is deceiving. Learners proposed recently performed worse on the Stagger concepts than they did on newer problems that are supposedly harder [2, 13]. We discuss these issues further in Sect. 5.

The main contribution of this chapter is the retrospective view of our results that have appeared in the literature since the mid-1990's. Specifically, we present a survey of the performance of learners for concept drift on the Stagger concepts. Results suggest that dynamic weighted majority [11, 13] and systems based on AQ11 [18, 20, 21, 23] outperformed other methods in terms of area under the performance curve after the first drift point. We hope our survey of these results will give researchers additional insights into the performance of methods for concept drift, insights that will lead to new methods for an important problem.

The organization of this chapter is as follows. In an effort to make the chapter self-contained, in Sect. 2, we review basic concepts of machine learning, online learning methods, and concept drift. In Sect. 3, we describe the AQ methods of concept learning. In Sect. 4, we describe our methods of concept drift based on the AQ algorithm, our ensemble methods for coping with concept drift, and the methods that other researchers have developed for this problem. As mentioned previously, we evaluated all of these methods on the Stagger concepts, which we present in Sect. 5. In Sect. 6, we present results from experimental studies that we have conducted since the mid-1990's using implementations of our methods and the methods of other researchers. We discuss these results in Sect. 7 and place them in a broader context. After describing avenues for future work in Sect. 8, we conclude the paper with Sect. 9.

2 Preliminaries

In this section, to introduce the notation we use in the chapter and in an attempt to make the chapter self-contained, we present some basic concepts of machine learning. We focus on supervised concept learning, evaluation, and concept drift. Readers who require additional preparation or would like a broader perspective of the field should consult an introductory book on artificial intelligence (e.g., [28]) or on machine learning (e.g., [1, 4, 9, 14, 25, 35]).

2.1 Concept Learning

A method of machine learning improves its performance on a task with experience [25]. Concept learning refers to a setting in which a learner must acquire a description for identifying members of classes. Such learning often takes place with the aid of a critic or a teacher, who gives the learner the correct categories of observations. We refer to this setting as *supervised concept learning* or *supervised machine learning*. Other settings include *discovery*, in which the learner is unsupervised, and *reinforcement learning*, in which the learner receives reward for the actions it takes.

In this chapter, we focus on supervised learning for classification tasks. The task is to learn to place observations in one of two or more categories or *classes*. As mentioned, there is a teacher that provides the correct classification for observations, and this supervision can occur in a number of ways. For example, the teacher may give the learner a batch of observations with their correct categories, or the teacher may present observations and their categories sequentially.

The learner attempts to acquire a mapping between the characteristics or attributes of the observations and the classes. Some characteristics may be important for this mapping, whereas others may not. Moreover, the learner should be able to generalize its experience to observations that it did not encounter during training.

Regardless of how interaction with the learner occurs, the teacher (or perhaps someone else) monitors and measures the learner's performance through testing. In the first scenario, in which the teacher gives the learner a batch of observations with their correct categories, the teacher may provide another batch of observations, ask the learner to classify them, and then score the learner's performance using, say, the number of observations correctly classified. In the second scenario, in which the teacher presents observations to the learner one by one, the teacher may determine whether the learner predicts correctly for each new observation before providing the correct classification. The teacher could also test the learner at certain intervals using a batch of observations, as in the first scenario.

In both scenarios, we expect the learner's performance, as measured by the percentage of correct predictions, to increase with more and more observations when accompanied with supervision from the teacher. However, a more interesting situation is when we evaluate two or more learners. In this case, we might examine how quickly each learns, how quickly each classifies, how well each classifies, and – because they are machine learners – how each represents concepts.

In more precise language, let $S = \{x, y\}_{t=1}^{T}$ such that $y_t = f(x_t)$ and where f is unknown. x is *observation*, y is a *class label*, and the pair $\langle x, y \rangle$ is an *example*. Each dimension of x corresponds to an *attribute* with a defined domain. x_i is an *attribute value* from that domain. y is also an attribute, but since it is the one we wish to predict, it is traditionally treated separately.

The domains of the attributes can be categorical or numeric. For classification tasks, the domain of y is categorical. (For regression tasks, which we do not discuss, it is numeric). For categorical attributes, let v_{ik} be the kth value of the ith attribute's domain. Herein, when we refer to v_{ik}, it is understood that $v_{ik} \equiv [x_i = v_k] \equiv [x_i = v_{ik}]$, for some value of k. For example, if the ith attribute is *color* with the domain $\{red, green, blue\}$, then $red \equiv [color = red]$. For the purposes of this chapter, we define the domain of y to be either a positive or negative example of a class of interest (i.e., $y \in \{-, +\}$). Put another way, the example is either a positive or negative example of the target concept.

A learning method consists of three components: the concept description language, the learning element, and the performance element. The concept description language is the formalism the method uses to build concept descriptions, models, or hypotheses to represent its knowledge of concepts. As examples, researchers have used probabilities, linear and nonlinear equations, propositional and first-order logic, rules, trees, and graphs. We use $\hat{f}$ to refer to a learner's model, which is the learner's approximation to the target function f.

The learning element builds a model from a set of examples. That is, from S, it produces $\hat{f}$. The performance element uses the model to make a prediction about an observation. That is, given x, it uses $\hat{f}$ to return $\hat{y} \in \{-,+\}$.

In the previous discussion, the learner built, maintained, and used a single model, $\hat{f}$. However, researchers have also developed *ensemble methods* that maintain multiple models (i.e., $\hat{f}^{(i)}$, for $i = 1, \ldots, M$). Generally, learning algorithms produce the same model from a given set of training examples, and such models make the same predictions. Consequently, ensemble methods must use some mechanism to produce different models. As examples, researchers have sampled the examples of the training set, sampled attributes of the training set, and used different learning methods to produce different models.

A learner may process all available examples at once, in a single batch, or it may process them online as they arrive. We review the AQ algorithm for batch learning in the next section, but algorithms that cope with concept drift are necessarily online. Most of the algorithms we review in the next section fall into this category.

Concept drift [19, 30] is a phenomenon in which examples have legitimate labels at one time and have different legitimate labels at another time. Geometrically, if we view a target concept as a cloud of points in a feature space, concept drift may entail the cloud changing its position, shape, and size. From the perspective of Bayesian decision theory, these transformations equate to changes to the form or parameters of the prior and class-conditional distributions.

Since all useful methods of online learning can acquire a target concept, one strategy of coping with concept drift is to use the method to build a model of a target concept, and when it changes, use the method to build a new model. However, it is difficult to detect when concepts change. A drop in predictive accuracy is an indicator of drift, and such drops in performance are correlated with concept drift, but they can also be due to sampling or the learner's instability.

At some level, all online methods of learning can cope with concept drift. When the target concept changes, given enough examples, such methods will eventually converge to the new target concept, but of course, the number of examples and the amount of time necessary to acquire the new target concept would be almost certainly impractical. Characterizing an online method's inherent ability to cope with concept drift is important. Most do not perform well, so it is natural to investigate what modifications we could make to a standard online algorithm to let it better cope with concept drift. On the other hand, we can also design completely new algorithms for concept drift, and these may or may not be useful for online learning of concepts that do not change over time. Once we have these algorithms, then it is important to conduct a proper empirical evaluation to characterize their performance, as we discuss in the next section.

2.2 Evaluation

Algorithms for supervised concept learning carry out inductive inference. They reason from examples to hypotheses about the characteristics of the categories to which the examples belong. As such, inference is uncertain, and thus, it is important to estimate this uncertainty.

Researchers have devised a number of schemes for evaluating inductive learners, and here, we focus on the evaluation of online learners that cope with concept drift. Evaluations divide into two broad categories: those that involve synthetic problems and those that involve real-world data sets.

Synthetic problems play an important role in research on systems for concept drift because researchers can design problems that test specific research hypotheses. Such hypotheses could involve learning certain classes of target concepts, learning with varying amounts of noise, drift, or overlap among concepts, and the like.

Instance generators for synthetic problems proceed through a series of target concepts. At each time step, the generator produces a random observation and assigns a label to it based on the time step and target concept. The random example is then presented to the learner.

To test the learner, protocols vary, but include measuring performance on the next example, on a randomly drawn test set for the time step, or on a single, randomly drawn test set for the period over which the target concept persists.

Researchers have used a variety of functions to represent target concepts. The Stagger concepts use conjunctive and disjunctive target concepts [30, 34], and we discuss this problem further in Sect. 5. Over the past eight years, the majority of the proposed instance generators have used shifting hyperplanes (e.g., [10, 12, 31, 33]).

Real-world tasks that have been used to evaluate learners for concept drift include calendar scheduling [26] web-page caching [10], and electricity price prediction [8]. These tasks are useful for comparing learners and establishing that they work well on real-world data sets, but to date, we are unaware of any real-world data set with ground truth sufficient to let researchers draw strong conclusions about the reasons for drops in a learner's accuracy, the types of drift that might occur in the data set, or even the presence of drift. For example, drops in a learner's performance could be due to sampling or instability, rather than to concept drift.

Over the years, we have used both synthetic problems and real-world data sets in our research. Since the Stagger concepts are the synthetic problem with the longest history in machine learning, we have used it most often in our evaluations. Indeed, in the mid-1990's, when we began investigating methods for concept drift based on Michalski's AQ algorithm, which we discuss in the next section, it was the only synthetic problem for concept drift.

3 The AQ Methods of Concept Learning

To develop AQ methods for concept drift, we based our efforts on AQ methods of both batch and incremental learning, which have been implemented in two systems: AQ15 [24] and AQ11 [23], respectively. AQ15 implements the AQ algorithm [22] for the propositional case. It uses a set of decision rules for each class as the concept representation. In the simplest case, a decision rule consists of a conjunction of conditions and an action. Each condition consists of a single attribute, a relational operator, and a set of attribute values. The action for classification tasks assigns a class label to the decision variable. Notice that a training example is a decision rule in which a single value is present for all attributes and all conditions consist only of the equality operator.

The performance element uses an instance's attribute values to evaluate each condition of each decision rule. If, for a given attribute, the instance and a condition have the same value, then the instance matches on that attribute and the condition evaluates to true; otherwise it evaluates to false. Using a strict matching scheme, if all of a rule's conditions match, then the rule matches, and the performance element predicts the class label that the action assigns to the decision variable. It is possible that no rule matches perfectly, so using a flexible matching scheme, the performance element returns the prediction of the rule that most closely matches. Researchers have a devised a number of schemes for flexibly matching decision rules.

The learning element takes as input a set of positive training examples and a set of negative training examples. It randomly picks one positive training example as the *seed*. The AQ algorithm forms a rule by using the other positive examples to generalize the seed until producing a rule that matches or *covers* a negative example. The algorithm removes from further consideration the positive examples that the rule covers, and repeats until all positive examples have been covered. The learning element returns the set of rules.

The AQ algorithm builds a set of decision rules for the negative class using the same procedure by exchanging the role of the negative and positive examples. For problems with more than two classes, the algorithm selects a class, treats its examples like positive examples, treats the examples of the other class like negative examples, and proceeds as described previously. It then repeats using each of the remaining classes as the positive class.

The AQ11 system [23] uses the AQ algorithm [22] to incrementally learn decision rules. That is, AQ11 uses new training examples to modify an existing set of decision rules. The concept description and performance elements of AQ11 are the same described previously.

AQ11's learning element consists of three steps [23]. The first step determines which of the new examples the current set of rules misclassifies. If, say, a positive rule covers a negative example, then as a second step, AQ11 uses the AQ algorithm [22] to specialize the rule so it no longer covers the negative example. Finally, in the third step, AQ11 combines the newly specialized rule with the new training examples and uses the AQ algorithm to generalize the rule until it intersects a negative rule or covers a negative example.

AQ11 [23] repeats this procedure for the negative class. For problems with more than two classes, it selects a class, treats it as if it were positive, combines the other classes and treats them as if they were negative, and applies the three-step procedure described in the preceding paragraph.

4 Methods for Concept Drift

Researchers have developed a number of learning methods for coping with concept drift. In this section, we review methods based on Michalski's AQ algorithm [22]: AQ-PM [20], AQ11-PM [21], AQ11-PM+WAH [18]. We also review other methods not based on AQ that we included in our study for the sake of comparison. These methods include Stagger [30], winnow [15], weighted majority [5, 16], dynamic weighted majority [11, 13], additive experts [12], streaming ensemble algorithm [31], and accuracy weighted ensemble [33]. Other methods for coping with concept drift include the FLORA systems [34], concept-drifting very fast decision trees [10], StreamMiner [6], ultra-fast decision trees [7], online rank [3], and paired learners [2].

4.1 Methods Based on the AQ Algorithm

AQ-PM [20] is based on the AQ algorithm [22] and tracks concept drift by maintaining examples from the stream that lie on the boundaries of the current concept description over a fixed window of time. It has the same concept description language and performance element as the AQ systems described previously.

Specifically, AQ-PM's learning element stores a set of decision rules and the examples from previous training episodes that lie on the boundaries of the rules [20]. These boundary or extreme examples document and enforce the important boundaries between concept descriptions.

When new examples arrive, AQ-PM produces a new set of decision rules from the stored examples and the new examples [20]. It then uses the new decision rules to select and store a new set of boundary examples. It also removes any examples that are older than a user-defined threshold. Put another way, it remembers boundary examples for a specified period of time.

Notice that the method of storing and using boundary examples is independent of the rule-learning mechanism. AQ-PM combines boundary examples with new examples and uses the batch version of the AQ algorithm to produce a new set of decision rules. However, one can also use an incremental method to modify a set of decision rules using the boundary examples and new training examples. This is the idea behind AQ11-PM [21].

AQ11-PM [21] is identical to AQ-PM [20], except the former learning method uses AQ11's incremental algorithm for inducing rules [23]. AQ11-PM stores a set of decision rules and boundary examples. When new examples arrive, it combines them with the boundary examples and uses AQ11 to incrementally learn a new set of decision rules. Using the examples that are newer than the threshold, it selects those examples that lie on the boundaries of the new decision rules.

Experimental results on the Stagger concepts [30, 34], which we discuss further in Sect. 6, suggest that AQ11-PM outperformed AQ-PM in terms of slope and asymptote. They also suggest that the performance of both AQ11-PM and AQ-PM may have suffered because of the methods' use of fixed windows for remembering and forgetting examples.

In an attempt to avoid using a fixed window for forgetting, we added Widmer and Kubat's [34] window adjustment heuristic (WAH) to AQ11-PM [21]. We called the new method AQ11-PM+WAH [18].

The window adjustment heuristic attempts to shorten the window when the target concept is changing, and lengthens the window when the concept is stable. If the concept is extremely stable, then the heuristic gradually reduces the size of the window by a fixed amount, which saves memory and readies the algorithm for the onset of any forthcoming instability.

The heuristic takes as input three thresholds for acceptable accuracy, minimum coverage, and maximum coverage. During learning, the heuristic computes the coverage, complexity, and accuracy of the current concept description. *Coverage* is the number of positive examples in the window that the positive rules cover. *Complexity* is the number of conditions in the positive rules. *Accuracy* is the percentage of correctly classified examples in the window.

If the accuracy of the rules is decreasing and the coverage or accuracy falls below their respective thresholds, then the heuristic shortens the current window by 20%. Otherwise, if accuracy is above its threshold and coverage is twice its threshold, then the heuristic reduces the size of the window by one unit. However, if accuracy and coverage of the rules are simply above their respective thresholds, then the heuristic leaves the window at its current length. Finally, if accuracy or coverage is below its respective threshold, then the heuristic increases the size of the window by one unit.

It is true that replacing AQ11-PM's static window with the window adjustment heuristic increases the number of parameters from one to three. However, given that it is unlikely that drift will occur at regular intervals, adding the heuristic should let the method cope with a broader range of learning tasks that involve drift.

4.2 Other Methods for Concept Drift

As mentioned previously, researchers have developed a number of methods for learning concepts that drift or change over time. Although we do not provide an extensive review of these methods here, we have attempted to do so elsewhere [13]. Consequently, we survey only those methods included in this study. These methods include Stagger [30], winnow [15], weighted majority [5, 16], dynamic weighted majority [11, 13], additive experts [12], streaming ensemble algorithm [31], accuracy weighted ensemble [33].

Stagger [29, 30] was the first learning method designed to cope with concept drift. As its concept description, Stagger maintains four counts for each feature for positive confirming, positive infirming, negative confirming, and negative infirming evidence of the target concept. It can also combine features using conjunction, disjunction, and negation operators.

When a new example arrives, the learning element updates the appropriate counts based on whether the example is a positive or negative example of the target concept. It also decays the counts to cope with concept drift. Then, using the counts, Stagger computes probabilistic measures of necessity and sufficiency of each feature for the target concept. Through a process called *chunking*, based on these probabilistic measures and thresholds, Stagger may assert or retract generalizations, specializations, and negations of features and chunks.

The performance element uses an observation's attribute values, the counts, and the measures of necessity and sufficiency to compute the probability of the instance being positive. If the probability is above the decision threshold, then Stagger predicts the positive class.

For this study, we used our implementation of Stagger, which we produced using WEKA [35]. We have not yet implemented chunking, so the results we present are preliminary. We plan to report results for Stagger with chunking in future publications.

Winnow [15] represents concepts as a linear discriminant function. It is similar to the perceptron, but uses a different function to update its weights. Given the weights w and the observation x, winnow predicts

$$\hat{y} = \begin{cases} + & \text{if } \sum_{i=1}^{n} w_i x_i > \theta \ ; \\ - & \text{otherwise} \ . \end{cases}$$

To set the weight vector, winnow uses two values to either *promote* or *demote* each weight [15]. It first sets the threshold θ and each weight to initial values (e.g., $\theta = n/2$, $w_i = 1.0$). For each new example, winnow first makes a prediction for the example, and if its prediction is correct, then it does nothing. However, if the prediction is incorrect, then it updates its weights depending on whether the misclassification was a false positive or a false negative. If winnow misclassifies a negative example, then it promotes each weight by multiplying it by α (e.g., $\alpha = 2.0$); otherwise, it demotes each weight by multiplying it by β (e.g., $\beta = 0.5$). We used the implementation of winnow in WEKA [35] and ran it with its default settings.

Weighted majority [16] is an ensemble method that maintains a collection of weighted *experts* or base learners. The algorithm initializes the weight for each learner to one, and makes a prediction for an observation by obtaining a prediction from each learner in the pool. It makes a global prediction by returning the weighted-majority vote. Learning entails obtaining a prediction from each expert for an example, and, for each learner that makes a mistake, the algorithm multiplies the learner's weight by β (e.g., $\beta = 0.5$).

Blum implemented weighted majority using pairs and triples of features as base learners [5]. We focus discussion on learners with pairs of features, since we used them in this study. Each expert uses a pair of features to predict, and there is one expert for each pair. Each expert maintains a history of the k most recent labels associated with its pair. To predict, if an expert's features are present in the observation being considered, then it predicts the majority class label in its history. During learning, if an expert's features match an example, then it stores the example's label in its history by replacing the oldest of the k labels.

Dynamic weighted majority (DWM) [11, 13] is an extension of the weighted majority algorithm [5, 16]. DWM uses as its concept description a dynamically sized, weighted collection of online base learners. Initially, DWM starts with a single base learner with a weight of one that predicts a default or random class. For each new example DWM receives, it obtains a prediction from each member of the ensemble, weights those predictions based on each learner's weight, and predicts the weighted-majority vote. If an individual learner's prediction is incorrect, then DWM multiplies its weight by β (e.g., $\beta = 0.5$). If DWM's global prediction is incorrect, it adds a new base learner to the ensemble with an initial weight of one. DWM also removes learners with weights below a threshold θ (e.g., $\theta = 0.01$). The method also has a parameter p that governs how frequently it updates its weights and adds and removes learners. For the results we present in this study, $p = 1$, so DWM updates its weights and adds and removes learners with the arrival of each new example.

We present results for two instantiations of DWM: DWM-NB and DWM-ITI [13]. DWM-NB uses an online version of naive Bayes (NB) as its base learner, whereas DWM-ITI uses the Incremental Tree Inducer (ITI) [32] as its.

An online version of naive Bayes stores counts for the number of examples processed, the number of examples of each class, and the number of attribute values given each class. Learning entails incrementing these counts as new examples arrive. To classify an observation, the performance element uses these counts to compute estimates of the prior and class-conditional probabilities, assumes conditional independence of the attributes, and uses Bayes rule to determine the most probable class. We used the implementation of naive Bayes in WEKA [35].

ITI is an incremental algorithm for inducing decision trees [32]. A decision tree is a rooted tree with internal nodes that correspond to attributes and external nodes that correspond to class labels. From an internal, attribute node, there are edges for each value the attribute takes. ITI produces decision trees with only binary splits, and at each node, it maintains a set of counts of class labels and attribute values.

To predict, ITI traverses from the root to a leaf node using the observation's attributes and their values to guide the traversal. Upon reaching a leaf node, ITI returns the class label stored in the node as its prediction for the observation.

When processing examples, ITI sorts an example to a leaf node, as described in the previous paragraph, and updates the counts at each node along the path from the root node to the leaf node [32]. When the counts at a node suggest that the split at that node could be better, ITI may restructure its decision tree by selecting other splitting attributes.

The additive experts algorithm for discrete case (AddExpD) [12] is an extension of DWM [13]. Rather than initializing a new learner's weight to one, AddExpD initializes a new learner's weight to the ensemble's total weight multiplied by the constant γ (e.g., $\gamma = 0.1$). Although this is a simple change to DWM, it has both theoretical and practical implications. Theoretically, unlike DWM's, AddExpD's weighting procedure submits to formal analysis, and we proved a worst-case bound on the number of mistakes that AddExpD makes when learning under concept drift. Practically, there are slight differences in performance between DWM and AddExpD, as we discuss in Sect. 6. In this study, we present results for AddExpD with naive Bayes as its base learner, which we refer to as AddExpD-NB.

The streaming ensemble algorithm (SEA) is also an ensemble method, but one that maintains a fixed-capacity collection of unweighted batch learners [31]. The performance element returns as its prediction the majority vote of the predictions of the classifiers in the ensemble. The learning element maintains a collection of m batch learners, processes batches of p examples, and keeps two classifiers in reserve: C_i and C_{i-1}. C_i is built from the ith batch of examples, whereas C_{i-1} is built from the $i-1$th batch.

After the ith batch is complete, consisting of p examples, SEA builds the classifier C_i [31]. If the ensemble is not full, then SEA adds C_{i-1} to the ensemble and sets C_{i-1} to C_i. If the ensemble is full, then SEA evaluates C_{i-1} and the members of the ensemble on the current batch using a measure of *quality*. If the quality of C_{i-1} is greater than the quality of the lowest quality member of the ensemble, then SEA replaces that member of the ensemble with C_{i-1}.

SEA measures *quality* using a heuristic that takes into account the performance of the new classifier, the best two classifiers in the ensemble, and the ensemble as a whole [31]. Although the authors claim that quality proved better than accuracy, we have not found this to be the case for the problems we have considered, including the one presented here. Consequently, we measured quality using accuracy on the new batch.

In this chapter, we present results for SEA [31] with naive Bayes as its base learner, and we refer to this learner as SEA-NB. We have evaluated SEA with other base learners, including J48, which is the implementation of C4.5 [27] in WEKA [35], but we found that for the Stagger concepts [30, 34], naive Bayes produced the best performance.

The accuracy weighted ensemble (AWE) maintains a fixed-capacity collection of weighted batch learners [33]. The performance element predicts the weighted-majority vote of the predictions of the members of the ensemble. The learning element maintains a collection of m batch learners and processes batches of p examples. Upon completing a batch, AWE uses cross-validation to measure the base learner's performance on the new batch. It uses this information to derive a weight for the classifier C_i, which is then builds from the entire batch.

Although one could use costs as weights, we used mean-squared error (see Eq. 8 in [33]). That is, $w_i = \text{MSE}_r - \text{MSE}_i$, where MSE_r is the mean-squared error of a classifier that predicts the majority class in S (i.e., the default rule), and MSE_i is the mean-squared error of the classifier C_i on the current batch.

We present results for AWE [33] with naive Bayes as the base learner, and refer to this learner as AWE-NB. As with SEA [31], we experimented with other base learners, but found that naive Bayes produced the best results for the Stagger concepts [30, 34], which we discuss in the next section.

5 Problems and Data Sets for Concept Drift

Researchers have proposed and used a number of problems and data sets for evaluating learners for concept drift. Synthetic problems, in particular, are widely used mainly because of the control they afford of the target concepts, the drift points, the degree of drift, class noise, and other such factors.

Real-world data sets have also been used, but most if not all of these data sets have little or no ground truth. As a consequence, it is difficult to understand the characteristics of the drift occurring in the data set, if it is occurring at all. It is then equally difficult to understand the changes one observes in a learner's performance.

What we call the *Stagger concepts* form a synthetic problem that Schlimmer and Granger originally used to evaluate Stagger [29, 30]. Since then, it has been used in a number of studies to evaluate learners for concept drift (e.g., [2, 11, 12, 18, 20, 21, 34]), and it is the problem we used in this study to evaluate and compare the methods described in the previous section.

The Stagger concepts consist of three attributes that each take three values: size, which can be small, medium, or large; shape, which can be circle, triangle, or square; and color, which can be red, blue, or green. The presentation of examples occurs over 120 time steps, with one example presented each time step. After each presentation, the learner is evaluated on a random sample of 100 examples. We measure the learner's performance on the test set using percent correct.

There are three target concepts, each of which persists for 40 time steps. For the first 40 time steps, the target concept is $color = red \wedge size = small$. For the next 40 time steps, the target concept is $color = green \vee shape = circle$. For the last 40 time steps, the target concept is $size = medium \vee size = large$.

We conducted 50 trials and computed the average accuracy at each time step. We also computed 95% confidence intervals. To obtain a single measure of accuracy, we computed the area under the performance curves (AUC). We computed the area under the entire curve and under the curve after the first drift point. We have found the area under the curve after the first drift point particularly useful because most learners do well on the first concept, and we are most interested in the accuracy of learners *after* drift occurs [2].

While we have found area to be a convenient measure, it may not make clear important aspects of performance, such as how quickly a learner responds to drift and acquires the next target concept. Consequently, we also plot performance curves for each method and analyze the slope and asymptote of these curves.

We should note that the version of the Stagger concepts we describe is different than that originally proposed by Schlimmer and Granger [29, 30]. Our version is due to Widmer and Kubat [34]. The target concepts of the two versions have the same functional form, but they differ in the features involved. This does not change the problem since all attributes are nominal and take the same number of values. For example, Schlimmer and Granger's [30] first target concept was $color = red \wedge shape = square$, whereas Widmer and Kubat's [34] was $color = red \wedge size = small$.

A second difference is the manner in which examples are presented to the learner. It appears that Schlimmer and Granger [30] presented examples of a target concept until the learner achieved 100% accuracy on the test set and then changed to the next target concept for a single run (see Fig. 2 in [30]). Widmer and Kubat presented 40 examples from each target concept and changed to the next regardless of the learner's performance. They also conducted 10 trials and plotted the average accuracy [34]. In contrast, we conducted 50 trials and plotted average accuracy and 95% confidence intervals. As mentioned, we also computed area under the entire curve and under the curve after the first drift point.

Some may perceive that the Stagger concepts are simple. While it is true that the size of the feature space is only 27 examples, it is not only the size of the feature space that makes a learning problem challenging. For the Stagger concepts, the first target concept is conjunctive, whereas the second is disjunctive. Furthermore, the first and second concepts share only a single positive example, so it is almost a concept reversal. Finally, the third concept involves internal disjunction, meaning that a single attribute can take two (or more) different values. Indeed, as we have shown elsewhere [2], controlling for differences in base learners and using slope, asymptote, and area under the performance curve, a number of recently proposed ensemble methods for concept drift performed better in terms of slope and asymptote on newer, supposedly harder problems, but performed worse on the Stagger concepts. We contend that the Stagger concepts remain useful for evaluating learners for concept drift.

Although our study involves only the Stagger concepts, other synthetic problems and real-world data sets exist for evaluating learners for concept drift. In terms of synthetic problems, drifting hyperplanes have become quite popular and numerous over the last eight years [10, 12, 31, 33].

Real-world data sets include the CAP data set for a calendar-scheduling task [26], electricity pricing [8], and web-page caching [10]. However, an issue with real-world data sets is that they have little or no ground truth, so it is difficult to draw scientific conclusions about differences in performance or whether drift occurs.

6 Experimental Study

Since the mid-1990's, we have designed, implemented, and evaluated a number of learning algorithms that cope with concept drift. We have also used implementations and implemented the algorithms of other researchers. We have reported results with these implementations on both synthetic problems and real-world data sets. In this section, we summarize the results we have obtained on the Stagger concepts [30, 34].

Table 1 lists the overall area under the curve and the area under the curve after the first drift point for thirteen algorithms evaluated on the Stagger concepts [30, 34]. We also included the references to the papers in which the methods and results on the Stagger concepts originally appeared. We ordered the results based on area under the curve after the first drift point because we are most concerned about performance after drift occurs, especially since most methods perform well on the first target concept. Sorting the methods by area under the entire curve does not change the ordering except in the case of AQ11-PM+WAH; it moves up one place from second to first.

As one can see, DWM-ITI was the best performing method, with an AUC after the first drift point of .882. The two methods based on AQ11 (i.e., AQ11-PM and AQ11-PM+WAH) and AQ11 itself achieved areas after the first drift point equal to or above

Table 1. Results on the Stagger concepts for thirteen methods. Measures are average area under the performance curve over 50 runs for the overall curve and for the curve after the first drift point. Entries are ordered by AUC after the first drift point.

Method	Area Under the Curve Overall	Area Under the Curve After	Original References
DWM-ITI	.889	.882	[11, 13]
AQ11-PM+WAH	.898	.875	[18]
AQ11	.890	.875	[18, 21]
AQ11-PM	.891	.874	[18, 21]
DWM-NB	.867	.859	[11, 13]
Winnow	.858	.853	—
AddExpD-NB	.856	.831	[12]
AWE-NB	.829	.808	[2]
Blum's WM	.827	.770	[11, 13]
Stagger	.822	.783	—
AQ-PM	.796	.744	[20]
SEA-NB	.776	.732	[2]
Naive Bayes	.647	.516	[2, 11, 13]

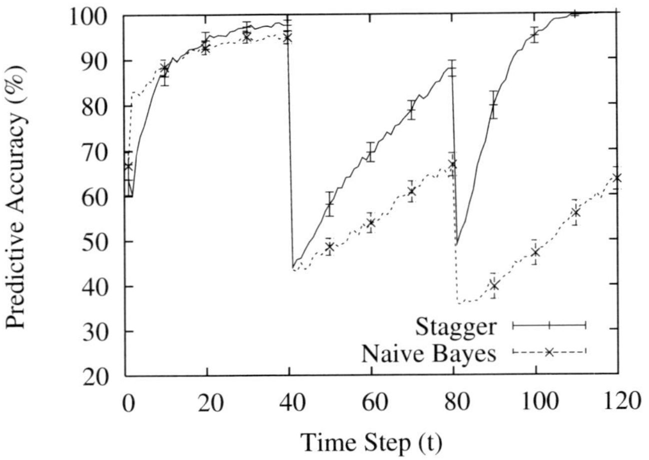

Fig. 1. Performance curves for naive Bayes and Stagger on the Stagger concepts. Measures are accuracy with 95% confidence intervals.

.875. For context, we also present results for naive Bayes, which performed relatively poorly, since it has no mechanisms to cope with concept drift.

As we mentioned in the previous section, area under the curve is a convenient measure, but it may not reveal all important aspects of performance, such as slope and asymptote. In Figs. 1–13, we present performance curves for these methods, and we compare each method to Stagger.

We compare all of the methods to Stagger because it was the first learner to cope with concept drift [30]. As we have mentioned, our implementation of Stagger is incomplete. We implemented Stagger's Bayesian weighting scheme, but we have not yet implemented its so-called chunking mechanisms, which let Stagger generalize, specialize, and invert expressions [29]. Without chunking Stagger performed comparably to the other methods, and outperformed AQ-PM [20] and SEA-NB [31]. We suspect that chunking will improve Stagger's performance, but doubt that it will let Stagger outperform DWM-ITI, for example. Naturally, this is a conjecture, and we plan to present results for Stagger with chunking in future publications.

While our results for Stagger are admittedly preliminary without chunking, we chose to include these results because, as mentioned, Stagger performed comparably to the other methods, and most crucially, because we are unaware of any direct comparisons to Stagger. Direct comparisons to Stagger are particularly important because Widmer and Kubat [34] modernized Schlimmer's original experimental design [29, 30]. As a consequence, it is difficult to make even indirect comparisons to the original study of Stagger and its results. Without direct comparisons, it is difficult to know if the methods for concept drift developed since 1986 are truly better than is Stagger for concept drift. We hope the results we present in this chapter are a step in this direction.

6.1 Performance of the AQ Methods for Concept Drift

Figure 1 shows performance curves for Stagger and naive Bayes. Naive Bayes acquires the first target concept as well as does Stagger, but then performs poorly on subsequent target concepts, since it has no mechanisms for coping with drift. Stagger outperformed naive Bayes overall, although it did not perform well on the second target concept, which is a disjunctive concept. Since our implementation of Stagger uses only Bayesian weighting, it was disadvantaged on the second concept since disjunction is difficult to represent using only weights of features. We suspect that Stagger's performance on the second target concept will improve when we implement chunking.

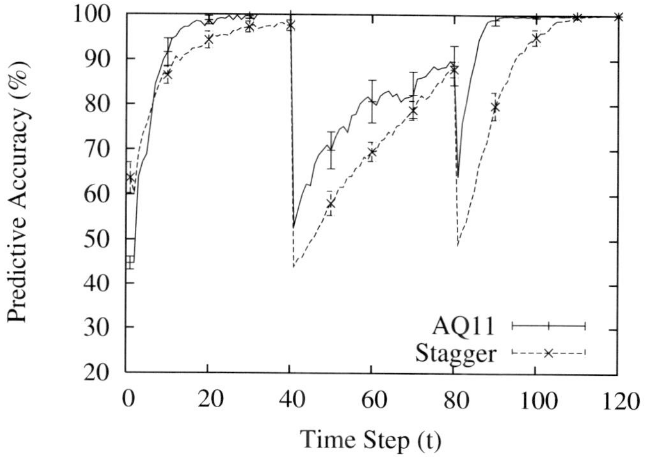

Fig. 2. Performance curves for AQ11 and Stagger on the Stagger concepts. Measures are accuracy with 95% confidence intervals.

Figure 2 contains the performance curves for AQ11 [23] and Stagger [30]. Even though AQ11 has no inherent mechanisms for concept drift, it outperformed Stagger. We have argued elsewhere [18] that AQ11 performed well on this task mainly because of its expressive language for representing concepts. For example, AQ11 can directly represent internal disjunction (e.g., $size = medium \vee large$), whereas other versions of AQ must represent internal disjunction with multiple rules.

We contend that AQ11's expressive language let it build more complex models. As we know, simpler models often perform well because they are less likely to overfit. But it appears that in this case AQ11's tendency to overfit may have let it respond well to concept drift. The expressiveness of its concept description language let it change its rules more quickly in response to change in the target concept.

What is not represented in the graph is the size of the rules that AQ11 produced. As we have remarked elsewhere [17], AQ11 occasionally produced rules with a large number of conditions from a single new example and a relatively simple set of rules. This type of instability may not be desirable for all problems, but it seems to have aided AQ11 in this instance. We are unaware of any other online learning method that performs this well on the Stagger concepts with no additional mechanisms for coping with concept drift.

Figure 3 shows the performance curves for AQ-PM [20], AQ11 [23], and Stagger [30]. AQ-PM performed comparably to Stagger on the first and second target concepts,

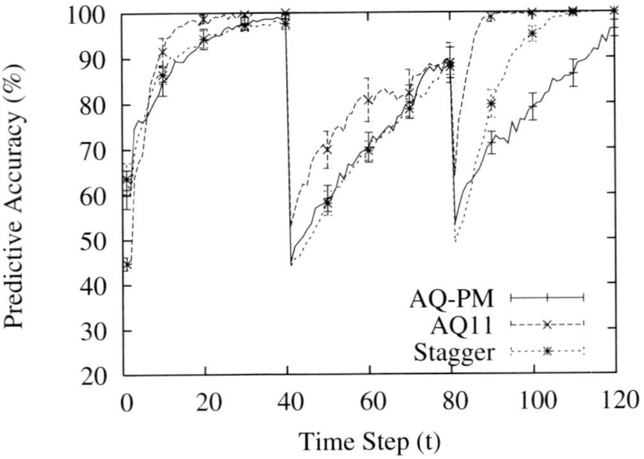

Fig. 3. Performance curves for AQ-PM, AQ11, and Stagger on the Stagger concepts. Measures are accuracy with 95% confidence intervals.

but performed worse on the third. AQ-PM also performed worse than did AQ11. We anticipate the main reason for AQ-PM's lackluster performance was its use of a fixed window of thirty examples. That is, AQ-PM maintained a window of thirty examples taken from the boundaries of concept descriptions, which it used with a new example to produce a new set of rules.

The curves in Fig. 4 illustrate the effects of adding partial instance memory to AQ11 [23]. We have reasoned elsewhere [17, 18] that the examples stored in partial memory decreased AQ11's propensity to overfit. While this improved performance on the second

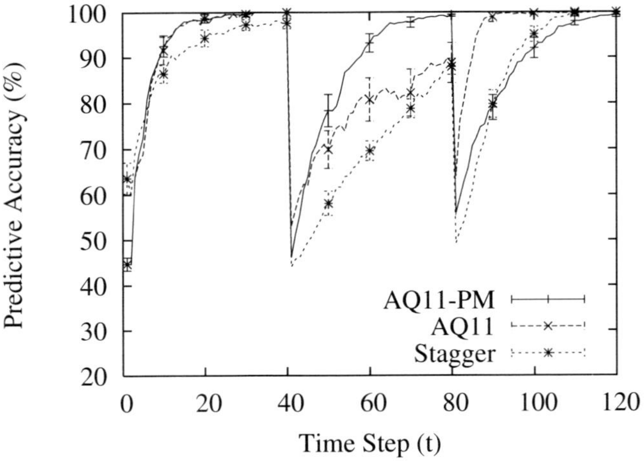

Fig. 4. Performance curves for AQ11-PM, AQ11, and Stagger on the Stagger concepts. Measures are accuracy with 95% confidence intervals.

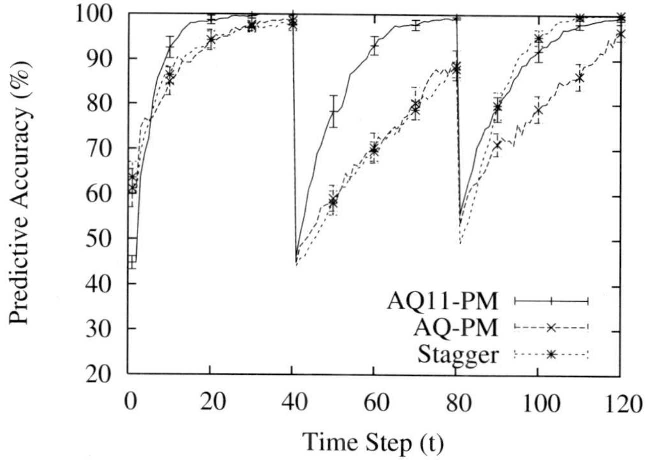

Fig. 5. Performance curves for AQ11-PM, and AQ-PM, and Stagger on the Stagger concepts. Measures are accuracy with 95% confidence intervals.

target concept, it decreased accuracy on the third. As shown previously in Table 1, both algorithms performed similarly in terms of area under the curve after the first drift point.

For context, in Fig. 5, we show the performance curves of AQ-PM and AQ11-PM. As one can see, AQ11-PM performed much better on this task than did AQ-PM in terms of slope and asymptote. Referring to Table 1, AQ11-PM achieved an AUC after the first drift point of .874, whereas AQ-PM achieved one of .744.

Both learners operated with fixed windows of thirty examples. Previously, we mentioned that AQ-PM's performance was harmed by this fixed window. We contend that

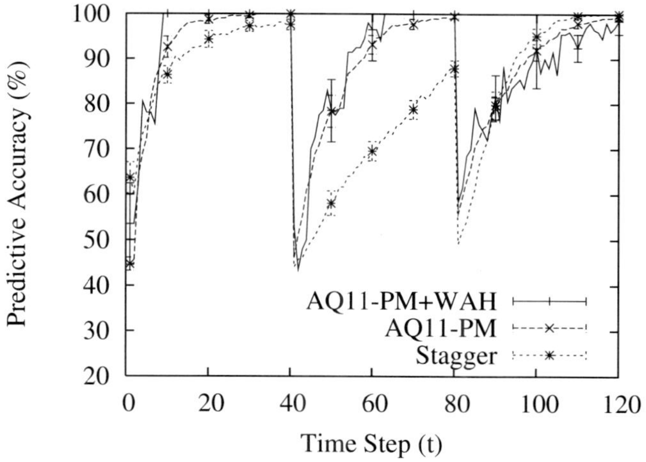

Fig. 6. Performance curves for AQ11-PM+WAH, AQ11-PM, and Stagger on the Stagger concepts. Measures are accuracy with 95% confidence intervals.

AQ11-PM was not as affected because of its more expressive language for representing rules.

Figure 6 shows the effect of adding Widmer and Kubat's window adjustment heuristic (WAH) [34] to AQ11-PM [18]. The curve for the version of AQ11-PM with the adjustment heuristic has higher variance than does the version with a fixed window. Both versions achieved the same AUC after the first drift point (see Table 1). AQ11-PM+WAH performed better on the second target concept, whereas AQ11-PM performed better on the third. In principle, the version with the dynamically-sized window should perform over a wider range of conditions, and we hope to explore this notion further in future work.

6.2 Performance of Other Methods for Concept Drift

As mentioned previously, we have developed a number of methods for concept drift that are not based on the AQ algorithm. Additionally, we have implemented a number of methods for concept drift that other researchers have proposed. In this section, we review the performance of these methods on the Stagger concepts. We relate how these methods performed with respect to Stagger and AQ11-PM.

We begin with winnow. Figure 7 shows its performance curve and those for AQ11-PM and Stagger. Aside for its performance on the first concept, winnow performed comparably to AQ11-PM. We used the implementation in WEKA [35], and we are unsure why this implementation of winnow was able to acquire the second and third concepts, but not the first. The first is conjunctive and is linearly separable, so winnow can represent it, at least in principle. The implementation in WEKA transforms nominal attributes into a binary representation, and this may have had an adverse effect, although it is unclear why this did not also affect winnow's performance on the disjunctive concepts.

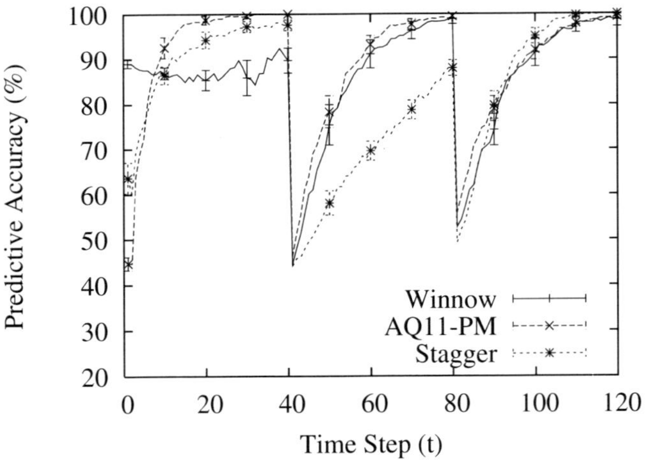

Fig. 7. Performance curves for Winnow, AQ11-PM, and Stagger on the Stagger concepts. Measures are accuracy with 95% confidence intervals.

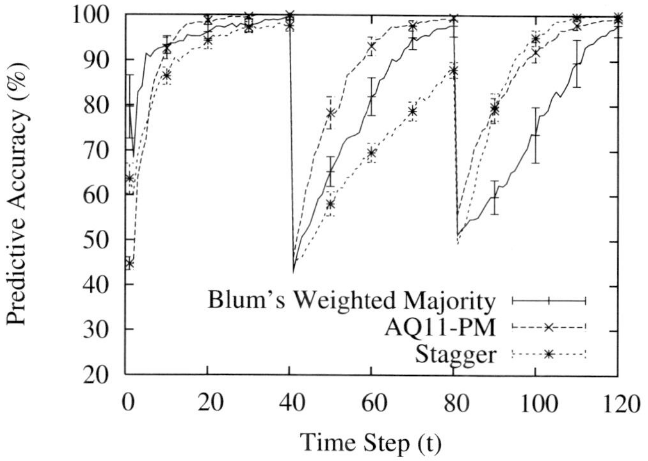

Fig. 8. Performance curves for Blum's weighted majority, AQ11-PM, and Stagger on the Stagger concepts. Measures are accuracy with 95% confidence intervals.

Figure 8 compares the performance of Blum's weighted majority with pairs of features [5] to AQ11-PM [21] and to Stagger on the Stagger concepts. We ran Blum's version to characterize its performance on this problem [11]. As one can see, Blum's weighted majority did not perform as well as did AQ11-PM on this problem. Through subsequent analysis, we concluded that the method's use of conjunctive pairs of features as experts led to its poor performance [11, 13].

In an attempt to replace pairs of features with more general learning methods, we developed dynamic weighted majority [11, 13], and we show its performance curve in Fig. 9 with naive Bayes as the base learner. DWM-NB outperformed Blum's weighted

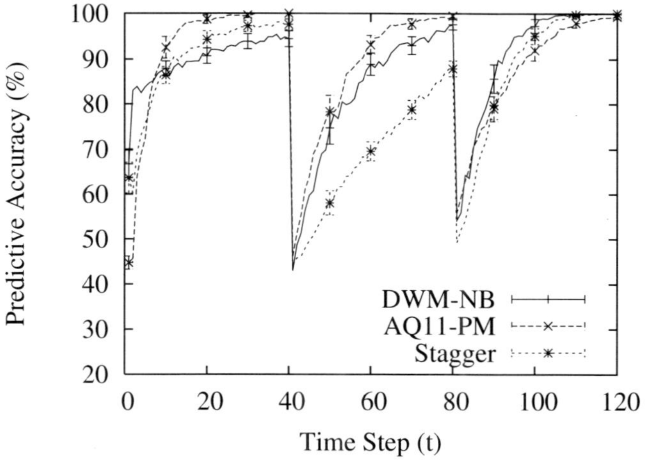

Fig. 9. Performance curves for DWM-NB, AQ11-PM, and Stagger on the Stagger concepts. Measures are accuracy with 95% confidence intervals.

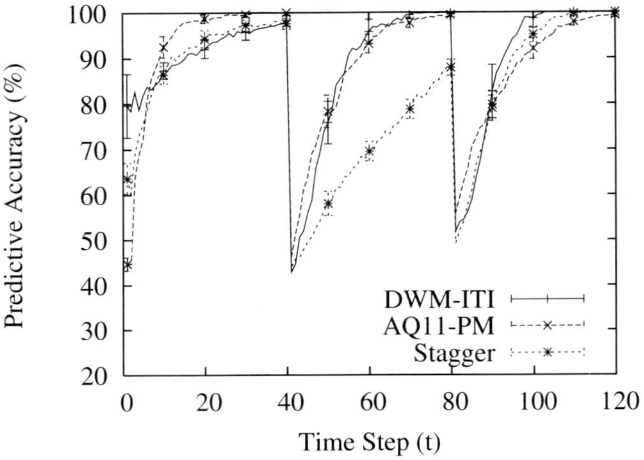

Fig. 10. Performance curves for DWM-ITI, AQ11-PM, and Stagger on the Stagger concepts. Measures are accuracy with 95% confidence intervals.

majority on the second concept, but did so mainly on the third. Although performance was better, naive Bayes was unable to well-represent the disjunctive concepts.

Consequently, we paired DWM with ITI [11], an incremental algorithm that learns decisions trees [32]. Although DWM-ITI's performance on the first concept was not as good as AQ11-PM's, it performed slightly better than did AQ11-PM on the second and third concept. Indeed, referring to Table 1, AQ11-PM had an AUC after the first drift point of .874, whereas DWM-ITI achieved an AUC of .882. In the future, we hope to use AQ11 and AQ11-PM as base learners for DWM.

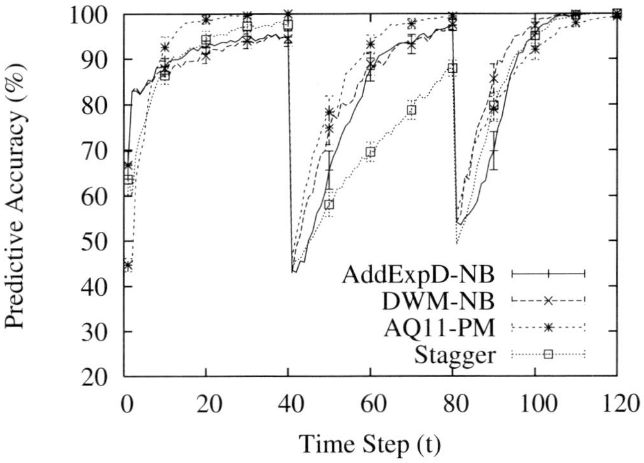

Fig. 11. Performance curves for AddExpD-NB, DWM-NB, AQ11-PM, and Stagger on the Stagger concepts. Measures are accuracy with 95% confidence intervals.

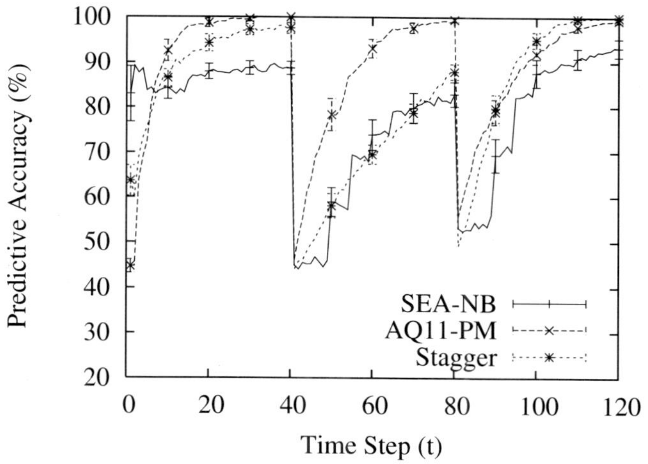

Fig. 12. Performance curves for SEA-NB, AQ11-PM, and Stagger on the Stagger concepts. Measures are accuracy with 95% confidence intervals.

As mentioned previously, AddExpD [12] can perform similarly to DWM [13]. Figure 11 illustrates. DWM-NB outperformed AddExpD-NB on this task, the former achieving an AUC .859 and the latter achieving .831, as shown in Table 1.

In Fig. 12, we present the performance curves for SEA-NB, AQ11-PM, and Stagger. As one can see, SEA-NB did not perform as well as did the other learners on this task. A problem is that it reads examples in batches and produces a classifier for each batch. It is easy to see the effect of this scheme in the "stair steps" in the performance curve. SEA also maintains a classifier in reserve, C_{i-1}, which impeded its ability to react quickly to drift when it occurred.

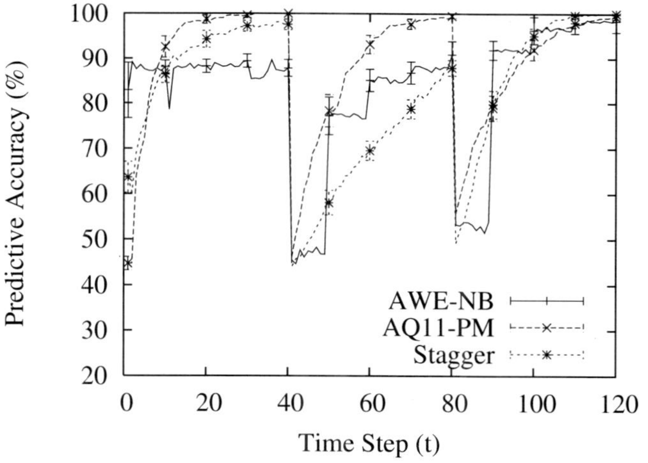

Fig. 13. Performance curves for AWE-NB, AQ11-PM, and Stagger on the Stagger concepts. Measures are accuracy with 95% confidence intervals.

Finally, in Fig. 13, we present the performance curves for AWE-NB, AQ11-PM, and Stagger. Recall that AWE is similar to SEA, but AWE weights its classifiers in the ensemble and does not have an extra classifier in reserve. As a result, AWE-NB outperformed SEA-NB on this task, but did not outperform AQ11-PM. Like SEA-NB, the performance curve for AWE-NB has "stair steps" that result from building new classifiers from batches of examples.

7 Discussion

The experimental results described in the previous section suggest that DWM-ITI and the AQ11 systems outperformed a number of other methods for concept drift on the Stagger concepts. One caveat, as we have mentioned, is that our implementation of Stagger only adjusted its weights and did not use chunking to generalize and specialize its features and chunks. This surely affected performance, and we plan to complete our implementation and report additional experimental results elsewhere.

Except for the results we present here and those found in the original papers describing Stagger [29, 30], we are unaware of any experimental studies that directly compare Stagger to other proposed methods for concept drift. It is quite possible that many algorithms proposed over the last decade do not outperform Stagger, and we intend to pursue this line of inquiry in future work.

We do believe that the performance of the AQ11 systems was due in part to its inductive bias. As evidence, DWM-ITI outperformed DWM-NB, and the only difference between these two learners is that DWM-ITI represented concepts using decision trees. We concluded that inductive bias mainly contributed to the observed differences performance. It is therefore possible that rules and AQ11's method of forming them gave those systems an advantage on this problem. The method of checking this conjecture is to use AQ11 as a base learner for DWM. As we discuss in the next section, we plan to pursue this and other ideas we have mentioned in future work.

8 Future Directions

Throughout the chapter, we have mentioned several possible avenues for future work. We plan to complete our implementation of Stagger by writing routines for chunking. Additionally, as originally designed, Stagger worked only with nominal attributes; it would be fruitful to design mechanisms for Stagger so it can handle numeric attributes. We also would like to use AQ11 as a base learner for DWM. This should let us better understand the role that inductive bias played in the performance of the AQ11 systems. We hope that the learner DWM-AQ11 will outperform the AQ11 systems and DWM-ITI. Finally, we are in the process of conducting studies to understand the problems for which unweighted (e.g., SEA) and weighted (e.g., DWM, AWE) ensembles are appropriate. We have reported initial results elsewhere [2].

9 Concluding Remarks

In this chapter, we have presented a survey of results on the Stagger concepts for a number of learning methods for concept drift, methods that have been previously published in the literature since 1986. Our early efforts in this area focused on systems built using Michalski's AQ algorithm, a covering algorithm that has been the foundation of numerous learning systems since its publication in 1969. We also described our more recent work on ensemble methods for concept drift, and the work of other researchers.

Empirical results on the Stagger concepts suggest that systems for concept drift based on the AQ11 algorithm outperformed the majority of the other systems included in the study. DWM-ITI performed the best in terms of area under the performance curve after the first drift point, and AQ11-PM+WAH did so based on area under the entire curve, although much of this is due to its excellent performance on the first target concept, which relates more to its ability to acquire concepts than to its ability to cope with concept drift.

Concept drift, shifting targets, time-changing data streams, and non-stationary environments are phenomena important for many real-world applications, from personal learning assistants to spam detection. Since 1986, researchers have introduced a number of learning methods, synthetic problems, and real-world data sets for concept drift. We anticipate that our survey of some of these learning methods and their performance on the Stagger concepts will contribute to researchers' understanding of the mechanisms important for learning in the presence of concept drift. Furthermore, we encourage researchers to conduct large-scale evaluations of methods of concept drift that involve multiple published algorithms, problems, and data sets, as is now common when evaluating batch learning algorithms for static concepts. These actions are required to place new methods in context with previous work. We hope our efforts in this regard will lead to better methods for coping with concept drift.

Acknowledgement. The author would like to thank Professor Ryszard S. Michalski for his years of advice, guidance, knowledge, and support. I tried to understand the beauty of his ideas, even though I did not always understand the ideas themselves. I did not always appreciate his lessons at the time he gave them, but a few years of experience have led me to comprehend more now what I then did not believe or understand. This research was conducted in the Department of Computer Science at Georgetown University.

References

1. Alpaydin, E.: Introduction to Machine Learning. MIT Press, Cambridge (2004)
2. Bach, S.H., Maloof, M.A.: Paired learners for concept drift. In: Proceedings of the Eighth IEEE International Conference on Data Mining. IEEE Press, Los Alamitos (2008)
3. Becker, H., Arias, M.: Real-time ranking with concept drift using expert advice. In: Proceedings of the Thirteenth ACM SIGKDD International Conference on Knowledge Discovery and Data Mining, pp. 86–94. ACM Press, New York (2007)
4. Bishop, C.M.: Pattern Recognition and Machine Learning. Springer, Heidelberg (2006)
5. Blum, A.: Empirical support for winnow and weighted-majority algorithms: Results on a calendar scheduling domain. Machine Learning 26, 5–23 (1997)
6. Fan, W.: StreamMiner: A classifier ensemble-based engine to mine concept-drifting data streams. In: Proceedings of the Thirtieth International Conference on Very Large Data Bases, pp. 1257–1260. Morgan Kaufmann, San Francisco (2004)

7. Gama, J., Medas, P., Rodrigues, P.: Learning decision trees from dynamic data streams. In: Proceedings of the 2005 ACM Symposium on Applied Computing (SAC 2005), pp. 573–577. ACM Press, New York (2005)
8. Harries, M., Sammut, C., Horn, K.: Extracting hidden context. Machine Learning 32(2), 101–126 (1998)
9. Hastie, T., Tibshirani, R., Friedman, J.: The Elements of Statistical Learning: Data Mining, Inference, and Prediction. Springer, Heidelberg (2001)
10. Hulten, G., Spencer, L., Domingos, P.: Mining time-changing data streams. In: Proceedings of the Seventh ACM SIGKDD International Conference on Knowledge Discovery and Data Mining, pp. 97–106. ACM Press, New York (2001)
11. Kolter, J.Z., Maloof, M.A.: Dynamic weighted majority: A new ensemble method for tracking concept drift. In: Proceedings of the Third IEEE International Conference on Data Mining, pp. 123–130. IEEE Press, Los Alamitos (2003)
12. Kolter, J.Z., Maloof, M.A.: Using additive expert ensembles to cope with concept drift. In: Proceedings of the Twenty-second International Conference on Machine Learning, pp. 449–456. ACM Press, New York (2005)
13. Kolter, J.Z., Maloof, M.A.: Dynamic weighted majority: An ensemble method for drifting concepts. Journal of Machine Learning Research 8, 2755–2790 (2007)
14. Langley, P.W.: Elements of Machine Learning. Morgan Kaufmann, San Francisco (1996)
15. Littlestone, N.: Learning quickly when irrelevant attributes abound: A new linear-threshold algorithm. Machine Learning 2, 285–318 (1988)
16. Littlestone, N., Warmuth, M.K.: The weighted majority algorithm. Information and Computation 108, 212–261 (1994)
17. Maloof, M.A.: Progressive partial memory learning. Ph.D. thesis, School of Information Technology and Engineering, George Mason University, Fairfax, VA (1996)
18. Maloof, M.A.: Incremental rule learning with partial instance memory for changing concepts. In: Proceedings of the International Joint Conference on Neural Networks, pp. 2764–2769. IEEE Computer Society Press, Los Alamitos (2003)
19. Maloof, M.A.: Concept drift. In: Wang, J. (ed.) Encyclopedia of Data Warehousing and Mining, pp. 202–206. Information Science Publishing, Hershey (2005)
20. Maloof, M.A., Michalski, R.S.: Selecting examples for partial memory learning. Machine Learning 41, 27–52 (2000)
21. Maloof, M.A., Michalski, R.S.: Incremental learning with partial instance memory. Artificial Intelligence 154, 95–126 (2004)
22. Michalski, R.S.: On the quasi-minimal solution of the general covering problem. In: Proceedings of the Fifth International Symposium on Information Processing, vol. A3, pp. 125–128 (1969)
23. Michalski, R.S., Larson, J.B.: Selection of most representative training examples and incremental generation of VL_1 hypotheses: The underlying methodology and the description of programs ESEL and AQ11. Technical Report UIUCDCS-R-78-867, Department of Computer Science, University of Illinois, Urbana (1978)
24. Michalski, R.S., Mozetic, I., Hong, J., Lavrac, H.: The multi-purpose incremental learning system AQ15 and its testing application to three medical domains. In: Proceedings of the Fifth National Conference on Artificial Intelligence, pp. 1041–1045. AAAI Press, Menlo Park (1986)
25. Mitchell, T.M.: Machine Learning. McGraw-Hill, New York (1997)
26. Mitchell, T.M., Caruana, R., Freitag, D., McDermott, J., Zabowski, D.: Experience with a learning personal assistant. Communications of the ACM 37(7), 80–91 (1994)
27. Quinlan, J.R.: C4.5: Programs for Machine Learning. Morgan Kaufmann, San Francisco (1993)

28. Russell, S.J., Norvig, P.: Artificial Intelligence: A Modern Approach, 2nd edn. Prentice Hall, Upper Saddle River (2003)
29. Schlimmer, J.C.: Concept acquisition through representational adjustment. Ph.D. thesis, Department of Information and Computer Science, University of California, Irvine (1987)
30. Schlimmer, J.C., Granger, R.H.: Beyond incremental processing: Tracking concept drift. In: Proceedings of the Fifth National Conference on Artificial Intelligence, pp. 502–507. AAAI Press, Menlo Park (1986)
31. Street, W.N., Kim, Y.: A streaming ensemble algorithm (SEA) for large-scale classification. In: Proceedings of the Seventh ACM SIGKDD International Conference on Knowledge Discovery and Data Mining, pp. 377–382. ACM Press, New York (2001)
32. Utgoff, P.E., Berkman, N.C., Clouse, J.A.: Decision tree induction based on efficient tree restructuring. Machine Learning 29, 5–44 (1997)
33. Wang, H., Fan, W., Yu, P.S., Han, J.: Mining concept-drifting data streams using ensemble classifiers. In: Proceedings of the Ninth ACM SIGKDD International Conference on Knowledge Discovery and Data Mining, pp. 226–235. ACM Press, New York (2003)
34. Widmer, G., Kubat, M.: Learning in the presence of concept drift and hidden contexts. Machine Learning 23, 69–101 (1996)
35. Witten, I.H., Frank, E.: Data Mining: Practical Machine Learning Tools and Techniques, 2nd edn. Morgan Kaufmann, San Francisco (2005)

Machine Learning Algorithms Inspired by the Work of Ryszard Spencer* Michalski

Krzysztof J. Cios[1] and Łukasz A. Kurgan[2]

[1] Virginia Commonwealth University, Richmond, USA, and IITiS, Polish Academy of Sciences, Poland
[2] University of Alberta, Edmonton, Canada

Abstract. In this chapter we first define the field of inductive machine learning and then describe Michalski's basic AQ algorithm. Next, we describe two of our machine learning algorithms, the CLIP4: a hybrid of rule and decision tree algorithms, and the DataSqeezer: a rule algorithm. The development of the latter two algorithms was inspired to a large degree by Michalski's seminal paper on inductive machine learning (1969). To many researchers, including the authors, Michalski is a "father" of inductive machine learning, as Łukasiewicz is of multivalued logic (extended much later to fuzzy logic) (Łukasiewicz, 1920), and Pawlak of rough sets (1991). Michalski was the first to work on inductive machine learning algorithms that generate rules, which will be explained via describing his AQ algorithm (1986).

1 Introduction

Machine learning (ML) is meant that machines/computers perform the learning instead of humans. The broadest definition of ML algorithms concerns the ability of a computer program to improve its own performance, in some domain, based on the past experience. Another, more specific, definition of ML is an ability of a program to generate a new data structure, different from the structure of the original data, such as a (production) IF… THEN… rule generated from numerical and/or nominal data (Kodratoff, 1988; Langley, 1996; Mitchell, 1997, Cios et al., 2007). ML algorithms are one of many data mining tools used for building models of data. However, the advantage of inductive ML algorithms is that they are one of only a few tools capable of generating user-friendly models. Namely, they generate models of the data in terms of the IF…THEN… rules that can be easily analyzed, modified, and used for training/learning purposes. This is in contrast to "black box" methods, such as neural networks and support vector machines, which generate models that are virtually impossible to interpret. Therefore, inductive ML algorithms (and their equivalent: decision trees) are preferred over other methods in fields where a decision maker needs to understand/accept the generated rules (like in medical diagnostics).

* Professor Michalski, after delivering talk on artificial intelligence at the University of Toledo, Ohio, in 1986, at the invitation of the first author, explained the origin of his second name: Spencer. Namely, he used the right of changing his name while becoming a United States citizen and adopted it after the well-known philosopher Herbert Spencer.

J. Koronacki et al. (Eds.): Advances in Machine Learning I, SCI 262, pp. 49–74.
springerlink.com © Springer-Verlag Berlin Heidelberg 2010

Michalski was involved in the development of algorithms that address both the supervised and unsupervised learning. Here we are concerned mainly with the supervised learning, although we also briefly comment on his work on clustering (the key unsupervised method). The supervised learning, also known as learning from examples, happens when the user/teacher provides examples (labeled data points) that describe concepts/classes. Thus, any supervised learning algorithm needs to be provided with a training data set, *S*, that consists of *M* training data pairs, belonging to *C* classes:

$$S = \{(\mathbf{x}_i, c_j) \mid i = 1,...,M; j = 1,...,C\}$$

where $\mathbf{x}_i$ is an n-dimensional pattern vector, whose components are called features/attributes, and c_j is a known class.

The mapping function f: c = f(x), is not known and a learning algorithm aims at finding/approximating this function. The training set represents information about some domain with the frequently used assumption that the features represent only properties of the examples but not relationships between the examples. A supervised ML algorithm searches the space of possible hypotheses, H, for the hypothesis (one or more) that best estimates the function f. The resulting hypotheses, or concept descriptions, are often written in the form of IF… THEN… rules.

The key concept in inductive ML is that of a hypothesis that approximates some concept. An example of a concept is, say, the concept of a hybrid car. We assume that only a teacher knows the true meaning of a concept and describes it by means of examples given to a learner (in our case a ML algorithm) whose task is to generate hypotheses that best approximate the concept. The concept of a hybrid car can be provided in terms of input-output pairs such as (gas&electric engine, hybridcar), (very low gas consumption, hybridcar), etc. We often assume that the terms concept and hypothesis are equivalent (which is not quite correct since the learner receives from a teacher only a finite set of examples that describe the concept so the generated hypotheses can only approximate it). Since hypotheses are often described in terms of rules we also use the term rule (and Michalski's notion of a cover, defined later) to denote the hypothesis.

Any supervised inductive ML process has two phases:

- Learning phase, where the algorithm analyzes training data and recognizes similarities among data objects to build a model that approximates f,
- Testing phase, when the generated model (say, a set of rules) is verified by computing some performance criterion on a new data set, drawn from the same domain.

Two basic techniques for inferring information from data are deduction and induction. Deduction infers information that is a logical consequence of the information present in the data. It is provably correct if the data/examples describing some domain are correct. Induction, on the other hand, infers generalized information/knowledge from the data by searching for some regularities among the data. It is correct for the data but only plausible outside of the given data. A vast majority of the existing ML algorithms are inductive. Learning by induction is a search for a correct rule, or a set of rules, guided by training examples. The task of the search is to find hypotheses that best describe the concept. We usually start with some initial hypothesis and then search for one that covers as many input data points (examples) as possible. We say that an example is covered by a rule when it satisfies all conditions of the IF… part of the rule. Still another view of inductive ML is one of designing a classifier, i.e., finding boundaries that encompass only examples belonging to a given class. Those boundaries can either partition the entire sample space into parts containing examples from one class only, and sometimes leave parts of the space unassigned to either of the classes (a frequent outcome).

A desirable characteristic of inductive ML algorithms is their ability (or inability) to deal with incomplete data. The majority of real datasets have records that include missing values due to a variety of reasons, such as manual data entry errors, incorrect measurements, equipment errors, etc. It is common to encounter datasets that have up to half of the examples missing some of their values (Farhanghfar et al., 2007). Thus, a good ML algorithm should be robust to missing values as well as to data containing errors, as they often have adverse effect on the quality of the models generated by the ML algorithms (Farhanghfar et al., 2008).

The rest of the chapter contains a review of Michalski's work in supervised learning and a review of our algorithms, which were inspired by his work, but first we briefly comment on Michalski's work in unsupervised learning. A prime example of unsupervised learning is clustering. However, there is a significant difference between the classical clustering and clustering performed within the framework of ML. The classical clustering is best suited for handling numerical data. Thus Michalski introduced the concept of conceptual clustering to differentiate it from classical clustering since conceptual clustering can deal with nominal data (Michalski, 1980; Fisher and Langley, 1986; Fisher, 1987). Conceptual clustering consists of two tasks: clustering itself which finds clusters in a given data set, and characterization (which is supervised learning) which generates a concept description for each cluster found by clustering. Conceptual clustering can be then thought of as a hybrid combining unsupervised and supervised approaches to learning; CLUSTER/2 by Michalski (1980) was the first well-known conceptual clustering system.

Table 1. Set of ten examples described by three features (F_1-F_3) drawn from two categories (F_4)

S	F_1	F_2	F_3	F_4 decision attribute
e_1	1	1	2	1
e_2	1	1	1	1
e_3	1	2	2	1
e_4	1	2	1	1
e_5	1	3	2	1
e_6	3	4	3	2
e_7	2	5	3	2
e_8	3	1	3	2
e_9	2	2	3	2
e_{10}	2	3	3	2

2 Generation of Hypotheses

The process of generating hypotheses is instrumental for understanding how inductive ML algorithms work. We first illustrate this concept by means of a simple example from which we will (by visual inspection) generate some hypotheses and later describe ML algorithms that do the same in an automated way. Let us define an *information system* (IS):

$$IS =< S, Q, V, f >$$

where

- S is a finite set of examples, $S = \{e_1, e_2, ..., e_M\}$ and M is the number of examples
- Q is a finite set of features, $Q = \{F_1, F_2, ..., F_n\}$ and n is the number of features
- $V = \cup V_{F_j}$ is a set of feature values where V_{F_j} is the domain of feature $F_j \in Q$
- $v_i \in V_{F_j}$ is a value of feature F_j
- $f = S \times Q \rightarrow V$ is an information function satisfying $f(e_i, F_i) \in V_{F_j}$ for every $e_i \in S$ and $F_j \in Q$

The set S is known as the *learning/training data*, which is a subset of the universe (that is known only to the teacher/oracle); the latter is defined as the Cartesian product of all feature domains V_{F_j} (j=1,2…n).

Now we analyze the data shown in Table 1 and generate a rule/hypothesis that describes class1 (defined by attribute F_4):

IF F_1=1 AND F_2=1 THEN class1 (or F_4=1)

This rule covers two (e_1 and e_2) out of five positive examples. So we generate another rule:

IF F_1=1 AND F_3=2 THEN class1

This rule covers three out of five positive examples, so it is better than the first rule; rules like this one are called strong since they cover a majority (large number) of positive (in our case class1) training examples. To cover all five positive examples we need to generate one more rule (to cover e_4):

IF F_1=1 AND F_3=1 THEN class1

While generating the above rules we paid attention so that none of the rules describing class1 covered any of the examples from class2. In fact, for the data shown in Table 1, this made it more difficult to generate the rules because we could have generated just one simple rule:

IF F_1=1 THEN class1

that would perfectly cover all class1 examples while not covering the class2 examples. The generation of such a rule is highly unlikely to describe any real data where hundreds of features may describe thousands of examples; that was why we have generated more rules to illustrate typical process of hypotheses generation.

As mentioned above, the goal of inductive ML algorithms is to automatically (without a human intervention) generate rules (hypotheses). After learning, the generated rules must be tested on unseen examples to assess their predictive power. If the rules fail to correctly classify (to calculate the error we assume that we know their "true" classes) a majority of the test examples the learning phase is repeated by using procedures like cross-validation. The common disadvantage of inductive machine learning algorithms is their ability to, often almost perfectly, cover/classify training examples, which may lead to the overfitting of data. A trivial example of overfitting would be to generate five rules to describe the five positive examples; the rules would be the positive examples themselves. Obviously, if the rules were that specific they would probably perform very poorly on new examples. As stated, the goodness of the generated rules needs to be evaluated by testing the rules on new data. It is important to establish a balance between the rules' generalization and specialization in order to generate a set of rules that have good predictive power. A more general rule (strong rule) is one that covers more positive training examples. A specialized rule, on the other hand, may cover, in an extreme case, only one example.

In the next section we describe rule algorithms also referred to as rule learners. Rule induction/generation is distinct from the generation of decision trees. While it is trivial to write a set of rules given a decision tree it is more complex to generate rules directly from data. However, the rules have many advantages over decision trees. Namely, they are easy to comprehend; their output can be easily written in the first-order logic format, or directly used as a knowledge base in knowledge-based systems; the background knowledge can be easily added into a set of rules; and they are modular and independent, i.e., a single rule can be understood without reference to other rules. Independence means that, in contrast to rules written out from decision trees, they do not share any common attributes (partial paths in a decision tree). Their disadvantage is that they do not show relationships between the rules as decision trees do.

3 Rule Algorithms

As already said, Michalski was the first one to introduce an inductive rule-based ML algorithm that generated rules from data. In his seminal paper (Michalski, 1969) he framed the problem of generating the rules as a set-covering problem. We illustrate one of early Michalski's algorithms, the AQ15, in which IF…THEN…rules are expressed in terms of variable-value logic (VL1) calculus (Michalski, 1974; Michalski et al. 1986). The basic notions of VL1 are that of a selector, complex, and cover. A selector is a relational statement:

$$(F_i \ \# \ v_i)$$

where # stands for any relational operator and v_i is one or more values from dom_i of attribute F_i.

A complex L is a logical product of selectors:

$$L = \cap \ (F_i \# v_i)$$

The cover, C, is defined as a disjunction of complexes:

$$C = \cup \ L_i$$

and forms the conditional part of a production rule covering a given data set.

Two key operations in the AQ algorithms are the generation of a *star* $G(e_i | E_2)$ and the generation of a *cover* $G(E_1 | E_2)$, where $e_i \in E_1$ is an element of a set E_1, and E_2 is another set such that $E_1 \cup E_2 = S$, where S is the entire training data set.

We use data shown in Table 1 to illustrate operation of the family of AQ algorithms.

First, let us give examples of information functions for data shown in Table1:

$$F_1 = 1$$
$$F_3 = 1 \text{ OR } 2 \text{ OR } 3.$$

The full form of the first information function is:

$$(F_1 = 1) \text{ AND } (F_2 = 2 \text{ OR } 3 \text{ OR } 4 \text{ OR } 5 \text{ OR } 1) \text{ AND } (F_3 = 1 \text{ OR } 2 \text{ OR } 3)$$

Similarly the second information function can be rewritten as:

$$(F_1 = 1 \text{ OR } 2 \text{ OR } 3) \text{ AND } (F_2 = 2 \text{ OR } 3 \text{ OR } 4 \text{ OR } 5 \text{ OR } 1) \text{ AND } (F_3 = 1\text{OR } 2 \text{ OR } 3)$$

A function *covers* an example if it matches all the attributes of a given example, or, in other words, it evaluates to TRUE for this example. Thus, the information function $(F_1 = 1)$ covers the subset $\{e_1, e_2, e_3, e_4, e_5\}$, while the function $(F_3 = 1 \text{ OR } 2 \text{ OR } 3)$ covers all examples shown in Table 1.

The goal of inductive machine learning, in Michalski's setting, is to generate information functions while taking advantage of a decision attribute (feature F_4 in Table 1). The question is whether the information function, IF_B, generated from a set of training examples will be the same as a true information function, IF_A (Kodratoff, 1988). In other words, the question is whether the ML algorithm (B) can learn what only the teacher (A) knows. To answer the question let us consider the function:

$$IF_A : (F_1 = 3 \text{ OR } 2) \text{ AND } (F_2 = 1 \text{ OR } 2)$$

that covers subset $\{e_8, e_9\}$. Next, we generate, via induction, the following information function:

$$IF_B : (F_1 = 3 \text{ OR } 2) \text{ AND } (F_2 = 1 \text{ OR } 2) \text{ AND } (F_3 = 3)$$

which covers the same two examples, but IF_B is different from IF_A. IF_B can be rewritten as:

$$IF_B = IF_A \text{ AND } (F_3 = 3)$$

We say that IF_B is a specialization of IF_A (it is less general). So in this case the learner learned what the teacher knew (although in a slightly different form). Note that frequently this is not the case.

In order to evaluate the goodness of the generated information functions we use criteria such as the sparseness function, which is defined as the total number of examples it can potentially cover minus the number of examples it actually covers. The smaller the value of the sparseness function the more compact the description of examples. Let us assume that we have two particular information functions IF_1 and IF_2, such that IF_1 covers subset E_1 and the other covers the remaining part of the training data, indicated by subset E_2. If the intersection of sets E_1 and E_2 is empty then we say that these two functions *partition* the training data. The goal of AQ algorithms, as well as all ML algorithms, is to find such partitions. Assuming that one part of the training data represents positive examples, and the remaining part represents negative examples, then IF_1 becomes the rule (hypothesis) covering all positive examples. To calculate the sparseness of a partition the two corresponding information sparsenesses can be added together and used for choosing among several generated alternative partitions (if they exist). Usually we start with the initial partition in which subsets E_1 and E_2 intersect and the goal is then to come up with information functions that result in the partition of training data. The task of the learner is to modify this initial partition so that all intersecting elements are incorporated into "final" subsets, say E_{11} and E_{22}, which form a *partition*:

$$E_{11} \cap E_{22} = \emptyset \text{ and } E_{11} \cup E_{22} = S$$

Michalski et al. (1986) proposed the following algorithm.

Given: Two disjoint sets of examples

1. Start with two disjoint sets E_{01} and E_{02}. Generate information functions, IF_1 and IF_2 from them and generate subsets, E_1 and E_2, which they cover.
2. If sets E_1 and E_2 intersect then calculate differences between sets E_1 and E_2 and the intersecting set

   $E_p = E_1 - E_1 \cap E_2$
   $E_n = E_2 - E_1 \cap E_2$

   and generate corresponding information functions, IF_p and IF_n; otherwise we have a partition; stop.
3. For all examples e_i from the intersection do:
   create sets $E_p \cup e_i$ and $E_n \cup e_i$ and generate information functions for each, IF_{pi} and IF_{ni}
4. Check if (IF_p, IF_{ni}) and (IF_n, IF_{pi}) create partitions of S
   a) if they do, choose the better partition, say in terms of sparseness (they become new E_1 and E_2), go to step 1 and take the next example e_i from the intersection
   b) if not, go to step 2 and check another example from the intersection

Result: Partition of the two sets of examples.

This algorithm does not guarantee that all examples will be assigned to one of the two subsets if a partition is not found.

We illustrate this algorithm using data from Table 1.

1. Assume that the initial subsets are $\{e_1, e_2, e_3, e_4, e_5, e_6\}$ and $\{e_7, e_8, e_9, e_{10}\}$.

Notice that we are not as yet using a decision attribute/feature F_4. We will use it later for dividing the training data into subsets of positive and negative examples. We only try to illustrate how to move the intersecting examples so that the resulting subsets create a partition.

The information functions generated from these two sets are:

IF_1 : (F_1 = 1 OR 3) AND (F_2 = 1 OR 2 OR 3 OR 4) AND (F_3 = 1 OR 2 OR 3)
IF2 : (F_1 = 3 OR 2) AND (F_2 = 5 OR 2 OR 1 OR 3) AND (F_3 = 3)

Function IF_1 covers set $E_1 = \{e_1, e_2, e_3, e_4, e_5, e_6, e_8\}$ and function IF_2 covers set $E_2 = \{e_7, e_8, e_9, e_{10}\}$.

2. Since $E_1 \cap E_2 = \{e_8\}$, which means that they intersect, hence we calculate:

$E_p = E_1 - \{e_8\} = \{e_1, e_2, e_3, e_4, e_5, e_6\}$
$E_n = E_2 - \{e_8\} = \{e_7, e_9, e_{10}\}$

and generate two corresponding information functions:

IF_p : (F_1 = 1 OR 3) AND (F_2 = 1 OR 2 OR 3 OR 4) AND (F_3 = 1 OR 2 OR 3)
IF_n : (F_1 = 2) AND (F_2 = 5 OR 2 OR 3 AND (F_3 = 3)

Note that IF_p is exactly the same as IF_1 and thus covers E_1, while IF_n covers only E_n.

3. Create the sums $E_p \cup e_i = E_1$ and $E_n \cup e_i = E_2$, where $e_i = e_8$, and generate the corresponding information functions. The result is $IF_{pi} = IF_1$ and $IF_{ni} = IF_2$.

4. Check if pairs (IF_p, IF_2) and (IF_n, IF_1) create a partition.

The first pair of information functions covers the subset $\{e_1, e_2, e_3, e_4, e_5, e_6, e_8\}$ and subset $\{e_7, e_8, e_9, e_{10}\}$. Since the two subsets still intersect, this is not a partition yet. The second pair covers the subsets $\{e_7, e_9, e_{10}\}$ and $\{e_1, e_2, e_3, e_4, e_5, e_6, e_8\}$; since they do not intersect, and sum up to S, this is a partition.

Note that in this example we have chosen the initial subsets arbitrarily and our task was just to find a partition. In inductive supervised ML, however, we use the decision attribute to help us in this task. We observe that we have generated information functions not by using any algorithm, but by visual inspection of Table 1.

The AQ search algorithm (as used in AQ15) is an irrevocable top-down search which generates a decision rule for each class in turn. In short, the algorithm at each step starts with selecting one positive example, called a seed, and generates all complexes (a star) that cover the seed but do not cover any negative examples. Then by using criteria such as the sparseness and the length of complexes (shortest complex first) it selects the best complex from the star, which is added to the current (partial) cover. The pseudocode, after Michalski et al. (1986), follows.

Given: Sets of positive and negative training examples
While partial *cover* does not cover all positive examples do:

1. Select an uncovered positive example (a seed)
2. Generate a star, that is determine maximally general complexes covering the seed and no negative examples
3. Select the best complex from the star, according to the user-defined criteria
4. Add the complex to the partial cover

While partial *star* covers negative examples do:

1. Select a covered negative example
2. Generate a partial star (all maximally general complexes) that covers the seed and excludes the negatie example
3. Generate a new partial star by intersecting the current partial star with the partial star generated so far
4. Trim the partial star if the number of disjoint complexes exceeds the predefined threshold, called maxstar (to avoid exhaustive search for covers which can grow out of control)

Result: Rule(s) covering all positive examples and no negative examples

Now we illustrate the generation of a cover using the decision attribute, F_4. However, we will use only a small subset of the training data set consisting of four examples: two positive (e_4 and e_5) and two negative (e_9 and e_{10}), to be able to show all the calculations. The goal is to generate a cover that covers all positive (class 1) examples (e_4 and e_5) and excludes all negative examples (e_9 and e_{10}). Thus, we are interested in generating a cover properly identifying subset $E_1 = \{e_4, e_5\}$ and rejecting subset $E_2 = \{e_9, e_{10}\}$; such a cover should create a partition of $S = \{e_4, e_5, e_9, e_{10}\}$.

Generation of a cover involves three steps:

For each positive example $e_i \in E_1$, where E_1 is a positive set:

1. Find $G(e_i | e_j)$ for each $e_j \in E_2$, where E_2 is a negative set
2. Find a star $G(e_i | E_2)$. It is THE conjunction of $G(e_i | e_j)$ terms found in step 1. When there is more than one such term (after converting it into a disjunctive form) select the best one according to some criteria, like the sparseness.
3. Finding a cover of all positive examples against all negative examples $G(E_1 | E_2)$. It is the disjunction of stars found in step 2. The final cover covers all positive examples and no negative examples.

Let us start with finding $G(e_4 | e_9)$. It is obtained by comparing the values of the features in both examples, skipping those which are the same, and making sure that the values of features in e_9 are different from those of e_4, and putting them in disjunction. Thus,

$$G(e_4 | e_9) = (F_1 \neq \text{black}) \text{ OR } (F_3 \neq \text{large})$$

Note that it is the most general information function describing e_4 since it makes sure that only example e_9 is not covered by this function.

Next, we calculate the star $G(e_i | E_2)$, for all $e_i \in E_1$, against all e_j from E_2. A star for e_i is calculated as the conjunction of all G(.)s and constitutes a cover covering e_i

$$G(e_i | E_2) = \cap G(e_i | e_j) \quad \text{for all } e_j \in E_2$$

Since we started with $e_i = e_4$ we will obtain a cover of e_4 against e_{10}, and combine it using the conjunction with the previous cover; this results in

$$G(e_4 | E_2) = ((F_1 \neq 2) \text{ OR } (F_3 \neq 3)) \text{ AND } ((F_1 \neq 2) \text{ OR } (F_2 \neq 3) \text{ OR } (F_3 \neq 3))$$

The expression is converted into the disjunctive form:

$$\begin{aligned} G(e_4 | E_2) = {} & ((F_1 \neq 2) \text{ AND } (F_1 \neq 2) \text{ OR } ((F_1 \neq 2) \text{ AND } (F_2 \neq 3)) \text{ OR} \\ & ((F_1 \neq 2) \text{ AND } (F_3 \neq 3)) \text{ OR } ((F_3 \neq 3) \text{ AND } (F_1 \neq 2)) \text{ OR} \\ & ((F_3 \neq 3) \text{ AND } (F_2 \neq 3)) \text{ OR}((F_3 \neq 3) \text{ AND } (F_3 \neq 3)) \end{aligned}$$

Next, by using various laws of logic it is simplified into:

$$\begin{aligned} G(e_4 | E_2) = (F_1 \neq 2) \quad & \text{OR} \quad (28) \\ ((F_1 \neq 2) \text{ AND } (F_2 \neq 3)) \; & \text{OR} \quad (23) \\ ((F_1 \neq 2) \text{ AND } (F_3 \neq 3)) \; & \text{OR} \quad (18) \\ ((F_3 \neq 3) \text{ AND } (F_2 \neq 3)) \; & \text{OR} \quad (23) \\ ((F_3 \neq 3) \quad & \end{aligned}$$

and after calculating the sparseness (shown in parentheses) the best is kept:

$$G(e_4 | E_2) = (F_1 \neq 2) \text{ AND } (F_3 \neq 3)$$

Then we repeat the same process for $G(e_5 \mid E_2)$:

$$G(e_5 \mid E_2) = ((F_1 \neq 2) \text{ OR } (F_2 \neq 2) \text{ OR } (F_3 \neq 3)) \text{ AND}((F_1 \neq 2) \text{ OR } (F_3 \neq 3))$$

which is next converted into the disjunctive form:

$$\begin{aligned} G(e_5 \mid E_2) = {} & ((F_1 \neq 2) \text{ AND } (F_1 \neq 2) \text{ OR } ((F_1 \neq 2) \text{ AND } (F_3 \neq 3)) \text{ OR} \\ & ((F_2 \neq 2) \text{ AND } (F_1 \neq 2)) \text{ OR } ((F_2 \neq 2) \text{ AND } (F_3 \neq 3)) \text{ OR} \\ & ((F_3 \neq 3) \text{ AND } (F_1 \neq 2)) \text{ OR } ((F_3 \neq 3) \text{ AND } (F_3 \neq 3)) \end{aligned}$$

and simplified to:

$$G(e_5 \mid E_2) = (F_1 \neq 2) \text{ AND } (F_3 \neq 3)$$

Finally in step 3 we need to combine the two stars (rules) into a cover:

$$G(E_1 \mid E_2) = (F_1 \neq 2) \text{ AND } (F_3 \neq 3)$$

From the knowledge of the feature domains we can write the final cover, or rule, covering all positive examples as:

$$G(E_1 \mid E_2) = (F_1 = 1 \text{ OR } 3) \text{ AND } (F_3 = 1 \text{ OR } 2)$$

The cover is actually written as:

$$<F_1 \neq 2> <F_3 \neq 3>$$

which reads

$$\text{IF } (F_1 = 1 \text{ OR } 3) \text{ AND } (F_3 = 1 \text{ OR } 2) \text{ THEN class positive}$$

As one can see the generation of a cover is computationally very expensive. In terms of a general set covering problem creating a cover $G(E_1 \mid E_2)$, while using the $\geq$ operators in the description of selectors, means that we are dividing the entire space, into subspaces in such a way that in one subspace we will have all the positive examples while all the negative examples will be included in another, nonintersecting subspace.

A substantial disadvantage of AQ algorithms is that they handle noise outside of the algorithm itself, by rule truncation.

4 Hybrid Algorithms

After reviewing Michalski's rule algorithms we concentrate on the description of our hybrid algorithm, the CLIP4 (Cover Learning (using) Integer Programming). The CLIP4 algorithm is a hybrid that combines ideas (like its predecessors the CLILP3 and CLIP2 algorithms) of Michalski's rule algorithms and decision trees. More precisely, CLIP4 uses a rule-generation schema similar to Michalski's AQ algorithms, as well as the tree-growing technique to divide training data into subsets at each level of a (virtual) decision tree similar to decision tree algorithms (Quinlan, 1993). The main difference between CLIP4 and the two families of algorithms is CLIP4's extensive use of our own algorithm for set covering (SC), which constitutes its core operation. SC is performed several times to generate the rules. Specifically, the SC algorithm is used to select the most discriminating features, to grow new branches of the tree, to select data subsets from which CLIP4 generates the least overlapping rules, and to generate final rules from the (virtual) tree leaves, which store subsets of the data. An important characteristic that distinguishes CLIP4 from the vast majority of ML algorithms is that it generates production rules that involve inequalities. This results in generating a small number of compact rules, especially in domains where attributes have large number of values and where majority of them are associated with the target class. In contrast, other inductive ML algorithms that use equalities would generate a large number of complex rules for these domains.

CLIP4 starts by splitting the training data in a decision-tree-like manner. However, it does so not by calculating any index of "good" splitting, like entropy, but it selects features and generates rules by solving an Integer Programming (IP) model. CLIP4 uses the training data to construct an IP model and then uses a standard IP program to solve it. CLIP4 differs from the decision tree algorithms is that it splits the data into subsets in several ways, not just in one "best" way. In addition, there is no need to store the entire decision tree in CLIP4. It keeps only the leaf nodes of the "tree" (the tree, in fact, does not exist). This results in the generation of simpler rules, a smaller number of rules, and a huge memory saving. Another advantage is that the solution of the IP model for splitting the data is relatively quick, as compared to the calculation of entropies. The solution returned from the IP model indicates the most important features to be used in the generation of rules. IP solutions may include preferences used in other machine learning algorithms (Michalski and Larson, 1978), like the *largest complex first* where IP solution can generate features that cover the largest number of positive examples. Or, the *background knowledge first*, where any background knowledge can be incorporated into the rules by including user-specified features, if it is known that they are crucial in describing the concept.

4.1 Our Set Covering Algorithm

As we mentioned above, several key operations performed by CLIP4 are modeled and solved by the set covering algorithm, which is a simplified version of integer programming (IP). IP is used for function optimization that is subject to a large number of constraints. Several simplifications are made to the IP model to transform it into the SC problem: the function that is the subject of optimization has all its coefficients set to one; their variables are binary, $x_i=\{0,1\}$; the constraint function coefficients are also binary; and all constraint functions are greater than or equal to one. The SC problem is NP-hard, and thus only an approximate solution can be found. First, we transform the IP problem into the binary matrix (BIN) representation that is obtained by using the variables and constraint coefficients. BIN's columns correspond to variables (features/attributes) of the optimized function; its rows correspond to function constraints (examples), as illustrated in Figure 1. CLIP4 finds the solution of the SC problem in terms of selecting a minimal number of columns that have the smallest total number of 1's. This outcome is obtained by minimizing the number of 1's that overlap among the columns and within the same row. The solution consists of a binary vector composed of the selected columns. All rows for which there is a value of 1 in the matrix, in a particular column, are assumed to be "covered" by this column.

$$
\begin{gathered}
Minimize: \\
x_1 + x_2 + x_3 + x_4 + x_5 = Z \\
Subject \ \ to: \\
x_1 + x_3 + x_4 \geq 1 \\
x_2 + x_3 + x_5 \geq 1 \\
x_3 + x_4 + x_5 \geq 1 \\
x_1 + x_4 \geq 1 \\
Solution: \\
Z = 2, \quad when \ x_1 = 1, x_2 = 0, x_3 = 1, x_4 = 0, x_5 = 0
\end{gathered}
$$

$$
\begin{gathered}
Minimize: \\
x_1 + x_2 + x_3 + x_4 + x_5 = Z \\
Subject \ \ to: \\
\begin{bmatrix} 1,0,1,1,0 \\ 0,1,1,0,1 \\ 0,0,1,1,1 \\ 1,0,0,1,0 \end{bmatrix} \cdot \begin{bmatrix} x_1 \\ x_2 \\ x_3 \\ x_4 \\ x_5 \end{bmatrix} \geq 1
\end{gathered}
$$

Fig. 1. A simplified set-covering problem and its solution (on the left); in the BIN matrix form (on the right)

To obtain a solution we use our SC algorithm, which is summarized as follows.

Given: BINary matrix. *Initialize*: Remove all empty (inactive) rows from the BINary matrix; if the matrix has no 1's, then return error.

1. Select active rows that have the minimum number of 1's in rows – *min-rows*
2. Select columns that have the maximum number of 1's within the *min-rows* – *max-columns*
3. Within *max-columns* find columns that have the maximum number of 1's in all active rows – *max-max-columns*. If there is more than one *max-max-column*, go to Step 4., otherwise go to Step 5.
4. Within *max-max-columns* find the first column that has the lowest number of 1's in the inactive rows
5. Add the selected column to the solution
6. Mark the inactive rows. If all the rows are inactive then terminate; otherwise go to Step 1.

Result: Solution to the SC problem.

In the above pseudocode, an active row is a row not covered by a partial solution, and an inactive row is a row already covered by a partial solution. We illustrate how the SC algorithm works in Figure 2 using a slightly more complex BIN matrix that the one shown in Figure 1. The solution consists of the second and fourth columns, which have no overlapping 1's in the same rows.

Before we describe the CLIP4 algorithm in detail, let us first introduce a necessary notation. The set of all training examples is denoted by S. A subset of positive examples is denoted by S_P and the subset of negative examples by S_N. S_P and S_N are represented by matrices whose rows represent examples and whose columns correspond to attributes. The matrix of positive examples is denoted as POS and their number by N_{POS}. Similarly for the negative examples, we have matrix NEG and number N_{NEG}. The following properties are satisfied for the subsets:

$$S_P \cup S_N = S, \quad S_P \cap S_N = \emptyset, \quad S_N \neq \emptyset, \text{ and } \quad S_P \neq \emptyset$$

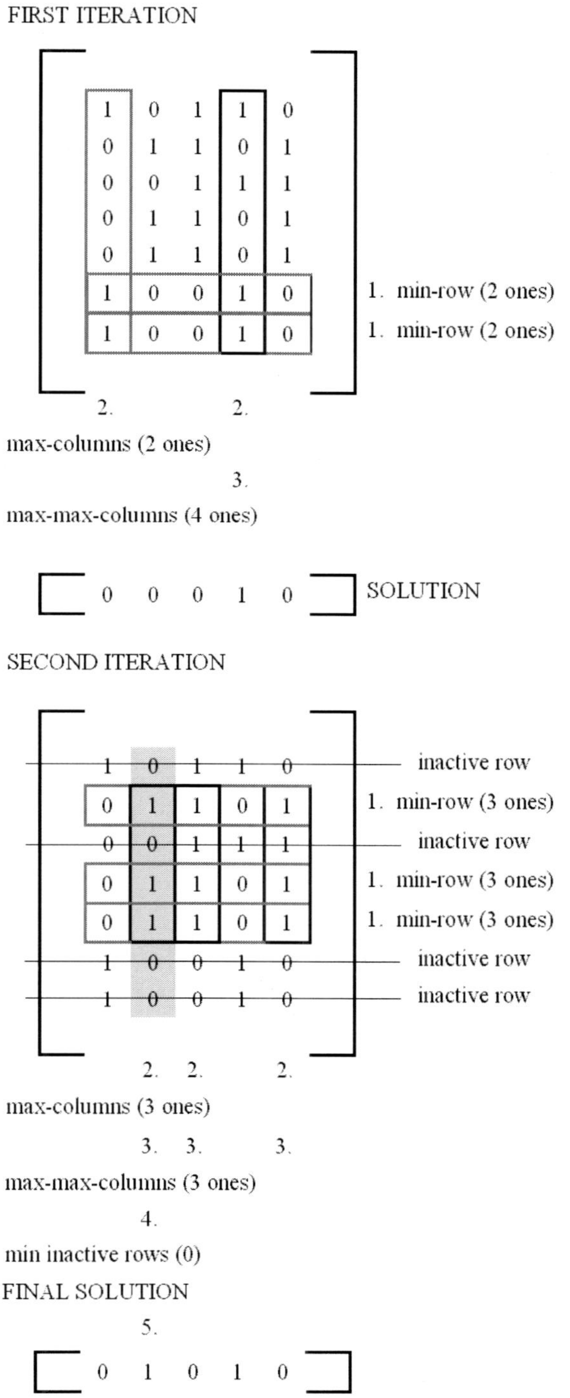

Fig. 2. Solution of the SC problem using the SC algorithm

The examples are described by a set of K attribute-value pairs:

$$e = \wedge_{j=1}^{K} [a_j \# v_j]$$

where a_j denotes the jth attribute with value $v_j \in d_j$, and # is a relation ($\neq$, =, <, $\approx$, $\leq$, etc.), where K is the number of attributes. An example e consists of a set of selectors $s_j = [a_j \neq v_j]$.

The CLIP4 algorithm generates rules in the form:

$$\text{IF } (s_l \wedge \ldots \wedge s_m) \qquad \text{THEN class} = \text{class}_i$$

where all selectors are only in the form $s_i = [a_j \neq v_j]$, that is they use only inequalities.

The positive examples from the matrix POS are described by a set of values $\text{pos}_i[j]$, where j=1,…,K is the column number and i is the example number (the row number in the POS matrix). The negative examples are described similarly by a set of $\text{neg}_i[j]$ values. CLIP4 uses binary matrices (BIN) that are composed of K columns, filled with either 1 or 0 as values. Each element of the BIN matrix is denoted by $bin_i[j]$, where i is a row number and j is a column number. These matrices are the result of operations performed by CLIP4 which are modeled and solved using the SC algorithm described above. The pseudocode of the CLIP4 algorithm is provided in Figure 3.

The algorithm consists of three key phases that are explained below:

Phase I: The POS data is partitioned into subsets of similar data in a decision-tree-like manner. Each node represents a data subset. Each level of the tree is built using one negative example to find selectors that distinguish between all positive examples and the negative example. The selectors are used to create new branches of the tree. During the tree growth, we use pruning to eliminate noise from the data (described later) and to avoid excessive growth, which reduces execution time.

The tree is grown in a top-down manner. At the ith tree level, N_i subsets, represented by matrices $\text{POS}_{i,j}$ (j=1,…,N_i), are generated using N_{i-1} subsets from the previous tree level and the single negative example neg_i. Each subset, represented by the matrix $\text{POS}_{i,j}$ (examples that constitute this matrix's elements are denoted as $\text{pos}^{i,j}$), is transformed into a BIN matrix of the same size as the $\text{POS}_{i,j}$ matrix, using the neg_i example, and then modeled and solved using the SC method. The solution is used to generate subsets for the next tree level, represented by the $\text{POS}_{i+1,j}$ matrices.

CLIP4 grows a virtual "tree", since any iteration consists of only one level: the most recent bottom tree level. Moreover, this tree is pruned so that even the one level that is kept has at most a few nodes. This results in a high memory efficiency. The data splits are made by generating not just one "best" division, based on some "best" feature (say in terms of the highest information gain, as in decision trees), but a set of divisions based on any feature that distinguishes between the positive and the one negative example. This mechanism of partitioning data assures generation of the (possibly) most general rules.

```
N0=1; Create POS0,1 consisting of entire SP; Create NEG consisting of entire SN     // Initialize
for negi, i=1 to NNEG do {                                                          // PHASE I STARTS
  for j=1 to Ni-1 do {                                                              //for each POSi-1,j matrix
    for k=1 to K do {                                                               //create new BINj matrix
      for l=1 to number of POSi-1,j rows do {
        if posl^(i-1,j)[k] = negi[k] then binl^j[k] = 0;
        if posl^(i-1,j)[k] ≠ negi[k] then binl^j[k] = 1;
        if posl^(i-1,j)[k] = '*' then binl^j[k] = 0;                                // missing value encountered
        if negi[k] = '*' then binl^j[k] = 0; }}                                     // missing value encountered
    SOLj = SolveSCProblem(BINj);}
  PruneMatrices(POSi-1,j , SOLj , j=1,...,Ni-1);
  ApplyGeneticOperators(POSi-1,j , SOLj , j=1,...,Ni-1);
  Ni=1;                                                                             //counter for POSi+1 matrices
  for j=1 to Ni-1 do {                                                              //for each POSi,j matrix
    if POSi-1,j was not pruned or redundant then {
      for k=1 to K do {                                                             //through entire solution vector
        if SOLj[k]=1 then {                                                         //then create new POSi,Ni matrix
          for l=1 to number of POSi-1,j rows do {
            if posl^(i-1,j)[k]≠negi[k] then add posl^(i-1,j) to POSi,Ni matrix; }}
    Ni=Ni+1; }}}
  for j=1 to Ni do {            //for each POSi matrix check if it large enough to be not considered as a noise
    if number of rows of POSi,j < NoiseThreshold then { remove POSi,j from the tree; Ni=Ni-1;}}
  EliminateRedundantMatrices(POSi,j j=1,...,Ni); }
Create BIN matrix that consist of N_NNEG columns and NPOS rows, and fill with zeros // PHASE II STARTS
for i=1 to N_NNEG do {                                                              // for all tree leaves
  for j=1 to NPOS do {
    for k=1 to number of rows of POS_(NNEG,i) do { if posj^(0,0) = posk^(NNEG,i) then binj[i]=1;}}}
SOL = SolveSCProblem (BIN);                                                         // select best leaf node subsets
for i=1 to N_NNEG do {                                                              // through entire solution vector
  create BINi=NEG;
  if SOL[i]=1 then {                                                                // back-project POSNneg,i matrix
    for j=1 to K do {
      for k=1 to NNEG do {
        if negk[j]='*' then bink^i[j]=0; else {
          for l=1 to number of rows of POS_(NNEG,i) do { if posl^(NNEG,i)[j]=negk[j] then bink^i[j]=0; }}}}}
  for j=1 to K do {                                                                 // convert BINi values to binary
    for k=1 to NNEG do { if bink^i[j]≠0 then bink^i[j]=0;}}
  SOLi = SolveSCProblem (BINi);
  for j=1 to K do {                                                                 // start generation of i-th rule
    if SOLi[j]=1 then {                                                             // add selectors to the rule
      for k=1 to NNEG do { if bin^i_k[j]=1 then add "aj≠neg k[j]" selector to the Rulei; }}}}
                                                                                    // PHASE III STARTS
best#covered=0; previous_best#covered=0;                                            // holds # ex. covered by the rule
while best#covered≥0 do {                                                           //until best rules are accepted
  for i=1 to N_NNEG do {                                                            // through all generated rules
    coversi=NPOS;                                                                   // # examples covered by Rulei
    for j=1 to NPOS do {                                                            // for all examples in POS
      for k=1 to K do {
        for l=1 to number of selectors in Rulei do {
          if al=k and vk=posj^(0,1)[k] then { coversi=coversi -1; j=j+1; }}}}        // example not covered by Rulei
    if best#covered<coversi then best#covered=coversi; best_rule=Rulei; }
  NPOS = NPOS – best#covered;
  if NPOS<StopThreshold or NPOS=0 then STOP;
  POS0,1 = POS0,1 – examples covered by best_rule;
  if (best#covered<previous_ best#covered/2 and best#covered<NPOS/2) then best#covered=-1;    // multiple rules
  previous_best#covered_examples=best#covered_examples; }
```

Fig. 3. Pseudocode of the CLIP4 algorithm

Phase II: A set of terminal subsets (tree leaves) is selected using two criteria. First, large subsets are preferred over small ones (according to Occam's razor which hopefully will result in the rules that are "strong" and more general), while the accepted subsets (between them) must cover the entire POS data. Second, we want to use the completeness criterion. To that end, we first perform a back-projection of one of the selected positive data subsets using the entire NEG data, and then convert the resulting matrix into a BIN matrix and solve it using the SC method. The solution is used to generate a rule, and the process is repeated for every selected positive data subset. Explanation of lines 24-41: $N_{N_{NEG}}$ is the number of tree leaves. The binary BIN matrix is used to select the minimal set of leaves that covers the entire POS data. Back-projection results in a matrix BIN_i that has 1's for the selectors that can distinguish between positive and negative examples. The back-projection generates one matrix for every terminal subset ($POS_{N_{NEG},i}$). It is computed from the NEG matrix, by setting a value from NEG to zero if the same value appears in the corresponding column in the $POS_{N_{NEG},i}$ matrix; otherwise it is left unchanged. The ith rule is generated by solving the SC problem for BIN_i and adding a selector for every 1 that is in any column indicated by the solution.

The rule induction generates a rule directly from two sets of data (NEG data and the selected subset of positive data). It does not require the use of logic or storing and traversing the entire tree, as in decision trees, but only a comparison of values between the two matrices. Thus, the only two data structures required to generate the rules are lists (vectors) and matrices.

Phase III: A set of best rules is selected from the generated rules. Rules that cover the most POS examples are chosen as being possibly the most general. If there is a tie between the "best rules", the rule that uses the minimal number of selectors is chosen.

To find the number of examples covered by a rule, we first find examples not covered by the rule, and subtract these from the total number of examples. We do this because the rules consist of selectors involving inequalities. The variable called $covers_i$ records the number of positive examples covered by the ith rule. After a rule is accepted, the positive examples covered by it are removed from the POS matrix (line 53). In this phase, more than one rule can be generated. The rules are accepted in the order from the strongest to the weakest. We use a heuristic for accepting multiple rules, which states that the next rule is accepted when it covers at least half the number of examples covered by the previously accepted rule and at least half the number of positive examples not covered by any rule so far. CLIP4 generates multiple rules in a single sweep through the data. The thresholds used in the pseudocode in Figure 3 are described later. We illustrate the high-level view of the process used by CLIP4 in Figure 4.

The basic characteristics of the CLIP4 algorithm are its completeness (the generated set of rules describes all positive examples), consistency (rules do not describe any of the negative examples) and use of a minimal number of selectors in the generated rules. CLIP4 is memory-efficient (because it stores only the bottom level of the tree) and is robust to noisy and missing-value data, as shown later in the chapter.

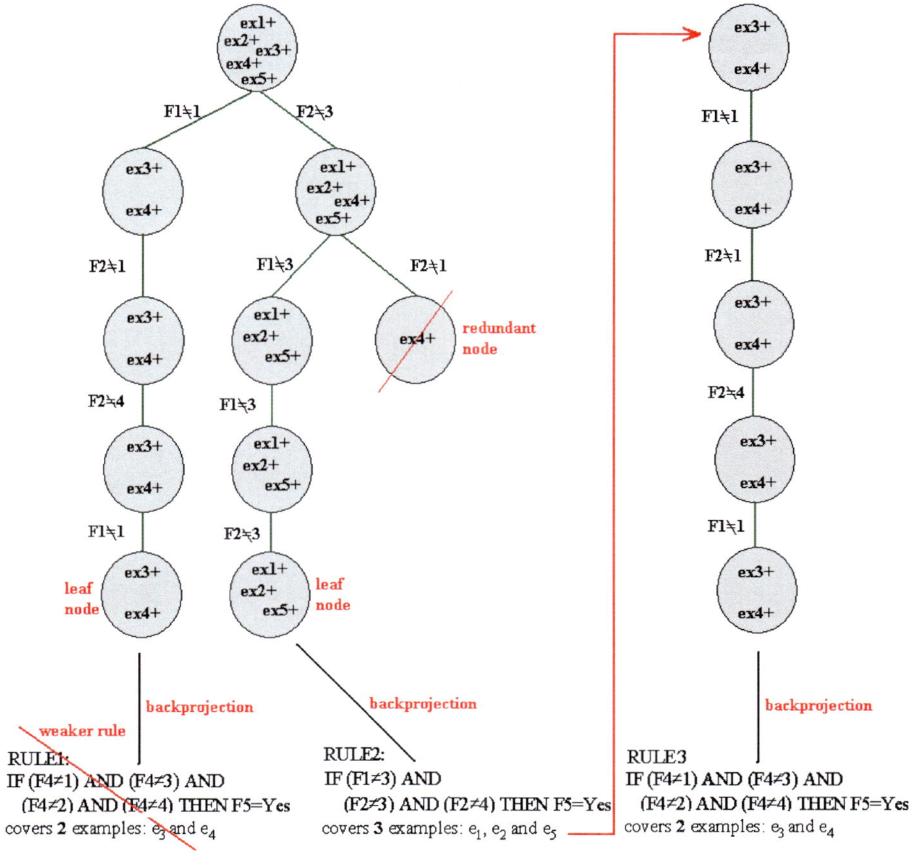

Fig. 4. A different view of the process used by the CLIP4 algorithm

4.2 Handling of Missing Values

Data with the missing values of the attributes are often encountered in real-world situations. If the number of examples with missing values is small, i.e., only a small fraction of the data is affected, we can simply discard them in a preprocessing step. However, in many situations – for example, in medical data - the number of examples (records) with missing values can exceed 50% of the data (Kurgan et al., 2005). In such situations, we need to use all examples and all attributes with values and omit from our computations only these attributes with missing values. The CLIP4 algorithm does not discard any examples with missing values nor does it fill them with any computed values. It simply takes advantage of the available values in the examples and ignores the missing values.

Attributes with missing values influence the first two phases of the algorithm. They are removed from processing by filling in the corresponding cell in the BIN matrix cell with a 0 when the new branch of the tree in Phase I is generated, and when the back-projection in Phase II is performed. The mechanism used in CLIP4 to deal with missing attribute values assures that they are not used in generated rules and that all existing attribute values, even in examples with missing attributes, are used during the rule generation process. The advantage of this mechanism is that the user simply supplies data to the algorithm and obtains his or her without needing to use some statistical methods to calculate missing values. This approach works well under the assumption that the remaining (complete) portion of the data is sufficient to infer the correct model.

Data shown in Table 2 is used to illustrate how CLIP4 deals with missing values. The missing values are denoted by * and class1 constitutes the positive class.

Table 2. Data with missing values

S	F_1	F_2	F_3	F_4	Class
e_1	1	2	3	*	1
e_2	1	3	1	2	1
e_3	*	3	2	5	1
e_4	3	3	2	2	1
e_5	1	1	1	3	1
e_6	3	1	2	5	2
e_7	1	2	2	4	2
e_8	2	1	*	3	2

CLIP4 generates two rules:

IF $F_1 \neq 3$ AND $F_1 \neq 2$ AND $F_3 \neq 2$ THEN class1
IF $F_2 \neq 2$ AND $F_2 \neq 1$ THEN class1

The first rule covers examples 1, 2, and 5, while the second covers examples 2, 3, and 4. Between them they cover all positive examples, including those with missing values, and none of the negative examples. The rules overlap since both cover the second example.

4.3 Classification

CLIP4 classifies examples by using sets of rules generated for all classes. Two classification outcomes are possible: an example is assigned to a particular class, or it is left unclassified. To classify an example, two principles are used:

- All rules that cover the example are found. If no rules cover the example then it remains unclassified. Such a situation may occur if the example has missing values for all attributes used by the rules
- For every class, goodness of the rules describing a particular class and covering the example is summed up. The example is assigned to a class that has the highest summed value. If there is a tie, then the example is left unclassified. For each generated rule, a goodness value, equal to the percentage of the positive training examples that it covers, is calculated. For instance, if an example is covered by two rules from class 1 with corresponding goodness values of 15% and 20%, but the example is also covered by two rules from class 2 with goodness values 50% and 10%, then it would be classified to Class 2 (the sum of goodness values for Class 1 is 35% vs. 60% for Class 2).

4.4 Thresholds

CLIP4 uses three thresholds to perform tree pruning and to remove "noisy" examples:

Noise Threshold determines which nodes (possibly containing noisy positive examples) are pruned from the tree grown in Phase 1. The threshold prunes every node that contains fewer examples than its value.

Pruning Threshold is used to prune nodes from the generated tree. It uses a goodness value, identical to the fitness function described later, to perform selection of the nodes. The threshold selects the first few nodes with the highest fitness value and removes the remaining nodes from the tree.

Stop Threshold stops the algorithm when fewer than the threshold number of positive examples remains uncovered. CLIP4 generates rules by partitioning the data into subsets containing similar examples, and removes examples covered by already generated rules. This approach has the advantage of eliminating small subsets of positive examples (which contain examples different than the majority already covered) from the subsequent rule generation process. If the user were to know the amount of noise in the data then the threshold could be set to this value (e.g., 5%). The noise and stop thresholds are specified as a percentage of the size of positive data and thus are easily scalable.

4.5 Use of Genetic Operators to Improve Accuracy on Small Training Data

CLIP4 uses genetic algorithms to improve the accuracy of generated rules. Its genetic module works by exploiting a single loop through a number of evolving populations. The loop consists of establishing the initial population of individuals and then selecting the new population from the old population, altering and evaluating the new population, and replacing the old one with the new. These operations are performed until a termination criterion is satisfied. CLIP4 uses the GA in Phase I to enhance the partitioning of the data and to obtain more "general" leaf node subsets. The components of the genetic module are as follows:

- *Population* and *individual*
  An individual/chromosome is defined as a node in the tree and consists of the $POS_{i,j}$ matrix (the jth matrix at the ith tree level) and $SOL_{i,j}$ (the solution to the SC problem obtained from the $POS_{i,j}$ matrix). A population is defined as a set of nodes at the same level of the tree.
- *Encoding* and *decoding scheme*
  There is no need for encoding using the individuals defined above since GA operators are used on the binary $SOL_{i,j}$ vector.
- *Selection* of the new population
  The initial population is the first tree level that consists of at least two nodes. CLIP4 uses the following fitness function to select the most suitable individuals for the next generation:

$$fitness_{i,j} = \frac{number\ of\ examples\ that\ constitute\ POS_{i,j}}{number\ of\ subsets\ that\ will\ be\ generated\ from\ POS_{i,j}\ at\ (i+1)^{th}\ tree\ level}$$

The fitness value is calculated as the number of rows of the $POS_{i,j}$ matrix divided by the number of 1's from the $SOL_{i,j}$ vector. The fitness function has high values for the tree nodes that consist of a large number of examples with a low branching factor. These two properties influence the generalization ability of the rules and the speed of the algorithm.

The mechanism for selecting individuals for the next population is as follows:

- All individuals are ranked using the fitness function
- Half of the individuals with the highest fitness are automatically selected for the next population (they will branch to create nodes for the next tree level)
- The second half of the next population is generated by matching the best with the worst individuals (the best with the worst, the second best with the second worst, etc.) and applying GA operators to obtain new individuals (new nodes in the tree). This mechanism promotes the generation of new tree branches that contain large number of examples.

An illustration of the selection mechanism used in the CLIP4 algorithm is shown in Figure 5.

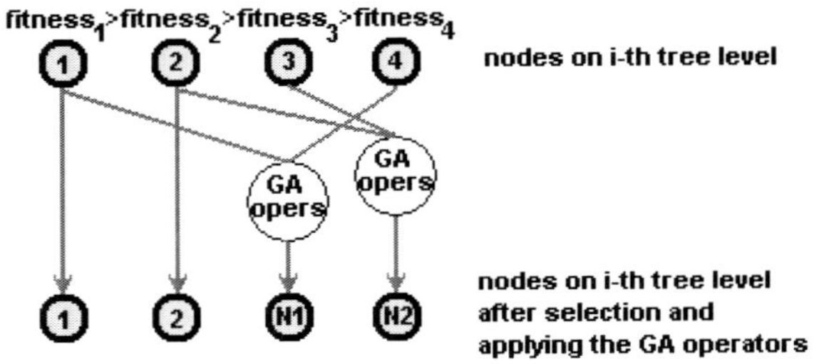

Fig. 5. Selection mechanism performed by the GA module

The GA module uses the crossover and mutation operators. Both are applied only to the $SOL_{i,j}$ vectors. The resulting $ChildSOL_{i,j}$ vector, together with the $POS_{i,j}$ matrix of the parent with the higher fitness value, constitutes the new individual. The selection of the $SOL_{i,j}$ matrix assures that the resulting individual is consistent with CLIP4's way of partitioning data. The crossover operator is defined as

$$ChildSOL_i = \max (Parent1SOL_i, Parent2SOL_i)$$

where $Parent1SOL_i$ and $Parent2SOL_i$ are the ith values of $SOL_{i,j}$ vectors of the two parent nodes.

CLIP4 uses the mutation with 10% probability to flip a value in the ChildSOL vector to 1, regardless of the existing value. For particular data, the probability of mutation can be established experimentally. Each 1 in the ChildSOL generates a new branch, except for 1's taken from the $SOL_{i,j}$ of the parent with higher fitness value, which are discarded because they would generate branches redundant with the branches generated by the parent.

The termination criterion checks whether the bottom level of the tree has been reached. The entire evolution process of the GA module is shown in Figure 6.

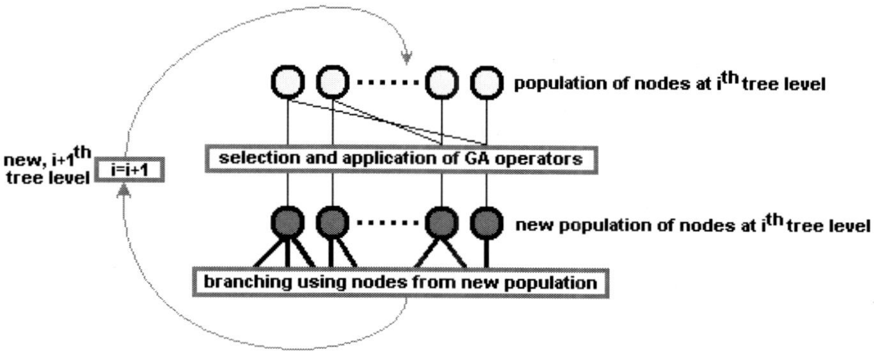

Fig. 6. The evolution process performed by the GA module

The GA module of the CLIP4 algorithm is used when a small data set covering a small portion of the search space is available; the reason is that then CLIP4 will generate rules that potentially cover a not yet described portion of the state space.

4.6 Pruning

CLIP4 uses prepruning to prune the tree during its process of tree generation. The prepruning stops the learning process even if some positive examples have still not been covered and while some negative examples are still covered. More precisely, CLIP4 prunes the tree grown in Phase 1 as follows:

- First, it selects a number (defined by the pruning threshold) of best nodes on the ith tree level. The selection is performed based on the goodness criterion identical to the fitness function described later. Only the selected nodes are used to branch into new nodes, and are passed to the $(i+1)^{th}$ tree level.
- Second, all redundant nodes that resulted from the branching process are removed. Two nodes are redundant if one mode contains positive examples that are identical to each other , or forms a subset of positive examples of the other node. The node with the smaller number of examples is pruned first.
- Third, after the redundant nodes are removed, each new node is evaluated using the noise threshold. If the node contains fewer examples than the number specified by the noise threshold, then the node is pruned.

The prepruning method used in the CLIP4 algorithm avoids some disadvantages of classical prepruning. Namely, it never allows negative examples to be covered by the rules, and it ensures the completeness condition of the rule generation process. This type of prepruning increases the accuracy of the generated rules and lowers the complexity of the algorithm.

4.7 Feature and Selector Ranking

The CLIP4 algorithm not only generates a data model in terms of rules but also ranks attributes and selectors. This aspect of the algorithm can be used for feature selection, which is defined as a process of finding a subset of original features that is optimal according to some defined criterion; in the case of classification, the criterion is to obtain the highest predictive accuracy. CLIP4 ranks attributes and selectors by assigning to them a goodness value that quantifies their relevance to a particular learning task. These rankings provide additional insight into the information hidden in the data.

The goodness of each attribute and selector is computed by means of a set of generated rules for a given class. All attributes with goodness value greater than zero are strongly relevant to the classification task. Strong relevancy means that the attribute cannot be removed from the attribute set without decreasing the accuracy of classification. The other attributes can be removed from the data. The attribute and selector goodness values are computed in the following manner:

- Each generated rule has a goodness value equal to the percentage value of the positive training examples it covers. Recall that each rule consists of one or more (attribute, value of the selector) pairs.
- Each selector has a goodness value equal to the goodness of the rule from which it comes. The goodness of the same selectors from different rules is summed, and then scaled to the (0,100) range; with 100 being the highest. Scaling of the goodness values is necessary because otherwise the summed goodness for a particular selector can grow to over 100, while only the ratio to the goodness's of other selectors is important.
- For each attribute, the sum of the scaled goodness of all attribute selectors is computed and then divided by the number of attribute values to obtain the goodness of the attribute.

The feature and selector ranking performed by CLIP4 algorithm can be used to:

- select only relevant attributes (features) and discard irrelevant ones; the user can discard all attributes that have a goodness of 0 and still have an equally accurate model of the data, and
- provide additional insight into data properties. The selector ranking can help in analyzing the data in terms of the relevance of the selectors to the classification task.

5 DataSqueezer Algorithm

The DataSqueezer algorithm was first published in (Kurgan 2003; Kurgan et al., 2006) and later it was incorporated into a larger data mining system called MetaSqueezer (Kurgan and Cios, 2004). The latter system was recently successfully used in analysis of clinical data concerning patients with cystic fibrosis (Kurgan et al., 2005); it was able to discover clinical information that is implicated in stabilization or improvement of patient's health. The DataSqueezer induces a set of production rules from a supervised training data by utilizing a very simple data reduction procedure via prototypical concept learning, inspired by the Find S algorithm (Mitchell, 1997). The main advantages of this method are its log-linear complexity resulting in very fast generation of the rule-based classification models and robustness to missing values. DataSqueezer is also relatively simple to implement.

Let us denote a training dataset by D, consisting of s examples and k attributes. The training dataset is divided into two disjoint subsets, D_P that consists of positive examples and D_N that includes the negative examples. Figure 7 shows the pseudocode of the algorithm. Vectors and matrices are denoted by capital letters, while elements of vectors/matrices are denoted with the same name but use small letters. The matrix of positive/negative examples is denoted as POS/NEG and their number as N_{POS}/N_{NEG}. The examples are represented by rows, and features/attributes by columns. Positive examples, POS, and negative examples, NEG, are denoted in the DataReduction procedure by $d_i[j]$ values, where j=1,…,k is the column number, and i is the example number (row number in matrix D, which denotes either POS or NEG). The algorithm also uses matrices that store intermediate results (G_{POS} for the POS, and G_{NEG} for the NEG), which have k columns. Each cell of the G_{POS} matrix is denoted by $gpos_i[j]$, where i is a row number and j is a column number; and similarly for the G_{NEG} matrix each cell is denoted by $gneg_i[j]$. The G_{POS} stores a reduced subset of the data from the POS matrix, and the G_{NEG} stores a reduced subset of the data from the NEG matrix. The G_{NEG} and G_{POS} matrices have an additional $(k+1)^{th}$ column that stores the number of examples from the NEG and POS matrices, described by a particular row in the G_{NEG} and G_{POS}, respectively. For example, $gpos_2[k+1]$ stores the number of examples from the POS that are described by the second row in the G_{POS} matrix.

DataSqueezer works in two steps. In step 1 it performs data reduction to generalize information stored in the original data. Data reduction is performed via the use of a prototypical concept learning procedure. This is performed for both positive and negative data and it results in generation of the G_{POS} and G_{NEG} matrices. This reduction

Given: POS, NEG, k (number of attributes), s (number of examples)
Step1.
1.1 G_{POS} = DataReduction(POS, k);
1.2 G_{NEG} = DataReduction(NEG k);

Step2.
2.1 Initialize RULES = []; i=1; *// where $rules_i$ denotes i^{th} rule stored in RULES*
2.2 create LIST = list of all columns in G_{POS}
2.3 within every G_{POS} column that is on LIST, for every non missing value a from selected column j compute sum, s_{aj}, of values of $gpos_i[k+1]$ for every row i, in which a appears and multiply s_{aj}, by the number of values the attribute j has
2.4 select maximal s_{aj}, remove j from LIST, add "j = a" selector to $rules_i$
2.5.1 **if** $rules_i$ does not describe any rows in G_{NEG}
2.5.2 **then** remove all rows described by $rules_i$ from G_{POS}, i=i+1;
2.5.3 **if** G_{POS} is not empty go to 2.2, **else** terminate
2.5.4 **else** go to 2.3
Output: RULES describing POS

DataReduction (D, k) // data reduction procedure for D=POS or D=NEG
DR.1 Initialize G = []; i=1; tmp = d_1; g_1 = d_1; $g_1[k+1]$=1;
DR.2.1 **for** j=1 to N_D // for positive/negative data; N_D is N_{POS} or N_{NEG}
DR.2.2 **for** kk = 1 to k *// for all attributes*
DR.2.3 **if** ($d_j[kk] \neq$ tmp[kk] **or** $d_j[kk]$ = '*')
DR.2.4 **then** tmp[kk] = '*'; *// '*' denotes missing" do not care" value*
DR.2.5 **if** (number of non missing values in tmp $\geq$ 2)
DR.2.6 **then** g_i = tmp; $g_i[k+1]$++;
DR.2.7 **else** i++; g_i = d_j; $g_i[k+1]$=1; tmp = d_j;
DR.2.8 return G;

Fig. 7. Pseudocode of the DataSqueezer algorithm

is related to the least generalization in the FindS method. The main difference is that FindS performs the least generalization multiple times for the entire positive set through a beam-search strategy, while the DataSqueezer performs it only once in a linear fashion by generalizing consecutive examples. It also generalizes the negative data. In step 2 the DataSqueezer generates rules by a greedy hill-climbing search on the reduced data. A rule is generated by using the search procedure starting with an empty rule, and adding selectors until the termination criterion is satisfied. The rule that is being generated consists of selectors generated using the G_{POS}, and is checked against the G_{NEG}. If the rule covers any data in the G_{NEG}, a new selector is added to make it more specific, and thus better able to distinguish between positive and negative data. The maximum depth of the search is equal to the number of features. Next, the examples covered by the generated rule are removed, and the process is repeated.

When applied to dataset given in Table 1 the DataSqueezer algorithm generates two rules:

If $F_1 = 1$ THEN Class = 1
If $F_3 = 3$ THEN Class = 2

Note that each rule was generated separately, i.e., the first rule was generated by using class 1 as positive and class 2 as negative, while the second rule was generated by using class 2 as positive and class 1 as negative. This means that each of these rules can be used independently to classify the data shown in Table 1. When applied to the Table 1 data, each of these rules perfectly separates the two classes.

The DataSqueezer algorithm handles data with missing values very well. Similarly to CLIP4, it uses all available information while ignoring missing values during the induction process, i.e., the missing values are handled as "do not care" values. The algorithm is robust to a very large number of missing values and was shown to successfully generate rules even from data that have more than half of missing values (Kurgan et al., 2005). DataSqueezer, like the other algorithms covered in this chapter handles only discrete-valued numerical and nominal attributes (the latter are automatically encoded into numerical values). We note that continuous attributes can be converted into discrete-values attributes using a discretization algorithm (Kurgan and Cios, 2004). The generated rules are independent of the encoding scheme because during the rule induction process no distances are calculated between the feature values.

DataSqueezer uses two thresholds that can be set by the user. By default these thresholds are set to zero. The pruning threshold is used to prune very specific rules, i.e., rules that cover only a few examples. The rule generation process is terminated if the first selector added to $rule_i$ has s_{aj} value equal to or smaller than the threshold's value. The algorithm induces rules by selecting maximal s_{aj} values (selectors that cover most of the POS examples) and removes examples that have already been covered. This approach has the benefit of leaving small subsets of positive examples that store different examples from the majority already covered. These potential outliers can be filtered out using this threshold. The generalization threshold is used to allow for rules that cover a small amount of negative data. It relaxes the requirement from line 2.5.1 (Figure 7) and allows for acceptance of rules that cover negative examples; the number is equal to or smaller than this threshold. It is a useful mechanism in the case of data with overlapping classes, or in the case of inconsistent examples present in the training dataset (examples that should have been but were not eliminated during preprocessing). Both thresholds should be set to small values preferably expressed as percentages of the POS data size.

6 Conclusions

The works of Michalski undoubtedly provided solid foundations and a strong motivation for many of the subsequent works on rule-based inductive ML algorithms. Some of our own methods, including the CLIP family of algorithms and DataSqueezer, were inspired by his work. We believe that ours and future generations of computer scientists will continues this line of research, which is fueled by an increasing demand for scalable methodologies that can generate human-readable models for current and future datasets.

References

Cios, K.J., Liu, N.: An algorithm which learns multiple covers via integer linear programming. Part I - The CLILP2 Algorithm. Kybernetes 24, 29–50 (1995)

Cios, K.J., Liu, N.: An algorithm which learns multiple covers via integer linear programming. Part I - The CLILP2 Algorithm. Kybernetes 24, 29–50 (1995); (The Norbert Wiener 1997 Outstanding Paper Award, `http://www.mcb.co.uk/literati/outst97.htm#k`)

Cios, K.J., Wedding, D.K., Liu, N.: CLIP3: cover learning using integer programming. Kybernetes 26(4-5), 513–536 (1997)

Cios, K.J., Pedrycz, W., Swiniarski, R., Kurgan, L.: Data mining: a knowledge discovery approach. Springer, Heidelberg (2007)

Cios, K.J., Pedrycz, W., Swiniarski, R.: Data mining methods for knowledge discovery. Kluwer, Dordrecht (1998)

Cios, K.J., Kurgan, L.: CLIP4: Hybrid inductive machine learning algorithm that generates inequality rules. Information Sciences 163(1-3), 37–83 (2004)

Farhangfar, A., Kurgan, L., Pedrycz, W.: A novel framework for imputation of missing values in databases. IEEE Transactions on Systems, Man, and Cybernetics, Part A 37(5), 692–709 (2007)

Farhangfar, A., Kurgan, L., Dy, J.: Impact of imputation of missing values on classification error for discrete data. Pattern Recognition 41(12), 3692–3705 (2008)

Kurgan, L.: Meta mining system for supervised learning, Ph.D dissertation, the University of Colorado at Boulder, Department of Computer Science (2003)

Kurgan, L., Cios, K.J.: Meta mining architecture for supervised learning. In: 7th International Workshop on High Performance and Distributed Mining, Proc. 4th International SIAM Conference on Data Mining, Lake Buena Vista, FL, pp. 18–26 (2004)

Kurgan, L., Cios, K.J.: CAIM discretization algorithm. IEEE Transactions on Data and Knowledge Engineering 16(2), 145–153 (2004)

Kurgan, L., Cios, K.J., Sontag, M., Accurso, F.: Mining the cystic fibrosis data. In: Zurada, J., Kantardzic, M. (eds.) Next generation of data-mining applications, pp. 415–444. IEEE Press - Wiley (2005)

Kurgan, L., Cios, K.J., Dick, S.: Highly scalable and robust rule learner: performance evaluation and comparison. IEEE Transactions on Systems, Man, and Cybernetics, Part B 36(1), 32–53 (2006)

Mitchell, T.M.: Machine learning. McGraw-Hill, New York (1997)

Kodratoff, Y.: Introduction to machine learning. Morgan Kaufmann, San Francisco (1988)

Langley, P.: Elements of machine learning. Morgan Kaufmann, San Francisco (1996)

Łukasiewicz, J.: O logice trójwartościowej (in Polish). Ruch Filozoficzny 5, 170–171 (1920); English translation: On three-valued logic. In: Borkowski, L. (ed.) Selected works by Jan Łukasiewicz, pp. 87–88. North–Holland, Amsterdam (1970)

Michalski, R.S.: On the quasi minimal solution of the general covering problem. In: Proc. 5th International Symposium on Information Processing (FCIP 1969), Bled, Yugoslavia, vol. A3, pp. 25–128 (1969)

Michalski, R.S.: Variable valued logic: system VLl. In: Proc. 1974 International Symposium on Multiple Valued Logic and Pattern Recognition, West Virginia University, Morgantown, pp. 323–346 (1974)

Michalski, R.S.: Knowledge acquisition through conceptual clustering: a theoretical framework and algorithm for partitioning data into conjunctive concepts. International Journal of Policy Analysis and Information Systems 4, 219–243 (1980)

Michalski, R.S., Mozetic, I., Hong, J., Lavrac, N.: The multipurpose incremental learning system AQ15 and its testing application to three medical domains. In: Proc. 5th National Conference on Artificial Intelligence, pp. 1041–1045. Morgan-Kaufmann, San Francisco (1986)

Mitchell, T.M.: Machine learning. McGraw-Hill, New York (1997)

Quinlan, J.R.: C4.5 programs for machine learning. Morgan-Kaufmann, San Francisco (1993)

Pawlak, Z.: Rough sets - theoretical aspects of reasoning about data. Kluwer, Dordrecht (1991)

Inductive Learning: A Combinatorial Optimization Approach

Janusz Kacprzyk and Grażyna Szkatuła

Systems Research Institute
Polish Academy of Sciences
ul. Newelska 6
01-447 Warsaw
Poland
kacprzyk@ibspan.waw.pl

Abstract. We propose an improved inductive learning method to derive classification rules correctly describing (at least) *most* of the positive examples and do not correctly describe (at least) *most* of the negative examples. We start with a pre-analysis of data to assign higher weights to those values of attributes which occur more often in the positive than in the negative examples. The inductive learning problem is represented as a modification of the set covering problem which is solved by an integer programming based algorithm using elements of a greedy algorithm or a genetic algorithm, for efficiency. The results are very encouraging and are illustrated on a thyroid cancer data set.

Keywords: Inductive learning, learning from examples, set covering problem, integer programming, genetic algorithm, greedy algorithm.

1 Introduction

Inductive learning (from examples) is a process of inferring a classification rule (concept description) of a class from *positive examples* and *negative examples*. Traditionally, the most relevant requirements to be satisfied by learning procedures are:

- *completeness*, i.e. that the classification rule must correctly describe *all* the positive examples,
- *consistency*, i.e. that the classification rule must not describe *any* negative examples,
- *convergence*, i.e. the classification rule must be derived in a *finite* number of steps,
- the classification rule of *minimal "length"* is to be found, e.g. with the minimum number of attributes (or, more generally being "simple").

In practice, due to imperfect data and other elements of the process, the first two requirements should rather be meant in a relaxed way as:

- *a partial completeness*, i.e. that the classification rule must correctly describe, for instance, (at least) *most* of the positive examples,
- *a partial consistency*, i.e. that the classification rule must describe, for instance, (at most) *almost none* of the negative examples.

J. Koronacki et al. (Eds.): Advances in Machine Learning I, SCI 262, pp. 75–93.
springerlink.com © Springer-Verlag Berlin Heidelberg 2010

Examples are described (cf. Michalski, 1983) by a set of K "attribute - value" pairs written as $e = \bigwedge_{j=1}^{K} [a_j \# v_j]$, where a_j denotes attribute j with value v_j and # is a relation exemplified by =, <, >, ≈,≥, etc. For instance, if the attributes are: *height, color of hair, color of eyes*, than the concept "look of a one women" may be described by [*height* ="high"]∧[*color of hair* ="blond"]∧[*color of eyes* ="blue"].

We propose here a modified inductive learning procedure based on Michalski's (1973, 1983) star-type methodology (see also, for instance, Michalski, Mozetic, Hong and Lovrac, 1986). It is based on elements of our previous work (cf. Kacprzyk and Szkatuła, 1994 – 2005c; see also Iwa ski and Szkatuła, 2001, Szkatuła and Kacprzyk, 2005). A pre-processing of data (examples) is performed based on how frequent the values of the particular attributes occur in the examples. These frequencies are used to derive weights associated with those values, and the problem is represented as a modification of the set covering problems. For efficiency it is solved by a modification of a greedy algorithm (IP2_GRE) or a genetic algorithm (IP2_GA).

The transformation of an inductive learning problem into a set covering problem, which is well known in combinatorial optimization, has some tradition. In our context, Iwa ski and Szkatuła (1991) may be the first, and then it was used in the CLIP algorithms proposed by Cios and his collaborators (cf. Cios and Liu, 1995; Cios, Wedding and Liu, 1997; Cios and Kurgan, 2004). In our papers, Kacprzyk and Szkatuła (1994 – 2005c) we extended those formulation by, first, redefining the problem to account for a milder requirement of a partial completeness and partial consistency, second, by using an intuitively appealing data preprocessing to obtain weights associated with selectors and complexes, and – third – by proposing the use of a different greedy and genetic algorithm. Here we will further elaborate upon these issues.

In Section 2 we present the inductive learning problem as a modification of the set covering problems. In Section 3 our IP2 algorithm is described. In Section 4 and 5 we present the two procedures implementing a greedy and genetic algorithm, IP2_GRE and IP2_GA, respectively. In Section 6 we show computation results for the thyroid cancer problem.

2 Problem Formulation of Inductive Learning

Suppose that we have a finite set of examples U and a finite set of attributes $A = \{a_1,...,a_K\}$. $V_{a_j} = \{v_{j,1}, v_{j,2},..., v_{j,L_j}\}$ is a domain of the attribute a_j, $j = 1, ..., K$, where L_j denotes number of values of the j-th attribute. $V = \bigcup_{j=1,...,K} V_{a_j}$.

$f{:}U \times A \to V$ is a function such that $f(e^n, a_j) \in V_{a_j}$ for every $a_j \in A$ and $e^n \in U$, $n = 1, 2, ..., N$. Each example $e^n \in U$ is described by K attributes $A = \{a_1, ..., a_K\}$, and is represented by

$$e^n = \bigwedge_{j=1}^{K} [a_j = f(e^n, a_j)] \tag{1}$$

where $f(e^n, a_j) = v_{j,t(j,n)}$, and $v_{j,t(j,n)} \in V_{a_j}$. Function $f(e^n, a_j)$ denotes that the attribute a_j taking on a value $v_{j,t(j,n)}$ for example e^n. The index $t(j,n)$ for $j \in \{1,2,...,K\}$ and $n \in \{1,2,...,N\}$ specifies which value of the *j*-th attribute is used in the *n*-th example.

An example e^n in (1) is composed of *K* "attribute-value" pairs (*selectors*), denoted $s_j = [a_j = v_{j,t(j,n)}]$. Conjunction of $l \le K$ "attribute-value" pairs, i.e.

$$\underset{j \in I}{\wedge} s_j = C^l \tag{2}$$

where $I \subseteq \{1,...,K\}$, $card(I) = l$ is called a *complex*.

A complex C^l *covers* an example e^n if all the conditions on attributes given as *j*-th selectors are covered by (equal to) the values of the respective attributes in the example, $\forall j \in I$. The set of all the examples described by the conjunction C^l will be denoted $[C^l]$.

Suppose that we have a finite set of examples *U,* a finite set of attributes $A \cup \{a_d\}$, $\{a_d\} \cap A = \varnothing$, a_d is a decision attribute and $V_{a_d} = \{v_{d,1}, v_{d,2}, ..., v_{d,L_d}\}$ is a domain of the attribute a_d. We have the sets $\{Y_{v_{d,l}} : l = 1,..., L_d\}$, where $Y_{v_{d,l}} = \{e \in U : f(e, a_d) = v_{d,l}\}$, $\forall v_{d,l} \in V_{a_d}$ and $Y_{v_{d,1}} \cup ... \cup Y_{v_{d,L_d}} = U$, $Y_{v_{d,i}} \cap Y_{v_{d,j}} = \varnothing$ for $i \ne j$. Thus, the decision attribute splits the set of examples into the non-empty, disjoint and exhaustive subsets, that we call the *decision classes.*

Let us consider class $Y_{v_{d,l}}$, for $v_{d,l} \in V_{a_d}$, and a set of *positive examples*:

$$S_P(Y_{v_{d,l}}) = \{e \in U : f(e, a_d) = v_{d,l}\} \tag{3}$$

and a set of *negative examples*:

$$S_N(Y_{v_{d,l}}) = \tag{4}$$
$$\{e \in U : f(e, a_d) \ne v_{d,l} \text{ and } \forall e' \in S_P(Y_{v_{d,l}})\ \exists a_j \in A,\ f(e, a_j) \ne f(e', a_j)\}$$

with $S_P(Y_{v_{d,l}}), S_N(Y_{v_{d,l}}) \ne \varnothing$, $S_P(Y_{v_{d,l}}) \cap S_N(Y_{v_{d,l}}) = \varnothing$, by assumption.

An implication R_k: $C^{l_k} \to [a_d = v_{d,l}]$, $l \in \{1,..., L_d\}$ is called the *k*-th *elementary rule* for the class $Y_{v_{d,l}}$, where $C^{l_k} = \underset{j \in I_k}{\wedge} [a_j = v_{j,t(j,k)}]$ is description of example in terms of condition attributes a_j, $j \in I_k$, $I_k \subseteq \{1,..., K\}$ and this example belongs to class $Y_{v_{d,l}}$. The index $t(j,k)$ specifies which value of the *j*-th attribute is used in the *k*-th rule. Each rule is characterised by the coefficient of its strength. The strength of a rule R_k, which depends upon the number of examples described by the conditional part of the rule C^{l_k}, belonging to a given class $Y_{v_{d,l}}$ is defined in the following manner: $q(C^{l_k}) = \dfrac{card(\{e : e \in [C^{l_k}] \text{ and } f(e, a_d) = v_{d,l}\})}{card(\{e : e \in U\})}$. It is evident that $0 \le q(C^{l_k}) \le 1$. The more examples are described by the rule, the greater the value of the rule strength coefficient (i.e. the more important the rule is).

In this paper we consider the classification rules to be the disjunction (via "$\cup$") of "elementary" rules consisting of complexes of type (2), i.e.

$$C^{I_1} \cup \ldots \cup C^{I_L} = \underset{j_1 \in I_1}{\wedge} s_{j_1} \cup \ldots \cup \underset{j_L \in I_L}{\wedge} s_{j_L} \rightarrow [\text{class} = \text{class } Y_{v_{d,l}}] \tag{5}$$

where: $I_1, \ldots, I_L \subseteq \{1, \ldots, K\}$ and "$\cup$" corresponds to the connective "or".

Suppose that we have a finite set of examples U, a finite set of attributes $\{a_1, \ldots, a_K\} \cup \{a_d\}$, a_d is a decision attribute and $V_{a_d} = \{v_{d,1}, v_{d,2}, \ldots, v_{d,L_d}\}$ is a domain of the attribute a_d. Let us class $Y_{v_{d,l}}$, for $v_{d,l} \in V_{a_d}$. Suppose now that we have P positive examples and N negative examples. The positive examples are therefore $e^m \in S_P(Y_{v_{d,l}})$, $m = 1, \ldots, P$, and are written as

$$e^m = [a_1 = v_{1,t(1,m)}] \wedge \ldots \wedge [a_K = v_{K,t(K,m)}] \tag{6}$$

where: $v_{j,t(j,m)} \in V_{a_j}$, $j = 1, \ldots, K$, while the negative examples are $e^n \in S_N(Y_{v_{d,l}})$, $n = 1, \ldots, N$, and are written as

$$e^n = [a_1 = v_{1,t(1,n)}] \wedge \ldots \wedge [a_K = v_{K,t(K,n)}] \tag{7}$$

where: $v_{j,t(j,n)} \in V_{a_j}$, $j = 1, \ldots, K$ and $S_P(Y_{v_{d,l}}) \neq \varnothing$, $S_N(Y_{v_{d,l}}) \neq \varnothing$, $S_P(Y_{v_{d,l}}) \cap S_N(Y_{v_{d,l}}) = \varnothing$.

The idea of data pre-processing proposed by Kacprzyk and Szkatuła (1994 – 1999), which leads to the formulation of a set covering problem, is as follows. For each attribute a_j in the examples, not each possible value occurs with the same intensity (frequency). Clearly, if a value occurs more frequently in the positive examples and less frequently in the negative examples, then it can be argued that it should rather appear in the rule sought. Then can be formalized as follows: first we introduce the following function, for each attribute a_j, $j = 1, \ldots, K$ and particular values $v_{j,l} \in V_{a_j}$, $l = 1, \ldots, L_j$, where the j-th attribute a_j taking on the value $v_{j,t(j,m)}$ for example $e^m \in S_P$ and the value $v_{j,t(j,n)}$ for example $e^n \in S_N$

$$g_j(v_{j,l}) = \frac{1}{P} \sum_{m=1}^{P} \delta(e^m, v_{j,l}) - \frac{1}{N} \sum_{n=1}^{N} \delta(e^n, v_{j,l}) \tag{8}$$

where: $\delta(e^m, v_{j,l}) = \begin{cases} 1 & \textit{for } v_{j,t(j,m)} = v_{j,l} \\ 0 & \textit{otherwise} \end{cases}$ and analogously for $\delta(e^n, v_{j,l})$.

Therefore, (8) expresses to what degree the particular values $v_{j,l} \in V_{a_j}$ of attribute a_j occurs more often in the positive than in the negative examples. We may therefore assume that this normalized value $g_j(v_{j,l})$ is used as a weight of value $v_{j,l} \in V_{a_j}$ of each attribute a_j.

Suppose now that we have a positive example e^m, such that $e^m = [a_1 = v_{1,t(1,m)}] \wedge ... \wedge [a_K = v_{K,t(K,m)}]$, and we consider a complex

$$C^I = [a_{j_1} = v_{j_1,t(j_1,m)}] \wedge ... \wedge [a_{j_m} = v_{j_m,t(j_m,m)}] \tag{9}$$

that corresponds to the set of indices $I = \{ j_1, ..., j_m \} \subseteq \{1, ..., K\}$; the set of indices $\{j_1, ..., j_m\}$ is clearly equivalent to a vector $x = [x_j]^T$, such that $x_j = 1$ if a selector $[a_j = v_{j,t(j,m)}]$ occurs in the complex (9), and 0 otherwise, $j = 1, ..., K$.

To the value $v_{j,t(j,m)} \in V_{a_j}$, of each attribute a_j a weight $g_j(v_{j,t(j,m)})$ is assigned, and an example e_W^m with weights is written as

$$e_W^m = \bigwedge_{j=1}^{K} [a_j = v_{j,t(j,m)}; g_j(v_{j,t(j,m)})] \tag{10}$$

An example e_W^m given by (10) is then composed of the weighted selectors, $s_j^W = [a_j = v_{j,t(j,m)}; g_j(v_{j,t(j,m)})]$, and a conjunction of them, i.e.

$$C_W^I = \bigwedge_{j \in I \subseteq \{1, ..., K\}} s_j^W =$$

$$= \bigwedge_{j=j_1, j_2, ..., j_m} s_j^W = [a_{j_1} = v_{j_1,t(j_1,m)}; g_{j_1}(v_{j_1,t(j_1,m)})] \wedge ... \wedge [a_{j_m} = v_{j_m,t(j_m,m)}; g_{j_m}(v_{j_m,t(j_m,m)})] \tag{11}$$

is called a *weighted complex*. Notice that for the above C_W^I the vector *x* has the elements $x_{j_1}, x_{j_2}, ..., x_{j_m} = 1$, while $x_j = 0$ for $j \in \{1,2,..., K\} \setminus \{j_1, j_2, ..., j_m\}$.

For a weighted complex C_W^I its *weighted length* is

$$d_W(C_W^I) = \sum_{j=j_1}^{j_m} (1 - g_j(v_{j,t(j,m)})) \cdot x_j + \sum_{j \in \{1,...,K\} \setminus \{j_1, j_2, ..., j_m\}} (1 - g_j(v_{j,t(j,m)})) \cdot x_j =$$

$$= \sum_{j=1}^{K} (1 - g_j(v_{j,t(j,m)})) \cdot x_j \tag{12}$$

which reflects the philosophy of data pre-processing introduced above, i.e. a higher relevance of those values of attributes which occur more often in positive than in negative examples.

The *length of the weighted classification rule* composed of *L* weighted complexes, $R_W \to [class = Y_{v_{d,l}}]$, $R_W = C_W^{I_1} \cup ... \cup C_W^{I_L}$ is

$$d_{R_W}(C_W^{I_1} \cup ... \cup C_W^{I_L}) = \max_{i=1, ..., L} d_W(C_W^{I_i}) \tag{13}$$

The problem of learning from examples was to find an optimal classification rule $R_W^* = C_W^{I_1^*} \cup ... \cup C_W^{I_L^*}$ such that

$$\min_{I_1, ..., I_L} d_{R_W}(C_W^{I_1} \cup ... \cup C_W^{I_L}) \tag{14}$$

i.e. which minimizes the weighted length of the classification rule.

Since the (exact) solution of problem (14) is very difficult, an auxiliary problem is solved (cf. Kacprzyk and Szkatula, 1996 – 1999), i.e. an $R_W^* = C_W^{I_1^*} \cup ... \cup C_W^{I_L^*}$ is sough such that

$$\min_{I_1} d_W(C_W^{I_1}), ..., \min_{I_L} d_W(C_W^{I_L}) \tag{15}$$

where the minimization is consecutively performed over the sets of indices $I_1, ..., I_L$ [cf. (5)]; the solution of (15) is in general very close to that of (14), while much easier to obtain.

If the requirements to be satisfied by learning procedures are:

- *a partial completeness*, i.e. that the classification rule must correctly describe *most* of the positive examples,
- *a partial consistency*, i.e. that the classification rule must not describe *most* of the negative examples,

the problem (15) can be represented as a modification of the set covering problem that is well known in combinatorial optimization.

3 Solution via the IP2 Method

Suppose that we have a finite set of examples *U*, a finite set of attributes $\{a_1,...,a_K\} \cup \{a_d\}$, a_d is a decision attribute and $V_{a_d} = \{v_{d,1}, ..., v_{d,L_d}\}$ is a domain of the attribute a_d. We have the decision classes $\{Y_{v_{d,l}} : l = 1,..., L_d\}$, where $Y_{v_{d,l}} = \{e \in U : f(e, a_d) = v_{d,l}\}$, $\forall v_{d,l} \in V_{a_d}$. The rules are iteratively induced for each class.

For a class $Y_{v_{d,l}}$, for $v_{d,l} \in V_{a_d}$, suppose that we have a set of positive examples S_P [cf. (3)], $e^m \in S_P$, $m = 1,..., P$ and a set of negative examples S_N [cf. (4)], $e^{P+n} \in S_N$, $n = 1, ..., N$, $S_N, S_P \neq \varnothing$, $S_P \cap S_N = \varnothing$, by assumption.

For $e^m \in S_P$, $e^m = [a_1 = v_{1,t(1,m)}; g_1(v_{1,t(1,m)})] \wedge ... \wedge [a_K = v_{K,t(K,m)}; g_K(v_{K,t(K,m)})]$ and all the negative examples $e^{P+n} \in S_N$, $n = 1, ..., N$, where $e^{P+n} = [a_1 = v_{1,t(1,P+n)}; g_1(v_{1,t(1,P+n)})] \wedge ... \wedge [a_K = v_{K,t(K,P+n)}; g_K(v_{K,t(K,P+n)})]$, we construct a 0-1 matrix $Z_{N*K} = [z_{nj}]$, $n = 1, ..., N$, $j = 1, ..., K$, defined as

$$z_{nj} = \begin{cases} 1 & for \quad v_{j,t(j,m)} \neq v_{j,t(j,P+n)} \\ 0 & for \quad v_{j,t(j,m)} = v_{j,t(j,P+n)} \end{cases} \tag{16}$$

The rows of this matrix correspond to the consecutive negative examples $e^{P+n} \in S_N$, $n = 1, ..., N$ and the columns correspond to the subsequent attributes $a_1, ..., a_K$. The value $z_{nj} = 1$ occurs if attribute a_j takes on different values in the positive and negative example, i.e. $v_{j,t(j,m)}$ in e^m are not equal to $v_{j,t(j,P+n)}$ in e^{P+n}; and $z_{nj} = 0$ otherwise.

In such a matrix there are clearly no rows with all the elements equal 0 since, by assumption, the sets of positive and negative examples are disjoint (and non-empty), $S_N, S_P \neq \emptyset$, $S_P \cap S_N = \emptyset$. Thus, for any positive and negative example there always exists at least one attribute taking on a different value in these examples.

Consider now the following inequality

$$\sum_{j=1}^{K} z_{nj} x_j \geq \gamma_n, \qquad n = 1, ..., N \tag{17}$$

where $\gamma = [\gamma_1, ..., \gamma_N]^T$ is a zero-one vector (of N elements), and $x = [x_1, ..., x_K]^T$ such that $x_j \in \{0,1\}$, for $j = 1, ..., K$ [cf. the remark for (9)].

Any vector x defined above which satisfies (17) determines therefore in a unique way [cf. (9)] some complex composed of selectors from the description of the example such that the conditions of partial completeness and partial consistence are satisfied. It describes at least one example from the set of positive examples, and it does not describe most of the examples from the set of negative examples. If vector x does not describe the n-th negative example, than $\gamma_n = 1$; and $\gamma_n = 0$ otherwise.

The minimization in problem (15) may be written using the inequality (17) as

$$\min_{x:Zx\geq\gamma} \sum_{j=1}^{K} (1 - g_j(v_{j,t(j,m)})) \cdot x_j \tag{18}$$

The minimization over the set of indices I_l may be replaced by the minimization with respect to x which yields [cf. (14)] an $R_W^* = C_W^{I_1^*} \cup ... \cup C_W^{I_L^*}$ such that

$$\min_{x:Z^1x\geq\gamma} d_W(C_W^{I_1}), ..., \min_{x:Z^Lx\geq\gamma} d_W(C_W^{I_L}) \tag{19}$$

Each minimization with respect to x in (19) is therefore equivalent to the determination of a 0-1 vector x^* which uniquely determines the complex of the shortest weighted length. On the other hand, the satisfaction of $Zx \geq \Lambda$ (if Λ is a unit vector) guarantees that such a complex would not describe the all negative examples. If a complex must describe *almost none* of the negative examples, problem (18) can be written as a modification of the set covering problem

$$\min_{x,\gamma} \sum_{j=1}^{K} c_j x_j \tag{20}$$

subject to

$$\sum_{j=1}^{K} z_{nj} x_j \geq \gamma_n , \qquad n = 1, ..., N \tag{21}$$

and an additional constraint

$$\sum_{n=1}^{N} \gamma_n \geq N - rel \tag{22}$$

where

$$c_j = (1 - g_j(v_{j,t(j,m)})) ,\ e^m \in S_P ,\ z_{nj} \in \{0,1\} ,\ \ x_j \in \{0,1\} ,\ \ j = 1, ..., K ,$$

$$\gamma = [\gamma_1, ..., \gamma_N]^T ,\ \gamma_n \in \{0,1\} ,\ \ n = 1, ..., N \tag{23}$$

given a parameter $rel \geq 0$, *rel* is an integer number.

The above problem is the same as the original set covering problem with the exception that no more then *rel* rows are uncovered. Then, it is clear that no more then *rel* rows can be deleted from the problem. Note here that we may, in deleting rows, lose some information about the problem that could have been better used. Note also that this reduction test cannot always be applied. In the set covering problem (SCP) [cf. Balas (1980), Balas and Ho (1980), Beasley (1987, 1990, 1996), etc.] there is only constrains (21) and vector $\gamma = [\gamma_1, ..., \gamma_N]^T$ is a unit vector.

Our modified set covering problem given by (20) - (23) is the problem of covering at least *N-rel* rows of an *N*-row, *K*-column, zero-one matrix (z_{nj}) by a subset of the columns at minimal cost c_j. We define $x_j = 1$ if column *j* with cost $c_j > 0$ is in the solution and $x_j = 0$ otherwise. Equations (21) and (22) ensures that the most rows (at least *N-rel* rows) are covered by at least one column and equation (23) is the integrally constraint. It always has a feasible solution (a unit vector *x* of *K* element), due to the required disjoints of the sets of positive and negative examples and the way the matrix *Z* was constructed.

So, we are looking for a 0-1 vector *x* at minimum cost and a 0-1 vector $\gamma = [\gamma_1, ..., \gamma_N]^T$ which determines the covered rows, $\gamma_n = 1$ if *n*-th row is covered by solution *x* and $\gamma_n = 0$ otherwise. By assumption, at least *N-rel* rows (given a parameter $rel \geq 0$) must be covered by solution *x*.

The set covering problem is a well-known combinatorial optimization problem and has been proven to be NP-complete, i.e. not currently solvable in polynomial time (cf. Johnson, 1974). A number of optimal and heuristic algorithms which try to find a "good" solution quickly have been presented in the literature in recent years. For many aspects of the formulation and various exact and heuristic solution methods, cf. Balas and Padberg (1979), Beasley and Jornsten (1992), Christofides and Korman (1975), Etcheberry (1977), Grossman and Wool A. (1995), Paixao J. (1984), Vercellis (1984).

For the solution of (19) we apply the IP2 algorithm with elements of a greedy algorithm (cf. Chvatal, 1979) implemented in the IP2_GRE procedure, and and a genetic algorithm (cf. Beasley, 1996; see also Homaifar, Lai and Qi, 1994, Powell and Skolnick, 1993, Schoenauer and Xanthakis, 1993) implemented in the IP2_GA procedure.

We assume that the classification rule must correctly describe *most of the examples*, at least with accuracy $A_{learning}$. The measure of classification accuracy $A_{learning}$ is the ratio of examples correctly classified to the total number of examples, in percentage.

Suppose that we have a finite set of examples *U*, a finite set of attributes $A=\{a_1,...,a_K\}\cup\{a_d\}$. Let us have a class $Y_{v_{d,l}}$, for $v_{d,l}\in V_{a_d}$, and a set of positive examples S_P [cf. (3)], and a set of negative examples S_N [cf. (4)].

The consecutive steps of the IP2 algorithm are as follows:

Step 1. Set the initial values: $S = S_P$, i.e. the whole set of examples is initially assumed to contain the positive ones, S_N is a set of negative examples, and $R_W^* = \varnothing$, i.e. the initial set of complexes is assumed empty, iteration $r = 0$, given parameter $rel \geq 0$.

Step 2. Iteration $r = r + 1$. Determine the weights *G* by analyzing (pre-processing) of the examples due to (8).

Step 3. Determine an appropriate starting point; a good starting point may be a so-called centroid [cf. Kacprzyk and Szkatuła (1996, 1999)] that is some (possibly none existing) example in which the attributes take on values that occur most often in the positive examples and seldom in the negative examples.

The positive examples $e^m \in S_P$, $m = 1, ..., P$, are written as

$$e^m=[a_1=v_{1,t(1,m)};g_1(v_{1,t(1,m)})]\wedge...\wedge[a_K=v_{K,t(K,m)};g_K(v_{K,t(K,m)})] \tag{24}$$

where $v_{j,t(j,m)}\in V_{a_j}$, $j=1,...,K$, while the negative examples $e^{P+n}\in S_N$, $n=1,...,N$ (for simplicity the upper index *P* is omitted below and later on when it does not lead to confusion), are written as:

$$e^n=[a_1=v_{1,t(1,n)};g_1(v_{1,t(1,n)})]\wedge...\wedge[a_K=v_{K,t(K,n)};g_K(v_{K,t(K,n)})] \tag{25}$$

where: $v_{j,t(j,n)}\in V_{a_j}$, $j=1,...,K$.

For each a_j, $j=1,...,K$, we determine such a value $v_{j,t(j,*)}\in V_{a_j}$ that [cf. (8)]

$$\max_{v_{j,l}\in V_{a_j}}\{g_j(v_{j,l})\} = \max_{v_{j,l}\in V_{a_j}}\left\{\frac{1}{P}\sum_{m=1}^{P}\delta(e^m,v_{j,l})-\frac{1}{N}\sum_{n=1}^{N}\delta(e^n,v_{j,l})\right\} \tag{26}$$

where: $v_{j,l}\in V_{a_j}$, $l=1,...,L_j$, V_{a_j} is the set of possible values of attribute a_j, and $\delta(e^m,v_{j,l})=\begin{cases}1 & \textit{for } v_{j,t(j,m)}=v_{j,l}\\ 0 & \textit{otherwise}\end{cases}$ and analogously for $\delta(e^n,v_{j,l})$.

Now, we form an example, called a centroid

$$e^{*}=[a_{1}=v_{1,t(1,*)};g_{1}(v_{1,t(1,*)})]\wedge\ldots\wedge[a_{K}=v_{K,t(K,*)};g_{K}(v_{K,t(K,*)})] \tag{27}$$

that contains the selectors with the most typical positive values of the particular attributes. Needless to say that this example may be artificial, i.e. nonexistent. The concept of a centroid is crucial for the efficiency of the algorithm. We introduce a similarity measure η yielding a degree of similarity between the centroid e^{*} and a positive example e^{m} [cf. (24)] as, e.g.:

$$\eta(e^{m},e^{*})=\frac{1}{K}\sum_{j=1}^{K}\delta(e^{m},v_{j,t(j,*)}) \tag{28}$$

or

$$\eta(e^{m},e^{*})=\sum_{j=1}^{K}g_{j}(v_{j,t(j,*)})\cdot\delta(e^{m},v_{j,t(j,*)}) \tag{29}$$

where $g_{j}(v_{j,t(j,*)})$ is the weight of the value $v_{j,t(j,*)}$ of attribute a_{j} obtained by using, e.g., (8). Finally, in the set of positive examples we find such an example $e^{p}\in S_{P}$ for which

$$\max_{e^{m}\in S_{P}}\eta(e^{m},e^{*}) \tag{30}$$

i.e. the closest positive example to e^{*}.

Then, as the starting point for the next iterations there is adopted a real (existing) example that is the closest, e.g., in the sense of the measure of similarity (28) or (29), to the one found by (30). This example, e^{p}, given as

$$e^{p}=[a_{1}=v_{1,t(1,p)};g_{1}(v_{1,t(1,p)})]\wedge\ldots\wedge[a_{K}=v_{K,t(K,p)};g_{K}(v_{K,t(K,p)})] \tag{31}$$

is used in the algorithm proposed in the next steps as the starting point.

Step 4. For the e^{p} we form the matrix $Z_{N*K}=[z_{nj}]$, $n=1,\ldots,N$, $j=1,\ldots,K$, due to (16) and a modification of the set covering problem given by (20) – (23), such that $\min_{x,\gamma}\sum_{j=1}^{K}c_{j}x_{j}$, with linear constrains $\sum_{j=1}^{K}z_{nj}x_{j}\geq\gamma_{n}$, $n=1,\ldots,N$ and $\sum_{n=1}^{N}\gamma_{n}\geq N-rel$ where $c_{j}=(1-g_{j}(v_{j,t(j,p)}))$, $z_{nj}\in\{0,1\}$, $x_{j}\in\{0,1\}$, $n=1,\ldots,N$, $j=1,\ldots,K$, $\gamma=[\gamma_{1},\ldots,\gamma_{N}]^{T}$ is a 0-1 vector, $rel\geq 0$, rel is an integer.

Step 5. To solve above problem we apply a greedy algorithm implemented in the IP2_GRE procedure or a genetic algorithm implemented in the IP2_GA procedure. The 0-1 vector $x^{*}=[x_{1}^{*},\ldots,x_{K}^{*}]^{T}$ found in arbitrary way determines in a unique way the complex $C_W^{I_r*}$ and the 0-1 vector $\gamma = [\gamma_1, ..., \gamma_N]^T$ determines the fulfilled constrains.

Step 6. Include complex $C_W^{I_r*}$ found in Step 5 into the classification rule sought R_W^* (i.e. that with the minimal weighted length), $R_W^* := R_W^* \cup C_W^{I_r*}; q(C_W^{I_r*})$, where $q(C_W^{I_r*})$ is the strength of the *r*-th elementary rule, and discard from the set of positive examples *S* all examples covered by complex $C_W^{I_r*}$.

Step 7. If the set of examples *S* remaining is *small enough*, STOP and the rule

$$R_W^* = C_W^{I_1*}; q(C_W^{I_1*}) \cup ... \cup C_W^{I_L*}; q(C_W^{I_L*}) \tag{32}$$

$I_1, ..., I_L \subseteq \{1, ..., K\}$, is the one sought; otherwise, return to Step 2.

4 The IP2_GRE Procedure

To solve problem (20) - (23) (i.e. Step 5 in the IP2 procedure) we apply a modification of the greedy algorithm implemented in the IP2_GRE procedure. Example e^p, found in Step 3 of the IP2 procedure, is used in this procedure as the starting point. For the e^p we form the matrix Z_{N*K}=$[z_{nj}]$, $n = 1, ..., N$, $j = 1, ..., K$, due to (16). We assume initially that $x^* = [x_1^*, ..., x_K^*]^T = [0, ...,0]^T$ and $\gamma^* = [\gamma_1^*, ..., \gamma_N^*] = [0, ...,0]^T$, $rel \geq 0$.

Step 1'. We assume the initial matrix to be *M* := *Z*.

Step 2'. We calculate the efficiency of element x_j, $j = 1, ..., K$ with respect to the matrix *M* which is defined as

$$E(x_j, M) = \sum_{n=1}^{N} m_{nj} / (1 - g_j(v_{j,t(j,p)})) \tag{33}$$

We choose the highest value $E(x_{j^*}, M)$, and set $x_{j^*}^*$=1.

Step 3'. We denote the *j**-th column of *M* as $m_{j^*} = [m_{1j^*}, m_{2j^*}, ..., m_{Nj^*}]^T$ and calculate new matrix *M* as

$$((M)^T \times diag(1 - m_{1j^*}, 1 - m_{2j^*}, ..., 1 - m_{Nj^*}))^T \tag{34}$$

and obtain the new vector $\gamma^* = [\gamma_1^*, ..., \gamma_N^*]$ as

$$[\max\{m_{1j^*}, \gamma_1^*\}, ..., \max\{m_{Nj^*}, \gamma_N^*\}]^T \tag{35}$$

Step 4'. If $\sum_{n=1}^{N} \gamma_n^* \geq N - rel$ then STOP, otherwise return to Step 2'.

The 0-1 vector $x^* = [x_1^*, ..., x_K^*]^T$ found in such a way determines in a unique way the complex C_W^* sought and the 0-1 vector $\gamma^* = [\gamma_1^*, ..., \gamma_N^*]$ determines the fulfilled constrains. Now, we can go to Step 6 of the IP2 algorithm.

5 The IP2_GA Procedure

For the solution of problem (20) - (23) (i.e. Step 5 in the IP2 procedure) we apply modification of genetic algorithm implemented in the IP2_GA procedure.

The basic steps of a simple genetic algorithm are shown below:

Set ***t*** = 1. Generate an initial population *P(t)* of possible solutions;
  Evaluate fitness of solutions in the population;
while a satisfactory solution has been found **do**
begin
  Select solutions from the population;
  Recombine solutions to produce new solutions;
  Evaluate fitness of new solutions;
  Set $t = t+1$. Replace some or all of the solutions in the population by the new solutions and create the new population P(t).
end;

We assume a K-bit binary string which represents the potential solution structure, where K is the number of variables in our problem (i.e. columns in the set covering problem). A value 1 for the j-th bit implies that column j is in the solution x^l, i.e. that x_j^l is in the solution, respectively.

In our procedure, IP2_GA, in each iteration all solutions are evaluated with respect to their completeness and consistency. We adopted a simple approach of using a penalty/evaluation function which assigns utility to candidate solutions. The fitness of an individual solution x is calculated simply by

$$eval(x) = f(x) - g_{\max} \frac{K}{N} \sum_{n=1}^{N} f_n(x) \tag{36}$$

where:

$$f(x) = \sum_{j=1}^{K} c_j x_j \,, \quad f_n(x) = \begin{cases} 0 & for \sum_{j=1}^{K} z_{nj} \cdot x_j > 0 \\ 1 & for \sum_{j=1}^{K} z_{nj} \cdot x_j = 0 \end{cases} \,, \quad g_{\max} = \max\{g_j : j = 1,..., K\} \,,$$

for n=1,...,N and x_j is the value of the j-th column in the string corresponding to the solution x and c_j is the cost of j-th column.

The initial population was randomly generated and the size of the population was proportional to the number of columns. We use the crossover operator which for two solutions forms two new solutions. Under the one point crossover operator, two structures in the population exchange portions of their binary representations. This can be implemented by choosing a point at random, called a crossover point, and exchanging the segments to the right of this point. After the new population has been selected, mutation is applied to each structure in the new population. The rate of mutation is generally set to be a small value. As to the selection method, we use a single wheel spin. Our procedure steps are as follows:

Step 1'. Set $t = 1$. Generate an initial population of random solutions $P(t) = \{x^1, x^2, ..., x^P\}$. Each solution is simply a binary string of length K. Evaluate the fitness $eval(x^l)$ of individuals in the population, $l = 1,2, ..., P$.

Step 2'. A mutation operator is applied to each solution in the population.

Step 3'. For the first solutions a crossover operator is applied. Two solutions are chosen and form two new solutions.

Step 4'. The new solution generated by the crossover and mutation may not be feasible because the constraints may not all be satisfied. We evaluate the fitness $eval(x^l)$ of new individuals in the population.

Step 5'. If a termination condition satisfied STOP, the best solution found is the one with the smallest fitness in the population; otherwise, go to Step 6'.

Step 6'. Select a new population $P(t+1)$ from population $P(t)$ and return to Step 2'

The 0-1 vector $x^* = [x_1^*, ..., x_K^*]^T$ found by this way determines in a unique way the complex C_W^* sought. Now, we can go to Step 6 of IP2 algorithm.

The IP2_GA algorithm described above is relatively simple and efficient. It requires a number of parameters, e.g. the population size, probabilities of applying genetic operators, etc.

6 Application of the IP2 Algorithm for the Thyroid Cancer Data

The method proposed was used with good results on a variety of medical data sets from one of clinics in Warsaw, Poland. However, due to space limitation and to make possible the comparison of the results obtained with those using other methods, we will show results for the thyroid cancer problem.

The medical data set on thyroid cancer [cf. Nakache and Asselian (1983)] were collected on patients with the thyroid cancer at Hospital Ambroise Paré, from 1960 to 1980. They concern 281 patients, all subject to surgery. The time of analysis was fixed for July, 1981. At that time, a patient could be dead or alive. The survival time was then fully known for those patients who died before the time of analysis. For those who were still alive, we only knew the inferior limit of their survival time, i.e. the time of analysis. Patients who survive may be completely cured and their survival pattern mayhave no relationship to those who die.

Each patient was described by the following 12 attributes (the sets of their possible discrete values in the parentheses):

sex,	{male, female}
age,	{<40, 40-60, 60-70, >70}
histology,	{well differentiated, poorly differentiated}
metastasis,	{yes, no}
enlargement,	{uni-lobe, uni-lobe+isthm, all the thyroid}
clinical lymph nodes,	{yes, no}
clinical aspect,	{unique nodule, multi nodules, important enlargement}
pathological lymph nodes,	{yes, no}
compressive syndromes,	{yes, no}
invasion,	{no, small, average, large}
survival time,	{in months}, length in month of survival time from the entrance in the study (between 1960 and 1980) to the time of analysis,
survival,	{survivor, non survivor at time of analysis}.

Two of the 12 attributes are important: survival time (in month) of a patient at the time of analysis, and survival or non survival at the time of analysis. Looking at each attribute separately, we had 229 survivors and 52 dead during the study. Among 52 patients there were 14 patients with missing values.

The aim of such a study is to identify prognostic elements of disease evolution and to define a prognostic rule for a new case coming from the same population and being in the same conditions.

The patients (training examples) have been divided into two classes. The class attribute is the survival time. The first class includes patients with the survival time over 7 years. The patients with the survival time below 7 years belong to the second class. Thus, 48 examples belong to the first and 29 examples belong to the second class.

The problem of learning from examples is formulated as to find two classification rules:

$$R_W^{1*} \to [class = class1],$$

$$R_W^{2*} \to [class = class2].$$

The R_W^{l*} is specified by "elementary" conditions depending on the attributes, $l = 1, 2$, for the classes:

class 1: the patients will be alive over 7 years,
class 2: the patient will dead during 7 years.

We assume that the classification rules for elements belonging to class l must correctly describe *most of the examples* belonging to class l, l=1,2, with the accuracy al least $A_{learning}$, by assumption. The measure of classification accuracy $A_{learning}$ is the ratio of examples correctly classified to the total number of examples, in percentage.

We assume $A_{learning}$ = 100% and $A_{learning}$ = 97.5% (i.e. may not describe not more than two training examples).

The IP2 method presented in this paper was applied to the data described above. We use IP2_GRE procedure with elements of a greedy algorithm and IP2_GA procedure with elements of a genetic algorithm (we used the mutation rate = 0.01, the crossover rate = 0.6, and the size of the population = 50).

The classification rules obtained are presented below (we add information on which procedure has produced a particular rule):

[*age* <=40] [*metastasis* = *no*] [*clinical aspect* = *unique nodule*] [*compressive syndromes* = *no*]; 0.364 ∪ [*metastasis* = *no*] [*pathological lymph nodes* = *no*] [*compressive syndromes* = *no*]; 0.468 ∪
[*histology* = *well differentiated*] [*metastasis* = *no*] [*clinical aspect* = *unique nodule*] [*invasion* = *no*]; 0.403
→ [*class* = *class 1*] (via IP2_GRE, $A_{learning}$ = 100 %)

[*metastasis* = *yes*] [*compressive syndromes* = *yes*]; 0.104 ∪
[*histology* = *poorly differentiated*] [*metastasis* = *yes*]; 0.156 ∪
[*clinical aspect* = *important enlargement*]; 0.078 ∪
[*histology* = *poorly differentiated*] [*compressive syndromes* = *yes*]; 0.104 ∪ [40 < *age* <=60] [*histology* = *poorly differentiated*] [*pathological lymph nodes* = *yes*]; 0.065 ∪ [*metastasis* = *yes*]; 0.234 ∪
[*pathological lymph nodes* = *yes*] [*compressive syndromes* = *yes*]; 0.039 ∪ [*clinical aspect* = *multi nodules*] [*clinical lymph nodes* = *yes*]; 0.052.
→ [*class* = *class 2*] (via IP2_GRE, $A_{learning}$ = 100 %)

[*age* <=40] [*metastasis* = *no*] [*compressive syndromes* = *no*]; 0.364 ∪.. [*metastasis* = *no*] [*pathological lymph nodes* = *no*] [*compressive syndromes* = *no*]; 0.468 ∪ [*histology* =*well differentiated*] [*metastasis* =*no*] [*invasion* = *no*]; 0.455
→ [*class* = *class 1*] (via IP_GRE, $A_{learning}$ = 97.5 %)

[*metastasis* = *no*] [*clinical lymph nodes* = *no*] [*invasion* = *no*]; 0.468 ∪
[*compressive syndromes* = *no*] [*invasion* = *average*]; 0.052 ∪
[*clinical aspect* = *unique nodule*] [*clinical lymph nodes* = *yes*]; 0.104
→ [*class* = *class 1*] (via IP_GA, $A_{learning}$ = 97.5 %).

[*metastasis* = *yes*] [*compressive syndromes* = *yes*]; 0.104 ∪
[*histology* = *poorly differentiated*] [*metastasis* = *yes*]; 0.156 ∪
[*clinical aspect* = *important enlargement*]; 0.078 ∪
[*histology* = *poorly differentiated*] [*compressive syndromes* = *yes*]; 0.104 ∪ [40 < *age* <=60] [*pathological lymph nodes* = *yes*]; 0.078 ∪
[*metastasis* = *yes*]; 0.234 ∪ [*pathological lymph nodes* = *yes*] [*compressive syndromes* = *yes*]; 0.039 ∪
[*clinical aspect* = *multi nodules*] [*clinical lymph nodes* = *yes*]; 0.052.
→ [*class* = *class 2*] (via IP_GRE, $A_{learning}$ = 97.5 %)

[40 < *age* <=60] [*pathological lymph nodes* = *yes*]0.078 ∪
[*metastasis* = *yes*]; 0.234 ∪
[*clinical aspect* = *important enlargement*]; 0.078 ∪
[*sex* = *male*] [*invasion* = *average*]; 0.039 ∪ [*invasion* = large]; 0.065 ∪
[*age* <=40] [*clinical aspect* = *multi nodules*]; 0.026
→ [*class* = *class 2*] (via IP_GA, $A_{learning}$ = 97.5 %)

The rules obtained are therefore quite legible to the human being, and the shortest is obtained via the IP2_GA.

Some data on the effectiveness and efficiency of the procedures are given below. The ratio of correct classification decisions to the total number of decisions made was taken as the measure of classification accuracy, in percentage.

Table 1. Some parameters describing the process of finding a classification rule for the first class

Algorithm	$A_{learning}$ by assumption	Number of iterations	Number of selectors in rule
IP2_GRE	100 %	3	11
	at least 97.5 %	3	9
IP2_GA	at least 97.5 %	3	7

Table 2. Some parameters describing the process of finding a classification rule for the second class

Algorithm	$A_{learning}$ by assumption	Number of iterations	Number of selectors in rule
IP2_GRE	100 %	8	15
	at least 97.5 %	8	14
IP2_GA	at least 97.5 %	6	9

Table 3. Some parameters describing the process of classification the patients into the first or the second class

Algorithm	$A_{learning}$, by assumption	Classification accuracy, achieved
IP2_GRE	100 %	100 %
	at least 97.5 %	98.7 %
IP2_GA	at least 97.5 %	98.7 %

As it can be see, the method proposed attains a very high value of accuracy. It makes it possible to involve the partial completeness and partial consistency. The use of combinatorial optimization in this case seems to be therefore a viable and well justified solution. Since the problem shown, and in general the medical problems considered, is not of a particularly high dimensionality, the use of such effective and efficient heuristics for solving the set covering problem as a greedy or genetic algorithm given very good results.

7 Concluding Remarks

We followed a direction in inductive learning which boils down to the transformation of an inductive learning problem into a set covering problem. As opposed to other approaches in this area exemplified by the CLIP family of inductive learning algorithms proposed by Cios and his collaborators (cf. Cios and Liu, 1995; Cios, Wedding and Liu, 1997; Cios and Kurgan, 2004), we proposed an extended formulation which, by some preprocessing of data that leads to weighted selectors, complexes, etc. makes it possible to redefine the problem to be solved to account for a milder requirement of a partial completeness and partial consistency.

We presented an improved inductive learning algorithm IP2 based on the solution of a set covering problem well known in discrete optimization. Due to inherent numerical difficulties while solving such combinatorial optimization problems, we proposed a heuristic approach using a greedy and genetic algorithm, and developed the respective procedures, to derive classification rules from the sets of positive and negative examples. In the implementations of the greedy and genetic algorithm we proposed tools to make it possible to redefine the problem to be solved to account for a milder requirement of a partial completeness and partial consistency.

The approach proposed was applied to many medical data sets but, to make a comparison with other methods possible, in this paper it was applied to the well known thyroid cancer problem. The computational results are very encouraging.

References

Balas, E.: Cutting planes from conditional bounds: a new approach to set covering. Mathematical Programming Study 12, 19–36 (1980)

Balas, E., Ho, A.: Set covering algorithms using cutting planes, heuristics and sub gradient optimisation: A computation study. Mathematical Programming 12, 37–60 (1980)

Balas, E., Padberg, M.W.: Set partitioning - A survey. In: Christofides, N. (ed.) Combinatorial Optimisation. Wiley, New York (1979)

Beasley, J.E.: An algorithm for set covering problem. European Journal of Operational Research 31, 85–93 (1987)

Beasley, J.E.: A Lagrangian heuristic for set covering problem. Naval Research Logistics Quarterly 37, 151–164 (1990)

Beasley, J.E.: A genetic algorithm for the set covering problem. European Journal of Operational Research 94, 392–404 (1996)

Beasley, J.E., Jornsten, K.: Enhancing an algorithm for set covering problem. European Journal of Operational Research 58, 293–300 (1992)

Christofides, N., Korman, S.: A computational survey of methods for the set covering problem. Management Science 21, 591–599 (1975)

Chvatal, V.: A greedy heuristic for the set-covering problem. Mathematics of Operational Research 4(3), 233–235 (1979)

Cios, K.J., Liu, N.: An algorithm which learns multiple covers via integer linear programming. Part I - The CLILP2 Algorithm. Kybernetes 24, 29–50 (1995)

Cios, K.J., Wedding, D.K., Liu, N.: CLIP3: cover learning using integer programming. Kybernetes 26, 513–536 (1997)

Cios, K.J., Kurgan, L.: CLIP4: Hybrid inductive machine learning algorithm that generates inequality rules. Information Sciences 163, 37–83 (2004)

Etcheberry, J.: The set covering problem: A new implicit enumeration algorithm. Operations Research 25, 760–772 (1977)

Grossman, T., Wool, A.: Computational experience with approximation algorithms for the set covering problem. Working paper, Theoretical Division and CNLS, Los Alamos National Laboratory (1995)

Homaifar, A., Lai, S., Qi, X.: Constrained optimization via genetic algorithms. Simulation 62, 242–254 (1994)

Iwański, C., Szkatuła, G.: Inductive learning supported by integer programming. Computers and Artificial Intelligence 10, 57–66 (1991)

Johnson, D.A.: Approximation algorithms for combinatorial problems. Journal of Computer System Science 9, 256–278 (1974)

Kacprzyk, J., Iwański, C.: Fuzzy logic with linguistic quantifiers in inductive learning. In: Zadeh, L.A., Kacprzyk, J. (eds.) Fuzzy Logic for the Management of Uncertainty, pp. 465–478. Wiley, New York (1992)

Kacprzyk, J., Szkatuła, G.: Machine learning from examples under errors in data. In: Bouchon-Meunier, B., Yager, R.R., Zadeh, L.A. (eds.) IPMU 1994. LNCS, vol. 945, pp. 1047–1051. Springer, Heidelberg (1994)

Kacprzyk, J., Szkatuła, G.: An algorithm for learning from erroneous and incorrigible examples. International Journal of Intelligent Systems 11, 565–582 (1996)

Kacprzyk, J., Szkatuła, G.: An improved inductive learning algorithm with a preanalysis of data. In: Raś, Z.W., Skowron, A. (eds.) ISMIS 1997. LNCS, vol. 1325, pp. 157–166. Springer, Heidelberg (1997)

Kacprzyk, J., Szkatuła, G.: Deriving IF-THEN rules for intelligent decision support via inductive learning. In: Kasabov, N., et al. (eds.) Progress in Connectionist-Based Information Systems (Proceedings of ICONIP 1997, ANZIIS 1997 and ANNES 1997 Conference), Dunedin, New Zealand, Singapore, vol. 2, pp. 818–821. Springer, Heidelberg (1997b)

Kacprzyk, J., Szkatuła, G.: IP1 - An improved inductive learning procedure with a preprocessing of data. In: Proceedings of IDEAL 1998, Hong Kong, pp. 385–392. Springer, Heidelberg (1998)

Kacprzyk, J., Szkatuła, G.: An inductive learning algorithm with a preanalysis of data. International Journal of Knowledge - Based Intelligent Engineering Systems 3, 135–146 (1999)

Kacprzyk, J., Szkatuła, G.: An integer programming approach to inductive learning using genetic and greedy algorithms. In: Jain, L.C., Kacprzyk, J. (eds.) New Learning Paradigms in Soft Computing, pp. 322–366. Physica-Verlag, Heidelberg (2002a)

Kacprzyk, J., Szkatuła, G.: An integer programming approach to inductive learning using genetic algorithm. In: Proceedings of CEC 2002 - The 2002 Congress on Evolutionary Computation, Honolulu, Hawaii, pp. 181–186 (2002b)

Kacprzyk, J., Szkatuła, G.: A softened formulation in inductive learning and application to coronary disease data. In: Atanassov, K.T., Kacprzyk, J., Krawczak, M., Szmidt, E. (eds.) Issues in the Representation and Processing of Uncertain and Imprecise Information. Fuzzy Sets, Intuitionistic Fuzzy Sets, Generalized Nets, and Related Topics, pp. 181–197. AOW EXIT, Warszawa (2005a)

Kacprzyk, J., Szkatula, G.z.: A softened formulation of inductive learning and its use for coronary disease data. In: Hacid, M.-S., Murray, N.V., Raś, Z.W., Tsumoto, S. (eds.) ISMIS 2005. LNCS (LNAI), vol. 3488, pp. 200–209. Springer, Heidelberg (2005)

Kacprzyk, J., Szkatuła, G.z.: An inductive learning algorithm with a partial completeness and consistence via a modified set covering problem. In: Duch, W., Kacprzyk, J., Oja, E., Zadrożny, S. (eds.) ICANN 2005. LNCS, vol. 3697, pp. 661–666. Springer, Heidelberg (2005)

Michalski, R.S.: Variable valued logic: system VLl. In: Proc. of 1974 International Symposium on Multiple Valued Logic and Pattern Recognition, West Virginia University, Morgantown, pp. 323–346 (1974)

Michalski, R.S.: A theory and methodology of inductive learning. In: Michalski, R., Carbonell, J., Mitchell, T.M. (eds.) Machine Learning. Tioga Press (1983)

Michalski, R.S., Mozetic, I., Hong, J., Lovrac, N.: The multi-purpose incremental learning system AQ15 and its testing application to three medical domains. In: Proceedings of 5th National Conference on Artificial Intelligence, Philadelphia, pp. 1041–1045. Morgan Kaufmann, San Francisco (1986)

Nakache, J.P., Asselain, B.: Medical data set proposed for the workshop on data analysis. In: EIASM Workshop (April 1983)

Paixao, J.: Algorithms for large scale set covering problems. PhD. Thesis, Department of Management Science, Imperial College (1984)

Powell, D., Skolnick, M.M.: Using genetic algorithms in engineering design optimization with non-linear constraints. In: Forrest, S. (ed.) Proceedings of the Fifth International Conference on Genetic Algorithms, pp. 424–430. Morgan Kaufmann, San Mateo (1993)

Schoenauer, M., Xanthakis, S.: Constrained GA optimization. In: Forrest, S. (ed.) Proceedings of the Fifth International Conference on Genetic Algorithms, pp. 573–580. Morgan Kaufmann, San Mateo (1993)

Szkatuła, G.: Machine learning from examples under errors in data (In Polish), Ph.D. thesis, SRI PAS Warsaw, Poland (1996)

Szkatuła, G., Kacprzyk, J.: An inductive learning algorithm with a partial completeness and consistency. In: Draminski, M., Grzegorzewski, P., Trojanowski, T., Zadrozny, S. (eds.) Issues in intelligent systems. Models and techniques, pp. 229–246. AOW EXIT, Warszawa (2005)

Vercellis, C.: A probabilistic analysis of the set covering problem. Annals of Operational Research 1, 255–271 (1984)

Part II
General Issues

From Active to Proactive Learning Methods

Pinar Donmez and Jaime G. Carbonell

Language Technologies Institute
School of Computer Science
Carnegie Mellon University
{pinard,jgc}@cs.cmu.edu

1 Active Learning

1.1 Motivation

In many machine learning tasks, unlabled data abounds, but expert-generated labels are scarce. Consider the process of learning to build a classier for the Sloan Digital Sky Survey (http://www.sdss.org/) so that each astronomical observation may be assigned its class (e.g. "pinwheel galaxy", "globular galaxy", "quasar", "colliding galaxies", "nebula", etc.). The SDSS contains 230 million astronomical objects, among which professional Astronomers have classified manually less than one tenth of 1 percent. Consider classifying web pages into subject-matter based taxonomies, such as the Yahoo taxomy or a Dewy library catalog system. Whereas there are many billions of web pages, less than .001% have reliable topic or subject categories.

In both cases, despite of the massive quantities of unlabeled data, some categories have very few reliably labeled examples, but more are needed to to train machine learning systems to classify with sufficent accuracy. However, obtaining expert labels can prove to be expensive, bounded (e.g. Astronomers will only give so much of their time), or both. Hence, the question becomes one of selecting the most informative instances to label; some labeled examples can be much more useful than others in training classiers to minimize errors. For instance, selecting an example representative of many others, or selecting an example in an unexplored region of the instance space, or selecting an example close to the decision boundary induced from the earlier labeled examples are all better strategies than random selection. Active learning is the process of selecting the best examples to label in order to improve classifer performance, where "best" is typially defined as minimizing the loss function [Miller et al., 1993] of the classifier or otherwise directly minimizing its estimated classification error.

Alternatives to active learning attempt to use unlabeled data directly in helping to training classifiers, without obtaining additional labels. These efforts range from transductive support vector machines [Joachims, 1999] to co-training [Blum and Mitchell, 1998] and other semi-supervised methods [He et al., 2007]. Although effective in some cases, transductive and semi-supervised methods cannot in general rival active learning since they do not take advantage of additional labeled examples. Combining ideas from both semi-supervised and active learning remains a largely unexplored area for future research.

J. Koronacki et al. (Eds.): Advances in Machine Learning I, SCI 262, pp. 97–120.
springerlink.com © Springer-Verlag Berlin Heidelberg 2010

1.2 Active Learning Methods in Literature

Although instance selection has been implicitly part of Machine Learning from its early years, such as in Michalski's Star methodology and related work [Michalski et al., 1983; Reinke and Michalski, 1985; Michalski et al., 1986], the systematic study of active learning has flourished and generated considerable interest in the past 15 years, including the emergence of a solid thoeretical foundation for the field [Cohn et al., 1995]. Whereas there have been several studies investigating the use of active learning in classification, rank learning, regression, and function optimization, in this chapter, we mainly focus on the dominant and most studied variant: active learning for classification.

The most straightforward strategy for selecting examples to label is random sampling among unlabeled instances, which is usually adopted as a baseline for comparative analysis since it often outpeforms other naive approaches (such as representative sampling). We group active learning methods into three major categories based on their underlying sampling criteria; namely, 1) uncertainty-based sampling strategies, 2) density-based strategies and 3) ensemble methods. We acknowledge that this is not an exhaustive list, rather a summary grouping of the best performing and most widely used active learning algorithms in the literature. Uncertainty sampling was proposed by [Lewis and Gale, 1994] by choosing to obtain a label for the example on which the current learner is most uncertain. Tong and Koller, 2000 [Tong and Koller, 2000] adopted the same idea and applied it to a support vector machine (SVM) classifier. They provided a theoretical motivation based on shrinking the version space (the margin) as much as possible; in other words, selecting the example that will minimize the maximum expected size of the residual version space. In case of linear discriminant functions, this corresponds to the selection of the unlabeled example with the smallest margin [Tong and Koller, 2000]. Despite the theoretical justification of the version-space reduction methods, their method wastes lableing effort attempting to eliminating areas of the parameter space that have no direct effect on the classification error rate. Hence, these methods are not immune to selecting outliers [McCallum and Nigam, 1998] since they have high uncertainty, but outliner labels seldom help the learner improve its generalization performance. Nevertheless, uncertainty sampling plays a key role in many studies in the literature, including our previous work which is explained in detail in Section 1.3.

The density-based sampling strategies incorporate the underlying data distribution into the selection mechanism, and thus indirectly borrow from semi-supervised methods cited above. Most of these methods propose ways to trade off an uncertainty measure with the density of the instance space in the immediate neighborhood of the sample. Xu et al., 2003 [Xu et al., 2003] uses k-means to cluster the samples lying inside the margin of an SVM classifier trained on the current labeled data. The cluster centroids are then selected for labeling. Shen and Zhai, 2005 [Shen and Zhai, 2005] adopts the same idea in an information retrieval scenario, and uses the k-medoid algorithm for the top relevant examples. Similar active learning schemes are proposed by [Tang et al., 2002] for natural language parsing, and by [Zhang and Chen, 2002] for content-based information retrieval. In [McCallum and Nigam, 1998], a naive Bayes classifier is trained on both labeled and unlabeled data using an EM algorithm. Under the assumption that unlabeled data dominates the labeled data, the training algorithm clusters the full dataset but labeled data is used only to seed the clusers. Clustering information, then, contributes to the selection such that an uncertainty measure is weighed with the density surrounding the example. Another active learning approach that utilizes combines uncertainty and density criteria is our paired sampling strategy described in [Donmez and Carbonell, 2008]. The difference between our method and the density-based methods introduced above is that our method relies on balanced sampling on both sides of a decision boundary rather than sampling disproportionately on one side. Furthermore, it exploits the natural grouping (clustering) of the data to effectively define a non-Eucledian local distance function, and maximizes a utility-based conditional entropy criterion for sampling. We have shown that this new density-sensitive method yields significantly superior performance over multiple datasets compared to other popular active learning sampling methods [Donmez and Carbonell, 2008].

Nguyen and Smeulders, 2004 [Nguyen and Smeulders, 2004] suggested a probabilistic framework where clustering is combined with a discriminative model. They assume that examples lying on the classification boundary, in other words most uncertain examples, are informative, but using information about the underlying data distribution helps to select better examples. Their method favors higher density examples lying close to the decision boundary. Those examples are assumed to have the largest contribution to the current error. Though their strategy works well in practice, especially by reducing the error quickly, it exhibits very slow additional learning after substantial sampling. On the other hand, standard uncertainty sampling initially has a slower learning rate, but gradually outperforms their method, as shown in our previous work [Donmez et al., 2007]. The reason is that density based methods sample from maximal-density unlabeled regions, and thus help establish the initial classification boundary where it affects the most remaining unlabeled data. On the other hand, uncertainty sampling fine-tunes a decision boundary by sampling the regions where the classifier is least certain, regardless of the distribution of the unlabeled data. Our algorithm, DUAL [Donmez et al., 2007], tackles this issue by proposing a principled ensemble-based approach that selects which sampling method to apply based on estimated residual classification error reduction. DUAL incorporates a robust combination of density weighted and uniformly weighted uncertainty sampling. We have empirically shown that DUAL combines the best of both worlds, and leads to superior performance across various domains [Donmez et al., 2007].

Another ensemble learning method is proposed by [Baram et al., 2003]. Baram et al., 2003 presented an online algorithm (COMB) that selects among three alternative active learning strategies using a variant of the multi-armed bandit algorithm to decide the strategy to be used at each iteration. Although there are similarities between DUAL and COMB, there is a major distinction: COMB aims to select which sampling method is optimal for a given dataset, irrespective of how much the algorithm has already learend, whereas DUAL focuses on selecting the operating range most suitable for the sampling method. Other ensembled active learning methods have appeared in literature. [Melville and Mooney, 2003] extends the query-by-committee algorithm by constructing diverse committees by employing artificial training examples. Co-testing [Muslea et al., 2000] is another ensemble active learning strategy inspired by the multi-view approach in co-training [Blum and Mitchell, 1998]. It utilizes two redundant views of the training data to create an ensemble and selects the unlabeled instance where two classifiers disagree.

1.3 Addressing Performance Barriers

1.3.1 DUAL: A Context-Sensitive Strategy Selection

DUAL [Donmez et al., 2007] is a principled ensemble-based sampling approach for active learning. It combines the uncertainty sampling and density-based sampling by selecting sampling methods based on estimated residual classification error reduction. Almost all active learning strategies that trade-off uncertainty vs. data density aim to balance the uncertainty of the sample with its representativeness, but do so in a fixed manner, rather than by dynamically selecting or reweighing, based on residual error estimation. In other words, prior methods do not estimate the learning rate of the classifier, and therefore have no criterion to switch sampling strategy if one starts to saturate in performance.

DUAL is a context-sensitive sampling method with a primary focus on improving active learning for both the onset of active learning and especially the later portion of the process, rather than traditional methods that concentrate primarily on the initial label selection and do not adapt. DUAL significantly improves upon the work of [Nguyen and Smeulders, 2004] by incorporating a robust combination of density weighted uncertainty sampling and standard (uniform) uncertainty sampling. Nguyen and Smeulders, 2004 [Nguyen and Smeulders, 2004] propose a Density Weighted Uncertainty Sampling criterion, or DWUS for short:

$$s = \arg\max_{i \in I_u} E[(\hat{y}_i - y_i)^2 \mid x_i] p(x_i) \tag{1}$$

where $E[(\hat{y}_i - y_i)^2 \mid x_i]$ and $p(x_i)$ are the expected error and density of a given data point x_i, respectively. I_u is the index for the unlabeled data. This criterion favors points that have the largest contribution to the current classification error. In contrast, one can use an uncertainty-based selection criterion within the same probabilistic framework as illustrated by the following formula:

$$s = \arg\max_{i \in I_u} E[(\hat{y}_i - y_i)^2 \mid x_i] \tag{2}$$

We refer to the above principle as Uncertainty Sampling. Consider Fig. 1, which displays the performance of DWUS and Uncertainty Sampling on two UCI datasets. After rapid initial gains, DWUS exhibits very slow additional learning while uncertainty sampling continues to exhibit more rapid improvement.[1] We have investigated the behaviour of unlabeled data in these two scenarios to explore the underlying reasons for such a pattern. We have observed that at the early iterations, many points are highly uncertain. Thus, DWUS can pick high density points which are lower down in the uncertainty *ranking* but have a high absolute uncertainty score. Later, points with high absolute uncertainty are no longer in dense regions. As a result, DWUS picks points that have moderate density but low uncertainty because such points are scored highly according to the criterion in Equation 1. Hence, it wastes effort picking instances whose selection does not have a large effect on error rate reduction.

In order to address this problem, our algorithm DUAL adopts a dynamically re-weighted mixture of density and uncertainty components. DUAL works as follows:

[1] Although a quick drop in classification error for DWUS is also observed in [Nguyen and Smeulders, 2004], they did not compare with uncertainty sampling.

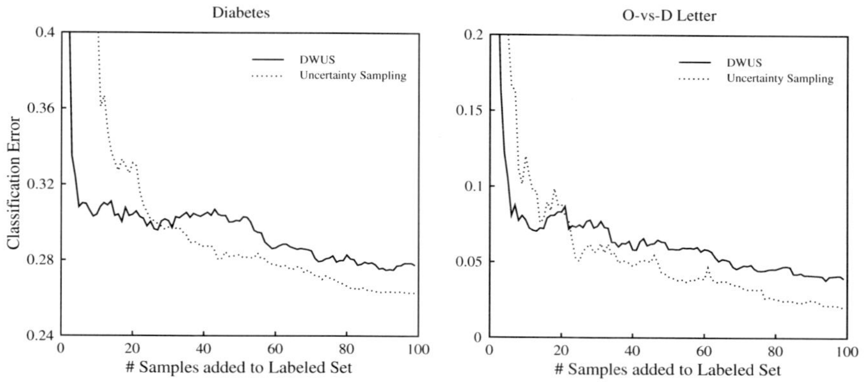

Fig. 1. Comparison of Density Weighted versus (standard) uniformly weighted Uncertainty Sampling on two UCI benchmark datasets

It starts executing DWUS[2] up until it estimates a cross-over point with uncertainty sampling by predicting a low derivative of the expected error, e.g. $\frac{\partial\epsilon(DWUS)}{\partial x_t} \leq \delta$. The derivative estimation need not be exact, requiring only the detection of diminishing returns which we explain soon. Then, it switches to execute a combined strategy of density-based and uncertainty-based sampling. In practice, we do not know the future classification error of DWUS, but we can approximate it by calculating the average expected error of DWUS on the unlabeled data. It will not give us the exact cross-over point, but it will provide a rough estimate of when we should consider switching between methods. The expected error of DWUS on the unlabeled data can be evaluated as follows:

$$\hat{\epsilon}_t(DWUS) = \frac{1}{n_t} \sum_{i \in I_u} E[(\hat{y}_i - y_i)^2 \mid x_i] \tag{3}$$

where $E[(\hat{y}_i - y_i)^2 \mid x_i]$ is the error expectation for a given unlabeled data point x_i:

$$E[(\hat{y}_i - y_i)^2 \mid x_i] = (\hat{y}_i - 1)^2 P(y_i = 1 \mid x_i) + (\hat{y}_i)^2 P(y_i = 0 \mid x_i) \tag{4}$$

Equation 3 is re-calculated at each iteration of active sampling. t is the iteration number, and n_t is the number of unlabeled instances at the t-th iteration and I_u is the set of indices of the unlabeled points. By monitoring the average expected error at every single iteration, we can estimate when DWUS' performance starts to saturate, i.e., $\frac{\partial\hat{\epsilon}(DWUS)}{\partial x_t} \leq \delta$. δ is assigned a fixed small value in our evaluations (see [Donmez et al., 2007] for details). When it is near zero, this is equivalent to detecting when a method is stuck in local minima/plateau in gradient descent methods. In fact, this principle is flexible enough to work with any two active learning methods where one is superior for labeling the initial data and the other is favorable later in the process. It generalizes to N sampling methods by introducing additional estimated

[2] Interested reader is referred to [Nguyen and Smeulders, 2004] for further details on the DWUS algorithm.

switchover points based on estimated derivative of expected error for each additional sampling strategy.

After we estimate the cross-over, we use a mixture model for active sampling:

$$x_s^* = \arg\max_{i \in I_u} \pi_1 * E[(\hat{y}_i - y_i)^2 \mid x_i] + (1 - \pi_1) * p(x_i) \tag{5}$$

It is desirable for the above model to minimize the expected future error. After labeling x_{US}^*, the expected error is:

$$f_{US} = \frac{1}{n} \sum_j E_{L+\{x_{US}^*, y\}}[(\hat{y}_j - y_j)^2 \mid x_j] \tag{6}$$

The subscript $L + \{x_{US}^*, y\}$ indicates that the expectation is calculated from the model trained on the data $L + \{x_{US}^*, y\}$. The appropriate weight in this scenario is inversely related with the expected error of uncertainty sampling. Thus, we can replace the weights by $\pi_1 = 1 - f_{US}$, and $1 - \pi_1 = f_{US}$, and obtain the following model:

$$x_s^* = \arg\max_{i \in I_u} (1 - f_{US}) * E[(\hat{y}_i - y_i)^2 \mid x_i] + f_{US} * p(x_i) \tag{7}$$

Achieving the minimum expected loss is guaranteed only for the extreme case where the expected error, f_{US}, of uncertainty sampling is equal to 0. However, correlating the weight of uncertainty sampling with its generalization performance increases the odds of selecting a better candidate after the cross-over.

In the real world, we do not know the true value of f_{US} so we need to approximate it. But, we approximate $\hat{\epsilon}(US)$ as the average expected error of uncertainty sampling

---

Algorithm 1. The DUAL Algorithm

Input: Labeled data L, Unlabeled data U, max number of iterations T, and δ.
Output: A set S of actively sampled data points.
Program
Initialize: $t = 0$ and $S = \{\}$.
while while(not switching point) **do**
  Run DWUS algorithm and compute $\frac{\partial \hat{\epsilon}(DWUS)}{\partial x_t}$.
  if $\frac{\partial \hat{\epsilon}(DWUS)}{\partial x_t} > \delta$ **then**
    $x_s^* = \arg\max_{i \in I_u} E[(\hat{y}_i - y_i)^2 \mid x_i] p(x_i)$
    Add the chosen point to set S: $S = S \cup x_s^*$
    t=t+1 (Increment counter t)
  else
    Hit the switching point.
  end if
end while
while while($t < T$) **do**
  Compute $E[(\hat{y} - y)^2 | x]$, $p(x)$ via DWUS, and $\hat{\epsilon_t}(US)$ via uncertainty sampling.
  $x_s^* = \arg\max_{i \in I_u} (1 - \hat{\epsilon_t}(US)) * E[(\hat{y}_i - y_i)^2 \mid x_i] + \hat{\epsilon_t}(US) * p(x_i)$
  Add the chosen point to set S: $S = S \cup x_s^*$
  t=t+1
end while

---

on the unlabeled portion of the data. This leads us to the following selection criterion for DUAL:

$$x_s^* = \arg\max_{i \in I_u} (1 - \hat{\epsilon}(US)) * E[(\hat{y}_i - y_i)^2 \mid x_i] + \hat{\epsilon}(US) * p(x_i) \tag{8}$$

$\hat{\epsilon}(US)$ is updated at every iteration t after the cross-over. Its calculation is exactly the same as in Equation 3. However, the data to sample from is restricted to the set of already labeled examples by active selection. This set is dynamically updated at each iteration with the actively sampled points. Consequently, in order to calculate the expected error of uncertainty sampling the algorithm never requests the label of a point that has not already been sampled during the active learning process. Such a restriction will prevent an exact estimate of the expected error. But, it is a reasonable alternative, and introduces no additional cost of labeling. The pseudo-code for the DUAL algorithm is given in Algorithm 1.

Experiments. To evaluate the performance of DUAL, we ran experiments on UCI benchmarks: splice, diabetes, image and letter. For the letter recognition dataset, we

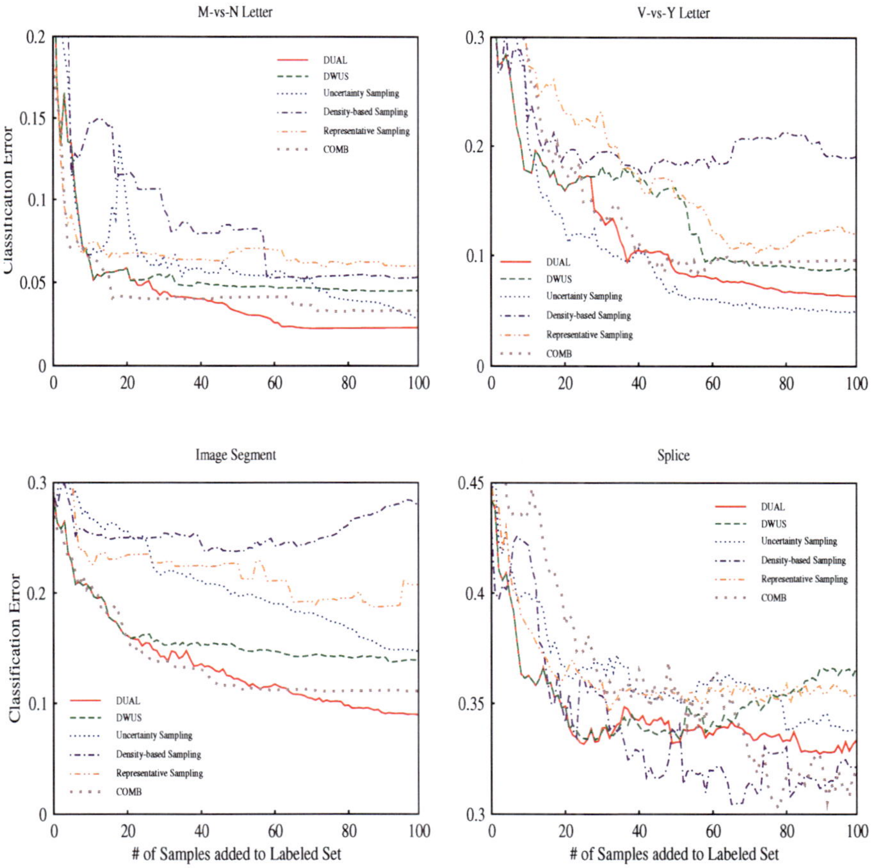

Fig. 2. Results on 4 different UCI benchmark datasets

picked three pairs of letters (M-vs-N,O-vs-D,V-vs-Y) that are most likely to be confused with each other. We compared the performance of DUAL with that of DWUS, uncertainty sampling, representative sampling [Xu et al., 2003], density-based sampling and the COMB method of [Baram et al., 2003]. Density-based sampling adopts the same probabilistic framework as DWUS but uses only the density information for active data selection: $x_s^* = \arg\max_{i \in I_u} p(x_i)$. COMB uses an ensemble of uncertainty sampling, sampling method of [Roy and McCallum, 2001], and a distance-based strategy choosing the unlabeled instance that is farthest from the current labeled set. COMB uses SVM with Gaussian kernel as the base classifier. For further implementation details on COMB, see [Baram et al., 2003].

Figure 2 presents the improvement in error reduction using DUAL over the other methods. We only display results on 4 datasets (See `www.cs.cmu.edu/~pinard/DUAL` for more thorough comparison). DUAL outperforms DWUS and representative sampling both with $p < 0.0001$ significance. DUAL outperforms COMB with $p < 0.0001$ significance on 4 out of 6 datasets, and with $p < 0.05$ on Image and M-vs-N data sets. We also calculate the error reduction of DUAL compared to the strong baseline DWUS. For instance, at the point in each graph after 3/4 of the sampling iterations after cross-over occurs, we observe 40% relative error reduction on O-vs-D data, 30% on Image, 50% on M-vs-N, 27% on V-vs-Y, 10% on Splice, and 6% on Diabetes dataset. These results are significant both statistically and also with respect to the magnitude reduction in relative residual error. DUAL is superior to Uncertainty sampling ($p < 0.001$) on 5 out of 6 datasets.

1.3.2 Density-Sensitive Paired Sampling

Another active learning algorithm we developed is called density-sensitive paired sampling [Donmez and Carbonell, 2008]. It relies on maximizing the likelihood of straddling the decision boundary with paired samples that potentially have the maximum information for the current learner. In order to sample points that are likely to be maximally informative to an active learner, we first seek to maximize the chance that we will sample on both sides of a decision boundary – sampling disproportionately on either side will not optimize boundary placement in the learning process. Maximizing the distance between two points is a step in the right direction, but Euclidean distance may not be the optimal measure; instead we investigate density-sensitive distance functions. The function we adopted in [Donmez and Carbonell, 2008] was proposed by [Chapelle and Zien, 2005] for clustering. The idea is to transform the data into a new space such that points in separate clusters are assigned low similarities, and vice versa. Then, the density-sensitive distance between any two points is approximated by the longest distance edge along each path that connects these two points. For details, see [Chapelle and Zien, 2005].

In order to maximize the likelihood of straddling the decision boundary, and to halve the computational time, we sample a pair of points to label at a time, in contrast to the typical active learning methods that select one point at each iteration. Figure 3 illustrates the motivation for paired sampling in active learning. Here we assume for simplicity the data is linearly separable. The dashed line shows the current decision boundary while the two solid lines define the region where the true boundary is expected to lie; namely the version space. The left figure in Figure 3 is an example of sampling a pair

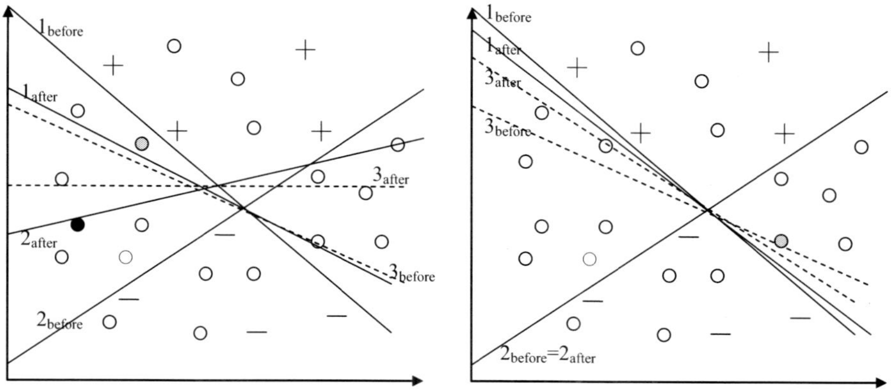

Fig. 3. Illustrative Example: The plus (minus) sign and circles indicate the positively (negatively) labeled points and unlabeled data, respectively. x_{after} and x_{before} indicate the line before and after data is sampled for labeling. The selected points are labeled either positive (shown in grey) or negative (shown in black). This example illustrates our motivation to sample two points with opposite labels at a time instead of a single point.

for labeling from opposite sides of the current boundary. It greatly reduces the version space since both points affect how the version space will be bounded. The figure on the right shows that only a single point is sampled for labeling. Clearly, the left figure has bigger potential to converge to the true hypothesis. These two scenarios illustrate why it is more advantageous to straddle the decision boundary in order to reduce the set of candidate hypotheses rapidly. With this goal in mind, we strive to sample two points with opposite class labels in a binary classification model. In multi-class scenarios, this is equivalent to sampling as many points as the number of classes at each iteration of active learning, seeking to maximize the chance of sampling each class once per round. Since the labels of the unlabeled data are unknown, we need to approximate the likelihood that any two points have opposite class labels, $P(y_i \neq y_j \mid x_i, x_j)$, for all $i, j \in I_u$ where I_u is the set of indices of the unlabeled points in the data. By the cluster assumption [Chapelle and Zien, 2005], points in different clusters are likely to have different labels. In the new representation of the data, points in different clusters are assigned low similarity. It is then reasonable to define $P(y_i \neq y_j \mid x_i, x_j)$ as proportional to the distance between x_i and x_j, i.e. $P(y_i \neq y_j \mid x_i, x_j) \propto \|x_i - x_j\|^2$. For an empirical analysis justifying this claim, see [Donmez and Carbonell, 2008].

In order to obtain a faster learning rate we need to select two points that are likely to have opposite labels *and* high uncertainty:

$$\begin{aligned}\{i^*, j^*\} &= \arg \max_{i \neq j \in I_u} P(y_i \neq y_j \mid x_i, x_j) * U(i, j) \\ &= \arg \max_{i \neq j \in I_u} \|x_i - x_j\|^2 * U(i, j) \end{aligned} \tag{9}$$

where $U(i, j)$ is a complex utility score. To compute $U(i, j)$, we focus on sampling the most uncertain points in high density regions. Hence, we propose an uncertainty weighted density measure:

$$\hat{p}(\boldsymbol{x}) = \sum_{k \in N_x} \exp(-\|\boldsymbol{x} - \boldsymbol{x}_k\|^2) * \min_{y_k \in \{-1,+1\}} \{P(y_k \mid \boldsymbol{x_k}, \hat{\boldsymbol{w}})\} \tag{10}$$

where N_x is the close neighborhood of the data point x, i.e. the set of indices of unlabeled data points whose distance to x is less than a small threshold. $\hat{p}(\boldsymbol{x})$ is a function that measures both the local density of a point and also the information content of its neighbors. The local density depends on the number of close neighbors as well as their proximity. Moreover, each neighbor's contribution to the density score is weighed by its uncertainty; hence, the effect of the neighbors with least uncertainty is reduced. Formally, we define the utility $U(i, j)$ of a pair of points as the sum of the density estimate for each point. By the definition of N_x, it includes the point $\boldsymbol{x}$ in consideration. Hence, Equation 10 includes the uncertainty of the point itself, $\min_{y \in \{-1,+1\}}\{P(y \mid \boldsymbol{x}, \boldsymbol{w})\}$, as a summand with weight equals to $\exp(-\|\boldsymbol{x} - \boldsymbol{x}\|^2) = 1$. We propose to give more flexibility to that uncertainty term by introducing a regularization coefficient. It quantifies a trade-off of the information content of an instance with the proximity weighted information content of its neighbors. This allows us to define the utility function as follows:

$$\begin{aligned} U(i,j) &= \log\{\hat{p}(\boldsymbol{x_i}) + \hat{p}(\boldsymbol{x_j})\} \\ &= \log\Bigg\{ \sum_{k \neq i \in N_{x_i}} \exp(-\|\boldsymbol{x}_i - \boldsymbol{x}_k\|^2) * \min_{y_k \in \{\pm 1\}} \{P(y_k \mid \boldsymbol{x_k}, \hat{\boldsymbol{w}})\} + \\ &\quad \sum_{r \neq j \in N_{x_j}} \exp(-\|\boldsymbol{x}_j - \boldsymbol{x}_r\|^2) * \min_{y_r \in \{\pm 1\}} \{P(y_r \mid \boldsymbol{x_r}, \hat{\boldsymbol{w}})\} \\ &\quad + s * \Big(\min_{y_i \in \{\pm 1\}} \{P(y_i \mid \boldsymbol{x_i}, \hat{\boldsymbol{w}})\} + \min_{y_j \in \{\pm 1\}} \{P(y_j \mid \boldsymbol{x_j}, \hat{\boldsymbol{w}})\} \Big) \Bigg\} \end{aligned} \tag{11}$$

Experiments. We conducted a set of experiments in order to evaluate our method on 5 real datasets and one artificial dataset, details of which can be found in [Donmez and Carbonell, 2008]. We compared our paired sampling approach with four other strategies: most uncertain, density-only, representative sampling [Xu et al., 2003], and random sampling. Figure 4 shows the results on the UCI Breast data comparing five methods. Our method has the steepest decrease as well as the lowest final error rate which is very close to the optimal achieved using the entire training data. A more thorough comparison of all methods on six datasets is given in Table 1. The first column in Table 1 shows the dataset and the corresponding iteration at which the error rates are compared. The percentage error reduction against the random sampling baseline is given in parenthesis. Lowest error rates are given in bold. Our method wins on the majority of the cases. Whenever it loses, there is only a slight difference between our method and the winner so our method is still comparable on cases where it is not the best.

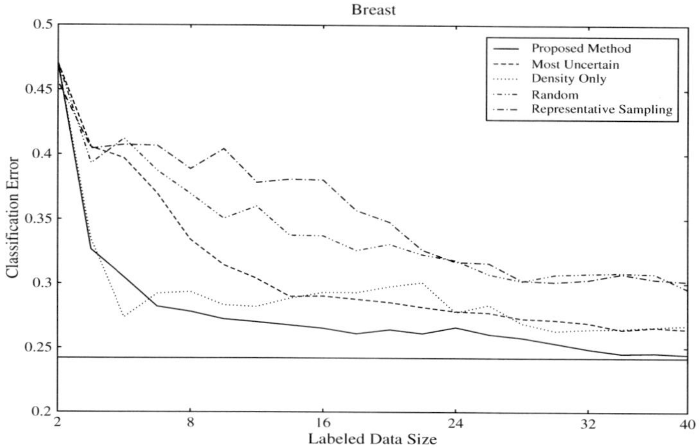

Fig. 4. Results on UCI Breast data. The solid horizontal line indicates the 10-fold cross-validation error using the entire data as the training data.

Table 1. Comparison of five different active learners on all datasets

Data	Proposed Method	Most Uncertain	Density Only	Representative	Random
Breast 5	**0.278 (-24.6%)**	0.334 (-9.04%)	0.293 (-20.5%)	0.380 (+2.9%)	0.369
Breast 11	**0.264 (-20%)**	0.285 (-13.6%)	0.297 (-10%)	0.347 (+5.1%)	0.330
Breast 17	**0.249 (-18.8%)**	0.269 (-12.3%)	0.264 (-14%)	0.302 (-1.6%)	0.307
Heart 5	**0.213 (-18.3%)**	0.245 (-6.1%)	0.220 (-15.7%)	0.216 (-17.2%)	0.261
Heart 11	**0.198 (-4.3%)**	0.208 (+0.4%)	0.220 (+6.2%)	0.205 (-0.9%)	0.207
Heart 17	0.166 (-13.5%)	**0.164 (-14.5%)**	0.219 (+14%)	0.20 (+4.1%)	0.192
Flare 5	0.465 (+5.2%)	0.454 (+2.7%)	0.454 (+2.7%)	0.478 (+8.1%)	**0.442**
Flare 11	**0.394 (-1.6%)**	0.451 (+10%)	0.422 (+2.9%)	0.417 (+1.7%)	0.410
Flare 17	**0.366 (-8.7%)**	0.449 (+11.9%)	0.401 (0%)	0.393 (-1.9%)	0.401
Face 5	0.350 (-1.9%)	0.468 (+31%)	0.420 (+17.6%)	**0.313 (-12.3%)**	0.357
Face 11	**0.210 (-23.3%)**	0.312 (+13.8%)	0.287 (+4.7%)	0.252 (-8%)	0.274
Face 17	**0.151 (-32.5%)**	0.196 (-12.5%)	0.189 (-15.6%)	0.202 (-9.8%)	0.224
Glass2 5	0.339 (-11%)	0.442 (+16%)	0.392 (+2.8%)	**0.326 (-14.4%)**	0.381
Glass2 11	0.317 (-7%)	0.341 (0%)	0.324 (-4.9%)	**0.31 (-9%)**	0.341
Glass2 17	**0.266 (-8.9%)**	0.292 (0%)	0.275 (-5.8%)	0.30 (+2.7%)	0.292
g50c 5	**0.169 (-46.3%)**	0.242 (-23.1%)	0.187 (-40.6%)	0.241 (-23.4%)	0.315
g50c 11	**0.110 (-37.8%)**	0.136 (-23.1%)	0.128 (-27.6%)	0.168 (-5%)	0.177
g50c 17	**0.079 (-34.1%)**	0.094 (-21.6%)	0.102 (-15%)	0.139 (+15.8%)	0.120

2 Proactive Learning

2.1 Why Proactively Learn?

Active learning, for all its popularity, operates in an idealized framework. For instance, we assume that there exists an oracle who provides accurate answers to all questions (error-free labeling), who never gets tired, and who charges uniformly regardless of question difficulty. In virtually all applications these assuptions break down: Human experts occasionally err; they can be reluctant to answer if tired or faced with difficult questions; different experts could charge differently for their services, perhaps based on quesion difficulty. If we use lab experiments to obtain labels, rather than human experts, these too can be fallible, producing inaccurate results, failing to produce results on occasion, and exhibiting different costs depending on the experimental setup.

Consider, for instance, proteomic classification [Cheng et al., 2005; Wang et al., 2007], where the task is to label proteins as to functional and structural family and subfamily membership based on their primary amino-acid sequence and other features (e.g. locus, density, homology, etc.). Although about three million proteins have been sequenced, only about 1.5% have been structurally resolved with certainty, and these are predominantly soluble proteins which are easier to crystalize and thefore ameanble to X-ray diffraction crystalography – the primary method for obtaining the structure of large proteins. If we need to obtain a structural label for a new protein, we have several options: 1) Ask a human expert, who will venture a educated guess if he or she can, or say "sorry" otherwise; 2) Check if the protein has very high sequence homology to a structurally resolved one, and if so infer an identical structure label, but this process will only provide labels for very sequence-proximate instances at best; or 3) Perform a lab experiment such as X-ray Crystallography – very expensive and uncertain to yield results, or NMR imaging for small proteins – less expensive but its reliability plummets with increasing protein size. The problem we have is not just to choose an instance to label, but also choose a labeling method (an "oracle") most appropriate for the instance. And, if we have a fixed budget, we must also calculate the cost of failure – experiments that do not yield results still incur costs. Hence, we cannot select the information-value best instance without regards to probability of successfully obtaining a label and without regard to cost. Ideally we want to jointly optimize the selection of instance and labeling method (i.e. oracle), which might yield a more certain or inexpensive choice over the information-value-only optimum.

There are many real learning problems with characteristics similar to protein labeling, perhaps more so than for the idealized oracle of active learning. Thus, we need to extend active learning to include a utility-based deision-theoretic approach that encompasses oracle errors, oracle reluctance, and variable costs. We call this extension *proactive learning*. In summary, proactive learning extends active learning as in Table 2.

Table 2. Active versus Proactive Learning

Oracle Property	Active Learning	Proactive Learning
Number	Individual (one oracle)	Plural (different oracles)
Accuracy	Infallible (always right)	Probabilistic (may err)
Reluctance	Indefatigable (always answers)	Reluctant (may refuse to answer)
Cost	Invariant (free or constant)	Variable (per oracle, per instance)

Proactive learning is a naescent field of investigation; the remainder of this paper provides a substantial exploratory foray, but much research remains to be done.

2.2 A Case-Study: Optimizing Proactive Learning with Two Oracles

In this section, we aim to show that a decision-theoretic approach [DeGroot, 2004] is beneficial for cost-effective sampling in cases with imperfect oracles. We mainly focus on three model scenarios embodying the notion of oracles with differing properties and costs. In each scenario, we target a single oracle property. In "Scenario 1", we assume there exist one reliable oracle and one reluctant oracle. The reliable oracle gives an answer every time it is invoked with a query, and the answer is always correct. The reluctant oracle, on the other hand, does not always provide an answer, but when it answers it does so correctly. We assume the probability of getting an answer from the reluctant oracle depends on the difficulty of the classification task. Not surprisingly, they charge different fees: the reliable oracle is more expensive than the potentially reluctant one.

Rather than fixing the number of instances to classify, as in standard active learning, proactive learning fixes a maximum budget envelope since instances and oracles may have variable costs. We formulate the problem as a joint optimization of which instance(s) to sample and which oracle to use to purchase their label(s), where the objective is to maximize the cumulative value of the information given a pre-defined budget:

$$\text{maximize } E[V(S)] - \lambda(\sum_k t_k * C_k) \quad \text{s.t. } \sum_k t_k * C_k = B \,, \; \sum_k t_k = |S|$$

where B is the budget, S is the set of instances to be sampled, and $E[V(S)]$ is the expected value of information of the sampled data to the learning algorithm. $V(S)$ is a value function that can be replaced with any active selection criterion. In our experiments, we adopt a density weighted uncertainty score proposed in [Donmez and Carbonell, 2008], which significantly outperforms other strong baselines. $k \in K$ denotes the chosen oracle from the set of oracles, K, and λ is a parameter controlling the relative importance of maximizing the information and minimizing the cost. For simplicity, we assumed $\lambda = 1$ in this work. C_k and t_k indicate the cost of the chosen oracle and the number of times it is invoked, respectively. UL is the set of unlabeled examples, $|S|$ is the total size of the sampled set[3]. The drawback of this approach is that it is difficult to optimize directly due to the fact that the maximization is over the entire set of potential sampling sequences, an exponentially large number. Thus, we cannot decide all the points to be sampled at once. A tractable alternative is a greedy approximation that will select the locally-optimal strategy at each round according to a joint utility function of the oracle-instance pair:

$$(x^*, k^*) = \arg \max_{x \in UL, k \in K} U(x,k) = \arg \max_{x \in UL, k \in K} \frac{P(ans \mid x,k) * V(x)}{C_k} \quad \text{where } k \in K \tag{12}$$

where the utility $U(x, k)$ is defined as the expected information value of that example at unit cost.

We need to simulate reliability and reluctance since there do not exist real-world datasets that have ground truth information on it. We assume the reliable (perfect) oracle resembles a system that has been trained on the entire dataset and has zero training error. Unlike the reliable oracle, a reluctant oracle has access only to a small portion of the data; therefore, it is not knowledgable for every point. We train a classifier on a small random subset of the entire data to obtain a posterior class distribution $P(y \mid x)$. We

[3] The extension of this formulation to more than two oracles is straightforward.

assume that the chance of obtaining an answer from the reluctant oracle is low when the uncertainty $\min_{y \in \mathcal{Y}} P(y \mid x)$ is high and vice versa.

We further assume that the reluctant oracle's knowledge level or response characteristics are unknown apriori. Hence, we estimate these properties in an exploration phase. We hypothesize that the oracle shows similar response characteristics for similar data instances since we model reluctance as a function of uncertainty which is similar for similar instances. Therefore, it is reasonable to estimate the answer probability of the reluctant oracle by inquiring the labels of the cluster centroids and propagating the information to nearby points. We cluster the unlabeled data using k-means clustering [Hartigan and Wong,]. The number of clusters depends on a pre-defined budget available for this phase and the cost of the reluctant oracle. For each cluster, if we obtain the label of the centroid, then we increase the answer probability of the points in this cluster. Similarly, we decrease the answer probability of the points in the clusters whose centroids we did not obtain the labels of. Initially, we assume the answer probability for each unlabeled point is 0.5, which indicates a random guess. Then, we update the answer probabilities of the reluctant (unreliable) oracle so that it changes as a function of the proximity of the point to the cluster centroid and oracle responsiveness:

$$\hat{P}(ans \mid x, reluctant) = \frac{0.5}{Z} * exp\left(\frac{h(x_{c_t}, y_{c_t})}{2} \ln \frac{max_d - \|x_{c_t} - x\|}{\|x_{c_t} - x\|}\right), \forall x \in C_t \tag{13}$$

where Z is a normalization factor. x_{c_t} is the centroid of the cluster C_t that includes x. $h(x_c, y_c) \in \{1, -1\}$ is an indicator function which is equal to 1 when we receive the label y_c for the centroid x_c, and -1 otherwise. $\|x_c - x\|$ is the Euclidean distance between the cluster centroid x_c and the point x and $max_d := \max_{x_{\acute{c}}, x} \|x_{\acute{c}} - x\|$ is the maximum distance between any cluster centroid and data point.

We substitute the estimated answer probability into the utility function, i.e. $\hat{U}(x, k) = \frac{\hat{P}(ans|x,k) * V(x)}{C_k}$. The joint sampling of the oracle-example pair can now be performed as shown in Algorithm 2. The algorithm works in rounds till the budget is exhausted. Each round corresponds to a single label acquisiton attempt where sampling persists until obtaining a label. However, we need to restrain from spending too much on a single attempt by adaptively penalizing the reluctant oracle every time it refuses to answer. At any given round, if the algorithm chooses the reluctant oracle and does not receive an answer, the utility of remaining examples with respect to this oracle decreases by the amount spent thus far at this round:

$$\hat{U}(x, reluctant) = \frac{\hat{P}(ans \mid x, reluctant) * V(x)}{C_{round}}$$

where C_{round} is the amount spent thus far in the given round. This penalization only applies to the reluctant oracle since the reliable oracle always provides the label. Algorithm 2 selects the maximum utility examples. This framework leads to an incrementally optimal solution in the sense that the most useful data is sampled at the minimum cost.

Algorithm 2. Proactive Learning: Scenario 1

Input: a classifier f, labeled data L, unlabeled data UL, entire budget B, clustering budget $B_C < B$, two oracles, each with a cost C_k, $k \in K = \{reliable, reluctant\}$
Output: f
- Cluster UL into $p = B_C / C_{reluctant}$ clusters
- Let x_{c_t} be the data point closest to its cluster centroid, $\forall t = 1, ..., p$
- Query the label y_{c_t} for each cluster centroid x_{c_t}
- Identify $\{x_{c_1}, ..., x_{c_g}\}$ for which we obtain the labels[4]
- Estimate $\hat{P}(ans \mid x, reluctant)$ via Equation 13
- Update $L = L \cup \{x_{c_t}, y_{c_t}\}_{t=1}^{g}$, $UL = UL \setminus \{x_{c_t}, y_{c_t}\}_{t=1}^{g}$
- cost spent so far $C_T = B_C$

while $C_T < B$ **do**
  - Train f on L
  - Initialize the cost of this round $C_{round} = 0$ and the set of queried examples $Q = \{\}$
  - $\forall k \in K, x \in UL$ estimate utility $\hat{U}(x, k)$

  repeat
    1. Choose $k^* = \arg\max_{k \in K} \max_{x \in UL \setminus Q} \{\hat{U}(x, k)\}$
    2. Choose $x^* = \arg\max_{x \in UL \setminus Q} \{\hat{U}(x, k^*)\}$
    3. Update $C_{round} = C_{round} + C_{k^*}$
    4. $Q = Q \cup \{x^*\}$
    5. Query the label y^* with probability $P(ans \mid x^*, k^*)$

  until label y^* is obtained
  - Update $C_T = C_T + C_{round}$
  - Update $L = L \cup (x^*, y^*)$ and $UL = UL \setminus (x^*, y^*)$

end while

In real applications, such as protein structure classification, there might also be fallible oracles which always answer, but the credibility of the answer is questionable[5]. We simulate this setting in "Scenario 2", where we assume two oracles; one reliable and one unreliable oracle. The reliable oracle is the perfect oracle that always provides the correct answer to any query. The unreliable oracle in this scenario is fallible that it may provide a random label for the queried example x with probability $1 - P(correct \mid x, k)$. We assume the probability of correct classification approaches 0.5 (random guess) if an example approaches the decision boundary. Similar to the previous scenario, we request the labels of the cluster centroids. Unlike the reluctant oracle, the fallible oracle provides the label together with its confidence[6]. The confidence is its posterior class probability for the provided label, $P(y \mid x)$. If the class posterior is within an uncertainty range, then we decide not to use the provided label since it is likely to be noisy. We decrease the correctness probability for the points in the cluster whose centroid has a class posterior in the uncertainty range and vice versa;

[5] Wrong answers might also be due to the noise in the data. But, we for now we assume the data is noise-free, and fallibility is only a property of the oracle.

[6] This could be a rather strong assumption. Section 2.3 discusses a more general case without requiring such information.

i.e. $\hat{P}(correct \mid x, fallible) = \frac{0.5}{Z} * exp\left(\frac{\tilde{h}(x_{c_t}, y_{c_t})}{2} \ln \frac{max_d - \|x_{c_t} - x\|}{\|x_{c_t} - x\|}\right) \forall x \in C_t$ where $\tilde{h}(x_{c_t}, y_{c_t}) = -1$ if $P(y \mid x_{c_t})$ is in the uncertainty range, and 1 otherwise. The pseudocode of the algorithm is given in Algorithm 3, where h denotes the number of high confident centroids.

Algorithm 3. Proactive Learning: Scenario 2

Input: a classifier f, labeled data L, unlabeled data UL, entire budget B, clustering budget $B_C < B$, two oracles, each with a cost C_k, $k \in K = \{reliable, fallible\}$
Output: f
- Cluster UL into $p = B_C / C_{fallible}$ clusters
- Let x_{c_t} be the data point closest to its cluster centroid, $\forall t = 1, ..., p$
- Query the label y_{c_t} for each cluster centroid x_{c_t}
- Identify $\{x_{c_1}, ..., x_{c_h}\}$ for which the fallible oracle has high confidence
- Estimate $\hat{P}(correct \mid x, fallible)$ as in Equation 13
- Update $L = L \cup \{x_{c_t}, y_{c_t}\}_{t=1}^{h}$, $UL = UL \setminus \{x_{c_t}, y_{c_t}\}_{t=1}^{h}$
- cost spent so far $C_T = B_C$

while $C_T < B$ **do**
1. Train f on L
2. $\forall k \in K, x \in UL$ estimate $\hat{U}(x, k)$
3. Choose $k^* = \arg\max_{k \in K} \max_{x \in UL} \{\hat{U}(x, k)\}$
4. Choose $x^* = \arg\max_{x \in UL} \{\hat{U}(x, k^*)\}$
5. Update $C_T = C_T + C_{k^*}$
6. Update $L = L \cup (x^*, y^*)$ and $UL = UL \setminus (x^*, y^*)$ where y^* is the correct label with probability $P(correct \mid x^*, k^*)$

end while

Thus far, we have only considered the settings where a uniform fee is charged for every query by an oracle, although each oracle may charge differently. Fraud detection in banking transactions is a good example for this setting. The customer records are saved in the bank database so it takes the same amount of time and effort, hence the same cost, to look up any entry in the database. However, it is possible that the costs are distributed non-uniformly over the set of instances. For instance in text categorization, it might be relatively easy for an annotator to categorize a short web page; hence the cost is modest. On the other hand, it is more time-consuming to assign a book into a category, which incurs a considerable reading time and therefore cost. We already discussed variable costs in protein classification, where less expensive NMR imaging may apply to smaller proteins. Another example for a non-uniform cost scenario is medical diagnosis. Some diseases such as herpes are easy to diagnose. Such diagnoses are not costly since there is usually a major definitive symptom, i.e. outbreak of blisters on the skin. On the other hand, diagnosing hepatitis can be more costly since it may require blood and urine tests, CT scans, or even a liver biopsy. In "Scenario 3", we explore the problem of deciding which instances to query for the labels when label acquisition cost varies with the instance. We assume two oracles one of which has a uniform and fixed cost for each query whereas the other charges according to the task

Algorithm 4. Proactive Learning: Scenario 3

Input: a classifier f, labeled data L, unlabeled data UL, entire budget B, two oracles, each with a cost C_k, $k \in K = \{unif, non - unif\}$
Output: f
cost spent so far $C_T = 0$
while $C_T < B$ **do**
1. Train f on L
2. $\forall k \in K, x \in UL$ calculate $\hat{U}(x, k)$ via Equation 14.
3. Choose $k^* = \arg \max_{k \in K} \max_{x \in UL} \{\hat{U}(x, k)\}$
4. Choose $x^* = \arg \max_{x \in UL} \{\hat{U}(x, k^*)\}$
5. Update $C_T = C_T + C_{k^*}$
6. Update $L = L \cup (x^*, y^*), UL = UL \setminus (x^*, y^*)$

end while

difficulty. We further assume that these oracles always provide an answer and both are perfectly reliable in their answers.

We model the cost of each example x as a function of the posterior class distribution $P(y \mid x)$ so that there is a positive correlation between the classification difficulty and the cost. We use the class posterior calculated similarly in the previous scenarios. The non-uniform cost $C_{non-unif}(x)$ per instance is then defined as follows:

$$C_{non-unif}(x) = 1 - \frac{\max_{y \in \mathcal{Y}} P(y \mid x) - 1/|\mathcal{Y}|}{1 - 1/|\mathcal{Y}|}$$

The cost increases as the instance approaches the decision boundary and vice versa. In other words, the oracle charges based on how difficult it is to classify, and how valuable it might potentially be to the learner. This sets up a challenging decision in terms of the utility-cost trade-off. The utility score in this scenario is calculated as the difference between the information value and the cost instead of the information value per unit cost[7]. This is to avoid infinitely large utility scores as a result of the division by small ϵ-cost. Thus, the revised utility score per oracle is given as follows:

$$\begin{aligned} \hat{U}(x, unif) &= V(x) - C_{unif} \\ \hat{U}(x, non - unif) &= V(x) - C_{non-unif}(x) \end{aligned} \tag{14}$$

where C_{unif} is the fixed cost of the uniform-cost oracle. The pseudocode of the algorithm is given in Algorithm 4. There is no clustering phase in Algorithm 4 since we assume we know the cost of every instance, which is realistic for many real-world applications.

Problem Setup. In order to simulate the reliability of the labeling source (oracle), we assume that a perfectly reliable oracle resembles by a strong classifier trained on the

[7] In general, if the cost and information value are not assessed in the same units, then they are normalized into the same range.

entire data, and having zero training error. An unreliable oracle, on the other hand, resembles a classifier trained on only a small subset of the entire data; hence a weak one. We randomly sampled a small subset from each dataset and trained a logistic regression classifier on this sample to output a posterior class distribution. Then, we identified the instances whose class posterior falls into the uncertainty range, i.e. $\min_y P(y \mid x) \in [0.45, 0.5]$. This range is used to filter the instances that the reluctant oracle does not answer or the fallible oracle outputs a random label. We varied the percentage of instances that fall into the uncertainty range $[.45, .5]$. The second column in Table 3 shows the different percentages used in our experiments. There is also a tradeoff between cost and unreliability to have an incentive to choose between oracles rather than exploiting a single one. The higher cost ratios for the fallibility scenario indicate more penalization for receiving a noisy label. See Table 3 for details.

Table 3. Oracle Properties and Costs. B_C is the clustering budget, B is the entire budget. Uncertain % is the percentage of the uncertain data points. Cost Ratio is the ratio of the cost of the reliable oracle to the cost of the unreliable one.

Scenario	Uncertain %	Cost Ratio	B_C	B
Scenario 1	45-50%	1:3	20	300
	50-55%	1:4	30	
	65-70%	1:5	50	
Scenario 2	45-50%	1:5	20	300
	50-55%	1:6	30	
	65-70%	1:7	50	

The other case we need to simulate involves the uniform and non-uniform cost oracles. The cost of each instance for the variable-cost oracle is defined as a function of the class posterior obtained on the randomly chosen subset. The cost of the uniform-cost oracle is chosen within the range of instance costs for the variable-cost oracle. We varied the fixed cost such that there is always an incentive to choose between oracles instead of fully exploiting a single one.

All the results are averaged over 10 runs. At each run, we start with one randomly chosen labeled example from each class which are assumed to be known a-prori, and therefore free. The rest of the data is considered unlabeled. The learner selects one example at each iteration to be labeled, and the learning function is tested on the remaining unlabeled set once the label is obtained. The learner pays the cost of each queried example regardless of whether the label is obtained. The budget is fixed at 300 in Scenario 1 and 2, and at 20 in Scenario 3[8]. A small budget is enough for the latter since the cost of individual instances can be very small depending on the posterior probability. The clustering budget, on the other hand, varies according to the unreliability, but is the same for each baseline under the same scenario (See Table 3).

[8] At each run, the total number of iterations to spend the entire budget may differ depending on how the budget is allocated between oracles. We rely on the minimum number of iterations attained over 10 runs for each experiment. This results in reporting different maximum elicitation costs smaller than the budget for different experiments

Evaluation. We conducted a thorough analysis to examine the performance of our method under various conditions. Due to the lack of existing work on cost-sensitive active learning with multiple oracles, we compared our method against active sampling with randomly chosen oracles and active sampling with a single oracle. We denote our method by *Joint*, the random sampling of oracles by *Random*. *Reliable*, *Reluctant*, and *Fallible* refer to the corresponding single oracle baseline.

We evaluated the performance of the proposed methods on various real-world benchmark datasets, but here we only report a representative subset of results. Figure 5 shows the results for the reluctance scenario on the Spambase dataset. Each plot indicates a different cost ratio. Our method outperforms the others on every case while the performance gap increases with the cost ratio. This is largely because the oracle differences leave more room for improvement via oracle selection in the latter case. When the unreliability gets higher, the reluctant oracle tend to spend almost the entire budget on a single label acquisition attempt. This leads to acquiring only a small amount of labeled data; hence, poor performance. As a result, we do not report the reluctant oracle baseline except for the 1 : 3 cost ratio.

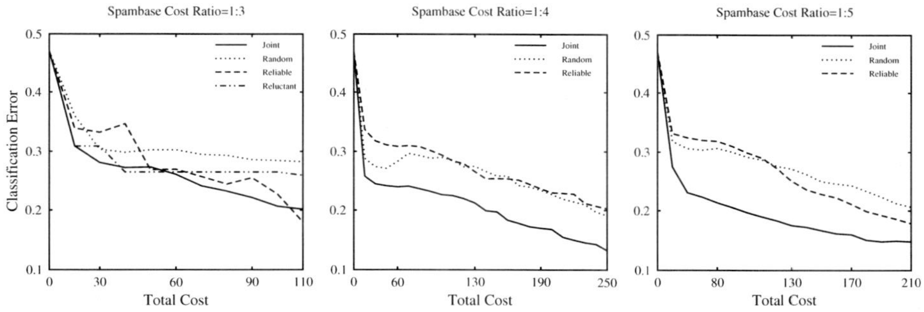

Fig. 5. Performance Comparison for Scenario 1 (Reluctance) on the Spambase dataset. The cost ratio is indicated above each plot.

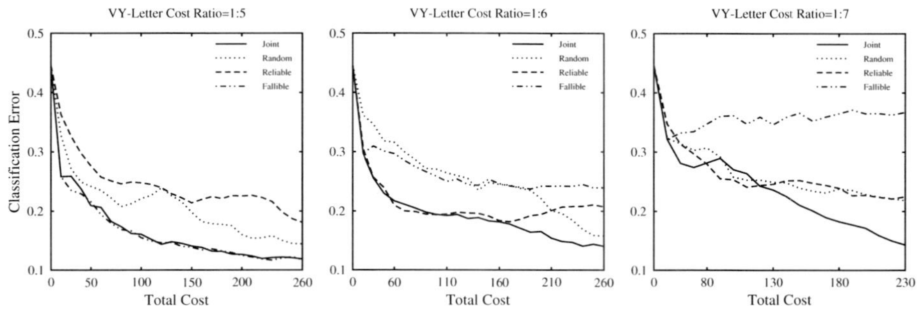

Fig. 6. Performance Comparison for Scenario 2 (Fallible) on the VY-Letter dataset. The cost ratio is indicated above each plot.

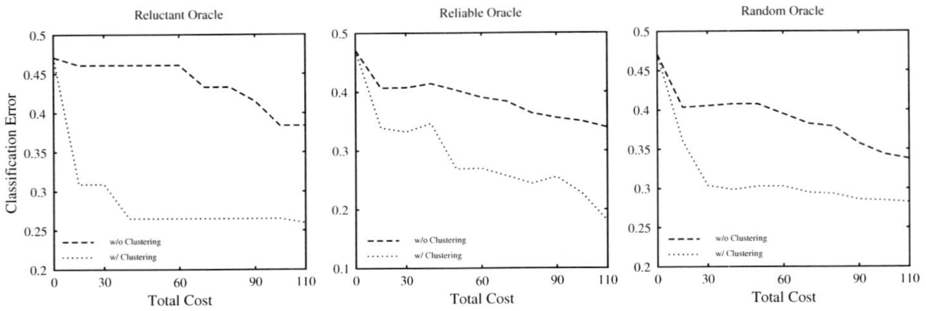

Fig. 7. Change in performance of each baseline with and without clustering on Spambase. The type of baseline is given in the title. The cost ratio is 1:3.

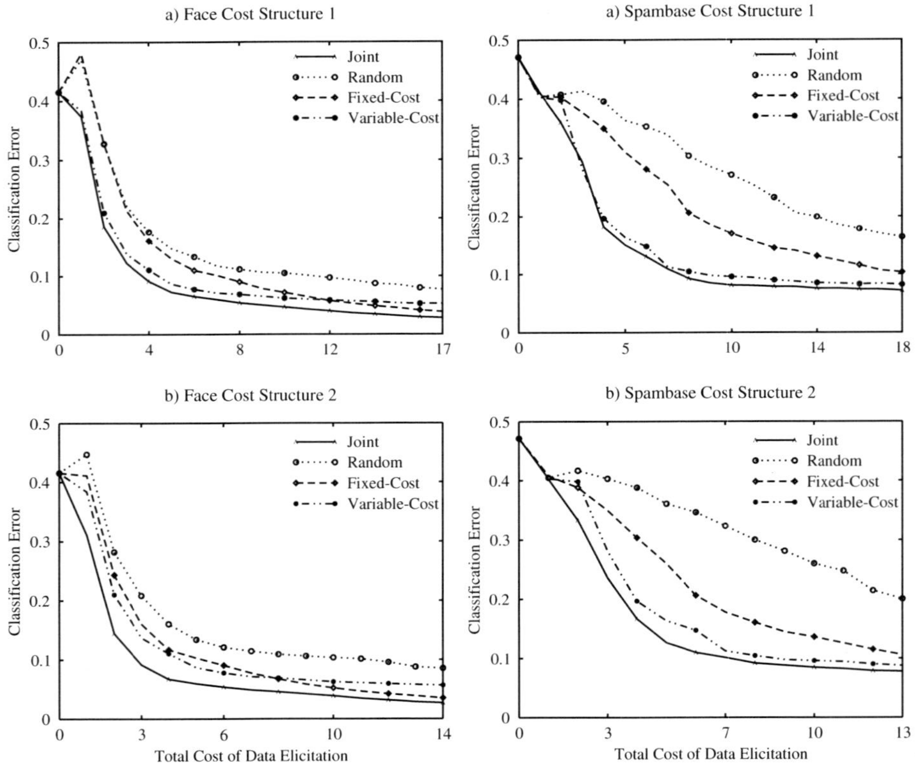

Fig. 8. Comparison of different algorithms under non-uniform cost structures (Scenario 3) on Face (left panel) and Spambase (right panel) datasets. a) Cost Structure 1: Fixed-Cost oracle has Cost1 b) Cost Structure 2: Fixed-Cost oracle has Cost2.

Figure 6 compares the performances for Scenario 2 on the VY-Letter dataset. The fallible oracle in this scenario performs poorly as unreliability increases since the labels are increasingly likely to be noisy. This pattern is especially evident in Figure 6 for the cost ratio 1 : 7. On the other hand, *Joint* strategy is quite effective for reducing the error in this scenario, indicating that it is capable of reducing the risk of introducing noisy data through strategic selection between oracles.

Figure 8 presents the evaluation results when the cost varies non-uniformly across the set of instances. We experimented with different assignments of the fixed cost, but we present two representative assignments for each dataset. The remaining assignments are not included since they are similar to those reported here. *Joint* is the best performer throughout the full operating range. The performance difference between *Joint* and each baseline is also statistically significant based on a paired two-sided t-test ($p < 0.01$).

In order to investigate if the initial clustering phase helps all the baselines, we re-ran each baseline excluding the clustering step. (See Figure 7). Every baseline significantly benefits from clustering, with the biggest boost in improvement occuring for the reluctant oracle. Hence, both the baselines and the ''Joint'' strategy benefit from the diversity-based sampling via clustering in their initial steps. Without pre-clustering, the reluctant oracle is prone to spend too much on a single elicitation attempt due to unsuccessful labeling requests. It can, however, maximize the chance of receiving a label through diversity sampling during the clustering step instead of getting stuck in one round for a single label.

2.3 Future Directions

Thus far, we explored simple cases each of which addresses a single oracle property with one perfect and one imperfect oracle. A natural next step is to extend it to multiple oracles with differing levels of reliability. With multiple oracles, the goal is to select the oracle-example pair with the maximum utility, but the estimation of the oracle properties becomes more challenging. For instance, oracles may be fallible at differing levels. A simple scenario is to assume that each oracle has a fixed probability of making a labeling mistake independent of the data instance to be classified. The goal, then, is to estimate each oracle's noise probability and use the estimates to select the least noisy oracle(s) for label acquisition. As before, noise estimation requires an exploration phase whereas querying the least noisy oracles is the exploitation phase. Hence, there is an exploration vs. exploitation tradeoff. This problem is analogous in many ways to multi-armed bandit problem.

The multi-armed bandit problem for a gambler is to decide the arm to pull in a K-slot machine to maximize her total reward in a series of trials. Each action does not just bring in reward, but also helps discover new knowledge about the slot machine that can be used to assist future actions. The reward of each lever is drawn from a specific distribution which is unknown to the gambler. Thus, she tries to acquire knowledge through repeated trials via balancing the exploration vs. exploitation tradeoff. There is quite a lot of work done in the literature to solve the multi-armed bandit problem. One popular bandit algorithm is the exponential weight algorithm introduced by [Auer et al., 2002]. The probability of choosing the lever k at round t is defined by:

$$p_k(t) = (1-\gamma)\frac{w_k(t)}{\sum_{K}^{j=1} w_j(t)} + \frac{\gamma}{K}$$

where $w_j(t+1) = w_j(t)exp(\gamma\frac{r_j(t)}{p_j(t)K})$ if the lever j has been pulled at time t with the observed reward $r_j(t)$, otherwise $w_j(t+1) = w_j(t)$. It has a theoretical guarantee that the regret of the algorithm is $O(\sqrt{KTlog(K)})$ with γ decreasing over time, where regret is defined the difference between the reward sum of the optimal strategy and the sum of the collected rewards. This strategy can be adopted to the proactive learning setting with multiple fallible oracles with the appropriate definitions of the reward function. In a completely unsupervised setting (absence of gold standard label information), the reward for an oracle output can be determined based on the majority vote from all and/or a subset of oracles. The majority voting principle has been used by [Sheng et al., 2008] in a simple context where each oracle has the same noise level; hence, there is no need for estimation.

More complexity arises when the error probability of each oracle is dependent on the data instance. This is probably a more realistic assumption than independence on the data. One possible solution is to leverage the fact that similar instances have similar labels. In other words, if an oracle makes a misprediction for some instance then it is likely to make mispredictions on similar instances. Then, it becomes important to query the oracles for instances that are most representative of the entire input space. A similar approach to out belief propagation framework can be quite effective to estimate the oracles' error probabilities without intensive labeling.

Another interesting problem occurs when each oracle has partial knowledge at only a certain region of the data distribution and its cost depends on its knowledge. The goal is to select the oracle who is an expert at the region of interest. Therefore, each oracle needs an estimate about the region at which it is knowledgeable. One possible way to attack this problem is to adopt a top-down strategy where each oracle is assumed to be an expert for the entire data space. The knon expertise region of each oracle shrinks over time based on inaccurate responses to labeling queries. How to calculate the shrinkage without massively repeated querrying remains an open question.

The journey to proactive machine learning is a long and exciting one. We have explored certain roads and not others. For instance, one could simplify the problem by assuming that oracle properties are fully known a-priori, or one could take detours into investigations such as referal networks among oracles, if the one queried does not know the answer with sufficient accuracy – but that would require a more complex cost model with referral fees and mode complex oracle network learning models.

References

[Auer et al., 2002] Auer, P., Cesa-Bianchi, N., Freund, Y., Schapire, R.E.: The nonstochastic multiarmed bandit problem. SIAM Journal on Computing 32(1), 48–77 (2002)

[Baram et al., 2003] Baram, Y., El-Yaniv, R., Luz, K.: Online choice of active learning algorithms. In: ICML 2003, pp. 19–26 (2003)

[Blum and Mitchell, 1998] Blum, A., Mitchell, T.: Combining labeled and unlabeled data with co-training. In: COLT 1998, pp. 92–100 (1998)

[Chapelle and Zien, 2005] Chapelle, O., Zien, A.: Semi-supervised classification by low density separation. In: Proceedings of the Tenth International Workshop on Artificial Intelligence and Statistics 2005 (2005)

[Cheng et al., 2005] Cheng, B.Y., Carbonell, J.G., Klein-Seetharaman, J.: Protein classification based on text document classification techniques. Proteins: Structure, Function and Bioinformatics 58(4), 955–970 (2005)

[Cohn et al., 1995] Cohn, D.A., Ghahramani, Z., Jordan, M.I.: Active learning with statistical models. In: Advances in Neural Information Processing Systems, vol. 7, pp. 705–712 (1995)

[DeGroot, 2004] DeGroot, M.H.: Optimal Statistical Decisions. Wiley Classics Library, Chichester (2004)

[Donmez and Carbonell, 2008] Donmez, P., Carbonell, J.G.: Paired sampling in density-sensitive active learning. In: International Symposium on Artificial Intelligence and Mathematics (2008)

[Donmez et al., 2007] Donmez, P., Carbonell, J.G., Bennett, P.N.: Dual strategy active learning. In: Kok, J.N., Koronacki, J., Lopez de Mantaras, R., Matwin, S., Mladenič, D., Skowron, A. (eds.) ECML 2007. LNCS (LNAI), vol. 4701, pp. 116–127. Springer, Heidelberg (2007)

[Hartigan and Wong,] Hartigan, J.A., Wong, M.A.: A k-means clustering algorithm. Applied Statistics, 28

[He et al., 2007] He, J., Carbonell, J., Liu, Y.: Graph-based semi-supervised learning as a generative model. In: IJCAI 2007 (2007)

[Joachims, 1999] Joachims, T.: Transductive inference for text classification using support vector machines. In: ICML 1999 (1999)

[Lewis and Gale, 1994] Lewis, D., Gale, W.: A sequential algorithm for training text classifiers. In: SIGIR 1994, pp. 3–12 (1994)

[McCallum and Nigam, 1998] McCallum, A., Nigam, K.: Employing em and pool-based active learning for text classification. In: ICML 1998, pp. 359–367 (1998)

[Melville and Mooney, 2003] Melville, P., Mooney, R.J.: Constructing diverse classifier ensembles using artificial training examples. In: IJCAI 2003, pp. 505–510 (2003)

[Michalski et al., 1983] Michalski, R.S., Carbonell, J.G., Mitchell, T.M.(eds.): Machine Learning: An Artificial Intelligence Approach. TIOGA Publishing Co. (1983)

[Michalski et al., 1986] Michalski, R.S., Carbonell, J.G., Mitchell, T.M. (eds.): Machine Learning: An Artificial Intelligence Approach, vol. II. Morgan Kaufmann Publishers, Inc., San Francisco (1986)

[Miller et al., 1993] Miller, J.W., Goodman, R., Smyth, P.: On loss functions which minimize to conditional expected values and posterior probabilities. IEEE Transactions on Information Theory 39(4), 1404–1408 (1993)

[Muslea et al., 2000] Muslea, I., Minton, S., Knoblock, C.: Selective sampling with naive co-testing: preliminary results. In: The ECAI 2000 workshop on Machine Learning for information extraction (2000)

[Nguyen and Smeulders, 2004] Nguyen, H.T., Smeulders, A.: Active learning with pre-clustering. In: ICML 2004, pp. 623–630 (2004)

[Reinke and Michalski, 1985] Reinke, R., Michalski, R.S.: Incremental learning of decision rules: A method and experimental results. Presented at the Machine Intelligence Workshop II (1985)

[Roy and McCallum, 2001] Roy, N., McCallum, A.: Toward optimal active learning through sampling estimation of error reduction. In: ICML 2001, pp. 441–448 (2001)

[Shen and Zhai, 2005] Shen, X., Zhai, C.: Active feedback in ad hoc information retrieval. In: SIGIR 2005, pp. 59–66 (2005)

[Sheng et al., 2008] Sheng, V.S., Provost, F., Ipeirotis, P.G.: Get another label? improving data quality and data mining using multiple, noisy labelers. In: KDD 2008, pp. 614–622 (2008)

[Tang et al., 2002] Tang, M., Luo, X., Roukos, S.: Active learning for statistical natural language parsing. In: ACL 2002 (2002)

[Tong and Koller, 2000] Tong, S., Koller, D.: Support vector machine active learning with applications to text classification. In: Proceedings of International Conference on Machine Learning, pp. 999–1006 (2000)

[Wang et al., 2007] Wang, Y., Wu, L.-Y., Chen, L., Zhang, X.-S.: Supervised classification of protein structures based on convex hull representation. International Journal of Bioinformatics Research and Applications 3(2), 123–144 (2007)

[Xu et al., 2003] Xu, Z., Yu, K., Tresp, V., Xu, X., Wang, J.: Representative sampling for text classification using support vector machines. In: Sebastiani, F. (ed.) ECIR 2003. LNCS, vol. 2633, pp. 393–407. Springer, Heidelberg (2003)

[Zhang and Chen, 2002] Zhang, C., Chen, T.: An active learning framework for content-based information retrieval. IEEE Trans. on Multimedia 4, 260–268 (2002)

Explicit Feature Construction and Manipulation for Covering Rule Learning Algorithms

Nada Lavrač[1,2], Johannes Fürnkranz[3], and Dragan Gamberger[4]

[1] Jožef Stefan Institute, Jamova 39, 1000 Ljubljana, Slovenia
nada.lavrac@ijs.si
[2] University of Nova Gorica, Vipavska 13, 5000 Nova Gorica, Slovenia
[3] TU Darmstadt, D-64289 Darmstadt, Germany
juffi@ke.informatik.tu-darmstadt.de
[4] Rudjer Bošković Institute, Bijenička 54, 10000 Zagreb, Croatia
dragan.gamberger@irb.hr

Abstract. Features are the main rule building blocks for rule learning algorithms. They can be simple tests for attribute values or complex logical terms representing available domain knowledge. In contrast to common practice in classification rule learning, we argue that the separation of feature construction and rule construction processes has a theoretical and practical justification. Explicit usage of features enables a unifying framework of both propositional and relational rule learning and we present and analyze procedures for feature construction in both types of domains. It is demonstrated that the presented procedure for constructing a set of simple features has the property that the resulting feature set enables the construction of complete and consistent rules whenever possible, and that the set does not include obviously irrelevant features. It is also shown that feature relevancy may improve the effectiveness of rule learning. It this work, we illustrate the relevancy concept in the coverage space, and show that the transformation from the attribute to the feature space enables a novel, theoretically justified way of handling unknown attribute values. The same approach enables that the estimated imprecision of continuous attributes can be taken into account, resulting in the construction of features that are robust to attribute imprecision.

1 Introduction

Michalski (1983), in his pioneering formulation of a formal framework for inductive learning, already noted that "*... determining [...] descriptors is a major part of any inductive learning problem*". Rule learning algorithms, most notably those that follow the separate-and-conquer or covering approach (Fürnkranz, 1999), which has been pioneered by Michalski's AQ (Michalski, 1969) and its successors (Michalski, 1980; Michalski and Larson, 1978; Michalski et al., 1986; Bergadano et al., 1992; Kaufman and Michalski, 2000; Wojtusiak et al., 2007), utilize object descriptors in the form of *features*, which can be combined into complex logical concept descriptions. Michalski's *variabled-value logic (VL1)* (Michalski, 1973a;b) is an elegant formalization of a subset of logic that is specifically tailored to rule learning algorithms.

Throughout this work we use the term *features* to denote simple boolean terms that can be inferred from the input data to be used for constructing complex logical expressions. However, different authors use a different terminology. Michalski (1973a;b)

J. Koronacki et al. (Eds.): Advances in Machine Learning I, SCI 262, pp. 121–146.
springerlink.com © Springer-Verlag Berlin Heidelberg 2010

introduced the term *selector*. Some rule learning systems, such as CN2 (Clark and Niblett, 1989), use the term *attribute test*. Others, like RIPPER (Cohen, 1995), prefer the more general term *condition*. In the descriptive induction framework of association rule learning, features correspond to *items*, and conjunctions of features correspond to *itemsets*. For example, items `apple` and `pear` correspond to features `Fruit = apple` and `Fruit = pear` in classification rule learning. Applications related to inductive logic programming and relational rule learning typically borrow the term *literal* from the logic programming terminology. Conjunctions of literals are also known as *templates* or *schemata*. The term *feature* as defined in this chapter, integrates properties of all these terms in the context of rule learning.

Note that in this work we clearly discriminate between the terms 'feature' and 'attribute'. This conflicts with the common practice in the machine learning community to use terms like *feature extraction*, *feature construction* or *feature selection* for approaches that aim at finding a suitable set of descriptors for the training examples by including expert knowledge, increasing the quality of learning or the expressiveness of the hypothesis language. As most learning algorithms, such as decision tree learners, focus on attributes, the term *feature* is in this context frequently used as a synonym of *attribute*.

The general form of a classification rule is IF *Conditions* THEN *Class*, or alternatively, *Class* ← *Conditions*. The *Conditions* part of the rule, also called the rule body, is a logical combination of features, where a feature is some test that checks whether the object to classify has (or does not have) some property. Because a complete hypothesis is built as a set of rules and each rule is built as a combination of features, features are indeed the basic building elements used in the rule construction process. In this way features and their properties significantly determine the form and quality of knowledge that can be induced by rule learning algorithms. Formally speaking, features, their properties and the way they are defined are a part of the hypothesis language definition.

As noted above, a *feature* describes a property of an instance, which can be used to form complex logical concept descriptions. The instance either does or does not have this property. Thus, features are always Boolean-valued, i.e., either *true* or *false*. Features can be simple literals that test a value of a single attribute, like $A_i > 3$, or they can represent complex logical and numerical relations, integrating properties of multiple attributes, like $A_k < 2 \cdot (A_j - A_i)$, as illustrated in Table 1.

Features are defined as logical tests of attribute values. As features are logical variables they are very different from attributes, which are either numerical variables (with values like 7 or 1.5) or nominal or discrete variables (with values like *red* or *female*). Also, in contrast to attributes, a feature cannot have a missing or unknown value. Note that features are different from binary attributes even for binary attributes that have values *true* or *false*. For example, in Table 2 for binary attribute A_m, acceptable features are $A_m = \textit{true}$ or $A_m = \textit{false}$, but also more complex features like $A_m \neq A_n$ or $A_m = \textit{true} \wedge A_n = \textit{true}$, where $\wedge$ denotes the logical conjunction.

Propositional learning is sometimes also called attribute-based (or attribute-value) learning with the intention to stress the difference to the clausal logic used in relational rule learning. But although the term attribute-based learning is justified for decision

Table 1. Illustration of simple and complex features in a domain with three instances described by three continuous attributes

Attributes			Features	
A_i	A_j	A_k	$A_i > 3$	$A_k < 2 \cdot (A_j - A_i)$
7	1.5	2	*true*	*false*
4	3	-4	*true*	*true*
1.07	2	0	*false*	*true*

Table 2. Examples of features defined on a set of three instances with two binary attributes

Attributes		Features			
A_m	A_n	$A_m = true$	$A_m = false$	$A_m \neq A_n$	$A_m = true \wedge A_n = true$
true	*true*	*true*	*false*	*false*	*true*
true	*false*	*true*	*false*	*true*	*false*
false	*false*	*false*	*true*	*false*	*false*

tree learning, rule learning is better described as feature-based learning, with features being the basic rule building blocks. This can, maybe, best be seen from the different definitions of information gain heuristics for decision tree learning (Quinlan, 1986) and rule learning (Quinlan, 1990), the former evaluating attributes, the latter evaluating features.

For many rule learning systems it is not so obvious that they are based on features because the process of transforming attributes into features is implicit, as it is integrated into the rule learning procedure (Clark and Niblett, 1989; Cohen, 1995). The main reason for this integration is the simplicity and memory usage efficiency. Explicit usage of features requires that relatively large feature *covering tables* are constructed (see Section 2.2). Nevertheless, there are rule learning algorithms that explicitly construct covering tables before starting the rule construction process. A classical example is the LINUS system for converting relational learning problems into the propositional form (Lavrač et al., 1991; Lavrač and Džeroski, 1994; Lavrač and Flach, 2001). In this framework also the concept of literal relevancy has been introduced by Lavrač et al. (1999).

In this work, we define the concept of explicit features both for propositional and (multi-)relational domains, describe procedures for transforming a set of attributes into a set of features, prove relevant relations that hold among features in two-class learning problem domains, and point out practical advantages of explicit feature covering table computation. Besides the concept of feature relevancy which enables a significant reduction of the number of features that effectively enter into the rule construction process, we describe also a novel approach to handling unknown or missing attribute values. In contrast to standard approaches which substitute an unknown attribute value with an adequate value estimate (Bruha and Franek, 1996), our approach is based on setting appropriate feature truth values which should prevent that unknown values are used for discriminating between the examples from different classes. We further extend the concept by claiming that known numerical attribute values should be, under specified conditions of expected attribute imprecision, treated as unknown values. The approach opens a possibility to handle continuous attributes more adequately for real world problem domains.

The organization of this chapter is as follows. In Section 2, propositional features are defined and an algorithm for the construction of a set of simple features is described. Section 3 presents relational features that are used for the representation of properties of structured objects. Feature relevancy, a general concept that is applicable to both propositional and relation features, is described in Section 4. This section includes a theoretical analysis of the concept of features, feature relevancy, and discriminating pairs of examples, including theorems defining relations among features, relations between features and complete and consistent rule sets, and properties of the presented algorithm for simple propositional feature construction. The paper concludes by presenting a novel approach to handling of unknown and imprecise attribute values during feature covering table construction in Section 5.

2 Propositional Features

The construction of features for a given set of training examples is the first step in the rule learning process. This step can be viewed as a transformation from the attribute space to the space of features. Although the construction of simple features is a straightforward and rather simple task, the construction of an appropriate set of more sophisticated features is harder. A part of the problem is that for a given set of attributes, the set of possible features may be very large. This is particularly true in situations when features include numerical relations based on more than one continuously valued attribute. If in such a situation the complexity of relations is not strictly limited, the number of potentially interesting features can explode. The second part of the problem is the fact that the appropriateness of the set of features is domain-dependent.

Discovering interesting features can be viewed as a separate knowledge discovery task. In the machine learning literature, automated feature construction is known as the *constructive induction* problem (Matheus, 1989; Mehra et al., 1989) which is out of the scope of this work. In this section, we present a straightforward approach to the construction of a set of simple features from a training set. It is assumed that more complex features representing background or domain expert knowledge, or the results of the constructive induction process, may always be added to this set of simple features.

2.1 Constructing Features from Attributes

An algorithm enabling the construction of simple features from attributes describing a two-class classification domain is presented in Figure 1 (Gamberger and Lavrač, 2002). The algorithm constructs features independently for each attribute. If an attribute is discrete then all distinct values in positive examples are detected, and features of the form $Att = value$ are constructed for all found values. Also, all distinct values for negative examples are detected, and features of the form $Att \neq value$ are constructed from them. Thus, the number of generated features for a discrete attribute is equal to the sum of distinct attribute values occurring in positive and negative examples.

procedure `ConstructFeatures(`E`)`

Input:
$E = Pos \cup Neg$: a set of positive and negative training examples

Procedure:
$FeatureSet := \emptyset$
for each attribute A_i **do**
  V := set of all different values of A_i appearing in positive examples
  W := set of all different values of A_i appearing in negative examples
  if A_i is a discrete attribute
    for each $v \in V$ **do** $FeatureSet := FeatureSet \cup \{A_i = v\}$
    for each $w \in W$ **do** $FeatureSet := FeatureSet \cup \{A_i \neq w\}$
  endif
  if A_i is a continuous attribute
    for each possible pair (v, w), $v \in V$ and $w \in W$
      if $v < w$ and does not exist $y \in V \cup W$ such that $v < y < w$ **do**
        $FeatureSet := FeatureSet \cup \{A_i < \frac{v+w}{2}\}$
      if $w < v$ and does not exist $y \in V \cup W$ such that $w < y < v$ **do**
        $FeatureSet := FeatureSet \cup \{A_i \geq \frac{v+w}{2}\}$
  endif
endfor

Output: $FeatureSet$

Fig. 1. Algorithm for the transformation of a set of attributes into a set of simple features

For a continuous attribute, features are generated by first identifying pairs of neighboring values, where neighboring means that there is no other value between them. From these pairs, the mean of the two neighboring values (*mean_value*) is first computed. If there are two neighboring values from different classes there are two options: first, if the smaller of the two values is from the positive class, feature $Att < mean\_value$ is constructed,and second, if the smaller of the values is from the negative class, feature $Att \geq mean\_value$ is constructed.[1] The number of features generated for a continuous attribute depends on the grouping of classes in the increasing value list but typically the number of constructed features is proportional to the number of examples.

For the domain presented in Table 3, consisting of two positive examples $\{p_1, p_2\}$ and three negative examples $\{n_1, n_2, n_3\}$, the feature construction algorithm generates the following features: for attribute A_1, features $A_1 = red$ and $A_1 = blue$ for values detected in the positive examples, and features $A_1 \neq green$ and $A_1 \neq red$ for A_1

[1] Note that this part of the feature construction algorithm can be implemented in a more efficient way, similar to the algorithm suggested by Fayyad and Irani (1992), by first sorting the attributes values, independently of the example class in an increasing value list, and then considering only neighboring pairs in this list (cf. Figure 2.1). The procedure becomes slightly more complicated when multiple examples with different class labels have identical attribute values.

Table 3. A domain consisting of two positive and three negative examples, described by three attributes A_1, A_2, and A_3 (left), and a covering table generated for this domain (right)

	Original Data			Covering Table					
Examples	A_1	A_2	A_3	$A_1 = red$	$A_1 \neq red$	$A_2 \geq 0.9$	$A_2 < 1.7$	$A_3 \geq 1.5$	...
p_1	red	1.2	4	*true*	*false*	*true*	*false*	*true*	...
p_2	blue	1.3	2	*false*	*true*	*true*	*false*	*true*	...
n_1	green	2.1	3	*false*	*true*	*true*	*true*	*true*	...
n_2	red	2.5	1	*true*	*false*	*true*	*true*	*false*	...
n_3	green	0.6	1	*false*	*true*	*false*	*false*	*false*	...

values detected in the negative examples. For attribute A_2 the algorithm generates features $A_2 \geq 0.9$ and $A_2 < 1.7$ as shown in Figure 2.1. For attribute A_3 the algorithm generates three features: $A_3 \geq 1.5$ for neighboring increasingly ordered value pair (1,2), $A_3 < 2.5$ for value pair (2,3), and $A_3 \geq 3.5$ for value pair (3,4).

It is important to note that, for continuous attributes, only features of the form $A_i \geq x$ and $A_i < x$ are generated. The form $A_i = value$ is typically not considered. The reason is that for most continuous attributes, it does not make much sense to test the exact value, e.g., to test if a person is exactly 20 years old or if a stick is exactly 1 meter long. It makes more sense to know if a person is younger or older than 20 years or if the stick is shorter or longer than 1 meter. Besides, there is always a problem of measurement precision, e.g., the question whether feature $Length = 1$ is true for sticks that are 1.1, 1.01, and 1.001 meters long. For feature $Length \geq 1$ the answer is positive for all three cases while for $Length = 1$ the answer depends on the acceptable tolerance that is typically not known. If we need to refer to the length of 1 meter, it is a better practice to use a conjunction of two simple features, like $(Length \geq 0.95) \wedge (Length < 1.05)$, because in this situation an accepted tolerance of 0.05 is explicitly defined. Such a complex feature can be automatically detected as relevant during rule body construction. In this situation the rule will give information that the length of 1 meter is relevant but also define the required measurement precision. Besides, if one knows in advance that the length of 1 meter is relevant it is possible to include such complex features as background knowledge.

There are some cases when it is justified to construct features of the form $A_i = v$ for continuous attributes. This is the case when attribute values are integers representing distinct, well specified and disjunctive concepts, e.g. floors in a building or school years as educational levels. In these cases, it is advisable to treat such attributes as being

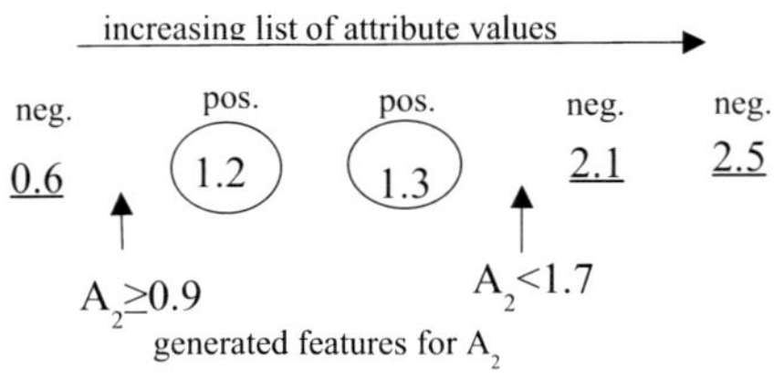

Fig. 2. Illustration of the feature construction process for continuous attribute A_2 of Table 3

both continuous and discrete, resulting in both types of features. An example may be attribute A_3 in Table 3 for which, in addition to the already mentioned features, one can construct also the following features: $A_3 = 4$, $A_3 = 2$, $A_3 \neq 3$, and $A_3 \neq 1$.

2.2 Covering Tables and Rule Induction

A *covering table* is a table which has examples as its rows and features as its columns. The right-most part of Table 3 presents a part of the covering table constructed for simple features generated for the domain presented in its left part. It can be noticed that the covering table has a much larger number of columns than the corresponding table presenting the attributes. Consequently, the conversion from the attribute to the feature space results in a significant expansion of the space complexity.

The covering table contains only values *true* and *false* as its elements. These truth values represent the covering properties of features on the set of available examples. Together with example classes, the covering table is the basic information necessary for the rule induction process. Actual rule construction from the available set of features can then be performed by any rule learning algorithms, regardless of whether exhaustive or heuristic search is used, what rule quality evaluation measures are used, and how rule pruning is performed. For all these steps we need only the information about the features contained in covering tables and the actual attribute values are no longer needed for rule construction.

A significant difference of such *explicit* feature construction is that we get all the features with the corresponding covering properties from the constructed covering table, while standard rule learning approaches construct features during the rule learning process (*implicit* feature construction). The price of explicit feature construction is reflected in the fact that we need to separate feature construction from the rule construction process and to save feature covering properties. The later represents the increase of the space complexity of rule learning algorithms. The advantages of explicit introduction of features are a possibility to use the same rule building blocks both for propositional and relational rule learning (described in Section 3), introduce feature relevancy and ensure that only most relevant features enter the rule learning process (Section 4), and to solve problems of handling unknown attribute values and imprecision of continuous attributes during feature covering table construction (Section 5).

3 Relational Features

Michalski (1980) first recognized that features (or selectors in his terminology) could not only be propositional tests for attribute values but could also represent relations between two or more attributes. In structured domains, entity- relationship diagrams representing the data model may provide the basis for defining features to be used by relational rules learning algorithms. For example, Figure 3(a) shows Michalski's East-West challenge problem (Larson and Michalski, 1977; Michalski, 1980). It can be described by the entity-relationship diagram presented in Figure 3(b) and data stored in three relational tables presented in Figure 3(c).

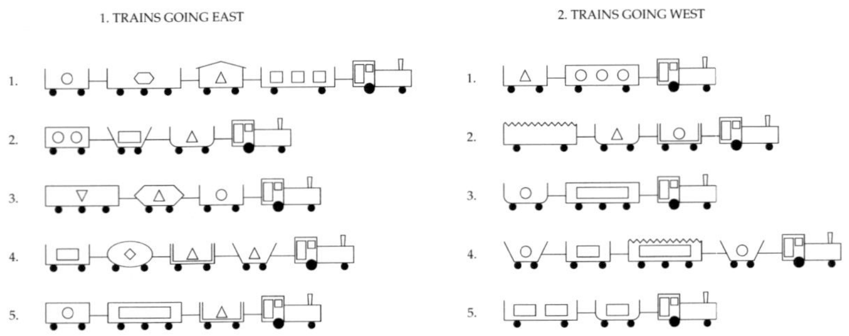

(a) Schematic illustration of the 10 examples.

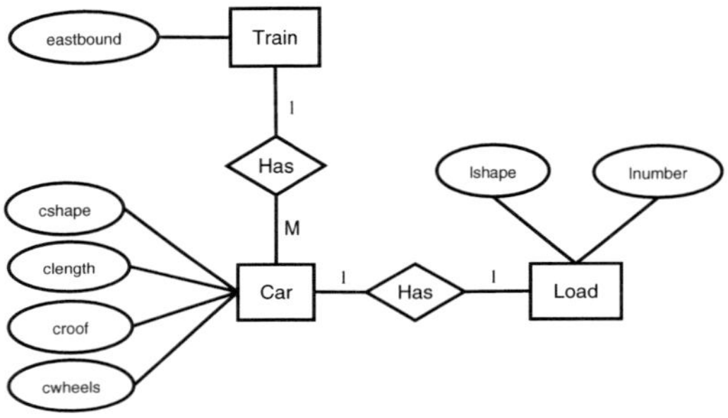

(b) The Entity-Relationship diagram.

TRAIN

trainID	eastbound
t1	true
...	...
t5	true
t6	false
...	...

LOAD

loadID	lshape	lnumber	car
l11	rect	3	c14
l12	tria	1	c13
l13	hexa	1	c12
l14	circ	1	c11
...	...	...	...

CAR

carID	cshape	clength	croof	cwheels	train
c11	rect	long	no_roof	2	t1
c12	rect	short	peak	2	t1
c13	rect	long	no_roof	3	t1
c14	rect	short	no_roof	2	t1
...	...	...	...	...	...

(c) A relational database representation.

Fig. 3. Michalski's East-West trains learning problem

Consider the following two clauses[2] that describe Eastbound trains.

```
eastbound(T) :-
   hasCar(T,C), clength(C,short), not croof(C,no_roof).
eastbound(T) :-
   hasCar(T,C1), clength(C1,short),
   hasCar(T,C2), not croof(C2,no_roof).
```

The first clause expresses that trains which have a short closed car are going East. The second clause states that trains which have a short car and a closed car are going East. The second clause is clearly more general than the first, covering all instances that the first one covers, and in addition instances where the short car is different from the closed car.

We say that the body of the first clause consists of a single *relational feature*, while the body of the second clause contains two distinct features. Formally, a feature is defined as a minimal set of literals such that no local (i.e., existential) variable occurs both inside and outside that set. The main point of relational features is that they localize variable sharing: the only variable which is shared among features is the global variable occurring in the rule head. This can be made explicit by naming the features:

```
hasShortCar(T)  :- hasCar(T,C), clength(C,short).
hasClosedCar(T) :- hasCar(T,C), not croof(C,no_roof).
```

Using these named features as background predicates, the second clause above can now be translated into a clause without local variables:

```
eastbound(T) :- hasShortCar(T), hasClosedCar(T).
```

This clause only refers to properties of trains, and hence could be captured extensionally by a single table describing trains in terms of those properties. It is therefore called a *semi-propositional* rule (it is not strictly a propositional rule, because propositional logic does not have variables).

3.1 Relational Feature Construction

LINUS (Lavrač et al., 1991; Lavrač and Džeroski, 1994; Lavrač and Flach, 2001) is a relational rule learning algorithm, performing *propositionalization* by transforming a multi-relational learning problem into a problem of learning from a single covering table.

Basic Relational Feature Construction with LINUS. The LINUS algorithm is based on the idea that the use of background knowledge can introduce features for learning relational rules (Lavrač et al., 1991). Through the construction of simple relational features, the learning problem can be transformed from a relational to a propositional form, and solved by a propositional learning algorithms.

[2] The two clauses are described in the Prolog programming language format, where :- stands for logical implication ←, and the comma stands for the logical conjunction.

Positive examples	Negative examples	Background knowledge	
`daughter(mary,ann).`	`daughter(tom,ann).`	`parent(ann,marry).`	`female(ann).`
`daughter(eve,tom).`	`daughter(eve,ann).`	`parent(ann,tom).`	`female(mary).`
		`parent(tom,eve).`	`female(eve).`
		`parent(tom,ian).`	

(a) Training examples and background knowledge for learning the `daughter` relation.

	Variables		Features					
`d(X,Y)`	`X`	`Y`	`f(X)`	`f(Y)`	`p(X,X)`	`p(X,Y)`	`p(Y,X)`	`p(Y,Y)`
true	`mary`	`ann`	*true*	*true*	*false*	*false*	*true*	*false*
true	`eve`	`tom`	*true*	*false*	*false*	*false*	*true*	*false*
false	`tom`	`ann`	*false*	*true*	*false*	*false*	*true*	*false*
false	`eve`	`ann`	*true*	*true*	*false*	*false*	*false*	*false*

(b) Resulting covering table.

Fig. 4. Learning the `daughter` relation problem with LINUS

As an illustration, consider a simple relational problem of learning family relations. The task is to define target relation `daughter(X,Y)`, which states that person `X` is daughter of person `Y`, in terms of background knowledge relations `female`, `male` and `parent`. All the variables are of type `person`, defined as `person` = {`ann,eve,ian,mary,tom`}. There are two positive and two negative examples of the target relation and some background knowledge, as shown in Figure 4(a).

Relational features are constructed as follows. All the possible applications of the background predicates on the arguments of the target relation are determined, taking into account argument types. Each such application introduces a new feature. In our example, all variables are of the same type `person`. The corresponding covering table is shown in Table 4(b), where `d` stands for `daughter`, `f` stands for `female` and `p` for `parent`. The tuples in the covering table are generalizations (relative to the given background knowledge) of the individual facts about the target relation.

In Table 4(b), 'Variables' stand for the arguments of the target relation, and 'Features' denote the newly constructed relational features. When learning function-free clauses, only the truth values of relational features are considered for learning.

Extended Relational Feature Construction with LINUS. The above approach is only feasible for a restricted class of structural learning problems. Thus, the hypothesis language of LINUS was initially restricted to function-free program clauses which are *typed* (each variable is associated with a predetermined set of values), *constrained* (all variables in the body of a clause also appear in the rule head) and *nonrecursive* (the predicate symbol of the head does not appear in any of the literals in the body), i.e., to function-free constrained deductive hierarchical database (DHDB) clauses.

We now describe a generalization of this approach, which is inspired by the entity-relationship data model and has been used by several ILP systems employing propositionalization (Flach and Lachiche, 1999; Lavrač and Flach, 2001). It is easy to define a simple general-to-specific search algorithm for constructing relational features.

In the definition of the feature bias, we distinguish between two types of predicates:

1. *structural* predicates, which "introduce" variables, and
2. *properties*, which are predicates which "consume" variables.

From the perspective of the entity-relationship data model, predicates defining properties correspond to attributes of entities, while structural predicates correspond to relationships between entities. A *relational feature* describing an individual is defined as a conjunction of structural predicates and properties as follows:

1. there is exactly one free variable which will play the role of the global variable in rules;
2. each structural predicate introduces a new existentially quantified local variable, and uses either the global variable or one of the local variables introduced by other structural predicates;
3. properties do not introduce new variables (this typically means that one of their arguments is required to be instantiated);
4. all variables are used either by a structural predicate or a property.

The actual relational feature construction can be restricted by parameters that define the maximum number of literals constituting a feature, maximum number of variables, and the number of occurrences of individual predicates. The following relational feature can be constructed in the trains example, allowing for 4 literals and 3 variables:

```
trainFeature42(T) :-
    hasCar(T,C), hasLoad(C,L), lshape(L,tria).
```

As shown above, a typical feature contains a chain of structural predicates, closed off by one or more properties. Properties can also establish relations between parts of the individual: e.g., the following feature expresses the property of "having a car whose shape is the same as the shape of its load".

```
trainFeature978(T) :-
    hasCar(T,C),cshape(C,CShape),
    hasLoad(C,L),lshape(L,LShape),
    CShape = LShape.
```

We demonstrate LINUS by applying it to the East-West trains example. Propositionalization is performed by providing LINUS with relational features defined as background predicates, such as "has a car with a triangular load". In the experiment below we simply provided LINUS with all the features that use at most two structural predicates introducing local variables, and at most two predicates describing properties. There are 190 such features, whose truth values are recorded in a single covering table (Table 4).

Taking Table 4 as input (using the first column as the class attribute), a propositional rule learner such as CN2 can be used to induce if-then rules like the following:

```
IF    f1(T) = true  ∧  f4(T) = true
THEN  eastbound(T) = true
```

Table 4. Covering table for the Eastbound trains learning problem

	Variable	Features					
eastbound(T)	T	f1(T)	f2(T)	f3(T)	f4(T)	...	f190(T)
true	t1	*true*	*false*	*false*	*true*	...	*false*
true	t2	*true*	*true*	*false*	*true*	...	*true*
	...	...	...	...	...	...	
false	t6	*false*	*false*	*true*	*true*	...	*true*
false	t7	*true*	*false*	*false*	*false*	...	*false*
	...	...	...	...	...	...	

Translated back to a Prolog rule format, the above rule has the following form:

```
eastbound(T) :- f1(T), f4(T).
```

The actual result of learning Eastbound trains by applying CN2 on the covering table with 10 rows and 190 columns describing the dataset from Figure 3(a) are the rules shown below (with feature names edited for clarity).

```
eastbound(T,true) :-
   hasCarHasLoadSingleTriangle(T),
   not hasCarLongJagged(T),
   not hasCarLongHasLoadCircle(T).
eastbound(T,false) :-
   not hasCarEllipse(T),
   not hasCarShortFlat(T),
   not hasCarPeakedTwo(T).
```

4 Feature Relevancy

In Section 2 we have presented an approach to feature construction in propositional tasks while in Section 3 we have discussed construction of relational features in multi-relational tasks. Regardless of the way features have been constructed, conditions for feature relevancy are the same. The feature relevancy concept is important because the detection and elimination of irrelevant features can lead to more effective and efficient rule learning in both frameworks. The presentation begins by discussing the notion of sufficiently rich feature sets.

4.1 Sufficiently Rich Feature Sets

A sufficiently rich feature set for a given training set of examples is a set of features from which a complete and consistent hypothesis can be built. Although the construction of complete and consistent rule sets is seldom the ultimate goal, we will investigate the theoretical properties of sufficiently rich feature sets for the construction of such rule sets, because they have some important consequences.

We will start by defining the concept of p/n pairs of examples.

Definition 1 (p/n pair). *A p/n pair is a pair of training examples p_i/n_j where $p_i \in Pos$ and $n_j \in Neg$, where Pos is a set of positive, and Neg a set of negative examples.*

Definition 2 (Discriminating feature[3]). *Let F denote a set of features. Feature $f \in F$ discriminates a p/n pair p_i/n_j iff feature f correctly classifies both examples, i.e., if feature f has value* true *for p_i and value* false *for n_j.*

From Definition 2 we can see that a good feature—that is *true* for many positive examples and *false* for many negative examples—discriminates many p/n pairs.

Theorem 1. *For training set E and a set of features F, a complete and consistent hypothesis H can be found using only features from set F if and only if for each possible p/n pair from training set E there exists at least one feature $f \in F$ that discriminates the p/n pair.*

Proof. Necessity: Suppose that a p/n pair exists that is not discriminated by any feature $f \in F$. Then no rule built of features from F will be able to discriminate between these two examples. Consequently, a description which is both complete and consistent can not be found.

Sufficiency: Assume that for each possible p/n pair from training set E there exists at least one feature $f \in F$ that discriminates the p/n pair. Take a positive example p_i. Select from F the subset of all features F_i that are *true* for p_i. We can now construct a conjunctive description R_i, $R_i = f_{i,1} \wedge \ldots \wedge f_{i,m}$, for $f_{i,1}, \ldots f_{i,m} \in F_i$. Each R_i will thus be *true* for p_i (by construction), and *false* for all $n_j \in Neg$, because, by assumption, for each n_j there must be at least one feature f_{ij} that is *true* for p_i and *false* for n_j. Clearly, $f_{ij} \in F_i$. This construction can be repeated for each positive example in Pos, resulting in a complete and consistent hypothesis $H = \{R_1, \ldots, R_k\}$ where k is the cardinality of the set of positive examples ($k = |P|$).

4.2 Concept of Feature Relevancy

Theorem 1 is important as it clearly shows that when deciding about the relevancy of features it is crucial to detect which p/n pairs can be discriminated (correctly classified) by the feature. The immediate consequence is the conclusion that features that do not discriminate any p/n pairs are useless for rule construction. Such features are called *totally irrelevant* and can be immediately removed from any feature set because they can not help in building rules that are *true* for positive examples and *false* for negative examples.

The relevancy concept and a possibility to detect and eliminate some features as irrelevant even before entering the rule construction process is important also for other reasons. The first is that it enables complexity reduction of the rule construction task. The second is that the elimination of irrelevant features is useful for overfitting avoidance by reducing the search space of hypotheses through the elimination of features and their combinations with low covering properties. The third is that through constructive

[3] In previous work, the term 'coverage' of p/n pairs was used instead of p/n pair 'discrimination'.

induction the learner can introduce many potentially interesting features while the relevancy approach can be used to eliminate—already in preprocessing—all the features that do not have a chance to be useful in the rule construction process.

The importance of considering feature relevancy goes beyond preprocessing and reducing the set of features that enter the rule construction process. Logical combinations of features may again be features, and complete rule bodies can also be complex features. Thus, the relevancy of features is also applicable to rules and complex features. The consequence is that feature relevancy can be integrated into the covering algorithm with the intention to reduce the hypothesis search space and to speed up the algorithm execution.

4.3 Total Irrelevancy

Definition 3 (Total irrelevancy). *Feature $f \in F$ which does not discriminate any p/n pair is totally irrelevant.*

Note that for checking whether a feature is totally irrelevant, we do not need to construct all p/n pairs.

Theorem 2. *A feature f is totally irrelevant iff it is* false *for all positive examples or* true *for all negative examples.*

Proof. A feature is totally irrelevant, if it does not discriminate any p/n pairs. This can be the case if it does not cover any positive examples. On the other hand, if it covers at least one positive example, it must also cover all negative examples because otherwise there would exist a p/n pair that is discriminated by f.

This situation can be illustrated in the *coverage space* which is a variant of the ROC space with non-normalized axes (Fürnkranz and Flach, 2005). A point in the ROC space (ROC: Receiver Operating Characteristic) (Provost and Fawcett, 2001) shows the classifier performance in terms of false alarm or *false positive rate* $FPr = \frac{|FP|}{|TN|+|FP|} = \frac{|FP|}{|N|}$ (plotted on the X-axis), and sensitivity or *true positive rate* $TPr = \frac{|TP|}{|TP|+|FN|} = \frac{|TP|}{|P|}$ (plotted on the Y-axis). A point (FPr, TPr) depicting rule R in the ROC space is determined by the covering properties of the rule. The ROC space is appropriate for measuring the success of the rule learning process. This is illustrated in the left-hand side of Figure 5.

An alternative to the ROC space is the so-called coverage space (see the right-hand side of Figure 5), where FPr on the X-axis is replaced by $|FP|$ and TPr on the Y-axis by $|TP|$.[4] The coverage space is equivalent to the ROC space when comparing the quality of subgroups induced in a single domain. The reminder of this chapter considers only this simpler coverage space representation.

Figure 6 shows the position of totally irrelevant features in the coverage space. Such features have either the number of true positive cases equal zero ($|TP| = p = 0$) or

[4] The covering space can be turned into the ROC space by simply normalizing the TP and FP axes to the [0,1]x[0,1] scale.

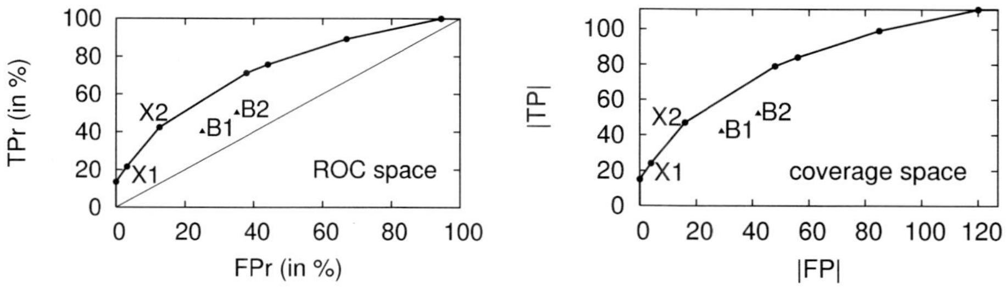

Fig. 5. The left-hand side figure shows the ROC space with a convex hull formed of seven rules that are optimal under varying TPr/FPr trade-offs, and two suboptimal rules B1 and B2. The right-hand side presents the positions of the same rules in the corresponding coverage space.

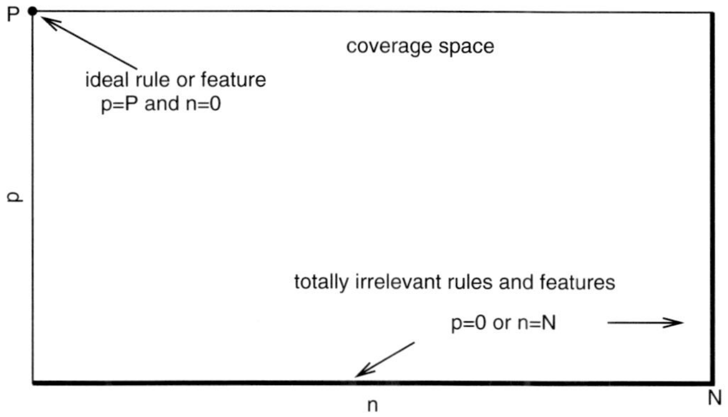

Fig. 6. The position of ideal rules or features and the position of totally irrelevant rules and features in the coverage space

the number of true negative cases equal zero ($|TN| = \bar{n} = 0$ or $|FP| = n = N$). The figure is also used to show the position of an ideal complete and consistent rule in point $(0, P)$ of the coverage space.

Note that one must not confuse feature irrelevancy with attribute irrelevancy. If we have a medical test T that perfectly recognizes cancer patients (i.e., T is *positive* if the patient has cancer, and is *negative* if the patient does not have cancer), one can be misled to believe that T is maximally relevant for the task of recognizing cancer, but totally irrelevant for the task of recognizing non-cancer. However, the correct interpretation is that T actually corresponds to two features: $T = positive$ is a relevant feature for predicting cancer and it is irrelevant for predicting non-cancer, whereas the opposite is true for $T = negative$.

Note that feature relevancy sheds additional light on the `ConstructFeatures` algorithm (Figure 1).

Theorem 3. *For a discrete attribute, all simple features consisting of a single attribute value that are not constructed by the* `ConstructFeatures` *algorithm are totally irrelevant.*

Proof. Assume a discrete attribute A.

The `ConstructFeatures` algorithm does not construct a feature $A = v_j$ if v_j does not occur in the positive examples, i.e., if v_j only occurs in the negative examples. The value of the feature $A = v_j$ is therefore *false* for all positive examples, and the feature is totally irrelevant by Theorem 2.

Analogously, a feature $A \neq v_j$ is not constructed if v_j does not occur in the negative examples. Thus, $A \neq v_j$ has the value *true* for all negative examples, and is therefore totally irrelevant.

A similar theorem regarding continuous attributes will be proved in the next section.

4.4 Relative Irrelevancy

The previous section showed that the quality of features depends on their usefulness for building rules, which in turn depends on their ability to correctly classify p/n pairs of examples. Recall that totally irrelevant features do not discriminate any p/n pair and can consequently be immediately excluded from the rule induction process. But, besides total irrelevancy, weaker forms of feature irrelevancy exist.

In the following, we denote with $PN(f)$ the set of p/n pairs discriminated by feature f, with $TP(f)$ the set of correctly covered/classified positive examples, and with $TN(f)$ the set of negative examples not covered, thus correctly classified by feature f.

Definition 4 (Relative irrelevancy). *If $PN(f) \subseteq PN(f_{rel})$ for two features $f, f_{rel} \in F$, we say that f_{rel} is* more relevant *than f and that feature f is* relatively irrelevant *because of the existence of feature f_{rel}. We denote this by $f_{rel} \succeq f$.*

Table 5 illustrates the concept of relative relevance. Let features f_1 and f_3 discriminate the following p/n pairs: $PN(f_1) = \{p_1/n_2, p_1/n_4, p_3/n_2, p_3/n_4\}$ and $PN(f_3) = \{p_1/n_1, p_1/n_2, p_1/n_4, p_3/n_1, p_3/n_2, p_3/n_4, p_5/n_1, p_5/n_2, p_5/n_4\}$, respectively. We say that f_3 is more relevant than f_1 because $PN(f_1) \subset PN(f_3)$.

Just as with total irrelevancy (Theorem 2), it is not necessary to actually construct p/n pairs for determining relative relevancy, even though the property of relative relevancy is based on the concept of p/n pairs.

Table 5. Feature f_3 is more relevant than f_1 and f_5 is more relevant than f_2 and f_4. Feature f_4 is also totally irrelevant because it is *true* for all negative examples.

	Features					
Examples	f_1	f_2	f_3	f_4	f_5	f_6
p_1	*true*	*false*	*true*	*false*	*true*	*false*
p_2	*false*	*true*	*false*	*true*	*true*	*false*
p_3	*true*	*false*	*true*	*false*	*false*	*true*
p_4	*false*	*true*	*false*	*true*	*true*	*false*
p_5	*false*	*false*	*true*	*false*	*true*	*false*
n_1	*true*	*false*	*false*	*true*	*false*	*true*
n_2	*false*	*true*	*false*	*true*	*false*	*true*
n_3	*true*	*true*	*true*	*true*	*true*	*false*
n_4	*false*	*true*	*false*	*true*	*true*	*false*

Theorem 4. *Feature f_{rel} is more relevant than feature f (i.e., $f_{rel} \succeq f$) iff $TP(f) \subseteq TP(f_{rel})$ and $TN(f) \subseteq TN(f_{rel})$. In terms of the sets of positive and negative examples covered by features f and f_{rel}, this is equivalent to $TP(f) \subseteq TP(f_{rel})$ and $FP(f_{rel}) \subseteq FP(f)$, where $FP(f)$ is the set of negative examples covered (incorrectly classified) by feature f.*

Proof. By definition, $f_{rel} \succeq f$ iff $PN(f) \subseteq PN(f_{rel})$. Suppose that $f_{rel} \not\succeq f$. Thus, there exists a p/n pair p_i/n_j that is discriminated by f, but not discriminated by f_{rel}. This implies that $p_i \in TP(f)$ and that $n_j \in TN(f)$, and that either $p_i \notin TP(f_{rel})$ or $n_j \notin TN(f_{rel})$. Both conclusions are contradictions to the assumptions that $TP(f) \subseteq TP(f_{rel})$ and $TN(f) \subseteq TN(f_{rel})$.

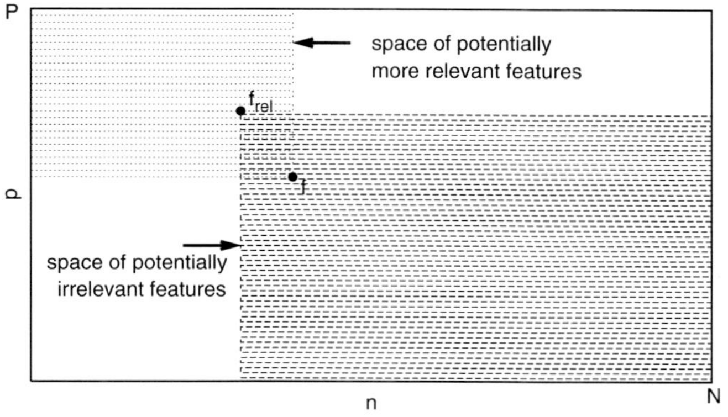

Fig. 7. The concept of relative relevancy in the coverage space

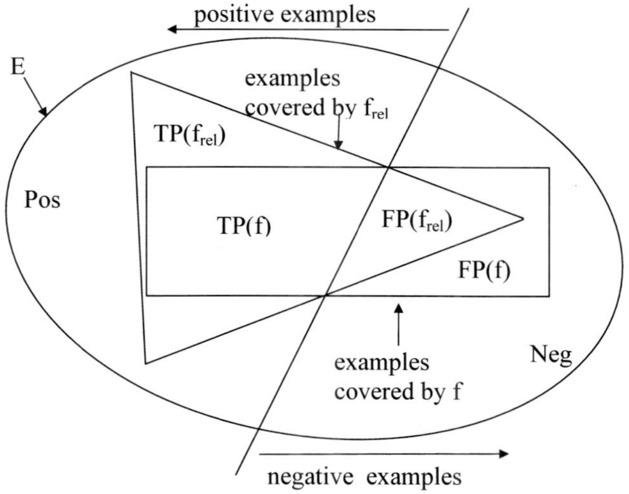

Fig. 8. Feature f_{rel} is more relevant than f iff $TP(f) \subseteq TP(f_{rel})$ and $TN(f) \subseteq TN(f_{rel})$. The later condition follows from $FP(f_{rel}) \subseteq FP(f)$.

A consequence of this theorem is that relative relevancy of features can be detected directly from covering tables. A more relevant feature should have value *true* at least for the positive examples for which the less relevant feature has value *true*, and for negative examples it should have value *false* at least in those rows of the covering table in which the less relevant feature is *false*.

Figure 7 illustrates the concept of relative relevancy in the coverage space. If $f_{rel} \succeq f$, then f_{rel} must lie in the rectangle that is spanned between the point $(0, P)$ and f. Note, however, that the coverage space does not actually show whether $TP(f) \subseteq TP(f_{rel})$ and $FP(f_{rel}) \subseteq FP(f)$, it only shows that the sizes of these sets $p(f) \leq p(f_{rel})$ and $n(f_{rel}) \leq n(f)$. Thus, the fact that f_{rel} lies in the rectangle spanned by $(0, P)$ and f does *not* imply that $f_{rel} \succeq f$. This is illustrated in Figure 8.

Relative irrelevancy is also important for understanding the `ConstructFeatures` algorithm (Figure 1). The following theorem has, in a different formulation, first been proved by Fayyad and Irani (1992).

Theorem 5. *For a continuous attribute, all simple features consisting of a single attribute value that are not constructed by the* `ConstructFeatures` *algorithm are relatively irrelevant.*

Proof. Assume a continuous attribute A. We will consider all features of the type $A < t$ and $A \geq t$ for all t being real numbers. Note that all such values between a pair of neighboring values (v, w) are equivalent in the sense that all $t : v \geq t < w$ cover the same positive and negative examples. Thus, the values $t = \frac{v+w}{2}$ that are considered by the algorithm are equivalent to all other values in this range.

Algorithm `ConstructFeatures` only adds values that are between pairs (v, w) where v originates from a positive example ($v \in V$) and w from a negative example ($w \in W$), or vice versa. Thus, the values that are ignored are those where both v and w occur in positive examples, or where both occur in negative examples.

Assume $v_1 < v_2 \in V$ are neighboring values both originating from positive examples, and f defined as $A \geq \frac{v_1+v_2}{2}$ is therefore ignored by `ConstructFeatures`. Let now $w \in W$ be the largest value from a negative example with $w < v_1 < v_2$ and let $v \in V$ be the smallest value with $w < v \leq v_1$. Note that feature f_{rel} defined as $A \geq \frac{v+w}{2}$ would be added by the algorithm. By construction, $TP(f_{rel}) \supset TP(f)$ because (at least) the positive example corresponding to the value v_1 will be covered by f_{rel} but not by f. Furthermore, f and f_{rel} cover the same set of negative examples, i.e., $FP(f) = FP(f_{rel})$. Thus, by Theorem 4, $f_{rel} \succeq f$.

Analogous arguments can be made for feature $A < \frac{v_1+v_2}{2}$, and for the case when both values originate from negative examples ($v_1, v_2 \in W$).

The importance of the concept of relative relevancy of features is due to the fact that a feature is not necessarily irrelevant because of its low number of true positive examples $p = |TP(f)|$ or low number of true negative examples $\overline{n} = |TN(f)|$, but because there exists another more relevant feature with better discrimination properties. A relevancy filter using the concept of relative relevancy of features eliminates only features that can be replaced by a more relevant feature. In this way it preserves the quality of rules that can be built from the reduced set of features. Besides that, and perhaps even more importantly from the point of view of avoiding overfitting, it ensures that rule learners will use only the best discriminating features available.

In order to demonstrate the quality-preserving property of relative relevancy, let F be a set of features such that it is possible to construct a complete and consistent rule set for the given training set. Furthermore, let F_{min} be the smallest among the subsets of F. According to Theorem 1, both F and F_{min} should be such that for every possible p/n pair from the training set there should be at least one feature in F_{min} that correctly discriminates the pair.

Theorem 6. *Let F' be a subset of F obtained by the elimination of all relatively irrelevant features from F. Then $|F'_{min}| = |F_{min}|$ where F'_{min} is the minimal subset of features from F' such that a complete and consistent hypothesis can be constructed.*

Proof. If all features that are elements of F_{min} are also elements of F' then F_{min} is identical to F'_{min} and their cardinality is equal. If there exists feature f, $f \in F_{min}$ and $f \notin F'$ then there exists feature $f_{rel} \in F_{min}$ so that $f_{rel} \succeq f$. We can now construct F'_{min} by replacing f with the corresponding f_{rel} for each feature $f \in F_{min}, f \notin F'$. By construction, $|F'_{min}| \leq |F_{min}|$. As F_{min} was minimal, $|F'_{min}| = |F_{min}|$ must hold. Moreover, from this set F'_{min} we must still be able to learn a complete and consistent solution because $f_{rel} \in F'_{min}$ discriminates at least the same p/n pairs as the corresponding $f \in F_{min}$.

In Theorem 6 we assumed that the starting set of features F allows to construct a complete and consistent hypothesis. This is not a necessary condition for the applicability of relative irrelevancy. Also for feature sets not satisfying this property, relative relevancy has the quality-preserving property reflected in the fact that the complexity of rules induced from reduced feature sets will not increase. This follows from the possibility to always substitute f by f_{rel}.

4.5 Constrained Relevancy

In order for a feature to be acceptable as a building block for a rule body that captures a genuine dependency between attribute values and classes, the feature itself must have appropriate covering properties on the training set. These can be defined in terms of user-defined support constraints.

Definition 5 (Constrained irrelevancy). *Feature f that has either $|TP(f)| < min_{TP}$ or $|TN(f)| < min_{TN}$ is irrelevant, for min_{TP} and min_{TN} being user defined constraints.*

For low values of min_{TP} and min_{TN}, feature f with $|TP(f)| < min_{TP}$ is *true* for a small number of target class examples, and feature g with $|TN(g)| < min_{TN}$ is *false* for a small number of non-target class examples. Such features discriminate a small number of p/n pairs, possibly by chance, so that it seems reasonable not to use features with either of these properties in the rule construction process. The part of the coverage space of features with constrained irrelevancy is represented by the shaded area of Figure 9.

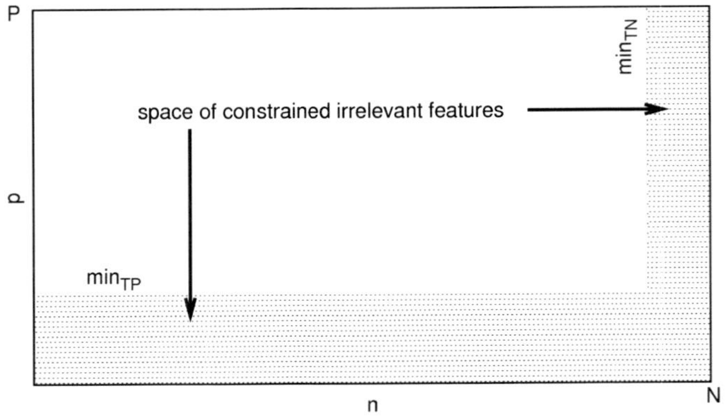

Fig. 9. The concept of constrained irrelevancy in the coverage space

Although the significance of rules is proportional to their distance from the diagonal in the coverage space, this property is not appropriate as a quality criterion for feature filtering. As logical combinations of features lying on the diagonal or very near to it can result in very significant conjunctions of features (rules), only relatively irrelevant features and constrained irrelevant features should be eliminated from the rule construction process.

As features are combined with conjunctions, a rule has at most as many correctly covered positive examples $|TP|$ as the smallest $|TP|$ value of the features forming the conjunction in the rule body. In contrast, the number of correctly classified negative examples $|TN|$ (number of negative examples not covered by the rule) will be at least as large as the largest $|TN|$ of the used features. This is the reason why min_{TP} is typically selected higher than min_{TN} and it can be as large as the minimal estimated number of examples that must be covered by a rule of acceptably high quality for the domain. The problem with constrained relevancy is that both min_{TP} and min_{TN} are user defined values and that no value can guarantee the actual relevancy of a feature.

5 Handling Unknown and Imprecise Attribute Values

In practical applications, not all attribute values are known, and learning algorithms have to be able to deal with missing information. A commonly used approach is to replace missing values with a default value. For example, CN2 (Clark and Niblett, 1989) replaces unknown values of discrete attributes by the most commonly occurring value (the *mode* value), and unknown values for continuous attributes by the average value (the *mean* value). This and similar techniques have drawbacks, especially when the number of unknown attributes is high.

Another problem encountered in practical rule learning applications is the problem of imprecise values. Imprecision is inherent to most non-integer continuous attributes. There are two main reasons for the necessity of imprecision handling. The first is that some continuous attributes are not or can not be measured with high precision (like some biological properties) or their values are significantly fluctuating (e.g., human blood pressure). In such cases, if the example values are very near to the decision values used in the features then the actual feature truth values for such examples are unreliable. The second reason is that although some attributes like income or human age can be known very precisely, building rules from features that have supporting examples very near to the decision values used in the features may be inadequate. The reasoning is that such features may lead to overfitting rules, or, in case of descriptive induction, may result in non-intuitive rules like that people of age 20 have significantly different properties from people of age 19 years and 11 months.

The approach presented in this section solves the problem of handling unknown/missing and imprecise values in the course of the covering table generation. It means that we do not have to replace the unknown and imprecise attribute values, but only to take care how to determine feature truth values in the cases when examples have unknown/imprecise values.

5.1 Unknown Attribute Values

The underlying principle in dealing with unknown values is that unknown attribute values should not affect the quality and potential usefulness of a feature for rule construction. We have seen that a good feature has a property of being *true* for many positive examples and *false* for many negative examples. From Theorem 1 we have learned that actually relevant is the set of discriminating p/n pairs. And an example pair is discriminated by the feature if it is *true* for the positive example p and *false* for the example n. From that it follows that if we do not know how to determine a correct truth value for a feature, an appropriate approach is to set its value to *false* if the unknown value occurs in a positive example and to value *true* if the unknown value is in a negative example. By reducing the number of discriminated p/n pairs we actually penalize features based on unknown attribute values, disabling them to appear in the constructed rules.

Table 6 shows examples of dealing with unknown values. The suggested approach is intuitively clear for positive examples: "If we do not know the attribute value, let us set all related feature values to *false*". In Table 6, by setting value *false* for example p_2 for feature $A_1 = red$, we prevent that p_2/n_3 is discriminated by this feature although n_3 has value *false*. This is inadequate for negative examples: by setting value *false* for unknown attribute values of negative examples, such negative example could form discriminating p/n pairs with all positive examples having *true* feature values. Therefore, for instance, in Table 6 we must set value *true* for feature $A_1 = red$ for example n_1, as otherwise the feature would discriminate the pair p_1/n_1, which is not wanted because the attribute value for n_1 is unknown.

The advantage of the proposed approach is that we do not have to change the rule induction process at all. Rules are constructed from completely specified features, rule quality measures may remain unchanged, and features can be combined in the rules in the same way as if all attribute values were known.

The approach can be directly applied also for complex features. For example, relation $A_k < 2 \cdot (A_j - A_i)$ from Table 1 will have an unknown value when any of attributes A_i, A_j, and A_k has an unknown value. The principle is easily extendable to features of any complexity combining both numerical and logical operations.

Table 6. In the case of an unknown attribute value, feature truth value is *true* for a negative example and *false* for a positive example

	Attributes			Features		
Examples	A_1	A_2	A_3	$A_1 = red$	$A_2 = green$	$A_3 < 1.6$ ...
p_1	red	?	1.2	*true*	***false***	*true*
p_2	?	green	?	***false***	*true*	***false***
n_1	?	blue	2.2	***true***	*false*	*false*
n_2	red	?	?	*true*	***true***	***true***
n_3	green	blue	2.0	*false*	*false*	*false*

At the end let us notice that the described approach to handling unknown attribute values can be applied also in rule induction algorithms that construct features during the rule construction process. But having explicit covering tables is a good strategy because it enables that the detection and elimination of irrelevant features before the rule induction process as the described technique for handling unknown attribute values can potentially result in many irrelevant features.

5.2 Imprecision of Continuous Attributes

An approach to dealing with inherent attribute imprecision, regardless of which type it is, is to treat attribute values near to the feature decision values as 'unknown' in order not to allow that such close values can affect feature quality. Such 'soft' unknown values are handled as standard unknown attribute vales as described in Section 5.1. It must be noted, however, that the actual attribute value is known, therefore it may happen that for a feature with some decision value it should be treated as an unknown value while for some other feature with a different decision value it may be treated as a regular known value. More precisely, the suggested approach to adequately dealing with continuous attributes is that in case of a feature decision value d we have to treat all attribute values v in the range $d - imp < v < d + imp$ as unknown attribute values, for imp being the user-estimated attribute imprecision boundary.

For illustration, in Table 6 there are attribute A_3 and feature $A_3 < 1.6$ that tests its value. The feature values in Table 6 for feature $A_3 < 1.6$ are determined assuming no or very small imprecision for attribute A_3. But if we assume that from the meaning of attribute A_3 or the way the data are collected we know that the expected imprecision of A_3 values is ± 0.5 then we have to take this into account when determining feature truth values. In this case it means that value 1.2 which is *true* for feature $A_3 < 1.6$ may be *false* in the case of imprecision of $+0.5$. A similar situation occurs with value 2.0 which is *false* for this feature but it may be *true* if the assumed imprecision of -0.5 really occurs. The appropriate approach is to treat both values as if they were unknown, resulting in value *false* for example p_1 and value *true* for example n_3.

The example presented in Table 7 demonstrates an artificial situation with two very similar continuous attributes such that from both ideal features discriminating all the 25 p/n pairs can be generated. But when the notion of imprecision of continuous attributes is introduced, the situation may change significantly. The right part of the table presents the covering properties of the same features but with imprecision 0.17. Now none of the

Table 7. Attribute values near the feature decision value are treated as unknown values. User defined constant *imp* defines the imprecision range. The example demonstrates the difference in feature covering properties for imprecision values 0.0 and 0.17, respectively. Attributes A_1 and A_2 are very similar but there are significant differences in feature covering properties when different expected attribute imprecision is taken into account.

	Attributes		Features ($imp = 0.0$)		Features ($imp = 0.17$)	
Examples	A_1	A_2	$A_1 < 1.95$	$A_2 < 1.95$	$A_1 < 1.95$	$A_2 < 1.95$
p_1	1.5	1.5	*true*	*true*	*true*	*true*
p_2	1.6	1.6	*true*	*true*	*true*	*true*
p_3	1.7	1.65	*true*	*true*	*true*	*true*
p_4	1.8	1.7	*true*	*true*	***false***	*true*
p_5	1.9	1.8	*true*	*true*	***false***	***false***
n_1	2.0	2.1	*false*	*false*	***true***	***true***
n_2	2.1	2.2	*false*	*false*	***true***	*false*
n_3	2.2	2.25	*false*	*false*	*false*	*false*
n_4	2.3	2.3	*false*	*false*	*false*	*false*
n_5	2.4	2.4	*false*	*false*	*false*	*false*

features is ideal. The first discriminates 9 and the second 16 p/n pairs. Although with assumed imprecision 0.0 both features seem equally good, with assumed imprecision 0.17 we have a clear preference for the second feature. By analyzing the attribute values it really seems that it is better to use attribute A_2 because it may turn out to be a more reliable classifier in an imprecise environment.

A good real world example for dealing with attribute imprecision are biological gene expression domains (Golub et al., 1999). In these domains there are thousands of attributes representing simultaneously measured gene expression values and it is known that the present technology does not enable precise and reliable measurement of gene expression values. In such situation, dealing with attribute imprecision as described in this section is suggested. Specially in cases with a very modest number of available training examples, the danger of overfitting is high and a good strategy is to try to prevent overfitting by selecting large attribute imprecision boundaries. In this way unreliable features supported by small differences between positive and negative examples will be excluded from the rule induction process.

Gene expression domains are interesting also because they contain attribute 'presence call values' representing discrete attributes constructed from actually measured continuous values by specialized software. There are three discrete attribute values: A (absent), M (marginal) and P (present). And although discrete attributes are out of the scope of this section, in this special situation when discrete values present an ordered set of levels, there is a possibility to treat marginal value M, lying between A and P values, as being in the area of imprecision, consequently treating the M value as value 'unknown'. The interpretation is that decision values between values A and P are somewhere in the M region, and that by selecting the imprecision boundary to be exactly equal to the whole M region, marginal values are considered to be unknown-valued. The motivation is the same as in the case of continuous attributes: we want that only large differences between positive and negative examples (it means only A and

Table 8. For attributes from gene expression domains there is a possibility to treat M (marginal) presence call values as unknown attribute values

	Attributes		Features			
Examples	A_1	A_2	$A_1 = A$	$A_1 = P$	$A_2 = A$	$A_2 = P$
p_1	A	A	*true*	*false*	*true*	*false*
p_2	M	P	***false***	***false***	*false*	*true*
p_3	P	M	*false*	*true*	***false***	***false***
n_1	P	P	*false*	*true*	*false*	*true*
n_2	A	M	*true*	*false*	***true***	***true***
n_3	M	M	***true***	***true***	***true***	***true***

P values) should affect the covering quality of features. The procedure is illustrated in Table 8.

6 Conclusions

We have defined propositional and relational features as the basic rule building blocks, and presented algorithms for propositional and relational feature construction. The most important advantage of explicit feature construction is the possibility to systematically handle the relevancy of features, by allowing to detect and eliminate many irrelevant features. For this purpose, we have shown that the notion of p/n pairs is crucial for detecting relevant feature covering properties. Also we have proved that the presented algorithm for the construction of propositional features constructs only potentially relevant features both for discrete and continuous attributes. Thus, feature sets constructed by this algorithm always allow the construction of complete and consistent rules if there are no contradictions among the training examples.

We have also demonstrated that explicit definition of features may be useful for systematic handling of unknown attribute values in the covering rule learning setting. In contrast to existing concepts which try to substitute the unknown value with some good approximation, we have solved handling of unknown values by appropriate definition of feature truth values. The approach is simple, applicable also to complex features based on many attributes, and conceptually well justified.

Finally, we proposed an approach to handling the inherent imprecision of continuous attributes, which is rarely analyzed in the rule learning framework, but which is essential for successful rule learning from real-life domains. Our proposal is to treat values near to the feature decision value as unknown attribute values. The approach relies on user estimation of expected imprecision levels and the possibility to effectively handle unknown attribute values through covering properties of constructed features.

Acknowledgments

This work was supported by Slovenian Ministry of Higher Education, Science and Technology project "Knowledge Technologies", German Science Foundation (DFG)

project FU 580/2 "Towards a Synthesis of Local and Global Pattern Induction (GLocSyn)", Croatian Ministry of Science, Education and Sport project "Machine Learning Algorithms and Applications", and EU FP6 project "Heartfaid: A knowledge based platform of services for supporting medical-clinical management of the heart failure within the elderly population".

References

Bergadano, F., Matwin, S., Michalski, R.S., Zhang, J.: Learning two-tiered descriptions of flexible concepts: The POSEIDON system. Machine Learning 8, 5–43 (1992)

Bruha, I., Franek, F.: Comparison of various routines for unknown attribute value processing: The covering paradigm. International Journal of Pattern Recognition and Artificial Intelligence 10(8), 939–955 (1996)

Clark, P., Niblett, T.: The CN2 induction algorithm. Machine Learning 3(4), 261–283 (1989)

Cohen, W.W.: Fast effective rule induction. In: Prieditis, A., Russell, S. (eds.) Proceedings of the 12th International Conference on Machine Learning (ML 1995), Lake Tahoe, CA, pp. 115–123. Morgan Kaufmann, San Francisco (1995)

Fayyad, U.M., Irani, K.B.: On the handling of continuous-valued attributes in decision tree generation. Machine Learning 8(2), 87–102 (1992)

Flach, P., Lachiche, N.: 1BC: A first-order bayesian classifier. In: Džeroski, S., Flach, P.A. (eds.) ILP 1999. LNCS (LNAI), vol. 1634, pp. 92–103. Springer, Heidelberg (1999)

Fürnkranz, J.: Separate-and-conquer rule learning. Artificial Intelligence Review 13(1), 3–54 (1999)

Fürnkranz, J., Flach, P.A.: ROC 'n' rule learning – Towards a better understanding of covering algorithms. Machine Learning 58(1), 39–77 (2005)

Gamberger, D., Lavrač, N.: Expert-guided subgroup discovery: Methodology and application. Journal of Artificial Intelligence Research 17, 501–527 (2002)

Golub, T., Slonim, D., Tamayo, P., Huard, C., Gaaseenbeek, M., Mesirov, J., Coller, H., Loh, M., Downing, J., Caligiuri, M., Bloomfield, C., Lander, E.: Molecular classification of cancer: Class discovery and class prediction by gene expression monitoring. Science 286, 531–537 (1999)

Idestam-Almquist, P.: Generalization of clauses under implication. Journal of Artificial Intelligence Research 3, 467–489 (1995)

Idestam-Almquist, P.: Generalization of clauses. PhD thesis, Stockholm University, Department of Computer and Systems Sciences (1993)

Kaufman, K.A., Michalski, R.S.: An adjustable rule learner for pattern discovery using the AQ methodology. Journal of Intelligent Information Systems 14, 199–216 (2000)

Larson, J.B., Michalski, R.S.: Inductive inference of VL decision rules. SIGART Newsletter 63, 38–44 (1977)

Lavrač, N., Džeroski, S.: Inductive Logic Programming: Techniques and Applications. Ellis Horwood (1994)

Lavrač, N., Flach, P.: An extended transformation approach to inductive logic programming. ACM Transactions on Computational Logic 2(4), 458–494 (2001)

Lavrač, N., Džeroski, S., Grobelnik, M.: Learning nonrecursive definitions of relations with LINUS. In: Kodratoff, Y. (ed.) EWSL 1991. LNCS, vol. 482, pp. 265–281. Springer, Heidelberg (1991)

Lavrač, N., Gamberger, D., Jovanoski, V.: A sudy of relevance for learning in deductive databases. Journal of Logic Programming 40(2/3), 215–249 (1999)

Matheus, C.J.: A constructive induction framework. In: Proceedings of the 6th International Workshop on Machine Learning, pp. 474–475 (1989)
Mehra, P., Rendell, L., Wah, B.: Principled constructive induction. In: Proceedings of the 11th International Joint Conference on Artificial Intelligence (IJCAI 1989), pp. 651–656. Morgan Kaufmann, San Francisco (1989)
Michalski, R.S.: On the quasi-minimal solution of the covering problem. In: Proceedings of the 5th International Symposium on Information Processing (FCIP 1969), Bled, Yugoslavia. Switching Circuits, vol. A3, pp. 125–128 (1969)
Michalski, R.S.: Pattern recognition and rule-guided inference. IEEE Transactions on Pattern Analysis and Machine Intelligence 2, 349–361 (1980)
Michalski, R.S.: A theory and methodology of inductive learning. Artificial Intelligence 20(2), 111–162 (1983)
Michalski, R.S.: AQVAL/1 — computer implementation of a variable-valued logic system VL_1 and examples of its application to pattern recognition. In: Proceedings of the 1st International Joint Conference on Pattern Recognition, pp. 3–17 (1973a)
Michalski, R.S.: Discovering classification rules using variable-valued logic system vl1. In: Proceedings of the 3rd International Joint Conference on Artificial Intelligence (IJCAI 1973), Stanford, CA, pp. 162–172 (1973b)
Michalski, R.S., Larson, J.B.: Selection of most representative training examples and incremental generation of vl1 hypotheses: the underlying methodology and the description of programs ESEL and AQ11. Technical Report 78-867, Department of Computer Science, University of Illinois at Urbana-Champaign (1978)
Michalski, R.S., Mozetič, I., Hong, J., Lavrač, N.: The multi-purpose incremental learning system AQ15 and its testing application to three medical domains. In: Proceedings of the 5th National Conference on Artificial Intelligence (AAAI 1986), Philadelphia, PA, pp. 1041–1045 (1986)
Provost, F., Fawcett, T.: Robust classification for imprecise environments. Machine Learning 42, 203–231 (2001)
Quinlan, J.R.: Induction of decision trees. Machine Learning 1, 81–106 (1986)
Quinlan, J.R.: Learning logical definitions from relations. Machine Learning 5, 239–266 (1990)
Wojtusiak, J., Michalski, R.S., Kaufman, K.A., Pietrzykowski, J.: The AQ21 natural induction program for pattern discovery: Initial version and its novel features. In: Proceedings of The 18th IEEE International Conference on Tools with Artificial Intelligence, Cincinnati, OH (2007)

Transfer Learning via Advice Taking

Lisa Torrey[1], Jude Shavlik[1], Trevor Walker[1], and Richard Maclin[2]

[1] University of Wisconsin, Madison WI 53706, USA
ltorrey@cs.wisc.edu, shavlik@cs.wisc.edu, twalker@cs.wisc.edu
[2] University of Minnesota, Duluth, MN 55812, USA
rmaclin@gmail.com

Abstract. The goal of transfer learning is to speed up learning in a new task by transferring knowledge from one or more related source tasks. We describe a transfer method in which a reinforcement learner analyzes its experience in the source task and learns rules to use as advice in the target task. The rules, which are learned via inductive logic programming, describe the conditions under which an action is successful in the source task. The advice-taking algorithm used in the target task allows a reinforcement learner to benefit from rules even if they are imperfect. A human-provided mapping describes the alignment between the source and target tasks, and may also include advice about the differences between them. Using three tasks in the RoboCup simulated soccer domain, we demonstrate that this transfer method can speed up reinforcement learning substantially.

1 Introduction

Machine learning tasks are often addressed independently, under the implicit assumption that each new task has no exploitable relation to the tasks that came before. *Transfer learning* is a machine learning paradigm that rejects this assumption and uses known relationships between tasks to improve learning. The goal of transfer is to improve learning in a *target task* by transferring knowledge from a related *source task*.

One context in which transfer learning can be particularly useful is *reinforcement learning* (RL), where an agent learns to take actions in an environment to receive rewards [26]. Complex RL tasks can require very long training times. However, when learning a new task in the same domain as previously learned tasks, there are opportunities for reducing the training times through transfer.

There are three common measures by which transfer might improve learning in RL. First is the initial performance achievable in the target task using only the transferred knowledge, before any further learning is done, compared to the initial performance of an ignorant agent. Second is the amount of time it takes to fully learn the target task given the transferred knowledge compared to the amount of time to learn it from scratch. Third is the final performance level achievable in the target task compared to the final level without transfer. Figure 1 illustrates these three measures.

Our transfer method learns *skills* from a source task that may be useful in a target task. Skills are rules in first-order logic that describe when an action should be successful. For example, suppose an RL soccer player has learned, in a source task, to pass to its teammates in a way that keeps the ball from falling into the opponents' possession. In the target task, suppose it must learn to work with teammates to score goals

J. Koronacki et al. (Eds.): Advances in Machine Learning I, SCI 262, pp. 147–170.
springerlink.com © Springer-Verlag Berlin Heidelberg 2010

against opponents. If this player could remember its passing skills from the source task, it should master the target task more quickly.

Even when RL tasks have shared actions, transfer between them is a difficult problem because differences in reward structures create differences in the results of actions. For example, the passing skill in the source task above is incomplete for the target task – in the target, unlike the source, passing needs to cause progress toward the goal in addition to maintaining ball possession. This indicates that RL agents using transferred information must continue to learn, filling in gaps left by transfer. Since transfer might also produce partially irrelevant or incorrect skills, RL agents must also be able to modify or ignore transferred information that is imperfect. Our transfer method allows this by applying skills as *advice*, with a learning algorithm that treats rules as soft constraints.

We require a human observer to provide a *mapping* between the source and target task. A mapping describes the structural similarities between the tasks, such as correspondences between player objects in the example above. It might also include simple advice that reflects the differences between the tasks. In our example, additional advice like "prefer passing toward the goal" and "shoot when close to the goal" would be helpful.

Our chapter's presence in this memorial volume is due to the way that our work touches on several topics of interest to Professor Ryszard Michalski. He contributed significantly to the area of rule learning in first-order logic [14], which we use to learn skills for transfer. He also did important work involving expert advice [2], which has connections to our advice-taking methods, and analogical learning [15], which is closely related to transfer learning.

The rest of the chapter is organized as follows. Section 2 provides background information on RL: an overview, and a description of our standard RL and advice-taking RL implementations. Section 3 presents RoboCup simulated soccer and explains how we learn tasks in the domain with RL. Section 4 provides background information on inductive logic programming, which is the machine-learning technique we use to learn skills. Section 5 then describes our transfer method, with experimental results in Section 6. Section 7 surveys some related work, and Section 8 reflects on some interesting issues that our work raises.

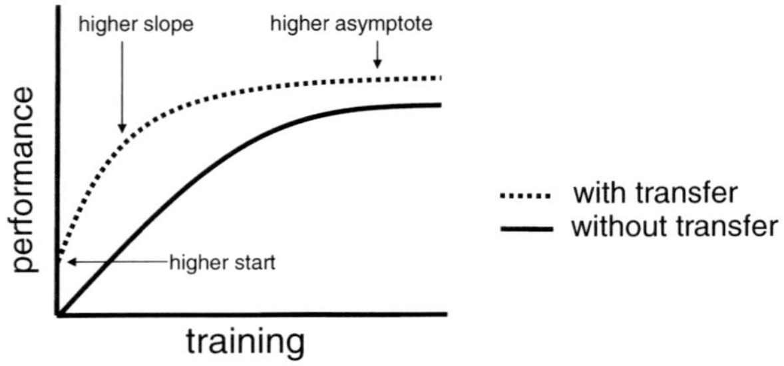

Fig. 1. Three ways in which transfer might improve reinforcement learning

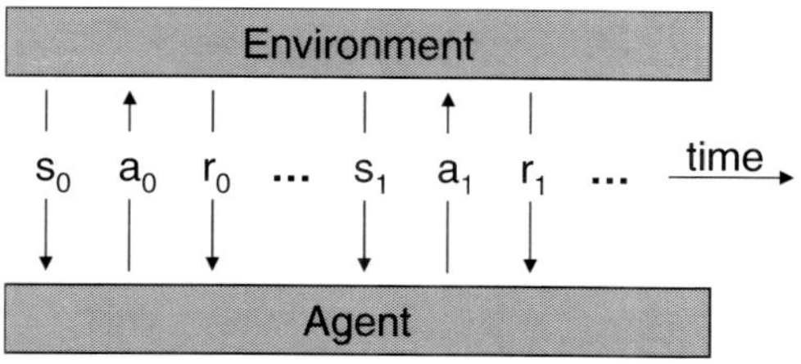

Fig. 2. A reinforcement learning agent interacts with its environment: it receives information about its state (s), chooses an action to take (a), receives a reward (r), and then repeats

2 Background on Reinforcement Learning

A reinforcement learning agent operates in a episodic sequential-control environment. It senses the *state* of the environment and performs *actions* that change the state and also trigger *rewards*. Its objective is to learn a policy for acting in order to maximize its cumulative reward during an episode. This involves solving a temporal credit-assignment problem, since an entire sequence of actions may be responsible for a single reward received at the end of the sequence.

A typical RL agent behaves according to the diagram in Figure 2. At time step t, it observes the current state s_t and consults its current policy π to choose an action, $\pi(s_t) = a_t$. After taking the action, it receives a reward r_t and observes the new state s_{t+1}, and it uses that information to update its policy before repeating the cycle. Often RL consists of a sequence of *episodes*, which end whenever the agent reaches one of a set of ending states (e.g. the end of a game).

Formally, a reinforcement learning domain has two underlying functions that determine immediate rewards and the state transitions. The reward function $r(s,a)$ gives the reward for taking action a in state s, and the transition function $\delta(s,a)$ gives the next state the agent enters after taking action a in state s. If these functions are known, the optimal policy $\pi^\star$ can be calculated directly by maximizing the *value function* at every state. The value function $V_\pi(s)$ gives the discounted cumulative reward achieved by policy π starting in state s (see Equation 1).

$$V_\pi(s_t) = r_t + \gamma r_{t+1} + \gamma^2 r_{t+2} + \ldots \tag{1}$$

The discount factor $\gamma \in [0,1]$. Setting $\gamma < 1$ gives later rewards less impact on the value function than earlier rewards, which may be desirable for tasks without fixed lengths.

During learning, the agent must balance between *exploiting* the current policy (acting in areas that it knows to have high rewards) and *exploring* new areas to find higher rewards. A common solution is the ε-greedy method, in which the agent takes random exploratory actions a small fraction of the time ($\varepsilon << 1$), but usually takes the action recommended by the current policy.

Often the reward and transition functions are not known, and therefore the optimal policy cannot be calculated directly. In this situation, one applicable RL technique is Q-learning [36], which involves learning a Q-function instead of a value function.

The Q-function, $Q(s,a)$, estimates the discounted cumulative reward starting in state s and taking action a and following the current policy thereafter. Given the optimal Q-function, the optimal policy is to take the action $argmax_a Q(s_t,a)$. RL agents in deterministic worlds can begin with an inaccurate Q-function and recursively update it after each step according to the rule in Equation 2.

$$Q(s_t,a_t) \longleftarrow r_t + \gamma\, max_a\, Q(s_{t+1},a) \tag{2}$$

In this equation, the current estimate of a Q-value on the right is used to produce a new estimate on the left. In the SARSA variant [26], the new estimate uses the actual a_{t+1} instead of the a with the highest Q-value in s_{t+1}; this takes the ε-greedy action selections into account. In non-deterministic worlds, a learning rate $\alpha \in (0,1]$ is required to form a weighted average between the old estimate and the new one. Equation 3 shows the SARSA update rule for non-deterministic worlds.

$$Q(s_t,a_t) \longleftarrow (1-\alpha)\, Q(s_t,a_t) + \alpha\, (r_t + \gamma\, Q(s_{t+1},a_{t+1})) \tag{3}$$

While these equations give update rules that look just one step ahead, it is possible to perform updates over multiple steps. In temporal-difference learning [25], agents can combine estimates over multiple lookahead distances.

When there are small finite numbers of states and actions, the Q-function can be represented in tabular form. However, some RL domains have states that are described by very large feature spaces, or even infinite ones when continuous-valued features are present, making a tabular representation infeasible. A solution is to use a function approximator to represent the Q-function (e.g., a neural network). Function approximation has the additional benefit of providing generalization across states; that is, changes to the Q-value of one state affect the Q-values of similar states.

Under certain conditions, Q-learning is guaranteed to converge to an accurate Q-function [37]. Although these conditions are typically violated (by using function approximation, for example) the method can still produce successful learning. For further information on reinforcement learning, there are more detailed introductions by Mitchell [16] and Sutton and Barto [26].

2.1 Performing RL with Support Vector Regression

Our implementation is a form of Q-learning called SARSA(λ), which is the SARSA variant combined with temporal-difference learning. We represent the state with a set of numeric features and approximate the Q-function for each action with a weighted linear sum of those features, learned via support-vector regression (SVR). To find the feature weights, we solve a linear optimization problem, minimizing the following quantity:

$$\text{ModelSize} + C \times \text{DataMisfit}$$

Here *ModelSize* is the sum of the absolute values of the feature weights, and *DataMisfit* is the disagreement between the learned function's outputs and the training-example outputs (i.e., the sum of the absolute values of the differences for all examples). The numeric parameter C specifies the relative importance of minimizing disagreement with the data versus finding a simple model.

Most Q-learning implementations make incremental updates to the Q-functions after each step the agent takes. However, completely re-solving the SVR optimization problem after each data point would be too computationally intensive. Instead, our agents perform batches of 25 full episodes at a time and re-solve the optimization problem after each batch.

Formally, for each action the agent finds an optimal weight vector w that has one weight for each feature in the feature vector x. The expected Q-value of taking an action from the state described by vector x is $wx+b$, where b is a scalar offset. Our learners use the ε-greedy exploration method.

To compute the weight vector for an action, we find the subset of training examples in which that action was taken and place those feature vectors into rows of a data matrix A. When A becomes too large for efficient solving, we begin to discard episodes randomly such that the probability of discarding an episode increases with the age of the episode. Using the current model and the actual rewards received in the examples, we compute Q-value estimates and place them into an output vector y. The optimal weight vector is then described by Equation 4.

$$Aw + b\overrightarrow{e} = y \tag{4}$$

where $\overrightarrow{e}$ denotes a vector of ones (we omit this for simplicity from now on).

Our matrix A contains 75% exploitation examples, in which the action is the one recommended by the current policy, and 25% exploration examples, in which the action is off-policy. We do this so that bad moves are not forgotten, as they could be if we used almost entirely exploitation examples. When there are not enough exploration examples, we create synthetic ones by randomly choosing exploitation steps and using the current model to score unselected actions for those steps.

In practice, we prefer to have non-zero weights for only a few important features in order to keep the model simple and avoid overfitting the training examples. Furthermore, an exact linear solution may not exist for any given training set. We therefore introduce *slack* variables s that allow inaccuracies on some examples. The resulting minimization problem is

$$\begin{aligned} \min_{(w,b,s)} \quad & ||w||_1 + \nu|b| + C||s||_1 \\ s.t. \quad & -s \leq Aw + b - y \leq s. \end{aligned} \tag{5}$$

where $|\cdot|$ denotes an absolute value, $||\cdot||_1$ denotes the one-norm (a sum of absolute values), and ν is a penalty on the offset term. By solving this problem, we can produce a weight vector w for each action that compromises between accuracy and simplicity. We let C decay exponentially over time so that solutions may be more complex later in the learning curve.

Several other parameters in our system also decay exponentially over time: the temporal-difference parameter λ, so that earlier episodes combine more lookahead distances than later ones; the learning rate α, so that earlier episodes tend to produce larger Q-value updates than later ones; and the exploration rate ε, so that agents explore less later in the learning curve.

2.2 Performing Advice Taking in RL

Advice taking is learning with additional knowledge that may be imperfect. It attempts to take advantage of this knowledge to improve learning, but avoids trusting it completely. Advice often comes from humans, but in our work it also comes from automated analysis of successful behavior in a source task.

We view advice as a set of soft constraints on the Q-function of an RL agent. For example, here is a vague advice rule for passing in soccer:

IF an opponent is near me AND
a teammate is open
THEN *pass* has a high Q-value

In this example, there are two conditions describing the state of the agent's environment: an opponent is nearby and there is an unblocked path to a teammate. These form the IF portion of the rule. The THEN portion gives a constraint on the Q-function that the advice indicates should hold when the environment matches the conditions.

In our advice-taking system, an agent can follow advice, only follow it approximately (which is like refining it), or ignore it altogether. We extend the support-vector regression technique described in Section 2.1 to accomplish this. Recall that Equation 5 describes the optimization problem for learning the weights that determine an action's Q-function. We incorporate advice into this optimization problem using a method called Knowledge-Based Kernel Regression (KBKR), designed by Mangasarian et al. [12] and applied to reinforcement learning by Maclin et al. [8].

An advice rule creates new constraints on the problem solution in addition to the constraints from the training data. In particular, since we use an extension of KBKR called Preference-KBKR [9], our advice rules give conditions under which one action is preferred over another action. Our advice therefore takes the following form:

$$Bx \leq d \Longrightarrow Q_p(x) - Q_n(x) \geq \beta, \tag{6}$$

This can be read as:

> If the current state satisfies $Bx \leq d$, then the Q-value of the preferred action p should exceed that of the non-preferred action n by at least β.

For example, consider giving the advice that action p is better than action n when the value of feature 5 is at most 10. The vector B would have one row with a 1 in the column for feature 5 and zeros elsewhere. The vector d would contain only the value 10, and β could be set to some small positive number.

Just as we allowed some inaccuracy on the training examples in Equation 5, we allow advice to be followed only partially. To do so, we introduce slack variables z and penalty parameters μ for trading off the impact of the advice with the impact of the training examples. Over time, we decay μ so that advice has less impact as the learner gains more experience.

The new optimization problem [9] solves the Q-functions for all the actions simultaneously so that it can apply constraints to their relative values. Multiple pieces of preference advice can be incorporated, each with its own B, d, p, n, and β, which makes it possible to advise taking a particular action by stating that it is preferred over all the other actions. We use the CPLEX commercial software to solve the resulting linear program. We do not show the entire formalization here, but it minimizes the following quantity:

$$\text{ModelSize} + C \times \text{DataMisfit} + \mu \times \text{AdviceMisfit}$$

We have also developed a variant of Preference-KBKR called ExtenKBKR [10] that incorporates advice in a way that allows for faster problem-solving. We will not present this variant in detail here, but we do use it for transfer when there is more advice than Preference-KBKR can efficiently handle.

3 RoboCup: A Challenging Reinforcement Learning Domain

One motivating domain for transfer in reinforcement learning is RoboCup simulated soccer. The RoboCup project [17] has the overall goal of producing robotic soccer teams that compete on the human level, but it also has a software simulator for research purposes. Stone and Sutton [24] introduced RoboCup as an RL domain that is challenging because of its large, continuous state space and nondeterministic action effects.

Since the full game of soccer is quite complex, researchers have developed several smaller games in the RoboCup domain (see Figure 3). These are inherently multi-agent games, but a standard simplification is to have only one agent (the one in possession of the soccer ball) learning at a time using a shared model built with data combined from all the players on its team.

The first RoboCup task we use is *M*-on-*N* KeepAway [24], in which the objective of the *M* reinforcement learners called *keepers* is to keep the ball away from *N* hand-coded players called *takers*. The keeper with the ball may choose either to hold it or to pass it to a teammate. Keepers without the ball follow a hand-coded strategy to receive passes. The game ends when an opponent takes the ball or when the ball goes out of bounds. The learners receive a +1 reward for each time step their team keeps the ball.

Our KeepAway state representation is the one designed by Stone and Sutton [24]. The features are listed in Table 1. The keepers are ordered by their distance to the learner *k0*, as are the takers.

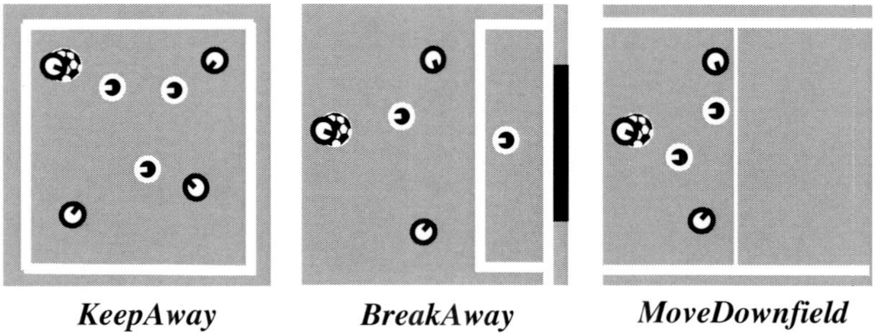

Fig. 3. Snapshots of RoboCup soccer tasks. In KeepAway, the keepers pass the ball around and keep it away from the takers. In BreakAway, the attackers attempt to score a goal against the defenders. In MoveDownfield, the attackers attempt to move the ball toward the defenders' side.

Table 1. Feature spaces for RoboCup tasks. The functions *minDistTaker(Keeper)* and *minAngleTaker(Keeper)* evaluate to the player objects *t0, t1*, etc. that are closest in distance and angle respectively to the given Keeper object. Similarly, the functions *minDistDefender(Attacker)* and *minAngleDefender(Attacker)* evaluate to the player objects *d0, d1*, etc.

KeepAway features	
distBetween(k0, Player)	Player ∈ {k1, k2, ...} ∪ {t0, t1, ...}
distBetween(Keeper, minDistTaker(Keeper))	Keeper ∈ {k1, k2, ...}
angleDefinedBy(Keeper, k0, minAngleTaker(Keeper))	Keeper ∈ {k1, k2, ...}
distBetween(Player, fieldCenter)	Player ∈ {k0, k1, ...} ∪ {t0, t1, ...}
MoveDownfield features	
distBetween(a0, Player)	Player ∈ {a1, a2, ...} ∪ {d0, d1, ...}
distBetween(Attacker, minDistDefender(Attacker))	Attacker ∈ {a1, a2, ...}
angleDefinedBy(Attacker, a0, minAngleDefender(Attacker))	Attacker ∈ {a1, a2, ...}
distToRightEdge(Attacker)	Attacker ∈ {a0, a1, ...}
timeLeft	
BreakAway features	
distBetween(a0, Player)	Player ∈ {a1, a2, ...} ∪ {d0, d1, ...}
distBetween(Attacker, minDistDefender(Attacker))	Attacker ∈ {a1, a2, ...}
angleDefinedBy(Attacker, a0, minAngleDefender(Attacker))	Attacker ∈ {a1, a2, ...}
distBetween(Attacker, goalPart)	Attacker ∈ {a0, a1, ...}
distBetween(Attacker, goalie)	Attacker ∈ {a0, a1, ...}
angleDefinedBy(Attacker, a0, goalie)	Attacker ∈ {a1, a2, ...}
angleDefinedBy(GoalPart, a0, goalie)	GoalPart ∈ {goalRight, goalLeft, goalCenter}
angleDefinedBy(topRightCorner, goalCenter, a0)	
timeLeft	

Note that we present these features as predicates in first-order logic. Variables are capitalized and typed (*Player*, *Keeper*, etc.) and constants are uncapitalized. For simplicity we indicate types by variable names, leaving out implied terms like *player(Player)*, *keeper(Keeper)*, etc. Since we are not using fully relational reinforcement learning, the predicates are actually grounded and used as propositional features during learning. However, since we transfer relational information, we represent them in a relational form here.

A second RoboCup task is *M*-on-*N* MoveDownfield, where the objective of the *M* reinforcement learners called *attackers* is to move across a line on the opposing team's side of the field while maintaining possession of the ball. The attacker with the ball may choose to pass to a teammate or to move ahead, away, left, or right with respect to the opponent's goal. Attackers without the ball follow a hand-coded strategy to receive passes. The game ends when they cross the line, when an opponent takes the ball, when the ball goes out of bounds, or after a time limit of 25 seconds. The learners receive symmetrical positive and negative rewards for horizontal movement forward and backward.

Our MoveDownfield state representation is the one presented in Torrey et al. [32]. The features are listed in Table 1. The attackers are ordered by their distance to the learner *a0*, as are the defenders.

A third RoboCup task is *M*-on-*N* BreakAway, where the objective of the *M* attackers is to score a goal against $N-1$ hand-coded *defenders* and a hand-coded *goalie*. The attacker with the ball may choose to pass to a teammate, to move ahead, away, left, or right with respect to the opponent's goal, or to shoot at the left, right, or center part of the goal. Attackers without the ball follow a hand-coded strategy to receive passes. The game ends when they score a goal, when an opponent takes the ball, when the ball goes out of bounds, or after a time limit of 10 seconds. The learners receive a +1 reward if they score a goal, and zero reward otherwise.

Our BreakAway state representation is the one presented in Torrey et al. [33]. The features are listed in Table 1. The attackers are ordered by their distance to the learner *a0*, as are the non-goalie defenders.

Our system discretizes each feature in these tasks into 32 tiles, each of which is associated with a Boolean feature. For example, the tile denoted by $distBetween(a0, a1)_{[10,20]}$ takes value 1 when *a1* is between 10 and 20 units away from *a0* and 0 otherwise. Stone and Sutton [24] found tiling to be important for timely learning in RoboCup.

The three RoboCup games have substantial differences in features, actions, and rewards. The goal, goalie, and shoot actions exist in BreakAway but not in the other two tasks. The move actions do not exist in KeepAway but do in the other two tasks. Rewards in KeepAway and MoveDownfield occur for incremental progress, but in BreakAway the reward is more sparse. These differences mean the solutions to the tasks may be quite different. However, some knowledge should clearly be transferable between them, since they share many features and some actions, such as the *pass* action. Furthermore, since these are difficult RL tasks, speeding up learning through transfer would be desirable.

4 Inductive Logic Programming

Inductive logic programming (ILP) is a technique for learning classifiers in first-order logic [16]. Our transfer algorithms uses ILP to extract knowledge from the source task. This section provides a brief introduction to ILP.

4.1 What ILP Learns

An ILP algorithm learns a set of first-order clauses, usually definite clauses. A definite clause has a *head*, which is a predicate that is implied to be true if the conjunction of predicates in the *body* is true. Predicates describe relationships between objects in the world, referring to objects either as constants (lower-case) or variables (upper-case). In Prolog notation, the head and body are separated by the symbol :- denoting implication, and commas denoting *and* separate the predicates in the body, as in the rest of this section.

As an example, consider applying ILP to learn a clause describing when an object in an agent's world is at the bottom of a stack of objects. The world always contains the object *floor*, and may contain any number of additional objects. The configuration of the world is described by predicates *stackedOn(Obj1, Obj2)*, where *Obj1* and *Obj2* are variables that can be instantiated by the objects, such as:

```
stackedOn(chair, floor).
stackedOn(desk, floor).
stackedOn(book, desk).
```

Suppose we want the ILP algorithm to learn a clause that implies *isBottomOfStack(Obj)* is true when $Obj = desk$ but not when $Obj \in \{floor, chair, book\}$. Given those positive and negative examples, it would learn the following clause:

```
isBottomOfStack(Obj) :-
    stackedOn(Obj, floor),
    stackedOn(OtherObj, Obj).
```

That is, an object is at the bottom of the stack if it is on the floor and there exists another object on top of it. On its way to discovering the correct clause, the ILP algorithm would probably evaluate the following clause:

```
isBottomOfStack(Obj) :-
    stackedOn(Obj, floor).
```

This clause correctly classifies 3 of the 4 objects in the world, but incorrectly classifies *chair* as positive. In domains with noise, a partially correct clause like this might be optimal, though in this case the concept can be learned exactly.

Note that the clause must be first-order to describe the concept exactly: it must include the variables *Obj* and *OtherObj*. First-order logic can posit the existence of an object and then refer to properties of that object. Most machine learning algorithms use *propositional* logic, which does not include variables, but ILP is able to use a more powerful and natural type of reasoning.

In many domains, the true concept is disjunctive, meaning that multiple clauses are necessary to describe the concept fully. ILP algorithms therefore typically attempt to learn a set of clauses rather than just one. The entire set of clauses is called a *theory*.

4.2 How ILP Learns

There are several types of algorithms for producing a set of first-order clauses, including Michalski's AQ algorithm [14]. This section focuses on the Aleph system [23], which we use in our experiments.

Aleph constructs a ruleset through *sequential covering*. It performs a search for the rule that best classifies the positive and negative examples (according to a user-specified scoring function), adds that rule to the theory, and then removes the positive examples covered by that rule and repeats the process on the remaining examples.

The default procedure Aleph uses in each iteration a heuristic search. It randomly chooses a positive example as the *seed* for its search for a single rule. Then it lists all the predicates in the world that are true for the seed. This list is called the *bottom clause*, and it is typically too specific, since it describes a single example in great detail. Aleph conducts a search to find a more general clause (a variablized subset of the predicates in the bottom clause) that maximizes the scoring function. The search process is top-down, meaning that it begins with an empty rule and adds predicates one by one to greedily maximize a scoring function.

Our rule-scoring function is the F(1) measure, which relies on the concepts of *precision* and *recall*. The *precision* of a rule is the fraction of examples it calls positive that are truly positive, and the *recall* is the fraction of truly positive examples that it correctly calls positive. The F(1) measure combines the two:

$$F(1) = \frac{2 * Precision * Recall}{Precision + Recall}$$

An alternative Aleph procedure that we also use is a randomized search [34]. This also uses a seed example and generates a bottom clause, but it begins by randomly drawing a legal clause of length N from the bottom clause. It then makes local moves by adding and removing literals from the clause. After M local moves, and possibly K repeats of the entire process, it returns the highest-scoring rule encountered.

5 Skill Transfer in RL via Advice Taking

Our method for transfer in reinforcement learning, called *skill transfer*, begins by analyzing games played by a successful source-task agent. Using the ILP algorithm from Section 4.2, it learns first-order rules that describe *skills*. We define a skill as a rule that describes the circumstances under which an action is likely to be successful [32]. Our method then uses a human-provided mapping between the tasks to translate skills into a form usable in the target task. Finally, it applies the skills as advice in the target task, along with any additional human advice, using the KBKR algorithm from Section 2.2.

Figure 4 shows an example of skill transfer from KeepAway to BreakAway. In this example, KeepAway games provide training examples for the concept "states in which

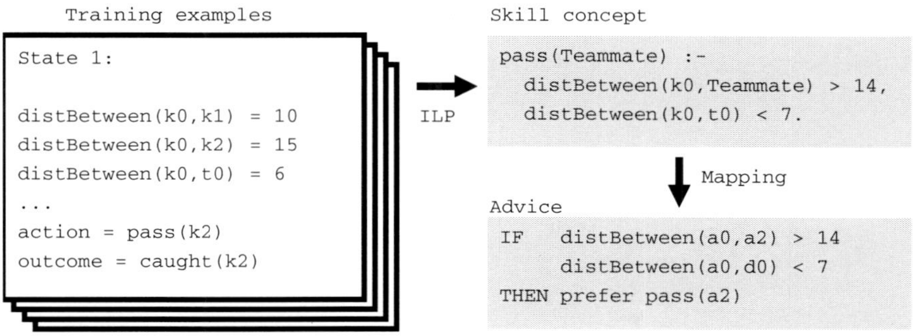

Fig. 4. Example showing how we transfer skills. We provide positive and negative source-task examples of *pass* actions to ILP, which learns a rule describing the *pass* skill, and we apply a mapping to produce target-task advice.

passing to a teammate is a good action," and ILP learns a rule representing the *pass* skill, which is mapped into advice for BreakAway.

We learn first-order rules because they can be more general than propositional rules, since they can contain variables. For example, the rule *pass(Teammate)* is likely to capture the essential elements of the passing skill better than rules for passing to specific teammates. We expect these common skill elements to transfer better to new tasks.

In a first-order representation, corresponding feature and action predicates can ideally be made identical throughout the domain so that there is no need to map them. However, we assume the user provides a mapping between logical objects in the source and target tasks (e.g., *k0* in KeepAway maps to *a0* in BreakAway).

The actions in the two tasks need not have one-to-one correspondences. If an action in the source does not exist in the target, we do not attempt to transfer a skill for it. The feature sets also do not need to have one-to-one correspondences, because the ILP search algorithm can limit its search space to only those feature predicates that are present in the target task. We therefore allow only feature predicates that exist in the target task to appear in advice rules. This forces the algorithm to find skill definitions that are applicable to the target task.

5.1 Learning Skills in a Source Task

For each action, we conduct a search with ILP for the rule with the highest F(1) score. To produce datasets for this search, we examine states from games in the source task and select positive and negative examples. Not all states should be used as training examples; some are not unambiguously positive or negative and should be left out of the datasets. These states can be detected by looking at their Q-values, as described below. Figure 5 summarizes the overall process with an example from RoboCup.

In a good positive example, several conditions should be met: the skill is performed, the desired outcome occurs (e.g. a pass reaches its intended recipient), the expected Q-value (using the most recent Q-function) is above the 10th percentile in the training

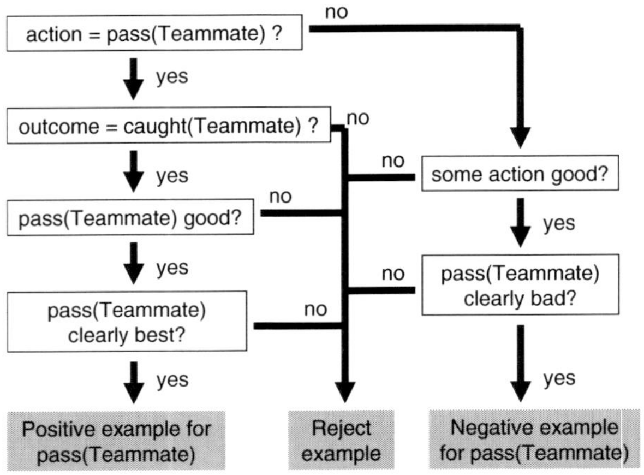

Fig. 5. Example of how our algorithm selects training examples for skills

set and is at least 1.05 times the predicted Q-values of all other actions. The purpose of these conditions is to remove ambiguous examples in which several actions may be good or no actions seem good.

There are two types of good negative examples. These conditions describe one type: some other action is performed, that action's Q-value is above the 10th percentile in the training set, and the Q-value of the skill being learned is at most 0.95 times that Q-value and below the 50th percentile in the training set. These conditions also remove ambiguous examples. The second type of good negative example includes states in which the skill being learned was taken but the desired outcome did not occur.

To make the search space finite, it is necessary to replace continuous features (like distances and angles) with finite sets of discrete features. For example, the rule in Figure 4 contains the Boolean constraint *distBetween(k0,t0)* < 7, derived from the continuous distance feature. Our algorithm finds the 25 thresholds with the highest information gain and allows the intervals above and below those thresholds to appear as constraints in rules. Furthermore, we allow up to 7 constraints in each rule. We found these parameters to produce reasonable running times for RoboCup, but they would need to be adjusted appropriately for other domains.

5.2 Mapping Skills for a Target Task

To convert a skill into transfer advice, we need to apply an object mapping and propositionalize the rule. Propositionalizing is necessary because our KBKR advice-taking algorithm only works with propositional advice. This automated process preserves the meaning of the first-order rules without losing any information, but there are several technical details involved.

First we instantiate skills like *pass(Teammate)* for the target task. For 3-on-2 BreakAway, this would produce two rules, *pass(a1)* and *pass(a2)*. Next we deal with any other conditions in the rule body that contain variables. For example, a rule might have this condition:

$$10 < \text{distBetween(a0, Attacker)} < 20$$

This is effectively a disjunction of conditions: either the distance to *a1* or the distance to *a2* is in the interval $[10, 20]$. Since disjunctions are not part of the advice language, we use tile features to represent them. Recall that each feature range is divided into Boolean tiles that take the value 1 when the feature value falls into their interval and 0 otherwise. This disjunction is satisfied if at least one of several tiles is active; for 3-on-2 BreakAway this is:

$$\text{distBetween(a0, a1)}_{[10,20]} + \text{distBetween(a0, a2)}_{[10,20]} \geq 1$$

If these exact tile boundaries do not exist in the target task, we add new tile boundaries to the feature space. Thus transfer advice can be expressed exactly even though the target-task feature space is unknown at the time the source task is learned.

It is possible for multiple conditions in a rule to refer to the same variable. For example:

$$\begin{array}{l}\text{distBetween(a0, Attacker)} > 15,\\ \text{angleDefinedBy(Attacker, a0, ClosestDefender)} > 25\end{array}$$

Here the variable *Attacker* represents the same object in both clauses, so the system cannot propositionalize the two clauses separately. Instead, it defines a new predicate that puts simultaneous constraints on both features:

```
newFeature(Attacker, ClosestDefender) :-
  Dist is distBetween(a0, Attacker),
  Ang is angleDefinedBy(Attacker, a0, ClosestDefender),
  Dist > 15, Ang > 25.
```

It then expresses the entire condition using the new feature; for 3-on-2 BreakAway this is:

$$\text{newFeature(a1, d0)} + \text{newFeature(a2, d0)} \geq 1$$

We add these new Boolean features to the target task. Thus skill transfer can actually enhance the feature space of the target task.

Each advice item produced from a skill says to prefer that skill over the other actions shared between the source and target task. We set the preference amount Δ to approximately 1% of the target task's Q-value range.

5.3 Adding Human Advice

Skill transfer produces a small number of simple, interpretable rules. This introduces the possibility of further user input in the transfer process. If users can understand the transfer advice, they may wish to add to it, either further specializing rules or writing their own rules for new, non-transferred skills in the target task. Our skill-transfer method therefore allows optional *user advice*.

For example, the passing skills transferred from KeepAway to BreakAway make no distinction between passing toward the goal and away from the goal. Since the new objective is to score goals, players should clearly prefer passing toward the goal. A user could provide this guidance by instructing the system to add a condition like this to the *pass(Teammate)* skill:

$$\text{distBetween(a0, goal)} - \text{distBetween(Teammate, goal)} \geq 1$$

Even more importantly, there are several actions in this transfer scenario that are new in the target task, such as *shoot* and *moveAhead*. We allow users to write simple rules to approximate skills like these, such as:

```
IF   distBetween(a0, GoalPart) < 10
AND  angleDefinedBy(GoalPart, a0, goalie) > 40
THEN prefer shoot(GoalPart) over all actions

IF   distBetween(a0, goalCenter) > 10
THEN prefer moveAhead over moveAway and the shoot actions
```

The advice-taking framework is a natural and powerful way for users to provide information not only about the correspondences between tasks, but also about the differences between them.

6 Results

We performed experiments with skill transfer in many scenarios with RoboCup tasks. Some are *close transfer* scenarios, where the tasks are closely related: the target task is the same as the source task except each team has one more player. Others are *distant transfer* scenarios, where the tasks are more distantly related: from KeepAway to BreakAway and from MoveDownfield to BreakAway. With distant transfer we concentrate on moving from easier tasks to harder tasks.

For each task, we use an appropriate measure of performance to plot against the number of training games in a learning curve. In BreakAway, it is the probability that the agents will score a goal in a game. In MoveDownfield, it is the average distance traveled towards the right edge during a game. In KeepAway, it is the average length of a game.

Section 6.1 shows examples of rules our method learned in various source tasks. Section 6.2 shows learning curves in various target tasks with and without skill transfer.

6.1 Skills Learned

From 2-on-1 BreakAway, one rule our method learned for the *shoot* skill is:

shoot(GoalPart) :-
  distBetween(a0, goalCenter) $\geq$ 6,
  angleDefinedBy(GoalPart, a0, goalie) $\geq$ 52,
  distBetween(a0, oppositePart(GoalPart)) $\geq$ 6,
  angleDefinedBy(oppositePart(GoalPart), a0, goalie) $\leq$ 33,
  angleDefinedBy(goalCenter, a0, goalie) $\geq$ 28.

This rule requires a large open shot angle, a minimum distance to the goal, and angle constraints that restrict the goalie's position to a small area.

From 3-on-2 MoveDownfield, one rule our method learned for the *pass* skill is:

pass(Teammate) :-
  distBetween(a0, Teammate) $\geq$ 15,
  distBetween(a0, Teammate) $\leq$ 27,
  angleDefinedBy(Teammate, a0, minAngleDefender(Teammate)) $\geq$ 24,
  distToRightEdge(Teammate) $\leq$ 10,
  distBetween(a0, Opponent) $\geq$ 4.

This rule specifies an acceptable range for the distance to the receiving teammate and a minimum pass angle. It also requires that the teammate be close to the finish line on the field and that an opponent not be close enough to intercept.

From 3-on-2 KeepAway, one rule our method learned for the *pass* skill is:

pass(Teammate) :-
  distBetween(Teammate, fieldCenter) $\geq$ 6,
  distBetween(Teammate, minDistTaker(Teammate)) $\geq$ 8,
  angleDefinedBy(Teammate, a0, minAngleTaker(Teammate)) $\geq$ 41,
  angleDefinedBy(OtherTeammate, a0, minAngleTaker(OtherTeammate)) $\leq$ 23.

This rule specifies a minimum pass angle and an open distance around the receiving teammate. It also requires that the teammate not be too close to the center of the field and gives a maximum pass angle for the alternate teammate.

Some parts of these rules were unexpected, but make sense in hindsight. For example, the shoot rule specifies a minimum distance to the goal rather than a maximum distance. Presumably this is because large shot angles are only available at reasonable distances anyway. This shows the advantages that advice learned through transfer can have over human advice.

6.2 Learning Curves

Figures 6, 7, and 8 are learning curves from our transfer experiments. One curve in each figure is the average of 25 runs of standard reinforcement learning. The other curves are RL with skill transfer from various source tasks. For each transfer curve we average 5 transfer runs from 5 different source runs, for a total of 25 runs (this way, the results include both source and target variance). Because the variance is high, we smooth the y-value at each data point by averaging over the y-values of the last 250 games.

These figures show that skill transfer can have a large overall positive impact in both close-transfer and distant-transfer scenarios. The statistical results in Table 2 indicate that in most cases the difference (in area under the curve) is statistically significant.

We use appropriate subsets of the human-advice examples in Section 5.3 for all of our skill-transfer experiments. That is, from KeepAway to BreakAway we use all of it, from MoveDownfield to BreakAway we use only the parts advising *shoot*, and for close-transfer experiments we use none.

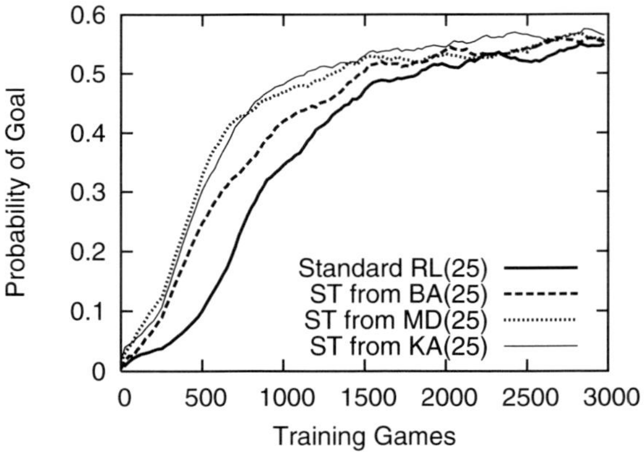

Fig. 6. Probability of scoring a goal while training in 3-on-2 BreakAway with standard RL and skill transfer (ST) from 2-on-1 BreakAway (BA), 3-on-2 MoveDownfield (MD) and 3-on-2 KeepAway (KA)

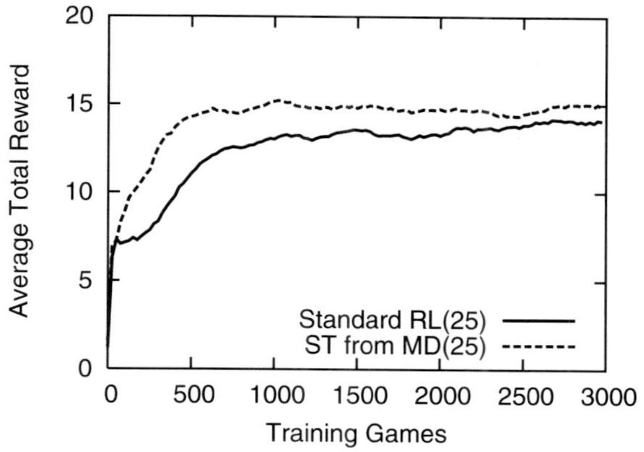

Fig. 7. Average total reward while training in 4-on-3 MoveDownfield with standard RL and skill transfer (ST) from 3-on-2 MoveDownfield (MD)

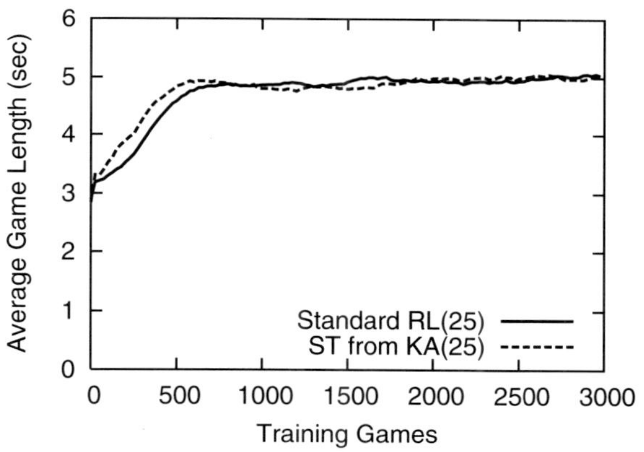

Fig. 8. Average game length while training in 4-on-3 KeepAway with standard RL and skill transfer (ST) from 3-on-2 KeepAway (KA)

Table 2. Statistical results from skill transfer (ST) experiments in BreakAway (BA), MoveDownfield (MD), and KeepAway (KA), comparing area under the curve to standard reinforcement learning (SRL)

Scenario	*Conclusion*	*p-value*	*95% confidence interval*
BA to BA	ST higher with 99% confidence	0.0003	63.75, 203.36
MD to BA	ST higher with 99% confidence	< 0.0001	153.63, 278.02
KA to BA	ST higher with 97% confidence	< 0.0001	176.42, 299.87
MD to MD	ST higher with 98% confidence	< 0.0001	3682.59, 6436.61
KA to KA	ST and SRL equivalent	0.1491	-114.32, 389.20

6.3 Further Experiments with Human Advice

To show the effect of adding human advice, we performed skill transfer without any (Figure 9). In the scenario shown, MoveDownfield to BreakAway, we compare learning curves for skill transfer with and without human advice. Our method still improves learning significantly when it includes no human advice about shooting, though the gain is smaller. The addition of our original human advice produces another significant gain.

To demonstrate that our method can cope with incorrect advice, we also performed skill transfer with intentionally bad human advice (Figure 10). In the scenario shown, KeepAway to BreakAway, we compare learning curves for skill transfer with our original human advice and with its opposite. In the bad advice the inequalities are reversed, so the rules instruct the learner to pass backwards, shoot when far away from the goal

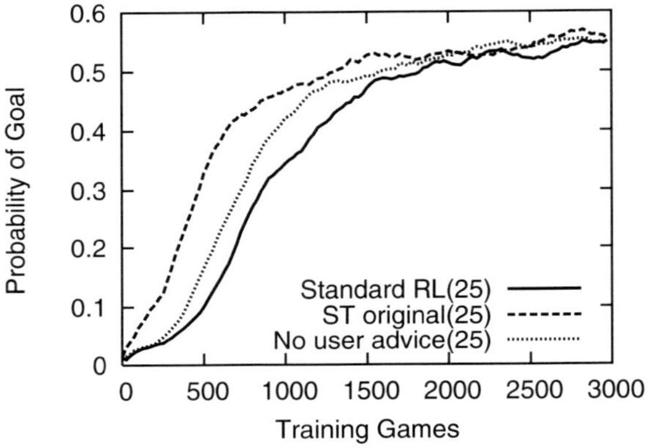

Fig. 9. Probability of scoring a goal while training in 3-on-2 BreakAway with standard RL and skill transfer (ST) from 3-on-2 MoveDownfield, with and without the original human advice

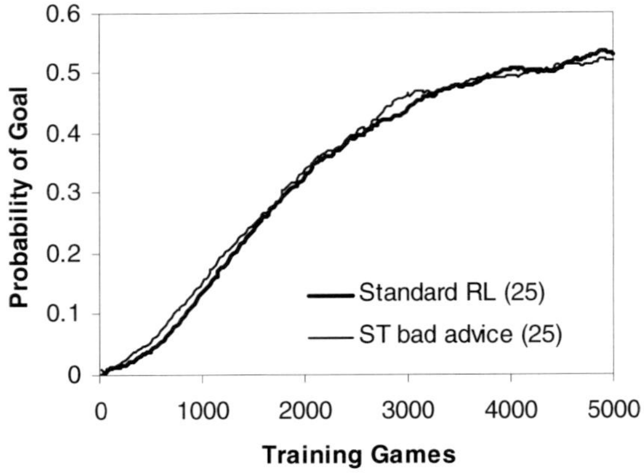

Fig. 10. Probability of scoring a goal while training in 3-on-2 BreakAway with standard RL skill transfer (ST) from 3-on-2 KeepAway that includes intentionally bad human advice

and at a narrow angle, and move when close to the goal. Our method no longer improves learning significantly with this bad advice, but since the KBKR algorithm can learn to ignore it, learning is never impacted negatively.

The robustness indicated by these experiments means that users need not worry about providing perfect advice in order for the skill-transfer method to work. It also means that skill transfer can be applied to reasonably distant tasks, since the source-task skills need not be perfect for the target task. It can be expected that learning with skill transfer will perform no worse than standard reinforcement learning, and it may perform significantly better.

7 Related Work

There is a strong body of related work on transfer learning in RL. We divide RL transfer into five broad categories that represent progressively larger changes to existing RL algorithms.

7.1 Starting-Point Methods

Since all RL methods begin with an initial solution and then update it through experience, one straightforward type of transfer in RL is to set the initial solution in a target task based on knowledge from a source task. Compared to the arbitrary setting that RL algorithms usually use at first, these *starting-point methods* can begin the RL process at a point much closer to a good target-task solution. There are variations on how to use the source-task knowledge to set the initial solution, but in general the RL algorithm in the target task is unchanged.

Taylor et al. [30] use a starting-point method for transfer in temporal-difference RL. To perform transfer, they copy the final value function of the source task and use it as the initial one for the target task. As many transfer approaches do, this requires a mapping of features and actions between the tasks, and they provide a mapping based on their domain knowledge.

Tanaka and Yamamura [27] use a similar approach in temporal-difference learning without function approximation, where value functions are simply represented by tables. This greater simplicity allows them to combine knowledge from several source tasks: they initialize the value table of the target task to the average of tables from several prior tasks. Furthermore, they use the standard deviations from prior tasks to determine priorities between temporal-difference backups.

Approaching temporal-difference RL as a batch problem instead of an incremental one allows for different kinds of starting-point transfer methods. In batch RL, the agent interacts with the environment for more than one step or episode at a time before updating its solution. Lazaric et al. [7] perform transfer in this setting by finding source-task samples that are similar to the target task and adding them to the normal target-task samples in each batch, thus increasing the available data early on. The early solutions are almost entirely based on source-task knowledge, but the impact decreases in later batches as more target-task data becomes available.

Moving away from temporal-difference RL, starting-point methods can take even more forms. In a model-learning Bayesian RL algorithm, Wilson et al. [38] perform transfer by treating the distribution of previous MDPs as a prior for the current MDP. In a policy-search genetic algorithm, Taylor et al. [31] transfer a population of policies from a source task to serve as the initial population for a target task.

7.2 Imitation Methods

Another class of RL transfer methods involves applying the source-task policy to choose some actions while learning the target task. While they make no direct changes to the target-task solution the way that starting-point methods do, these *imitation methods* affect the developing solution by producing different function or policy updates. Compared to the random exploration that RL algorithms typically do, decisions based on a source-task policy can lead the agent more quickly to promising areas of the environment. There are variations in how the source-task policy is represented and in how heavily it is used in the target-task RL algorithm.

One method is to follow a source-task policy only during exploration steps of the target task, when the agent would otherwise be taking a random action. Madden and Howley [11] use this approach in tabular Q-learning. They represent a source-task policy as a set of rules in propositional logic and choose actions based on those rules during exploration steps.

Fernandez and Veloso [5] instead give the agent a three-way choice between exploiting the current target-task policy, exploiting a past policy, and exploring randomly. They introduce a second parameter, in addition to the ε of ε-greedy exploration, to determine the probability of making each choice.

7.3 Hierarchical Methods

A third class of RL transfer includes *hierarchical methods*. These view the source as a subtask of the target, and use the solution to the source as a building block for learning the target. Methods in this class have strong connections to the area of hierarchical RL, in which a complex task is learned in pieces through division into a hierarchy of subtasks.

An early approach of this type is to compose several source-task solutions to form a target-task solution, as is done by Singh [22]. He addresses a scenario in which complex tasks are temporal concatenations of simple ones, so that a target task can be solved by a composition of several smaller solutions.

Mehta et al. [13] have a transfer method that works directly within the hierarchical RL framework. They learn a task hierarchy by observing successful behavior in a source task, and then use it to apply the MaxQ hierarchical RL algorithm [4] in the target task. This removes the burden of designing a task hierarchy through transfer.

Other approaches operate within the framework of *options*, which is a term for temporally-extended actions in RL [18]. An option typically consists of a starting condition, an ending condition, and an internal policy for choosing lower-level actions. An RL agent treats each option as an additional action along with the original lower-level ones.

In some scenarios it may be useful to have the entire source-task policy as an option in the target task, as Croonenborghs et al. [3] do. They learn a relational decision tree to represent the source-task policy and allow the target-task learner to execute it as an option. Another possibility is to learn smaller options, either during or after the process of learning the source task, and offer them to the target. Asadi and Huber [1] do this by finding frequently-visited states in the source task to serve as ending conditions for options.

7.4 Alteration Methods

The next class of RL transfer methods involves altering the state space, action space, or reward function of the target task based on source-task knowledge. These *alteration methods* have some overlap with option-based transfer, which also changes the action space in the target task, but they include a wide range of other approaches as well.

One way to alter the target-task state space is to simplify it through state abstraction. Walsh et al. [35] do this by aggregating over comparable source-task states. They then use the aggregate states to learn the target task, which reduces the complexity significantly.

There are also approaches that expand the target-task state space instead of reducing it. Taylor and Stone [29] do this by adding a new state variable in the target task. They learn a decision list that represents the source-task policy and use its output as the new state variable.

While option-based transfer methods add to the target-task action space, there is also some work in decreasing the action space. Sherstov and Stone [21] do this by evaluating in the source task which of a large set of actions are most useful. They then consider only a smaller action set in the target task, which decreases the complexity of the value function significantly and also decreases the amount of exploration needed.

Reward shaping is a design technique in RL that aims to speed up learning by providing immediate rewards that are more indicative of cumulative rewards. Usually it requires human effort, as many aspects of RL task design do. Konidaris and Barto [6] do reward shaping automatically through transfer. They learn to predict rewards in the source task and use this information to create a shaped reward function in the target task.

7.5 New RL Algorithms for Transfer

A final class of RL transfer methods consists of entirely new RL algorithms. Rather than making small additions to an existing algorithm or making changes to the target task, these approaches address transfer as an inherent part of RL. They incorporate prior knowledge as an intrinsic part of the algorithm. Our KBKR algorithm falls into this category of methods [32].

Price and Boutilier [19] propose a temporal-difference algorithm in which value functions are influenced by observations of expert agents. They use a variant of the usual value-function update that includes an expert's experience, weighted by the agent's confidence in itself and in the expert. They also perform extra backups at states the expert visits to focus attention on those areas of the state space.

There are several algorithms for case-based RL that accomodate transfer. Sharma et al. [20] propose one in which Q-functions are estimated using a Gaussian kernel over stored cases in a library. Cases are added to the library from both the source and target tasks when their distance to their nearest neighbor is above a threshold. Taylor et al. [28] use source-task examples more selectively in their case-based RL algorithm. They use target-task cases to make decisions when there are enough, and only use source-task examples when insufficient target examples exist.

8 Conclusion

We have described a method for transferring knowledge in reinforcement learning that learns logical rules to represent skills and uses them as advice for a new task. This approach can provide significant benefits in target tasks, as evidenced by our results in a complex RL domain. Our work has connections to Professor Michalski's interests in rule learning, advice, and analogical reasoning. As Michalski did, we emphasize the value of logic as a means of representing knowledge. We believe that first-order logic is a powerful mechanism for transfer.

An inherent aspect of transfer learning is recognizing the correspondences between tasks. Knowledge from one task can only be applied to another if it is expressed in a way that the target-task agent understands. If the task representations are not identical, a *mapping* is needed to translate between task representations. We assume a human-provided mapping so far, but learning a mapping is also an interesting task.

If a transfer method actually decreases performance, then *negative transfer* has occurred. One of the major challenges in developing transfer methods is to produce positive transfer between appropriately related tasks while avoiding negative transfer between tasks that are less related. Ideally, a transfer method would produce positive transfer between appropriately related tasks while avoiding negative transfer when the tasks are not a good match. In practice, these goals are difficult to achieve simultaneously. Approaches that have safeguards to avoid negative transfer often produce a smaller effect from positive transfer due to their caution.

Another challenge that we have encountered in RL transfer learning is that differences in reward structures between the source and target task make it difficult to transfer even shared actions. Changing the game objective or adding a new action changes the meaning of a shared skill. This means that it is important to continue learning in the target task and to avoid relying on source-task skills too much. We have also addressed this issue through human guidance, by allowing additional advice that points out differences between tasks.

Acknowledgements

This chapter was written while the authors were partially supported by DARPA grants HR0011-07-C-0060 and FA8650-06-C-7606.

References

1. Asadi, M., Huber, M.: Effective control knowledge transfer through learning skill and representation hierarchies. In: International Joint Conference on Artificial Intelligence, Hyderabad, India (2007)

2. Bloedorn, E., Michalski, R., Wnek, J.: Multistrategy constructive induction: AQ17-MCI. In: International Workshop on Multistrategy Learning (1993)
3. Croonenborghs, T., Driessens, K., Bruynooghe, M.: Learning relational skills for inductive transfer in relational reinforcement learning. In: International Conference on Inductive Logic Programming, Corvallis, OR (2007)
4. Dietterich, T.: Hierarchical reinforcement learning with the MAXQ value function decomposition. Journal of Artificial Intelligence Research 13, 227–303 (2000)
5. Fernandez, F., Veloso, M.: Probabilistic policy reuse in a reinforcement learning agent. In: Conference on Autonomous Agents and Multi-Agent Systems, Hakodate, Japan (2006)
6. Konidaris, G., Barto, A.: Autonomous shaping: Knowledge transfer in reinforcement learning. In: International Conference on Machine Learning, Pittsburgh, PA (2006)
7. Lazaric, A., Restelli, M., Bonarini, A.: Transfer of samples in batch reinforcement learning. In: International Conference on Machine Learning, Helsinki, Finland (2008)
8. Maclin, R., Shavlik, J., Torrey, L., Walker, T.: Knowledge-based support vector regression for reinforcement learning. In: IJCAI Workshop on Reasoning, Representation, and Learning in Computer Games, Edinburgh, Scotland (2005)
9. Maclin, R., Shavlik, J., Torrey, L., Walker, T., Wild, E.: Giving advice about preferred actions to reinforcement learners via knowledge-based kernel regression. In: AAAI Conference on Artificial Intelligence, Pittsburgh, PA (2005)
10. Maclin, R., Shavlik, J., Walker, T., Torrey, L.: A simple and effective method for incorporating advice into kernel methods. In: AAAI Conference on Artificial Intelligence, Boston, MA (2006)
11. Madden, M., Howley, T.: Transfer of experience between reinforcement learning environments with progressive difficulty. Artificial Intelligence Review 21, 375–398 (2004)
12. Mangasarian, O., Shavlik, J., Wild, E.: Knowledge-based kernel approximation. Journal of Machine Learning Research 5, 1127–1141 (2004)
13. Mehta, N., Ray, S., Tadepalli, P., Dietterich, T.: Automatic discovery and transfer of MAXQ hierarchies. In: International Conference on Machine Learning, Helsinki, Finland (2008)
14. Michalski, R.: A theory and methodology of inductive learning. Artificial Intelligence 20(2), 111–161 (1983)
15. Michalski, R.: Toward a unified theory of learning: Multistrategy task-adaptive learning. In: Buchanan, B.G., Wilkins, D.C. (eds.) Readings in Knowledge Acquisition and Learning: Automating the Construction and Improvement of Expert Systems. Morgan Kaufmann, San Francisco (1993)
16. Mitchell, T.: Machine Learning. McGraw-Hill, New York (1997)
17. Noda, I., Matsubara, H., Hiraki, K., Frank, I.: Soccer server: A tool for research on multiagent systems. Applied Artificial Intelligence 12, 233–250 (1998)
18. Perkins, T., Precup, D.: Using options for knowledge transfer in reinforcement learning. Technical Report UM-CS-1999-034, University of Massachusetts, Amherst (1999)
19. Price, B., Boutilier, C.: Implicit imitation in multiagent reinforcement learning. In: International Conference on Machine Learning, Bled, Slovenia (1999)
20. Sharma, M., Holmes, M., Santamaria, J., Irani, A., Isbell, C., Ram, A.: Transfer learning in real-time strategy games using hybrid CBR/RL. In: International Joint Conference on Artificial Intelligence, Hyderabad, India (2007)
21. Sherstov, A., Stone, P.: Action-space knowledge transfer in MDPs: Formalism, suboptimality bounds, and algorithms. In: Conference on Learning Theory, Bertinoro, Italy (2005)
22. Singh, S.: Transfer of learning by composing solutions of elemental sequential tasks. Machine Learning 8(3-4), 323–339 (1992)
23. Srinivasan, A.: The Aleph manual (2001)
24. Stone, P., Sutton, R.: Scaling reinforcement learning toward RoboCup soccer. In: International Conference on Machine Learning, Williamstown, MA (2001)

25. Sutton, R.: Learning to predict by the methods of temporal differences. Machine Learning 3, 9–44 (1988)
26. Sutton, R., Barto, A.: Reinforcement Learning: An Introduction. MIT Press, Cambridge (1998)
27. Tanaka, F., Yamamura, M.: Multitask reinforcement learning on the distribution of MDPs. Transactions of the Institute of Electrical Engineers of Japan 123(5), 1004–1011 (2003)
28. Taylor, M., Jong, N., Stone, P.: Transferring instances for model-based reinforcement learning. In: Daelemans, W., Goethals, B., Morik, K. (eds.) ECML PKDD 2008, Part II. LNCS (LNAI), vol. 5212, pp. 488–505. Springer, Heidelberg (2008)
29. Taylor, M., Stone, P.: Cross-domain transfer for reinforcement learning. In: International Conference on Machine Learning, Corvallis, OR (2007)
30. Taylor, M., Stone, P., Liu, Y.: Value functions for RL-based behavior transfer: A comparative study. In: AAAI Conference on Artificial Intelligence, Pittsburgh, PA (2005)
31. Taylor, M., Whiteson, S., Stone, P.: Transfer learning for policy search methods. In: ICML Workshop on Structural Knowledge Transfer for Machine Learning, Pittsburgh, PA (2006)
32. Torrey, L., Shavlik, J., Walker, T., Maclin, R.: Skill acquisition via transfer learning and advice taking. In: Fürnkranz, J., Scheffer, T., Spiliopoulou, M. (eds.) ECML 2006. LNCS (LNAI), vol. 4212, pp. 425–436. Springer, Heidelberg (2006)
33. Torrey, L., Walker, T., Shavlik, J., Maclin, R.: Using advice to transfer knowledge acquired in one reinforcement learning task to another. In: Gama, J., Camacho, R., Brazdil, P.B., Jorge, A.M., Torgo, L. (eds.) ECML 2005. LNCS (LNAI), vol. 3720, pp. 412–424. Springer, Heidelberg (2005)
34. Železný, F., Srinivasan, A., Page, D.:
35. Walsh, T., Li, L., Littman, M.: Transferring state abstractions between MDPs. In: ICML Workshop on Structural Knowledge Transfer for Machine Learning, Pittsburgh, PA (2006)
36. Watkins, C.: Learning from delayed rewards. PhD thesis, University of Cambridge (1989)
37. Watkins, C., Dayan, P.: Q-learning. Machine Learning 8, 279–292 (1992)
38. Wilson, A., Fern, A., Ray, S., Tadepalli, P.: Multi-task reinforcement learning: A hierarchical Bayesian approach. In: International Conference on Machine Learning, Corvallis, OR (2007)

Part III
Classification and Beyond

Determining the Best Classification Algorithm with Recourse to Sampling and Metalearning

Pavel Brazdil and Rui Leite

LIAAD-INESC Porto L.A./Faculty of Economics,
University of Porto, Rua de Ceuta, 118-6,
4050-190 Porto, Portugal
pbrazdil@liaad.up.pt, rleite@liaad.up.pt

Abstract. Currently many classification algorithms exist and no algorithm exists that would outperform all the others. Therefore it is of interest to determine which classification algorithm is the best one for a given task. Although direct comparisons can be made for any given problem using a cross-validation evaluation, it is desirable to avoid this, as the computational costs are significant. We describe a method which relies on relatively fast pairwise comparisons involving two algorithms. This method is based on a previous work and exploits sampling landmarks, that is information about learning curves besides classical data characteristics. One key feature of this method is an iterative procedure for extending the series of experiments used to gather new information in the form of sampling landmarks. Metalearning plays also a vital role. The comparisons between various pairs of algorithm are repeated and the result is represented in the form of a partially ordered ranking. Evaluation is done by comparing the partial order of algorithm that has been predicted to the partial order representing the supposedly correct result. The results of our analysis show that the method has good performance and could be of help in practical applications.

1 Introduction

Currently many classification algorithms exist and no algorithm exists that would outperform all the others. Therefore it is of interest to determine which classification algorithm is the best one for a given task. One possible way of solving the problem would be by evaluating all possible candidates using a well-known cross-validation (CV) procedure. We could then order all candidates using the mean success rate and pick the first item in the ordering. However, this procedure is slow, particularly if it involves many algorithms.

In this article we describe a method which relies on relatively fast pairwise comparisons involving two algorithms. This method is based on our previous work and exploits performance information of so called *sampling landmarks* representing accuracy estimates on simplified versions of the data [7, 19]. A series of sampling landmarks represents in effect a partial learning curve.

Earlier approaches have also explored data characteristics, captured in the form of various statistical and information–theoretic measures [3, 16]. Although methods based

J. Koronacki et al. (Eds.): Advances in Machine Learning I, SCI 262, pp. 173–188.
springerlink.com © Springer-Verlag Berlin Heidelberg 2010

on sampling landmarks usually perform better than those that rely on data characteristics [9, 10, 11], it was shown [10] that using both types of information may represent the best solution. So, the method described here uses both sampling landmarks and some dataset characteristics.

One key feature of this method is an iterative procedure for building up the characterization in the form of sampling landmarks. The characterization involves conducting experiments on a new dataset. The plan of these experiments is built up gradually, by taking into account the results of all previous experiments – both on previously used datasets and on the new dataset. Metalearning plays a vital role in determining what should be done and which algorithm is recommended.

The comparisons between various pairs of algorithm are repeated and the result is represented in the form of a partially ordered ranking. If two or more alternatives have a comparable performance, no ordering exists between them in the ranking.

In this article we also discuss how results of this form can be evaluated. The objects of study here are the predicted partial order of items which is compared to similar partial order representing the supposedly correct result. The results of this evaluation show that the method has good performance and could be of help in practical applications.

This article is organized as follows. In Section 2 we describe the problem of determining the best of N classification algorithms in more detail. In section 3 we review the method for determining which of the two given algorithms is better. This method exploits sampling landmarks and data characteristics. Section 4 discusses the problem of evaluation and presents the results.

2 Determining the Partial Order of Classification Algorithms

The issue of identifying the best classification algorithm(s) can be resolved in various ways. One simple possibility involves the following steps:

1. Pick a pair of classification algorithms and determine which one of the two is better using method T.
2. Repeat this for all pairs of algorithms.
3. Process the partial order obtained in order to identify the best element(s).

Let us refer to this method as $TestAll_T$, where the suffix $_T$ represents the test method used. It is a parameter of the method that needs to be supplied by the user. Let us analyze some of the steps in more detail, starting with step 1.

Step 1. Suppose we have chosen algorithms A_i and A_j. One simple method which is often used involves running a cross-validation (CV) procedure and applying a statistical significance test. Let us refer to this method here as CV_{ST} (comparison using CV accompanied by a significance test). Later on in this paper we describe an alternative method (SAM, Selection of Algorithms using Metalearning) which does this faster and produces also good results. Here we will use the term T to represent all the methods of this kind, including CV_{ST} and SAM.

Method T (standing for CV_{ST}, SAM) can be seen as function applied to algorithms A_i and A_j which returns result $+1(-1)$, if A_i is significantly better (worse) than A_j. The value 0 is returned when it is not possible to determine whether A_i is better than A_j, or vice versa.

Step 2. involves applying T to all pairs of algorithms. The results for N algorithms can be represented using a graph where algorithms are represented by nodes and the relation *is significantly better than* by links.

Link $A_i \rightarrow A_j$ is drawn if and only if $T(A_i, A_j) = +1$. The graph results in a partial order of items. The function T may return 0, in which case no link is drawn and consequently the order of these items is undetermined. Some authors use also the term *quasi-linear ranking*, instead of partial order of items [5].

Fig. 1 shows an example of a partial order of 6 algorithms representing the result for Abalone dataset, obtained by the method CV_{ST} discussed earlier. More details about the classification algorithms used here can be found later (Section 4).

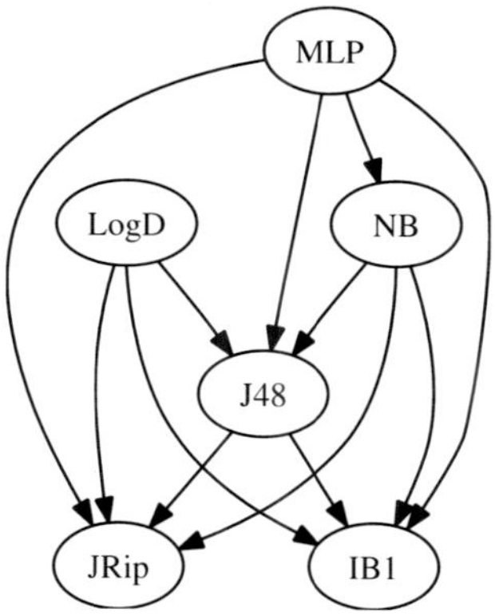

Fig. 1. An example of partial order of 6 classification algorithms obtained by $TestAll_{CV_{ST}}$ for Abalone dataset

Step 3. The aim is to analyze the partially ordered set of items and identify the top-most element(s). This seems straightforward, but there is one issue that requires some clarification.

Let us first consider an example that involves three elements identified as A_1, A_2 and A_3. Suppose we have completed steps 1 and 2. The following figure shows some of many possible outcomes:

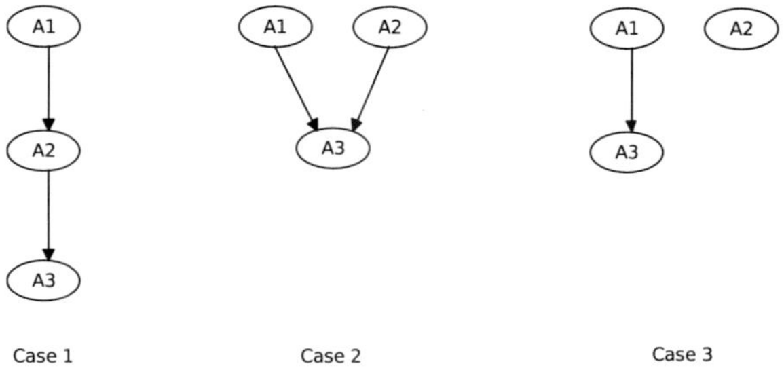

Fig. 2. Several possible outcomes after evaluation and pairwise comparisons of three items

Regards case 1 there is no doubt as to which algorithm should be positioned at the topmost level (i.e., A_1). Case 2 is also relatively straightforward. As we have no way of distinguishing between A_1 and A_2, both elements share the topmost level (level 1), while A_3 appears on level 2.

Case 3 is more problematic, as we can use different criteria to position elements into levels. Basically, we can consider the following strategies:

- L, considering losses (incoming arrows),
- W, considering wins (outgoing arrows),
- LW, considering both losses and wins.

If we adopt strategy L, the items A_1 and A_2 have the least number of losses (zero losses) and hence they should both share the topmost level (level 1) As item A_2 has suffered one loss (represented by an incoming arrow if we use a graph), it should be placed on level 2. The corresponding graph can be redrawn to take this into account. W note that Fig. 2 case 3 conforms already to this scheme.

Strategy W gives a somewhat different result. A_1 is placed on level 1, as it has scored one win, while A_2 and A_3 appear on level 2 (no wins). Strategy WL gives yet another result. Suppose wins are represented by +1, losses by -1 and draws by 0. If we use this scheme, A_1 scores +1, A_2 gets 0 points, while A_3 gets -1 point.

Here we have adopted strategy L, although we had no particular reason for preferring this strategy to the others.

Method for reordering elements into levels. The general method for reordering elements into levels using strategy L is quite straightforward. Let us assume that the graph is represented in the usual way, that is using a set of nodes, N and set of oriented links, L. In the first step the aim is to identify nodes with the least number of incoming arrows (i.e., the least number of losses).

Let us consider the example of 6 partially ordered algorithms shown earlier in Fig. 1. As MLP has no incoming arrows, it is placed at the topmost level (level 1). Then all nodes identified in this step are marked and so are all the links originating from these nodes. In our example this includes node MLP and the links $MLP \rightarrow LogD$, $MLP \rightarrow NB$ and $MLP \rightarrow J48$. The whole process is then repeated. As $LogD$ and NB have no incoming links they are placed on level 2. The process continues like this until termination.

3 Predicting the Relative Performance of a Pair of Algorithms

The aim in this section is to describe the method SAM which uses both data characteristics and performance on samples to predict which of the two given algorithm (A_p and A_q) should be used for a new dataset (i.e. dataset for which we seek a recommendation). The approach described follows the general strategy of metalearning [2]. The method relies on two kinds of information:

- Information about various datasets used in the past. This information is of two kinds: (1) data characteristics, and (2) information about accuracies of the two classifications algorithms on samples of data from these datasets. Following [2] this information is referred to as *metaknowledge*.

– Information about the new dataset. This is again of two kinds, similar to above, but the information about accuracies of the two classifications algorithms is only *partial*. It includes accuracies relative to *some data samples* only.

The method is based on an intuition that performance of the two algorithms on a set of samples sheds some light on the final outcome of the learning. Let us analyze Fig. 3 that illustrates this. Its shows the measured accuracy on three samples of data s_1, s_2 and s_3. Here $s_1 \cdots s_m$ are used to represent random samples of increasing size. The two curves have been extended on the basis of other similar curves retrieved from the metadatabase. It enables to determine that algorithm A_q will achieve higher accuracy at the end, that is, when the whole dataset is used. The details regards this method are described in the following.

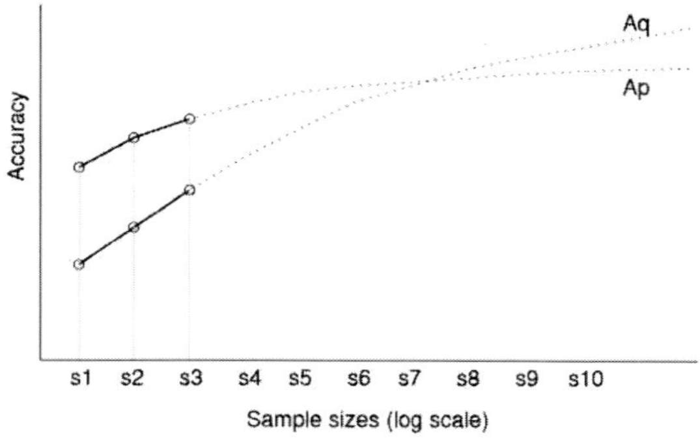

Fig. 3. Example of two extended learning curves determining which algorithm is better

Metaknowledge about Past Datasets. As has been mentioned before, the method described here relies on the existence metaknowledge about various datasets dealt with in the past. Let us refer to these datasets as $d_1 \cdots d_n$. As has been mentioned, the information is of two kinds: data characteristics and information about how various classifications algorithms performed on samples of data from these datasets.

Regards data characteristics here we consider the set presented in Table 1. We have chosen this set as it was used in other studies [18] and it was shown that this set

Table 1. Data Characteristics Measures Used

Id	Measure
DC_1	Number of examples
DC_2	Proportion of symbolic attributes
DC_3	Proportion of missing values
DC_4	Proportion of numerical attributes with outliers
DC_5	Class entropy
DC_6	Average mutual information between attributes and the class
DC_7	First canonical correlation

achieves similar results in metalearning tasks as a much larger set suggested in previous work [16].

Besides this, we need information about how various classifications algorithms performed on an exhaustive set of samples of data from these datasets. In other words, we need information about full learning curves.

Metaknowledge about the New Dataset. This metaknowledge is again of two kinds. First, it includes data characteristics, such as those shown earlier 1.

Besides, the method requires some information about how the chosen two classifications algorithms performed on *some* samples of data from the new dataset. Normally only some of these meta-attributes are sufficient for the prediction. In other words, only a part of each learning curve is sufficient for making quite good prediction.

In the next section we describe method SAM_F that uses a fixed set of samples. Further on (in Section 3.2) we describe a more general approach (SAM), that uses an automatic mechanism to determine which samples should be used.

3.1 Using Data Characteristics and Characteristics Based on a Fixed Set of Samples

This method described here, referred to as SAM_F, is based on previous work [10, 12]. As we shall see in the next section it forms a part of the more general method SAM.

The method requires as input a fixed set of samples for each algorithm. Let S_p (S_q) represent the samples required to characterize algorithm A_p (A_q). So, for instance, the samples passed as input can be $S_p =< s_1, s_2, s_3 >$ and $S_q =< s_1, s_2 >$. The method encompasses the following steps:

1. Compute the data characteristics for the new dataset d.
2. Characterize the new dataset d by conducting experiments with algorithm A_p on S_p samples and measuring accuracies. Repeat this for A_q on S_q samples. In other words build two partial learning curves.
3. Compute the distances between the information relative to dataset d and stored information relative to all other datasets $d_1 \cdots d_n$ (see the subsection below for details).
4. Identify the subset of k nearest datasets.
5. For each of the k nearest datasets identified and retrieved in the previous step, *adapt* each pair of learning curves to the new partial learning curves build for dataset d. Adaptation is done by rescaling each retrieved learning curve in order to minimize the square distance from this curve to the respective partial learning curve for dataset d.
6. For each pair of adapted curves decide which algorithm achieves higher performance on the adapted and extended learning curves.
7. Identify the algorithm that is better on the new dataset d, by considering the results on k pairs of nearest datasets.

More details on the distance function used in Step 3. The distance function used consists of two parts. One that calculates the distance using data characteristic measures and another that uses the estimates of accuracies of the algorithms on samples It is defined by the following equation:

$$d(d_i, d_j) = w_{DC} * d_{DC}(d_i, d_j) + w_S * d_S(d_i, d_j) \tag{1}$$

where w_{DC} and w_S are weights (normally set to 0.5 each). The distance relative to dataset characteristics is calculated using the equation shown below. Each measure is first rescaled to fit the interval [0,1].

$$d_{DC}(d_i, d_j) = \sum_{k \in DC} \frac{\mid DC_t(i) - DC_t(j) \mid}{max(DC_k) - min(DC_k)} \tag{2}$$

where DC contains the indices of classical data characteristics. The distance relative to accuracies on samples is calculated using the equation:

$$d_S(d_i, d_j) = \underbrace{\sum_{s \in S_p} \mid A_{p,i,s} - A_{q,j,s} \mid}_{Ap} + \underbrace{\sum_{s \in S_q} \mid A_{q,i,s} - A_{q,j,s} \mid}_{Aq} \tag{3}$$

where S_p (S_q) contains the indices of meta-features used to characterize the new case (dataset) using accuracies of algorithm A_p (A_q) on specific samples. The symbol $A_{p,i,j}$ represents the meta-feature corresponding to the estimate of accuracy of algorithm A_p on a sample of size s_j extracted from dataset d_i.

The disadvantage the method just described is that the user has to decide how many samples should be used and what their sizes should be. This is not easy to determine.

If we use the performance on more samples, or larger samples, it is reasonable to expect that better decisions will be made, as the shape of the learning curves will be better defined. However the computational costs involved in computing the meta-attributes (i.e. performance of the algorithms on these samples) will rise too. If we use fewer samples, or smaller samples, the quality of the decision will decrease. The advantage is that the computational costs will also decrease. A good decision involves striking a good compromise. In the next section we describe a method that resolves this issue.

3.2 Method that Determines which Samples to Use

Method SAM described here seeks a good compromise between trying to achieve a good performance and keep the computational costs to a minimum. In this respect it differs from an ordinary feature selection method (e.g. forward selection) that would only try to improve the accuracy of the method.

The desired sequence of samples is identified using a hill climbing approach. At each stage the existing set of characteristics is considered and a plan of possible extensions is built up. These are ordered by costs (i.e., time needed to compute them).

The aim is to improve the capacity of the method to predict which of the two algorithms is better for a given task. The plan formulated earlier is followed to determine which of the extensions defined earlier can improve the meta–accuracy of the method by a certain pre-defined amount Δ. As the alternatives are ordered by increasing costs, the first alternative that is expected to provide this improvement is accepted. The method continues like this until no further improvement can be made. An overview of the method is presented in the following:

1. Compute the data characteristics for the new dataset d.
2. Initialize the meta-characteristics of dataset d based on samples of data (e.g., with $S_{p,1} =< s_1 >$ and $S_{q,1} =< s_1 >$). Call SAM_F to determine which algorithm (A_p or A_q) is better and store the prediction on a stack.
3. Generate a plan concerning how to extend the existing meta-characteristics. Consider both classical characteristics and characteristics based on samples of data. Order the alternatives by increasing costs (i.e., time needed to compute them). (more details on this are given later).
4. For each alternative defined in the plan conduct virtual experiments with SAM_F to estimate the corresponding meta–accuracy As soon as an alternative is identified that is likely to improve the previous meta-accuracy by a given amount Δ. If no alternative has been identified, go to step 6 to terminate. (more details on this are given later).
5. Call SAM_F with the alternative identified to determine which algorithm (A_p or A_q) is better. Store the prediction on a stack and go back to step 3.
6. Retrieve the last prediction from the stack regards which algorithm (A_p or A_q) is better and output it.

More details on step 4. Virtual experiments are conducted by exploiting solely the information stored in the meta-database. Therefore these experiments are rather fast to execute. Method leave–one–out is followed. In each cycle one dataset is left out, and method SAM_F is called with the proposed set of meta-features. The aim is to obtain a prediction for the case in question and make comparisons with correct value. After all cycles of leave–one–out, it is possible to determine the estimate of meta–accuracy.

4 Evaluation of Method SAM Used for Pairwise Comparisons

In this section we describe the evaluation methodology and also the experiments whose aim was to evaluate the method SAM used in pairwise comparisons.

Evaluation Methodology. To evaluate the method proposed we need to compare, for each case (dataset), the predicted class with the true classification determined by a usual cross–validation evaluation procedure on each dataset for the two given algorithms. A statistical test (t–test) is used to compute the statistical significance. Instead of using the usual accuracy measure, a different measure is used that is more suited for our classification task with 3 possible outcomes. The errors are called *penalties* and are calculated as follows:

- If the classification is +1 (or -1), while the true class is -1 (or +1), this counts as a complete failure, and the penalty is 1.
- If the method classifies the dataset as 0, while the true class is +1 (or -1) then we consider that the method partially failed. The penalty is 0.5.

- If a particular method (e.g. SAM) classifies some case as +1 (or -1), while the true class is 0 (the given algorithms are not significantly different) then, from a practical point of view the method did not fail, because any decision is a good one. Therefore the penalty is 0.

The corresponding accuracy, referred to as *meta–accuracy*, is computed using the formula $1 - \sum_{d \in D} \frac{penalty(d)}{|D|}$, where D is the collection of datasets.

Evaluation Set-up. In this empirical study we have used the following 6 base-level classification algorithms, all implemented within Weka [20] machine learning tools:

- J48 (a version of C4.5 implemented in Weka),
- JRip - rule set learner (RIPPER [4]),
- LogD, logistic discriminant [8],
- MLP - multi-layer perceptron,
- IB1 - instance-based learner,
- NB, naive bayes.

Using this setting we get 15 classification problems, one for each pair of algorithms. In total 40 datasets were used in the evaluation. Some come from UCI [1], others from the project METAL [15].

The datasets were characterized using 6 data characteristic measures presented in Table 1.

Regards meta–attributes based on performance of classification algorithms on samples, samples $s_1 \cdots s_{10}$ were used. The calculation of the size of s_i was determined using the following formula $2^{6+0.5*i}$. For s_1, the size of the first sample, this gives $2^{6.5}$ 91 cases after rounding. For s_2 this gives 2^7, i.e. 128 cases etc. Each sample included randomly chosen cases subject to the restriction that s_i be a subset of s_{i+1} for all i. This restriction helps to restrict the variance of the resulting classifier to a minimum.

For each decision problem the method was evaluated using a leave-one-out methodology. In each case we measure meta-accuracy and computational cost. Computational costs are expressed as a ratio of times. The time required by the particular method (e.g. SAM) is normalized, by dividing it by the time required to obtain the decision by cross-validation. Cross-validation represents a slower, but, in principle, also a more reliable method.

In the experiments the following setting were used $\Delta = 0.07$ and $\Delta_{DC} = 0.001$ Since the data characteristic measures were previously shown to be less predictive in comparison with performance on samples meta-attributes the value of Δ_{DC} was set to a much smaller value than the value of Δ.

Regards the number of neighbours, here we have used $k = 23$ which was shown to be a good setting in previous experiments.

Results. Table 2 shows the results including the meta–accuracies. The decision problem in question is presented in column 1. Default meta–accuracy, determined as usual by the most frequent class, is shown in column 2. Meta-accuracy is shown in column 3.

As can be seen it is rather high on average (92.35) indicating that in a high proportion of cases method SAM recommended the right decision. The method is much more precise than the method that uses classical dataset characteristics (its average meta–accuracy was 81.15).

Regarding the costs of the method, measured by the time spent on computing required metafeatures, the results are summarized in Table 3. The results are presented in the form of ratios. The time spent by each method is related to the time taken by cross-validation. The values presented here are average values (geometric mean). The method described here is about 7 times faster ($= 1/0.139$) than cross–validation. The method that relies on classical data characteristics, SAM_{DC}, is faster, but its accuracy is much lower.

It is interesting to analyse which sequences of samples were generated by SAM. Here we present some typical patterns:

- $S_{IB1} =< s_1, s_2, s_4 >$ and $S_{NB} =< s_1, s_2, s_5 >$ (abalone),
- $S_{IB1} =< s_1, s_2, s_6 >$ and $S_{NB} =< s_1, s_2, s_5 >$ (acetylation),
- $S_{J48} =< s_1, s_8 >$ and $S_{JRip} =< s_1, s_8 >$ (abalone),
- $S_{J48} =< s_1, s_8 >$ and $S_{JRip} =< s_1, s_8 >$ (byzantine),

We note the sequences generated are typically quite short. The sequences shown above consist of at most three samples. Furthermore we note that the sequences may include gaps and nonconsecutive samples. For instance, in the first example shown that involves the algorithms $IB1$ and NB on *abalone* dataset, the samples s_1 and s_2 are not followed by s_3, but rather by s_4. The estimate of accuracy on s_3 is not sufficiently informative and does not allow to required increase the meta–accuracy. Therefore a larger sample (s_4) is used. Finally, we not also different algorithms are characterized using different samples. This is yet another interesting feature of the method.

Table 2. Meta–accuracy of the method (average)

Algorithm Pair	Default Accuracy	Meta-Accuracy
IB1–J48	80.00	92.50
IB1–JRip	72.50	92.50
IB1–LogD	66.67	97.22
IB1–MLP	84.21	92.11
IB1–NB	67.50	92.50
J48–JRip	75.00	90.00
J48–LogD	77.78	97.22
J48–MLP	63.16	86.84
J48–NB	85.00	90.00
JRip–LogD	75.00	86.11
JRip–MLP	68.42	89.47
JRip–NB	82.50	92.50
LogD–MLP	80.00	97.14
LogD–NB	94.44	94.44
MLP–NB	94.74	94.74
Mean	77.79	92.35

Table 3. Time of the method expressed as a proportion of time used by cross-validation

	SAM_{DC}	SAM
Geometric mean	0.007	0.139

A similar analysis can be made regards classical data characteristics. We have observed that most of the times these measures were not selected by SAM and when they were selected the decision concerning the best algorithm remained basically unaffected.

The measures seemed to have been useful for a relatively small proportion of datasets. The figures were in the range between 0% and 18.4%, but in many cases it was either 0% or 2.8% More details about this study can be found in [12].

5 Evaluation of the Method for Determining the Best of N Algorithms

Evaluation involves comparing a prediction to some baseline. In a general case both are partially ordered rankings. The prediction is represented by a partially ordered ranking returned by the variant of the method $TestAll_{SAM}$ which calls SAM. Figure 4 shows an example of a partially ordered ranking obtained for Abalone dataset. The

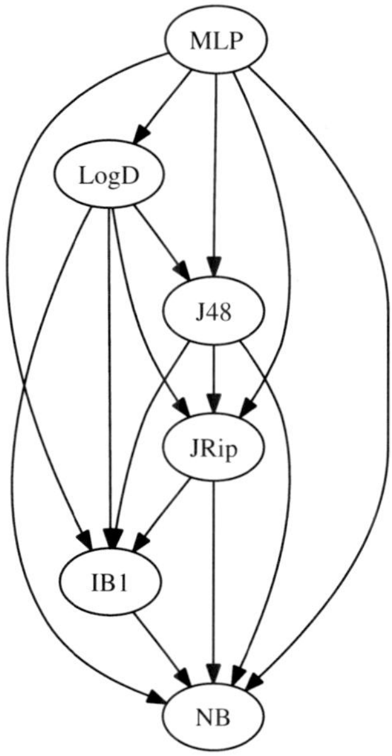

Fig. 4. An example of predicted partial order of 6 algorithms obtained for Abalone dataset obtained using $TestAll_{SAM}$

baseline is represented by the partially ordered ranking returned by the variant of method $TestAll_{CV_{ST}}$ which calls CV_{ST}. An example of a partially ordered ranking obtained for Abalone dataset using this method was shown earlier in Figure 1.

It should be stressed that the method $TestAll_{CV_{ST}}$ is not entirely reliable, as Type I and Type II errors may occur. However, the aim here is to use this method as a baseline for comparison purposes. The question is how well does the method $TestAll_{SAM}$ perform when compared to the baseline. The problem that we wish to discuss next is what metric should be used that would permit to evaluate the quality of predicted ranking with respect to the supposedly correct ranking.

Here we will focus on just the first level in both rankings. This makes sense, as normally we are interested to identify those algorithms that are considered to be the best ones (in general, there may be more than one). We want to have a way of evaluating such situations.

5.1 Evaluation That Focuses on the Top Level in Both Rankings

The aim of this section is to describe an evaluation method that is oriented towards just the first level in both rankings. Before describing our proposal, let us analyze a few situations that may occur (see Table 4).

Cases 1 and 2 are straightforward. As the predictions agree with the correct value, the measure of success (MS) is 1. Case 3 is again straightforward. The prediction involves only one item (A_1) and as this disagrees with the baseline prediction (A_2), the measure of success is 0.

In case 4 the prediction and the correct value are not identical. Our strategy here is to adopt a "benevolent position" and still consider that the measure of success is 1 in this case. This can be justified by the fact that if A_1 and A_2 are equivalent, it is good enough for the method to propose one of them (A_1 in this case).

Case 5 involves a prediction of two items (A_1, A_2), while only A_1 is correct. Here the predictions that include more than one item are considered as *uncertain predictions*. These should be accompanied by probabilities. If we have no specific information available, the best is to assume that the predictions are equi-probable. (i.e. here both $1/2$).

The method can be formalized as follows. First, let us introduce some notation. Let $\hat{Y}_k$ represent the set of elements predicted for case k. Let Y_k represent the set of elements that are correct for case k. So, for instance, if we consider the data shown in

Table 4. Examples of some possible situations involving the top level

Case	Prediction $TestAll_{SAM}$	Correct value $TestAll_{CV_{ST}}$	Measure MS
1	A_1	A_1	1
2	A_1, A_2	A_1, A_2	1
3	A_2	A_1	0
4	A_1	A_1, A_2	1
5	A_1, A_2	A_1	0.5

Table 5. Evaluation results for 6 algorithms on different datasets

	Prediction	Correct Value	MS
abalone	MLP	LogD MLP	1
acetylation	J48	IB1 J48 JRip NB	1
adult	JRip	J48	0
allbp	J48	J48 JRip MLP	1
allhyper	JRip	J48	0
ann	J48	J48 JRip	1
byzantine	IB1	IB1 MLP	1
car	MLP	MLP	1
cmc	JRip	J48 JRip LogD MLP	1
contraceptive	JRip	J48 JRip LogD MLP	1
injury_severity	J48	J48 MLP	1
internetad	JRip	J48 JRip NB	1
isolet	MLP	MLP	1
...	...	...	...

the table above, $\hat{Y}_5$ returns A1,A2 and $\hat{Y}_6$ returns A1. The method for calculating MS relative to case k can be expressed as follows:

$$MS_k = \sum_{i,j} I(\hat{Y}_k[i], Y_k[j]) * p(\hat{Y}_k[i])$$

where $I(x, y)$ returns 1 if $x = y$ and 0 otherwise and $p(\hat{Y}_k[i])$ represents the probability accompanying the prediction i for case k.

5.2 Evaluation Results for 6 Algorithms

The evaluation results for 6 algorithms results obtained for some of the datasets used in our tests are shown in Table 5. As can be seen, our method does the right recommendation in most cases as is indicated by the measure of success, $MS = 1$. The average value of the measure of success, MS, on all datasets considered was 0.9.

6 Discussion

6.1 Other Approaches for Evaluation of Rankings

In section 5.1 that has focused on the top level in both rankings. However, sometimes we may want to consider more levels in the rankings. We may want to distinguish situations when the prediction is not far from the ideal from others.

One possibility is to adopt a measure of correlation (e.g. Spearman correlation) or perhaps better, *weighted rank measure* that attributes more weight to items in top positions [6].

Another possibility is to assume that the evaluation is carried out with respect to the top level in the correct ranking and the aim is to determine how far down in the predicted ranking are the correct items. A measure of this kind, *estimated rank of best algorithm*, was used in [13]. Alternatively we could assume that the evaluation should be carried out with respect to the top level in the predicted ranking and the aim is to determine how far down in the correct ranking are the predicted elements. A measure of this type, referred to as *average rank of predicted algorithm*, was used in [14].

6.2 How Is the Method Described Related to Planning?

Approaches to planning (see e.g. [17]) can be divided into two groups depending on what the objective is. The first group include planners that generate a full plan which is then followed step by step in the execution phase. The second approach just determines the first action (or a few actions) to execute. The rest of the plan is not fully determined. The system then executes the actions and continues in this mode until the final goal has been attained. The latter approach is useful in situations where there is incomplete knowledge of the environment and when the execution of some actions brings in more information.

The method presented here follows the second approach referred to above. It tries to establish the next action to execute and executes this action. It continues like this until the stopping condition is satisfied. In our case the aim of the action it to gather more information about one of the two algorithms in question. This involves training each algorithm on a sample of a particular size and getting the estimates of performance on a test sample. This process terminates when enough information is available and a decision can be made as to which of two algorithms is better.

The problem tackled here is a somewhat different from typical problems handled by many planning systems. Solutions are characterized by training and test times, or in general by costs. So normally, considering that many algorithms exist, these will have different benefits (e.g. accuracies) and different costs. So, on one hand we can identify a low cost solutions (fast algorithms) which normally do not perform very well. On the other hand we may have solutions with achieve higher performance, but incur also higher costs.

In the work presented here we were interested in a compromise between the two. Our aim is to obtain a solution that approaches the performance of the best possible solution (selection by cross-validation), but incurs much lower costs (i.e., is N times faster).

7 Conclusions

In this paper we have described method SAM that exploits both classical data characteristics and estimates of accuracy on data samples to determine which of two given classification algorithms is better on a new dataset.

Here we have shown how we can extend our previous method, by including the measures of both kinds. The metafeatures (meta-attributes) used by the method are either data characteristics of the dataset in question, or performance estimates of the given classification algorithms on samples of specified sizes.

The method automatically establishes how many samples are needed and their sizes. Besides the method also determines which data characteristic measures are useful for the decision concerning which is the best algorithm.

Experimental evaluation has shown that method SAM used to determine the best algorithm in pairwise comparisons achieves good performance when compared to the other methods. The average meta-accuracy of this method was 92.35%.

The method $TestAll_{SAM}$ that was used to predict the best classification algorithm out of 6 given algorithms achieved also very good results. The method described did the right recommendation in most cases. The average value of the measure of success, MS, on all datasets considered was 90%.

The methods described represent thus a significant improvement over previous methods for dealing with the problem of predicting the relative performance of learning algorithms in a systematic manner.

References

1. Asuncion, A., Newman, D.J.: UCI machine learning repository (2007)
2. Brazdil, P., Giraud-Carrier, C., Soares, C., Vilalta, R.: Metalearning: Applications to Data Mining. Springer, Heidelberg (2009)
3. Brazdil, P., Soares, C., Costa, J.: Ranking learning algorithms: Using IBL and meta-learning on accuracy and time results. Machine Learning 50, 251–277 (2003)
4. Cohen, W.W.: Fast effective rule induction. In: Prieditis, A., Russell, S. (eds.) Proc. of the 12th International Conference on Machine Learning, Tahoe City, CA, July 9-12, pp. 115–123. Morgan Kaufmann, San Francisco (1995)
5. Cook, W.D., Kress, M., Seiford, L.W.: A general framework for distance-based consensus in ordinal ranking models. European Journal of Operational Research 96(2), 392–397 (1996)
6. Costa, J.P., Soares, C.: A weighted rank measure of correlation. Australian and New Zealand Journal of Statistics 47(4), 515–529 (2005)
7. Fürnkranz, J., Petrak, J.: An evaluation of landmarking variants. In: Proceedings of the ECML/PKDD Workshop on Integrating Aspects of Data Mining, Decision Support and Meta-Learning (IDDM 2001), pp. 57–68. Springer, Heidelberg (2001)
8. le Cessie, S., van Houwelingen, J.C.: Ridge estimators in logistic regression. Applied Statistics 41(1), 191–201 (1992)
9. Leite, R., Brazdil, P.: Predicting relative performance of classifiers from samples. In: ICML 2005: Proceedings of the 22nd international conference on Machine learning, pp. 497–503. ACM Press, New York (2005)
10. Leite, R., Brazdil, P.: An iterative process for building learning curves and predicting relative performance of classifiers. In: Neves, J., Santos, M.F., Machado, J.M. (eds.) EPIA 2007. LNCS (LNAI), vol. 4874, pp. 87–98. Springer, Heidelberg (2007)
11. Leite, R., Brazdil, P.: An iterative process of building learning curves and predicting relative performance of classifiers. In: Brazdil, P., Bernstein, A. (eds.) Proceedings of the Planning to Learn Workshop (PlanLearn 2007), held at ECML/ PKDD 2007, pp. 31–40 (2007)
12. Leite, R., Brazdil, P.: Selecting classifiers using metalearning with sampling landmarks and data characterization. In: Brazdil, P., Bernstein, A., Hunter, L. (eds.) Proceedings of the Planning to Learn Workshop (PlanLearn 2008), held at ICML/COLT/UAI 2008, Helsinki, Finland, pp. 35–41 (2008)
13. Ler, D., Koprinska, I., Chawla, S.: A new landmarker generation based on correlativity. In: Proceedings of the IEEE International Conference on Machine Learning and Applications, pp. 178–185. IEEE Press, Louisville (2004)
14. Ler, D., Koprinska, I., Chawla, S.: Utilizing regression-based landmarkers within a meta-learning framework for algorithm selection. In: Proceedings of the Workshop on Meta-Learning, associated with 22nd International Conference on Machine Learning, Bonn, Germany, pp. 44–51 (2005)
15. Metal project site (1999), `http://www.metal-kdd.org/`

16. Michie, D., Spiegelhalter, D.J., Taylor, C.C.: Machine Learning, Neural and Statistical Classification. Ellis Horwood (1994)
17. Russell, S.J., Norvig, P.: Artificial Intelligence: A Modern Approach. Pearson Education, London (2003)
18. Soares, C.: Learning Rankings of Learning Algorithms. PhD thesis, Department of Computer Science, Faculty of Sciences, University of Porto (2004)
19. Soares, C., Petrak, J., Brazdil, P.: Sampling-based relative landmarks: Systematically test-driving algorithms before choosing. In: Brazdil, P.B., Jorge, A.M. (eds.) EPIA 2001. LNCS (LNAI), vol. 2258, pp. 88–94. Springer, Heidelberg (2001)
20. Witten, I., Frank, E., Trigg, L., Hall, M., Holmes, G., Cunningham, S.: Weka: Practical machine learning tools and techniques with Java implementations (1999)

Transductive Learning for Spatial Data Classification

Michelangelo Ceci, Annalisa Appice, and Donato Malerba

Dipartimento di Informatica,
Università degli Studi di Bari,
Via Orabona 4, I-70126 Bari, Italy
{ceci,appice,malerba}@di.uniba.it

Abstract. Learning classifiers of spatial data presents several issues, such as the heterogeneity of spatial objects, the implicit definition of spatial relationships among objects, the spatial autocorrelation and the abundance of unlabelled data which potentially convey a large amount of information. The first three issues are due to the inherent structure of spatial units of analysis, which can be easily accommodated if a (multi-)relational data mining approach is considered. The fourth issue demands for the adoption of a transductive setting, which aims to make predictions for a given set of unlabelled data. Transduction is also motivated by the contiguity of the concept of positive autocorrelation, which typically affect spatial phenomena, with the smoothness assumption which characterize the transductive setting. In this work, we investigate a relational approach to spatial classification in a transductive setting. Computational solutions to the main difficulties met in this approach are presented. In particular, a relational upgrade of the naïve Bayes classifier is proposed as discriminative model, an iterative algorithm is designed for the transductive classification of unlabelled data, and a distance measure between relational descriptions of spatial objects is defined in order to determine the *k*-nearest neighbors of each example in the dataset. Computational solutions have been tested on two real-world spatial datasets. The transformation of spatial data into a multi-relational representation and experimental results are reported and commented.

1 Introduction

Several applications, such as fleet management, environmental and ecological modeling, remote sensing, are the source of a huge amount of spatial data, which are stored in spatial databases of Geographic Information Systems (GISs). A GIS is a software system that provides the infrastructure for editing, storing, analyzing and displaying spatial objects. Popular GISs (e.g. ArcView, MapInfo and Open GIS) have been designed as a toolbox that allows planners to explore spatial data by zooming, overlaying, and thematic map coloring. They are provided with functionalities that make the spatial visualization of individual variables effective, but overlook complex multi-variate dependencies. GIS vendors and researchers now recognize this limitation and have begun to address it by adding spatial data interpretation capabilities to the systems.

Wessel [48] describes a prototypical GIS extended with a knowledge-base, expressed in Description Logic, and some reasoning capabilities which can be useful for the interpretation of stored maps. However, this system has a limited range of applicability for a

J. Koronacki et al. (Eds.): Advances in Machine Learning I, SCI 262, pp. 189–207.
springerlink.com © Springer-Verlag Berlin Heidelberg 2010

variety of reasons. First, providing the GIS with operational definitions of some spatial concepts (e.g. morphological environments) is not a trivial task. Generally only declarative and abstract definitions, which are difficult to compile into database queries, are available. Second, the operational definitions of some spatial objects are strongly dependent on the data model adopted for the GIS. Finding relationships between density of vegetation and climate is easier with a raster data model, while determining the usual orientation of some morphological elements is simpler in a topological data model [15]. Third, different applications of a GIS will require the recognition of different spatial objects in a map. Providing the system in advance with all the knowledge required for its various application domains is strenuous.

The solution to these difficulties can be found in *spatial data mining*, which investigates how interesting, but not explicitly available knowledge can be extracted from spatial data. Several works on spatial data mining have already been reported in the literature. They concern several tasks, both predictive (e.g., regression [32] and localization [42]) and descriptive (e.g., clustering [19, 40] and discovery of association rules [24, 1], co-location [44], subgroups [22], emerging patterns [5], and spatial trends [13]).

In this chapter, we focus on the classification task, where the goal is to produce a model that can predict the class of unlabeled examples, by training on examples whose label, or class, is supplied. Some examples of applications of classification models to spatial data are the recognition of complex geo-morphologies for topographic map interpretation [33] and the land use and land cover mapping on the basis of remotely sensed data [34].

In the literature there are already several works on spatial classification. Ester *et al.* [14] have proposed an extension of decision trees based on neighborhood graph to consider both thematic attributes of the classified objects and relations with neighboring objects. Koperski [23] has investigated a method which deals with both spatial and hierarchical relations between spatial objects, and takes thematic attributes of neighboring objects into account. Shekhar *et al.* [43] compares two major approaches for incorporating spatial dependence into classification/prediction problems, namely the Markov random fields and the spatial autoregression model.

The formulation of a spatial classification method cannot leave out of consideration the following issues.

Issue 1: Heterogeneity of spatial objects. Each unit of analysis involves spatial objects of different types, such as a town and a highway. In spatial databases, objects of different types are organized in separate *layers*, each of which has a distinct set of attributes and possibly a geometry attribute represented in the vector mode. For instance, a town can be described in terms of economic and demographic factors, as well as a polygon corresponding to its administrative boundary, while a highway is described by the average speed limit, traffic and driving safety conditions, as well as a polyline corresponding to its path. To deal with object heterogeneity, the design of a spatial classification method should not be strictly bound to process objects in one specific layer.

Issue 2: The implicit definition of spatial relationships among objects. Spatial objects have a locational property which implicitly defines several spatial relationships between objects, such as topological, distance-based and directional. Topological relationships (e.g. 'intersects') are invariant under homomorphisms, such as rotation, translation and scaling. Their semantics is precisely defined by means of the 9-intersection model [11]. Distance relationships between two spatial objects can be easily defined in case of punctual representation of the objects (e.g., Euclidean metric) while are more elaborated for linear or surfacic representations. Directional relations can be expressed by the angle formed by two points with respect to the origin of the reference system, or by an extension of Allen's interval algebra which is based on projection lines [38]. In a spatial database, implicit binary spatial relationships correspond to *spatial joins* between layers. The relational nature of spatial rules makes the computation of these spatial joins crucial for the development of effective and efficient data analysis methods.

Issue 3: Spatial autocorrelation. By picturing the spatial variation of some observed variables in a map, we may observe regions where the distribution of values is smoothly continuous with some boundaries possibly marked by sharp discontinuities. In this case, a variable is correlated with itself through space. Formally, *spatial autocorrelation* (or *spatial dependence*, as it is typically called in statistics) is defined as the property of random variables taking values, at pairs of locations a certain distance apart, that are more similar (positive autocorrelation) or less similar (negative autocorrelation) than expected for randomly associated pairs of observations [29]. Informally, spatial positive (negative) autocorrelation occurs when the values of a given property are highly uniform (different) among similar spatial objects in the neighborhood (see Fig. 1). Spatial autocorrelation clearly indicates a violation of the independence assumption of observations usually made in statistics and data mining. Therefore, knowledge discovery techniques that ignore spatial autocorrelation typically perform poorly in the presence of spatial data. When properties of some spatial objects are investigated, attributes of spatially related objects must be taken into account as well, since they may have some kind of influence. This leads to distinguish between the *reference* (or target) *objects*, which are the main subject of analysis and identify the units of analysis, and the

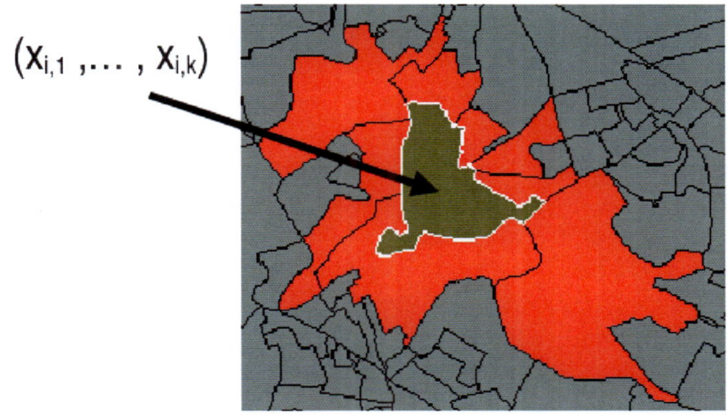

Fig. 1. An example of positive autocorrelation. The classification of the brown area as 'well served' is based on some features measured in that area (e.g., number of communal establishments (schools, hospitals, etc.) and in the nearby areas.

task-relevant objects, which are spatial objects "in the neighborhood" that can help to account for the spatial variation.

Issue 4: Using both labelled and unlabelled data. Learning algorithms designed for spatial classification tasks may require large sets of labeled data. However, the common situation is that only few labeled training data are available since manual annotation of the many objects in a map is very demanding. In this situation, it is important to exploit the large amount of information potentially conveyed by unlabeled data to better estimate the data distribution and to build more accurate classification models.

The first three issues reported above are due to the inherent structure of the units of analysis, which cannot be easily accommodated into a classical double-entry table, whose columns correspond to elementary (nominal, ordinal or numeric) single-valued attributes. In fact, spatial datasets can be naturally modeled as a set of relations $R_1, \ldots, R_n$, such that each relation R_i has a number of elementary attributes $A_1^i, \ldots A_{m_i}^i$ and possibly a geometry attribute G_i (in which case it is a layer). Relationships are either explicitly modeled by means of foreign key constraints or implicitly represented by spatial joins.

(Multi-)relational data mining (MRDM) offers the most suitable setting for spatial data mining tasks. Indeed, MRDM tools can be applied directly to data distributed over several relations to find relational patterns which involve multiple relations [10]. Relational patterns can be expressed either in SQL or in first-order logic (or predicate calculus). The latter reprentation formalism explains why many MRDM algorithms originate from the field of inductive logic programming (ILP) [37, 8, 28].

The earliest proposal of a relational approach to spatial classification has been reported in [33], where authors propose to exploit the expressive power of first-order logic to represent not only spatial relations, but also background knowledge. Ceci and Appice [3] have subsequently proposed the combination of two relational methods, one for the extraction of spatial association rules and the other for the Bayesian classification. Both proposals, however, do not deal with the problem of learning from both labelled and unlabelled data.

To deal with the fourth issue, two learning settings have been proposed in the literature: the *semi-supervised* setting and the *transductive* setting [41]. The former is a type of inductive learning [35], since the learned function is used to make predictions on any possible example. The latter asks for less - it is only interested in making predictions for the given set of unlabeled data.

Transduction [47] seems to be the most suitable setting for spatial classification tasks, for at least two reasons. First, in spatial domains observations to be classified are already known in advance: they are spatial objects on maps already available in a GIS. Second, transduction is based on a (semi-supervised) smoothness assumption according to which if two points x_1 and x_2 in a high-density region are close, then the corresponding outputs y_1 and y_2 should also be close [6]. In spatial domains, where closeness of points corresponds to some spatial distance measure, this assumption is implied by (positive) spatial autocorrelation.

Several transductive learning methods have been proposed in the literature for support vector machines ([2] [16] [20] [7]), for k-NN classifiers ([21]) and even for general classifiers ([27]). However, all of these transductive learning algorithms assume (un)labeled input examples are represented as rows of a classical double-entry table (or database relation). As previously said, this tabular representation of data, also known as *propositional* or *feature-vector* representation, turns out to be too restrictive for spatial classification tasks.

These considerations motivate the investigation of a transductive setting for spatial classification which is based on a relational approach. To devise an algorithm for transductive learning from spatial data represented in as a set of relations $R_1, \ldots, R_n$, we face several problems. First, we have to choose a strategy for the classification of unlabeled data and make it suitable for relational data. Second, we have to choose a classifier which can handle relational data representing different spatial objects and their spatial relationships. Third, we have to define a distance measure between examples described by several relations in order to define a meaningful neighborhood of an object. Solutions to these problems are described in the following sections. They have been implemented in a new classification algorithm based on an extension of the naïve Bayes classifier. Information on the potential uncertainty of classification conveyed by naïve Bayes classifiers is especially useful to prevent drastic changes in the prediction due to small changes in the attribute values of a test case. It is also useful when missing (or imprecise) information may prevent a new object from being classified at all.

This paper is organized as follows. The next section presents both the background of this work and the most relevant related literature. Section 3 introduces a formalization of the transductive learning problem and describes the proposed solution, which is based on an iterative k-NN based re-classification of labelled and unlabelled examples in order to identify "borderline" cases, i.e. examples for which the classification is more uncertain. Experimental results on two real-world spatial datasets are reported and discussed in Section 4. Finally, Section 5 concludes and presents ideas for further work.

2 Background and Related Work

For spatial classification, two methods have been reported in the literature which resort to a relational approach. Malerba et al. [33] proposed to exploit the expressive power of predicate logic to represent both spatial relations and background knowledge. For this purpose, the ILP system ATRE [30] has been integrated in the data mining server of a prototypical GIS, named INGENS, which allows, among other things, to mine classification rules for geographical objects stored in an object-oriented database. Training is based on a set of examples and counterexamples of geographic concepts of interest to the user (e.g., ravine or steep slopes). The symbolic representation of the training examples is automatically extracted from maps, although it is still controlled by the user who can select a suitable level of abstraction and/or aggregation of data by means of a data mining query language [31].

The main weakness of this approach is that learned rules are quite brittle, since conditions in the body must perfectly match against the object description in order to classify it. Ceci and Appice [3] have investigated an associative classification framework, where association rules discovered from training datasets are used by a naïve Bayes classifier which operates on relational representations of spatial data.

Given an object E to be classified, a classical naïve Bayes classifier assigns E to the class C_i that maximizes the *posterior probability* $P(C_i|E)$. By applying the Bayes theorem, $P(C_i|E)$ is expressed as follows:

$$P(C_i|E) = \frac{P(C_i) \cdot P(E|C_i)}{P(E)}. \tag{1}$$

In fact, the decision on the class that maximizes the posterior probability can be made only on the basis of the numerator, that is $P(C_i) \cdot P(E|C_i)$, since $P(E)$ is independent of the class C_i.

To work on relational representations, Ceci and Appice proposed to consider a set $\mathfrak{R}$ of association rules, expressed as first order definite clauses, which are mined on the training set and can be used to define a suitable decomposition of the likelihood $P(E|C_i)$ *à la* naive Bayes in order to simplify the probability estimation problem. In particular, if $\mathfrak{R}(E) \subseteq \mathfrak{R}$ is the set of first order definite clauses whose body covers E, the probability $P(E|C_i)$ is defined as follows:

$$P(E|C_i) = P(\bigwedge_{R_j \in \mathfrak{R}(E)} antecedent(R_j)|C_i). \tag{2}$$

The straightforward application of the naïve Bayes independence assumption to all literals in $\bigwedge_{R_j \in \mathfrak{R}(E)} antecedent(R_j)$ is not correct, since it may lead to underestimating $P(E|C_i)$ when several similar clauses in $\mathfrak{R}(E)$ are considered for the class C_i. To prevent this problem the authors resort to the logical notion of factorization [39]. Details are reported in [3].

For relational classification in the transductive setting, two methods have been reported in the literature. Krogel and Scheffer ([26]) investigate a transformation (known as *propositionalization*) of a relational description of gene interaction data into a classical double-entry table, and then study transduction with the well-known transductive support vector machines. Therefore, transduction is not explicitly investigated on relational representations and it is based on propositionalization, which is fraught with many difficulties in practice ([9, 17]). Taskar *et al.* ([45]) build, on the framework of Probabilistic Relational Models, a *generative* probabilistic model which captures interactions between examples, either labelled or unlabelled. However, given sufficient data, a *discriminative* model generally provides significant improvements in classification accuracy over generative models ([46]). This motivates our interest in designing classifiers based on discriminative models.

For spatial classification, there is no work, at best of our knowledge, concerning a transductive setting. Only recently, some work on semi-supervised learning for spatial classification has been reported [43]. Therefore, we intend this contribution to be a further step toward the investigation of methods which originate from the intersection of these three promising research areas, namely transduction, relational data mining and spatial data mining.

3 Problem Formalization and Proposed Solution

The transductive spatial classification problem can be formalized as follows:
Given

- a spatial database (*SDB*),
- a set *S* of reference spatial objects

- a set $TS \subseteq S$ of reference spatial objects tagged with a class label $Y \in \{C_1, \ldots, C_L\}$ (training reference objects),
- a set $WS = S - TS$ of unlabelled reference spatial objects (working reference objects),
- some sets R_k, $1 \leq k \leq m$ of task-relevant spatial objects

Find: the most accurate prediction of Y for each reference spatial object in WS.

Informally, the learner receives full information (including labels) on the reference spatial object in TS and partial information (without labels) on the reference spatial object in WS and is required to predict the class values only of the examples in WS. The original formulation of the problem of function estimation in a transductive (*distribution-free*) setting requires TS to be sampled from S without replacement. This means that, unlike the standard inductive setting, the examples in the training (and working) set are supposed to be mutually dependent. Vapnik also introduced a second (*distributional*) transduction setting, in which the learner receives training and working sets, which are assumed to be drawn i.i.d. from some unknown distribution. As shown in [47](Theorem 8.1), error bounds for learning algorithms in the distribution-free setting also apply to the more popular distributional transductive setting. Therefore, also considering the spatial autocorrelation issue, in this work we focus our attention on the first setting.

In order to take advantage of results obtained in the area of MRDM, we boil down the spatial database *SDB* to a relational database D with schema Σ. This requires the materialization of spatial queries which compute complex and massive spatial joins (or operators) between reference objects and task relevant objects. For example, spatial joins operations are used to compute the "interact" (non disjoint) relationship between spatial objects (either reference or task-relevant) and the result is stored in database relations. Hence, the schema Σ includes additional relations with respect to those originally present in *SDB* in order to model these spatial relationships.

Reference objects, task-relevant objects and (spatial) interactions among them are tuples stored in separate relations of D. In particular, D includes a *target* relation T whose tuples represent the reference objects in S. Each set R_i is stored in a distinct relation of D. The interaction between different objects is materialized in new relations of Σ (e.g., the relation "Towns_CrossedBy_Roads" describes the instances of the crossing relation between towns and roads). Foreign key constraints (FK) state how tuples in one relation relate to tuples in another and define the inherent *spatial structure* of data. For example, the relation "Towns_CrossedBy_Roads" has two foreign keys: one for the relation "Towns" and one for the relation "Roads". Foreign keys can be used to retrieve all task-relevant objects of D which are (spatially) related to a reference object and are relevant to discriminate between the values of Y (target attribute). Indeed, an example in S is represented as one tuple $t \in T$ and all tuples related to t in D according to FK.

3.1 The Transductive Learning Algorithm

The core of the transductive learning method is described in Algorithm 1. In keeping with the main idea expressed in [20], the algorithm iteratively refine the classification by changing the class of "borderline" training and working examples, i.e., of those examples whose classification is more uncertain. It starts with a given classification and, at each iteration, alternates a step during which examples are reclassified and a step during which the class of borderline examples is changed.

Algorithm 1. Top level transductive algorithm description

```
1: transductiveClassifier(initialClassification, TS, WS)
2: classification1 ← initialClassification;
3: changedExamples ← φ;
4: i ← 0;
5: repeat
6:    prevClassification ← classification1;
7:    prevChangedExamples ← changedExamples;
8:    classification2 ← reclassifyExamplesKNN(classification1, TS, WS);
9:    (classification1, changedExamples) ← changeClass(classification2);
10: until ( (++i ≥ MAX_ITERS) OR
    (computeOverlap(prevChangedExamples, changedExamples) ≥ MAXOVERLAP))
11: return prevClassification
```

The initial classification of $E \in WS \cup TS$ is obtained according to the classification function *preclass* defined as follows:

$$preclass(E) = \begin{cases} class(E) & \text{if } E \in TS \\ BayesianClassification(E) & \text{if } E \in WS, \end{cases}$$

where $BayesianClassification(E)$ is the initial probabilistic classifier built from the training set TS according to formula 2.

The examples are then reclassified by means of a version of the k-NN algorithm [36], tailored for transductive inference in MRDM. The idea is to classify each example $E \in \{TS \cup WS\}$ on the basis of a k-sized neighborhood $N_k(E) = \{E_1, \ldots, E_k\}$, consisting of the k examples of $TS \cup WS$ closest to E with respect to a dissimilarity measure d. More precisely, each example E is associated with an L-dimensional class probability vector $\mathbf{y}' = (y_1(E), \ldots, y_L(E))$, where $y_i(E)$ estimates $P(class(E) = C_i)$. Estimation is based of the following formula:

$$P(class(E) = C_i) = \frac{|\{E_j \in N_k(E) | C_{E_j} = C_i\}|}{k}, \tag{3}$$

where C_{E_j} is the class value associated to E_j at the previous step (at the first step, C_{E_j} is the class label returned by $preclass(E_j)$). The following properties hold for this estimate:

i) $P(class(E) = C_i) \geq 0$ for each $i = 1, \ldots, L$;
ii) $\sum_{i=1,\ldots,L} P(class(E) = C_i) = 1$;
iii) $P(class(E) = C_i)$ is estimated according to the transductive inference principle, as both training and working examples are taken into account in its computation.

The *changeClass* procedure is in charge of changing the classification of the borderline examples. Unlike what was originally proposed in [20], where examples on the border are identified by means of support vectors, we consider the examples for which the entropy of the decision made by the classifier is maximum. The entropy of an example E is computed from the probabilities associated with each class C_i:

$$Entropy(E) = -\sum_{i=1,\dots,L} P(class(E) = C_i) \times log(P(class(E) = C_i)). \quad (4)$$

Examples are ordered according to the entropy function and the class label of at most the first k examples, having $Entropy(E) > MINENTROPY$, is changed. In particular, each selected example E is assigned the most likely class C_i for E among those remaining after the the old class of E has been excluded. The threshold k, which is the same used for k-NN, is necessary in order to avoid changing the class of too many examples, in the extreme case, the class of all examples in a cluster.

Two distinct stopping criteria are used. The first criterion stops the execution of the algorithm when the maximum number of iterations (*MAX_ITERS*) is reached. This guarantees the termination of the algorithm. In any case, our experiments showed that this criterion is rarely attained when the parameter *MAX_ITERS* is as small as 10. The second criterion stops execution when a cycle processes the same examples as the previous one. For this purpose, the overlap between two sets of examples is determined. The *computeOverlap* function returns the ratio between the cardinality of the intersection between the sets of examples and that of their union.

3.2 Generating the First Order Definite Clauses

Differently from the original work in [3], the set $\mathfrak{R}$ of first order definite clauses used in formula 2 do not correspond to association rules. They are directly constructed by means of a depth-bounded breadth-first search in the space of definite clauses built by navigating foreign key paths. A *foreign key path* is an ordered sequence of relations $\vartheta = \{T_{i_1}, T_{i_2}, \dots T_{i_s}\}$, such that $T_{i_j} \in \Sigma$ $(j = 1 \dots s)$ and each relation T_{i_j} has a foreign key to the relation $T_{i_{j-1}}$ $(j = 2, \dots, s)$ or $T_{i_{j-1}}$ has a foreign key to relation T_{i_j}. To formally define the set of first order definite clauses $\mathfrak{R}$ in terms of foreign key paths, some definitions are introduced.

Definition 1 (Structural predicate). *A binary predicate p is a structural predicate associated to the pair of relations* $\langle T_i, T_j \rangle \in S$ *if a foreign key in* T_i *exists which references a relation* $T_j \in S$*, or vice-versa. The first argument of p represents the primary key of* T_j *and the second argument represents the primary key of* T_i*, or vice-versa.*

Definition 2 (Property predicate). *A binary predicate p is a property predicate associated to a relation* $T_i \in S$ *and an attribute Att of* T_i *if the first argument of p represents the primary key of* T_i *and the second argument represents a value observed for Att in* T_i*. The attribute Att is neither the primary key of* T_i *nor a foreign key in* T_i*.*

Hence, we can formally define a first order definite clause associated to a foreign key path.

Definition 3 (First order definite clause associated to a foreign key path). *A first order definite clause associated to the foreign key path $\vartheta = \{T_{i_1}, \ldots, T_{i_{s-1}}\}$ is a clause in the form:*

$$p_0(A_1, y) \leftarrow p_1(A_1, A_2), p_2(A_2, A_3), \ldots, p_{s-1}(A_{s-1}, A_s), p_s(A_s, c).$$

or

$$p_0(A_1, y) \leftarrow p_1(A_1, A_2), p_2(A_2, A_3), \ldots, p_{s-1}(A_{s-1}, A_s).$$

where:

1. *p_0 is a property predicate associated to both the target relation T and the target attribute Y.*
2. *p_1 is a structural predicate associated to the target relation T and relation T_{i_1} of ϑ.*
3. *p_k, $k = 2, \ldots, s-1$, is a structural predicate associated to the pair of relations $\langle T_{i_{k-1}}, T_{i_k} \rangle$ of ϑ.*
4. *p_s is an optional property predicate associated to both relation $T_{i_{s-1}}$ and an attribute Att of $T_{i_{s-1}}$.*

The set $\mathfrak{R}$ is constructed by searching the first order definite clauses R_j associated with a foreign key path ϑ, such that the antecedent of R_i covers at least one training example. The length of the foreign key path ϑ is upper bounded by a user-defined constant *MAX_LENGTH_PATH*.

The property predicate p_s is associated to either a discrete attribute or a continuous attribute *Att* of the relation $T_{i_{s-1}}$. In the former case, p_s checks a condition in the form "$Att = v$", where v is a value in the range of *Att*, while in the latter case, p_s checks a condition in the form "$Att \in [v1, v2]$", where $[v_1, v_2]$ denotes an interval.

3.3 The Relational Dissimilarity Measure

The re-classification of training and working examples is based on a dissimilarity measure d. The classical k-NN method assumes that examples correspond to points in the m-dimensional space $\mathbb{R}^m$ and the nearest neighbors of the example to classify are defined in terms of the standard Euclidean distance. However, in the proposed relational algorithm, examples cannot be associated to points of $\mathbb{R}^m$. This motivates the need for a different notion of a distance (dissimilarity) measure that applies to relational data.

The dissimilarity measure between each pair of examples E_1 and E_2 is computed by first converting the first order definite clauses in $\mathfrak{R}$ into a set of boolean features and then using these features as input of some propositional dissimilarity measure. Actually, the conversion into boolean features concerns only those first order definite clauses $R_j \in \mathfrak{R}$ whose property predicate p_s is defined on a continuous attribute *Att*. In particular, the range of values of *Att* is discretized[1] into *Nb* disjoint intervals (or bins), where *Nb* is a user-defined parameter, and the predicate p_s is defined on the basis of the returned sets of bins.

[1] In this work, we use an equal-width discretization strategy.

Special care is needed for this conversion. Indeed, if we let p_s simply modeling the condition $T_{i_{s-1}}.Att \in [v_1, v_2]$, where $[v_1, v_2]$ is a bin, we incur in a harmful information loss on the order relation of continuous values. To overcome this problem, we follow the idea formulated in [12] and transform the rule R_j into R'_j, such that:

1. $p'_i = p_i$, for each $i = 0, \ldots s-1$, while
2. the property predicate p'_s expresses the condition $T_{i_{s-1}}.Att \leq v_2$.

The advantage of R'_j with respect to R_j is that R'_j models as closer two examples E_1 and E_2, whose *Att* values belong to two consecutive bins, rather than two examples, whose *Att* values belong to distant bins.

Once the new set $\Re' = \{R'_j\}$ is constructed, boolean features are derived from $\Re'$, in order to represent examples by means of a single relation V. This is a form of propositionalization [25] which allows us to use dissimilarity measures defined for classical propositional representations. The schema of V includes $|\Re'|$ attributes, that is one for each first order definite clause R'_j. Each row of V corresponds to an example $E \in \{TS \cup WS\}$. If the antecedent of the first order clause R'_j covers E, then the j-th value of the row in V corresponding to E is set to *true*, *false* otherwise.

The similarity between the pair of examples E_1 and E_2 can be computed by means of the Kendall, Sokal-Michener similarity measure [12], computed on their feature vector representation $V(E_1)$ and $V(E_2)$ stored in V, that is:

$$s(E_1, E_2) = \frac{cardinality(V(E_1) \quad XNOR \quad V(E_2))}{|\Re'|}, \tag{5}$$

where $cardinality(\bullet)$ returns the number of *true* values occurring in the input boolean vector. The similarity coefficient computed in Equation (5) takes values in the unit interval. In particular, $s(E_1, E_2) = 1$, if the two vectors match perfectly, while $s(E_1, E_2) = 0$, if the two vectors are orthogonal. The dissimilarity between the pair of examples is then computed as follows:

$$d(E_1, E_2) = 1 - s(E_1, E_2). \tag{6}$$

4 Experimental Results

In this section, we evaluate the proposed approach on two real-world spatial datasets concerning Greater Manchester (UK) and Munich (Germany). A description of the two datasets is reported below.

Dataset 1: Greater Manchester. Data have been collected within the context of the European project SPIN![2] They concern Greater Manchester, one of the five counties of North West England. Greater Manchester is divided into 10 metropolitan districts, each of which is in turn decomposed into census sections (wards), for a total of 214 wards. Census data are available at ward level and provide socio-economic statistics (e.g. mortality rate) as well as some measures of the deprivation of each ward according to

[2] Web site of the project: http://www.ais. fraunhofer.de/KD/SPIN/project.html

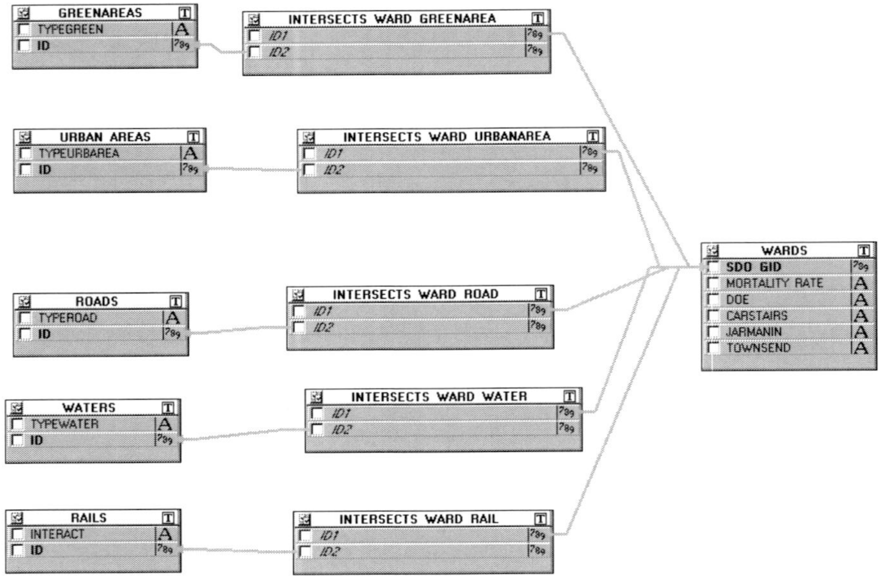

Fig. 2. Schema of the relational database storing data on deprived areas of Greater Manchester

information provided by Census combined into single index scores. We consider the Jarman Underprivileged Area Score (designed to estimate the need for primary care), the indices developed by Townsend and Carstairs (used to perform health-related analyses), and the Department of the Environment's (DoE) index (used in targeting urban regeneration funds). The higher the indices values the more deprived the ward. Layers available concern the boundaries of the 1998 census wards as well as the urban areas (115 lines), the green areas (9 lines), the road network (1,687 lines), the rail network (805 lines) and the water network (716 lines). The original spatial database boils down to a relational database where spatial objects of each layer are stored as single tuples of a database relation together with information on the object type. For instance, type information of an urban area can be either "large urban area" or "small urban area". Topological relationships between wards and objects in all these layers are materialized as database relations. The number of instances of the "non disjoint" spatial relationship is 5,313. The schema of the relational database is reported in Fig. 2. In this work, the reference spatial objects are the wards, that is, *WARDS* is the target relation, the task-relevant objects are all spatial objects in the six layers described above, and the the goal of the classification task is to predict a dichotomized value of the Jarman index (low=[0.001, 0.01] or high=]0.01,0,18]), which defines the class label Y. Information used for prediction purposes is given by other deprivation indices (Townsend index, Carstairs index and DoE index), the mortality rate and spatial objects that geometrically intersects with wards. An assumption of positive spatial autocorrelation seems sensible for this dataset, since deprived wards tend to spatially cluster.

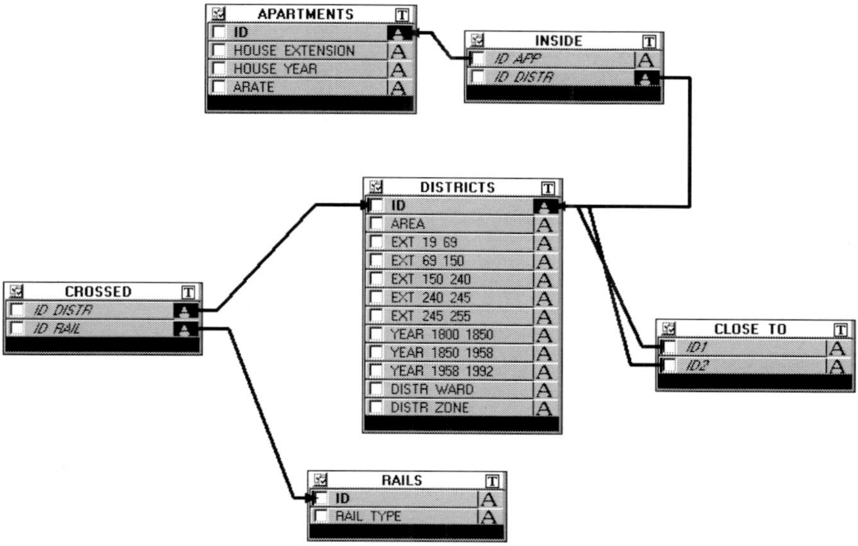

Fig. 3. Schema of the relational database storing rental data on Munich

Dataset 2: Munich. This dataset[3] has been collected in 1998 by Infratest Sozialforschung to develop the 1999 Munich rental guide. It concerns the monthly rent (expressed in German Marks) per square meter for flats in Munich. In particular, the data describe 2,180 geo-referenced flats located in the 446 subquarters of Munich obtained by first dividing the Munich metropolitan area up into three areal zones and then by decomposing each of these zones into 64 districts. The vectorized boundaries of districts as well as the map of public train stops (56 subway (U-Bahn) stops, 15 rapid train (S-Bahn) stops and 1 railway station) within Munich are available for this study. Trains stops are distinguished into three types, namely U-Bahn, S-Bahn or railway station, while flats are described by means of their "monthly rent per square meter", "floor space in square meters" and "year of construction". The schema of the relational database derived from the original spatial database is reported in Fig. 3. In this work, the reference spatial objects are the flats, that is, *APARTMENTS* is the target relation, the task relevant objects are bothe the districts and the train stops, and the the goal of the classification task is to predict a dichotomized value of the 'monthly rent per square meter" (low = [2.0,14.0] or high=]14.0,35.0]), which defines the class label Y. An assumption of positive spatial autocorrelation is sensible for this dataset as well. The spatial arrangement of data is defined by both the "close_to" relation between Munich metropolitan districts, the "inside" relation between flats and districts, and the "inside" relation between public train stops and districts. All these topological relations are materialized into as many database relations.

The experiments aimed at validating the actual advantage of the transductive algorithm over the basic inductive algorithm when few labeled examples are available. The

[3] Available at http://www.di.uniba.it/~ceci/mic Files/munich_db.tar.gz

basic inductive algorithm is that used for the initial classification of unlabelled examples (see line 2 of Algorithm 1). The empirical comparison is based on the misclassification error of the two algorithms. To estimate the misclassification error rate, a K-fold cross validation is performed. For each trial, both algorithms are trained on a single fold and tested on the hold-out $K-1$ folds, which form the working set. The comparative statistics is computed by averaging the misclassification error over the K-folds. The construction of the K folds is a two-stepped process. First, the target relation is partitioned into K blocks $B_1,\ldots,B_K$ of nearly-equal size. Then, for each block B_i, the subset of tuples in other non-target relations which are related, according to the foreign key constraints, to the tuples in B_i are extracted.

It is noteworthy that, unlike the standard cross-validation approach, here only one fold is used for the training set. In this way we can simulate datasets with a small percentage of labelled cases (the training set) and a large percentage of unlabelled data (the working set), which is the usual situation for a transductive setting. Therefore, high misclassification error rates with respect to those published in previous works [3,4] should not be surprising, due to the small amount of labelled examples that we consider.

Since the performance of the transductive algorithm may vary significantly depending on the size (k) of the neighborhood, experiments for different k are performed in order to set the optimal value. In theory, we should experiment with each value of k ranging in the interval $[1,N]$, where N is the number of reference objects in $TS \cup WS$. However, as observed in [49], it is not necessary to consider all possible values of k during cross-validation to obtain the best performance. This can be well approximated by means of cross-validation on no more than ten values of k. A similar consideration has also been reported in [18], where it is shown that the search for the optimal k can be substantially reduced from $[1,N]$ to $[1,\sqrt{N}]$, without degrading accuracy of learning algorithm too much. Hence, we have decided to consider in our experiments only those values of k such that $k \in \{\eta i | i = 1,\ldots,q\}$, where $\eta = \sqrt{N}/q$ is the step value and q is the number of steps.

The classifiers mined in this study are obtained by setting *MAX_ITERS=10*, *MAX_LENGTH_PATH=3*, *MINENTROPY*=0.65 and *MAXOVERLAP*=0.5. The step value η differs for each dataset.

The error rates of the transductive algorithm and its inductive counterpart on the two datasets are reported in Tables 1 and 2. The results are obtained according to both a 10-fold cross validation (CV) of the data and a 20-fold CV of the same data. For experiments on Greater Manchester data, we set $k \in \{4,7,9,11,14\}$, while for experiments on Munich data we set $k \in \{9,18,27,36,45\}$. The percentage of error loss refers to the difference of the error rate of the transductive algorithm with respect to the inductive algorithm. A positive (negative) value is in favor of the transductive (inductive) algorithm.

In both datasets, results confirm a lower error rate for the transductive setting with respect to the inductive one. The improvement depends on the k value and this result is more evident in the case of 10-fold CV. In 20-fold CV, there is an error propagation through algorithm iterations due to the presence of few training examples. A deeper analysis of results of 10-fold CV reveals that the error curve is U-shaped: it first decreases (the minimum is obtained when k=11 for dataset 1 and k= 36 for dataset 2), and

Table 1. Dataset 1: Transductive vs. inductive algorithm

Experiment	Transductive					Inductive
	k=4	k=7	k=9	k=11	k=14	
Avg 10-CV Error (%)	23.38	21.10	19.79	18.04	19.08	22.71
error loss (%)	*-2.97*	*7.06*	*12.84*	*20.56*	*15.99*	
Avg 20-CV Error (%)	33.87	34.41	33.82	33.20	33.28	34.31
error loss (%)	*0.00*	*-1.60*	*0.15*	*1.96*	*1.75*	

Table 2. Dataset 2: Transductive vs. inductive algorithm

Experiment	Transductive					Inductive
	k=9	k=18	k=27	k=36	k=45	
Avg 10-CV Error (%)	28.99	28.61	28.36	28.30	28.15	31.23
error loss (%)	*7.17*	*8.41*	*9.19*	*9.40*	*9.86*	
Avg 20-CV Error (%)	37.25	36.30	36.73	36.67	36.78	37.79
error loss (%)	*1.44*	*3.94*	*2.81*	*2.98*	*2.68*	

then slightly increases when k approximates $\sqrt{|N|}$. This poses the problem of automatically determining the best k value.

Finally, we observe that the error rate is also affected by the number of bins for the discretization of continuous values. In these experiments the number of bins (Nb) for discretization is set to 10 for dataset 1, and 40 for dataset 2. Some results, which we omit here due to space limitations, empirically prove that the higher the error loss the lower the number of bins.

5 Conclusions

In this work we have investigated the main issues met when facing a spatial classification learning problem. The search for a solution to these issues has motivated the proposal of a relational approach to spatial classification in a transductive setting. Indeed, our proposal is to transform the original spatial database into a relational database with multiple relations, which represent both the reference objects (i.e., the main subject of analysis), and the related task-relevant objects and their spatial relationships. The target relation associated to the reference objects includes the class attribute, whose value is null for unlabelled objects. A distinct database relation is associated to each type of spatial object and to each type of spatial relationship (topological, distance-based or directional) explicitly computed on the original spatial database.

An initial probabilistic classifier is induced from these relational data by processing only training (i.e., labeled) examples. The inductive learning algorithm is a relational upgrade of the naïve Bayes classifier and provides a discriminative model on the basis of which unlabelled reference objects (i.e., examples in the working set) can be preliminarily classified. This initial classification of the examples in the working set is then iteratively refined by a transductive learning algorithm. Each iteration consists in a k-NN classification of unlabeled examples and a subsequent reclassification of some "borderline" examples, i.e. examples for which the classification is more uncertain. Uncertainty is measured as the entropy of the class probabilities.

Neighbors are determined by computing a distance measure on a propositionalized representation of unlabelled examples. Propositional features are obtained by transforming a set of first order definite clauses into boolean features. Definite clauses are directly constructed by means of a depth-bounded breadth-first search in the space of definite clauses built by navigating foreign key paths.

The transductive classifier has been compared to its inductive counterpart on two spatial datasets. Experimental results are in favor of the transductive algorithm. The improvement in accuracy seems more significant than that observed by Joachims [21] who compared SVMs in both the inductive and transductive setting.

We believe that our investigation indicates that learning in a transductive setting is an important direction for further research in spatial classification, since there is a clear contiguity of the concept of positive autocorrelation, which typically affect spatial phenomena, with the smoothness assumption which characterize the transductive setting. In principle, we expect that a strong spatial autocorrelation should counterbalance the lack of labelled data, if a relational classifier which takes spatial autocorrelation into account is learned in a transductive setting. As future work, we intend to extend the empirical investigation in order to corroborate our intuition.

Acknowledgements. This work is in partial fulfillment of research objectives of the Strategic Project PS121 "Telecommunication Facilities and Wireless Sensor Networks in Emergency Management" funded by "Regione Puglia" as well as of the ATENEO-2008 project "Knowledge Discovery in Relational Domains" funded by the University of Bari, Italy.

References

1. Appice, A., Ceci, M., Lanza, A., Lisi, F.A., Malerba, D.: Discovery of spatial association rules in georeferenced census data: A relational mining approach. Intelligent Data Analysis 7(6), 541–566 (2003)
2. Bennett, K.P.: Combining support vector and mathematical programming methods for classification. In: Schölkopf, B., Burges, C.J.C., Smola, A.J. (eds.) Advances in kernel methods: support vector learning, pp. 307–326. MIT Press, Cambridge (1999)
3. Ceci, M., Appice, A.: Spatial associative classification: propositional vs. structural approach. Journal of Intelligent Information Systems 27(3), 191–213 (2006)
4. Ceci, M., Appice, A., Malerba, D.: Mr-SBC: A multi-relational naïve bayes classifier. In: Lavrač, N., Gamberger, D., Todorovski, L., Blockeel, H. (eds.) PKDD 2003. LNCS (LNAI), vol. 2838, pp. 95–106. Springer, Heidelberg (2003)

5. Ceci, M., Appice, A., Malerba, D.: Discovering emerging patterns in spatial databases: A multi-relational approach. In: Kok, J.N., Koronacki, J., Lopez de Mantaras, R., Matwin, S., Mladenič, D., Skowron, A. (eds.) PKDD 2007. LNCS (LNAI), vol. 4702, pp. 390–397. Springer, Heidelberg (2007)
6. Chapelle, O., Schölkopf, B., Zien, A. (eds.): Semi-supervised learning. MIT Press, Cambridge (2006)
7. Chen, Y., Wang, G., Dong, S.: Learning with progressive transductive support vector machines. Pattern Recognition Letters 24, 1845–1855 (2003)
8. De Raedt, L.: Interactive theory revision. Academic Press, London (1992)
9. De Raedt, L.: Attribute-value learning versus inductive logic programming: the missing links. In: Page, D.L. (ed.) ILP 1998. LNCS, vol. 1446, pp. 1–8. Springer, Heidelberg (1998)
10. Džeroski, L., Lavrač, N.: Relational data mining. Springer, Berlin (2001)
11. Egenhofer, M.J., Franzosa, R.: Point-set topological spatial relations. International Journal of Geographical Information Systems 5(2), 161–174 (1991)
12. Esposito, F., Malerba, D., Tamma, V., Bock, H.: Similarity and dissimilarity. In: Bock, H.H., Diday, E. (eds.) Analysis of Symbolic Data: Exploratory Methods for Extracting Statistical Information from Complex Data, pp. 139–152. Springer, Heidelberg (2000)
13. Ester, M., Gundlach, S., Kriegel, H., Sander, J.: Database primitives for spatial data mining. In: Proceedings of the International Conference on Database in Office, Engineering and Science, BTW 1999, Freiburg, Germany (1999)
14. Ester, M., Kriegel, H., Sander, J.: Spatial data mining: A database approach. In: Scholl, M.O., Voisard, A. (eds.) SSD 1997. LNCS, vol. 1262, pp. 47–66. Springer, Heidelberg (1997)
15. Frank, A.: Spatial concepts, geometric data models, and geometric data structures. Computers and Geosciences 18(4), 409–417 (1992)
16. Gammerman, A., Azoury, K., Vapnik, V.: Learning by transduction. In: Proc. of the 14th Annual Conference on Uncertainty in Artificial Intelligence, UAI 1998, pp. 148–155. Morgan Kaufmann, San Francisco (1998)
17. Getoor, L.: Multi-relational data mining using probabilistic relational models: research summary. In: Knobbe, A., Van der Wallen, D.M.G. (eds.) Proceedings of the 1st Workshop on Multi-Relational Data Mining, Freiburg, Germany (2001)
18. Góra, G., Wojna, A.: RIONA: A classifier combining rule induction and k-NN method with automated selection of optimal neighbourhood. In: Elomaa, T., Mannila, H., Toivonen, H. (eds.) ECML 2002. LNCS (LNAI), vol. 2430, pp. 111–123. Springer, Heidelberg (2002)
19. Han, J., Kamber, M., Tung, A.K.H.: Spatial clustering methods in data mining. In: Han, J., Kamber, M. (eds.) Geographic Data Mining and Knowledge Discovery, pp. 188–217. Taylor and Francis, Abington (2001)
20. Joachims, T.: Transductive inference for text classification using support vector machines. In: Proceedings of the 16th International Conference on Machine Learning (ICML 1999), pp. 200–209. Morgan Kaufmann, San Francisco (1999)
21. Joachims, T.: Transductive learning via spectral graph partitioning. In: Proc. of the 20th International Conference on Machine Learning, ICML 2003. Morgan Kaufmann, San Francisco (2003)
22. Klösgen, W., May, M.: Spatial subgroup mining integrated in an object-relational spatial database. In: Elomaa, T., Mannila, H., Toivonen, H. (eds.) PKDD 2002. LNCS (LNAI), vol. 2431, pp. 275–286. Springer, Heidelberg (2002)
23. Koperski, K.: Progressive Refinement Approach to Spatial Data Mining. PhD thesis, Computing Science, Simon Fraser University, British Columbia, Canada (1999)
24. Koperski, K., Han, J.: Discovery of spatial association rules in geographic information databases. In: Egenhofer, M.J., Herring, J.R. (eds.) SSD 1995. LNCS, vol. 951, pp. 47–66. Springer, Heidelberg (1995)

25. Krogel, M., Rawles, S., Zelezny, F., Flach, P., Lavrac, N., Wrobel, S.: Comparative evaluation of approaches to propositionalization. In: Horváth, T., Yamamoto, A. (eds.) ILP 2003. LNCS (LNAI), vol. 2835, pp. 197–214. Springer, Heidelberg (2003)
26. Krogel, M.-A., Scheffer, T.: Multi-relational learning, text mining, and semi-supervised learning for functional genomics. Machine Learning 57(1-2), 61–81 (2004)
27. Kukar, M., Kononenko, I.: Reliable classifications with machine learning. In: Elomaa, T., Mannila, H., Toivonen, H. (eds.) ECML 2002. LNCS (LNAI), vol. 2430, pp. 219–231. Springer, Heidelberg (2002)
28. Lavrač, N., Džeroski, S.: Inductive logic programming: techniques and applications. Ellis Horwood, Chichester (1994)
29. Legendre, P.: Spatial autocorrelation: Trouble or new paradigm. Ecology 74, 1659–1673 (1993)
30. Malerba, D.: Learning recursive theories in the normal ILP setting. Fundamenta Informaticae 57(1), 39–77 (2003)
31. Malerba, D., Appice, A., Ceci, M.: A data mining query language for knowledge discovery in a geographical information system. In: Meo, R., Lanzi, P.L., Klemettinen, M. (eds.) Database Support for Data Mining Applications. LNCS (LNAI), vol. 2682, pp. 95–116. Springer, Heidelberg (2004)
32. Malerba, D., Ceci, M., Appice, A.: Mining model trees from spatial data. In: Jorge, A.M., Torgo, L., Brazdil, P.B., Camacho, R., Gama, J. (eds.) PKDD 2005. LNCS (LNAI), vol. 3721, pp. 169–180. Springer, Heidelberg (2005)
33. Malerba, D., Esposito, F., Lanza, A., Lisi, F.A., Appice, A.: Empowering a GIS with inductive learning capabilities: The case of INGENS. Journal of Computers, Environment and Urban Systems 27, 265–281 (2003)
34. McIver, D., Friedl, M.: Estimating pixel-scale land cover classification confidence using nonparametric machine learning methods. IEEE Transactions on Geoscience and Remote Sensing 39(9), 1959–1968 (2001)
35. Michalski, R.S.: A theory and methodology of inductive learning. Artificial Intelligence 20(2), 111–161 (1983)
36. Mitchell, T.: Machine Learning. McGraw Hill, New York (1997)
37. Muggleton, S.: Inductive logic programming. Academic Press, London (1992)
38. Mukerjee, A., Joe, G.: A qualitative model for space. In: Proceedings of AAAI 1990, pp. 721–727. AAAI Press, Menlo Park (1990)
39. Robinson, J.A.: A machine oriented logic based on the resolution principle. Journal of the ACM 12, 23–41 (1965)
40. Sander, J., Ester, M., Kriegel, H., Xu, X.: Density-based clustering in spatial databases: The algorithm gdbscan and its applications. Data Mining and Knowledge Discovery 2(2), 169–194 (1998)
41. Seeger, M.: Learning with labeled and unlabeled data. Technical report, University of Edinburgh (2001)
42. Shekhar, S., Schrater, P.R., Vatsavai, R.R., Wu, W., Chawla, S.: Spatial contextual classification and prediction models for mining geospatial data. IEEE Transactions on Multimedia 4(2), 174–188 (2002)
43. Shekhar, S., Vatsavai, R., Chawla, S.: Spatial classification and prediction models for geospatial data mining. In: Miller, H.J., Han, J. (eds.) Geographic Data Mining and Knowledge Discovery, 2nd edn., pp. 117–147. Taylor & Francis, Abington (2009)
44. Shekhar, S., Zhang, P., Huang, Y.: Spatial data mining. In: Maimon, O., Rokach, L. (eds.) The Data Mining and Knowledge Discovery Handbook, pp. 833–851. Springer, Heidelberg (2005)

45. Taskar, B., Segal, E., Koller, D.: Probabilistic classification and clustering in relational data. In: Nebel, B. (ed.) Proceedings of the Seventeenth International Joint Conference on Artificial Intelligence (IJCAI 2001), pp. 870–878 (2001)
46. Vapnik, V.: The Nature of Statistical Learning Theory. Springer, New York (1995)
47. Vapnik, V.: Statistical Learning Theory. Wiley, New York (1998)
48. Wessel, M.: Some practical issues in building a hybrid deductive geographic information system with a dl component. In: Proceedings of the 10th International Workshop on Knowledge Representation meets Databases (KRDB 2003), Hamburg, Germany, September 15-16. CEUR Workshop Proceedings, vol. 79. CEUR-WS.org (2003)
49. Wettschereck, D.: A study of Distance-Based Machine Learning Algorithms. PhD thesis, Oregon State University (1994)

Beyond Sequential Covering – Boosted Decision Rules

Krzysztof Dembczyński[1], Wojciech Kotłowski[1], and Roman Słowiński[2]

[1] Poznań University of Technology, 60-965 Poznań, Poland
kdembczynski@cs.put.poznan.pl, wkotlowski@cs.put.poznan.pl
[2] Poznań University of Technology, 60-965 Poznań, Poland,
and Systems Research Institute, Polish Academy of Sciences, 01-447 Warsaw, Poland
rslowinski@cs.put.poznan.pl

Abstract. From the beginning of machine learning, rule induction has been regarded as one of the most important issues in this research area. One of the first rule induction algorithms was AQ introduced by Michalski in early 80's. AQ, as well as several other well-known algorithms, such as CN2 and Ripper, are all based on sequential covering. With the advancement of machine learning, some new techniques based on statistical learning were introduced. One of them, called boosting, or forward stagewise additive modeling, is a general induction procedure which appeared to be particularly efficient in binary classification and regression tasks. When boosting is applied to induction of decision rules, it can be treated as generalization of sequential covering, because it approximates the solution of the prediction task by sequentially adding new rules to the ensemble without adjusting those that have already entered the ensemble. Each rule is fitted by concentrating on examples which were the hardest to classify correctly by the rules already present in the ensemble. In this paper, we present a general scheme for learning an ensemble of decision rules in a boosting framework, using different loss functions and minimization techniques. This scheme, called ENDER, is covered by such algorithms as SLIPPER, LRI and MLRules. A computational experiment compares these algorithms on benchmark data.

1 Introduction

We consider the classification problem that consists in assigning objects described by several attributes to pre-defined decision classes. Using a set of training examples, our task is to build a classifier that assigns new objects to decision classes as accurately as possible. There are several learning algorithms introduced for this task. In the following, we consider classifiers based on decision rules.

Decision rule is a logical statement of the form: *if* [*condition*], *then* [*decision*]. If an object satisfies the condition part of the rule, then the decision is taken; otherwise no action is performed. A rule can be treated as a simple classifier that gives a constant response for the objects satisfying the conditions, and abstains from the response for all other objects. The main advantage of decision rules is their simplicity and human-readable form.

Induction of decision rules has been widely considered in the early machine learning approaches [1, 2, 3, 4]. The most popular algorithms were based on a *sequential covering* procedure (also known as separate-and-conquer approach). According to [4], this

J. Koronacki et al. (Eds.): Advances in Machine Learning I, SCI 262, pp. 209–225.
springerlink.com © Springer-Verlag Berlin Heidelberg 2010

procedure can be described as follows: "learn a rule that covers a part of the given training examples, remove the covered examples from the training set (the separate part) and recursively learn another rule that covers some of the remaining examples (the conquer part) until no examples remain".

Apart from the sequential covering, some other approaches to rule induction exist. For instance, the apriori-based algorithms are also used for induction of predictive rules [5, 6]. There are several rule-based approaches of lazy learning type [7, 8], possibly combined with instance-based methods [9, 10]. Other algorithms are based on Boolean reasoning and mathematical programming [11]. Let us also notice that decision rule models are strongly associated with rough set approaches to knowledge discovery [12, 13, 14, 15, 16, 17].

Recently, some new algorithms have been introduced that follow another approach to rule induction. They treat decision rules as subsidiary, base classifiers in the ensemble. More precisely, decision rule classifiers are built using boosting [18] or forward stagewise additive modeling [19]. The examples of such algorithms are SLIPPER [20], LRI [21], RuleFit [22], and MLRules [23]. We will refer to this family of algorithms as *boosted decision rules*.

In this paper, we discuss the relation between sequential covering and boosted decision rules. As we will see, these two procedures are quite similar and the latter can be seen as a generalization of the former. We will show a general scheme for rule induction that we called ENDER. Different rule learning algorithms fall into this scheme as special cases.

The paper is organized as follows. Section 2 states formally the problem of classification. In Section 3, the decision rule is defined. Sequential covering is briefly described in Section 4. We show how the boosted decision rules are related to sequential covering in Section 5. Two instances of boosting algorithms for rule induction, SLIPPER and MLRules, are also described there. Several extensions of boosted decision rules are discussed in Section 6. Experimental comparison of different rule induction algorithms is given in Section 7. The last Section concludes the paper.

2 Classification Problem

We consider the classification problem in which the aim is to predict for an object an unknown class label of attribute y taken from set $\{1,\ldots,K\}$, knowing joint values of attributes $\mathbf{x} = (x_1, x_2, \ldots, x_n) \in \mathscr{X}$. The task is to find function $f(\mathbf{x})$, called *classifier*, that predicts accurately the value of y. The accuracy of a single prediction is measured by *loss function* $L(y, f(\mathbf{x}))$ which determines the penalty for predicting $f(\mathbf{x})$ when the actual value is y. The overall accuracy of function $f(\mathbf{x})$ is measured by the expected loss, called *risk*, over joint distribution $P(y,\mathbf{x})$:

$$R(f) = \mathbb{E}_{y\mathbf{x}} L(y, f(\mathbf{x})).$$

Therefore, the optimal risk-minimizing classification function, called *Bayes classifier*, is given by:

$$f^* = \arg\min_f \mathbb{E}_{y\mathbf{x}} L(y, f(\mathbf{x})).$$

Since $P(y,\mathbf{x})$ is generally unknown, the learning procedure uses only a set of training examples $\{y_i,\mathbf{x}_i\}_1^N$ to construct f to be the best possible approximation of f^*. Usually, this is performed by minimization of the *empirical risk*:

$$R_{emp}(f) = \frac{1}{N}\sum_{i=1}^{N} L(y_i, f(\mathbf{x}_i)),$$

where f is chosen from a restricted family of functions.

The most natural loss function is the *0-1 loss*:

$$L_{0-1}(y, f(\mathbf{x})) = \begin{cases} 0, & \text{if } y = f(\mathbf{x}), \\ 1, & \text{if } y \neq f(\mathbf{x}). \end{cases} \tag{1}$$

The expected value of this loss function is simply a misclassification error of $f(\mathbf{x})$ defined by $\Pr(y \neq f(\mathbf{x}))$. Thus, Bayes classifier has the following form:

$$f^*(\mathbf{x}) = \arg\min_{f(\mathbf{x})} \mathbb{E}_{y|\mathbf{x}} L_{0-1}(y, f(\mathbf{x})) = \arg\max_{k\in\{1,\ldots,K\}} \Pr(y = k|\mathbf{x}). \tag{2}$$

By minimizing the 0-1 loss function, we estimate K regions in the attribute space, such that in the k-th region, the k-th class is observed with the highest probability.

For the sake of simplicity, we will confine our considerations to the two-class problem, referred to as *binary classification*. Let us assume that $y \in \{-1,1\}$ and the classifier gives continuous responses, $f(\mathbf{x}) \in \mathbb{R}$. In this case, the loss function can be expressed by $L(yf(\mathbf{x}))$, where $yf(\mathbf{x})$ is called *margin*. The positive margin means no error, and its magnitude tells, intuitively, what is the credibility of assigning an object to a given class. The *margin 0-1 loss* function is therefore:

$$L_{0-1}(yf(\mathbf{x})) = [\![yf(\mathbf{x}) < 0]\!],$$

where $[\![\pi]\!]$ is the Boolean test equal to 1 if predicate π is true, and 0 otherwise.

3 Decision Rule

Decision rule is a simple logical pattern having the form:

"if $[condition]$ then $[decision]$".

Its main advantage is simplicity and human-interpretable form that can model interactions between several attributes for which the condition is defined. As an example of a decision rule, let us consider a rule generated from the well-know CREDIT-G benchmark data set concerning the problem of predicting a risk-level of bank customers:

if CHECKING STATUS = *no checking account*
and OTHER PAYMENT PLANS ≠ *bank*
and AGE ∈ [22.5, 66.5]
and CREDIT AMOUNT ≤ 10924.5,
then CUSTOMER = *good*.

From a machine learning perspective, such a rule can be treated as a simple classifier that gives a constant response to examples satisfying the condition part, and abstains from the response for all other examples. Formally, a decision rule can be defined as follows.

Let $X_j \subset \mathscr{X}$ be a domain of attribute $j \in \{1,\ldots,n\}$. The condition part of the rule consists of a conjunction of elementary expressions of the general form

$$x_j \in S_j, \tag{3}$$

where x_j is the value of object $\mathbf{x}$ on attribute j and S_j is a subset of X_j. In particular, elementary expressions are of the form $x_j \geq s_j$, $x_j \leq s_j$, for quantitative attributes, and $x_j = s_j$, $x_j \neq s_j$, for qualitative attributes, where s_j is taken from a domain of attribute j. Let Φ be the set of elementary expressions constituting the condition part of the rule, and let $\Phi(\mathbf{x})$ be a function indicating whether object $\mathbf{x}$ satisfies the conjunction of elementary expressions Φ. In other words, $\Phi(\mathbf{x})$ defines an arbitrary axis-parallel region in the attribute space. We say that a rule *covers* object $\mathbf{x}$ if it belongs to this region, i.e. $\Phi(\mathbf{x}) = 1$. The number of training examples covered by the rule is referred to as *rule coverage*. The decision, or response, denoted by α, is a real value assigned to the region defined by Φ. Therefore, we define a decision rule as:

$$r(\mathbf{x}) = \alpha\Phi(\mathbf{x}).$$

Note that the decision rule takes only two values, $r(\mathbf{x}) \in \{\alpha, 0\}$, depending whether $\mathbf{x}$ satisfies the condition part or not.

Having defined the decision rule, we now need an algorithm that induces such rules from a set of training examples and combines them into a powerful classifier.

4 Sequential Covering

Almost all early algorithms for rule learning were based on sequential covering. The most popular are AQ [1], CN2 [2], Ripper [3], and LEM [14]. Sequential covering relies on learning a rule in each step, that covers a part of a given set of training examples, removing the covered examples from the training set, and repeating this step until no example remains. This procedure is repeated separately for each class. In each turn, the rules cover examples from one class only. These examples are called *positive*, while the others, *negative*.

This approach is more formally presented as Algorithm 4.1. It starts with an empty set of decision rules and successively adds rules to it, until all positive examples are covered. A single rule is built by the FindBestRule procedure in order to cover positive examples only. This can be done by specializing the current rule by adding new elementary expressions to its condition part. In order to move towards the goal of finding a rule that covers no negative examples, the algorithm selects the elementary expressions that optimize the purity measure of the rule. The purity measure is a function of the numbers of positive and negative examples. For instance, this can be the fraction of positive examples in the set of covered examples.

This approach to rule induction encounters several problems. For example, the set of decision rules does not usually cover the entire attribute space. Moreover, several

Algorithm 4.1. Sequential covering

input : set of training examples $\{y_i, \mathbf{x}_i\}_1^N$,
k – label of positive class.
output: set of decision rules R_k for k-th class.
$R_k = \emptyset$
$Pos = \{\mathbf{x}_i : y_i = k, i = 1, \ldots, N\}$ // positive examples
$Neg = \{\mathbf{x}_i : y_i \neq k, i = 1, \ldots, N\}$ // negative examples
$m = 0$
while $Pos \neq \emptyset$ **do**
  $r_m(\mathbf{x}) = \text{FindBestRule}(Pos, Neg)$
  $Pos = Pos - \{\mathbf{x}_i : r_m(\mathbf{x}) = 1\}$ // remove covered examples
  $R_k = R_k \cup r_m(\mathbf{x})$
  $m = m + 1$
end

rules may overlap. Thus, there is a problem how to classify new examples by a set of rules. In many rule induction algorithms, the rule generation phase and the classification phase are considered separately. First, the algorithm generates a set of rules. Next, one of several possible strategies is used to classify new examples. Usually, different voting schemes are considered. The weights of the rules used in the voting are simple statistics computed over examples covered by the rules. A popular choice is a relative number of covered examples or an estimate of conditional class probability within the rule. The latter suggests that the rules are treated as independent ones, what is not true taking into account how the sequential covering procedure works.

5 Boosted Decision Rules

The problem described above can be easily solved, if we first assume some form of the ultimate rule-based classifier. Let the classification function be a linear combination of M rules:

$$f_M(\mathbf{x}) = \alpha_0 + \sum_{m=1}^{M} r_m(\mathbf{x}), \tag{4}$$

where α_0 is a constant value, which can be interpreted as a default rule, covering the whole attribute space. This is an ensemble of simple and interpretable base classifiers. Single decision rule $r_m(\mathbf{x})$ defines an increase or decrease of $f_M(\mathbf{x})$ by α_m if a condition $\Phi_m(\mathbf{x})$ is met. In other words, this is a weighted majority voting. The prediction is made by $\text{sgn}(f_M(\mathbf{x}))$.

Having such a form of the classifier, we can proceed in a similar way to sequential covering. We add the rules one by one in order to minimize the empirical risk. In each subsequent iteration, a new rule is added by taking into account previously generated rules. Let $f_{m-1}(\mathbf{x})$ be a classification function after $m-1$ iterations, consisting of the first $m-1$ rules and the default rule. In the m-th iteration, a decision rule can be obtained by solving:

$$r_m(\mathbf{x}) = \arg\min_{r(\mathbf{x})} R_{emp}(f_{m-1}(\mathbf{x}) + r(\mathbf{x})) = \arg\min_{\Phi,\alpha} \sum_{i=1}^{N} L(y_i, f_{m-1}(\mathbf{x}_i) + \alpha\Phi(\mathbf{x}_i)). \tag{5}$$

This procedure corresponds to sequential covering, since the value of the margin 0-1 loss function decreases down to 0 for all correctly covered training examples, and there is no need for another rule to cover them again. This corresponds to removing such objects from the training set. All the rules will obtain the same absolute value of rule response $|\alpha|$, and the sign decides for which class the rule votes. The classification is then a simple majority voting.

Instead of the margin 0-1 loss, one can use, however, another loss function, like exponential loss or logit loss. These are surrogates (upper-bounding the 0-1 loss) commonly used in classification tasks. These functions are convex and differentiable, which makes the minimization process easier to cope with. The *exponential loss* is defined as:

$$L_{\exp}(yf(\mathbf{x})) = \exp(-yf(\mathbf{x})). \tag{6}$$

This loss function is used in AdaBoost [18]. The *logit loss*

$$L_{\log}(yf(\mathbf{x})) = \log(1 + \exp(-2yf(\mathbf{x}))) \tag{7}$$

is commonly used in statistics. These two loss functions are sensitive to the value of $yf(\mathbf{x})$. The Bayes classifier for both of them has the form [19]:

$$f^*(\mathbf{x}) = \frac{1}{2}\log\frac{\Pr(y=1|\mathbf{x})}{\Pr(y=-1|\mathbf{x})}, \tag{8}$$

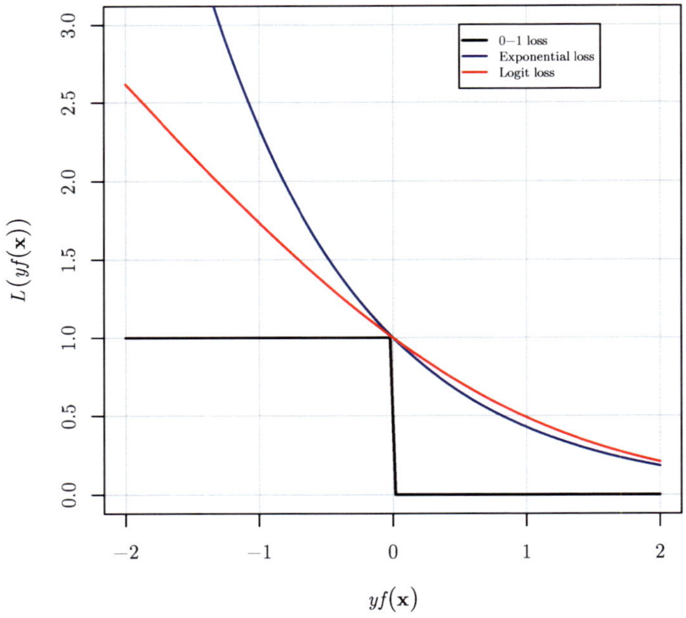

Fig. 1. Loss functions $L(yf(\mathbf{x}))$ expressed through the margin $yf(\mathbf{x})$ for binary classification

Algorithm 5.1. Ensemble of decision rules – ENDER

input : set of training examples $\{y_i, \mathbf{x}_i\}_1^N$,
$L(y, f(\mathbf{x}))$ – loss function,
M – number of decision rules to be generated.
output: ensemble of decision rules $f_M(\mathbf{x})$.

$\alpha_0 = \arg\min_\alpha \sum_{i=1}^N L(y_i, \alpha)$
$f_0(\mathbf{x}) = \alpha_0$;
for $m = 1$ *to* M **do**
  $\Phi_m = \arg\min_\Phi \mathscr{L}_m(\Phi)$
  $\alpha_m = \arg\min_\alpha \sum_{i=1}^N L(y_i, f_{m-1}(\mathbf{x}_i) + \alpha\Phi_m(\mathbf{x}_i))$
  $r_m(\mathbf{x}) = \alpha_m \Phi_m(\mathbf{x})$
end

which is the logit transform of the conditional probabilities. Expression (8) can be inverted to give:

$$\Pr(y = 1|\mathbf{x}) = \frac{1}{1 + \exp(-2f^*(\mathbf{x}))}. \tag{9}$$

Therefore, minimization of these loss functions on the training set can be seen as estimation of conditional probabilities $\Pr(y = 1|\mathbf{x})$. The sign of $f(\mathbf{x})$ estimates in turn the class with a higher probability. Characteristics of these loss functions are shown in Figure 1.

Using these loss functions in (5) results in the boosting-like algorithm for rule induction, we presented in [24] and called ENDER. We outline its main steps in Algorithm 5.1. First, the default rule is computed:

$$\alpha_0 = \arg\min_\alpha \sum_{i=1}^N L(y_i, \alpha). \tag{10}$$

In each next iteration, we proceed in two steps:

1. Find Φ_m by minimizing a functional $\mathscr{L}_m(\Phi)$ derived from (5) that does not depend of α:

$$\Phi_m = \arg\min_\Phi \mathscr{L}_m(\Phi). \tag{11}$$

2. Find α_m, the solution to the following line-search problem:

$$\alpha_m = \arg\min_\alpha \sum_{i=1}^N L(y_i, f_{m-1}(\mathbf{x}_i) + \alpha\Phi_m(\mathbf{x}_i)). \tag{12}$$

The algorithm finds iteratively the best rules minimizing the empirical risk. $\mathscr{L}(\Phi)$, which denotes a purity measure, is greedily minimized by a heuristic that builds the condition part of the decision rule. It works in a similar way to the FindBestRule procedure. The difference is that it does not stop when no negative example is covered, but when $\mathscr{L}_m(\Phi_m)$ cannot be decreased. The response of the rule is the weight in a voting procedure. It is computed by taking into account the previously generated rules, thus the dependencies between rules are not ignored. The entire procedure is very intuitive and, formally, well-founded. Below, we will state that several rule learning algorithms fall into this ENDER scheme.

In the case of the exponential loss, the ENDER algorithm works as follows. The response of the decision rule is computed according to:

$$\alpha_m = \arg\min_{\alpha} \sum_{i=1}^{N} e^{-y_i(f_{m-1}(\mathbf{x}_i)+\alpha\Phi_m(\mathbf{x}_i))} = \frac{1}{2}\log\frac{\sum_{y_i=1}\Phi(\mathbf{x}_i)e^{-f_{m-1}(\mathbf{x}_i)}}{\sum_{y_i=-1}\Phi(\mathbf{x}_i)e^{f_{m-1}(\mathbf{x}_i)}}. \tag{13}$$

The solution of α_m can be put to (5), and after straightforward transformations, we get the following form of the purity measure:

$$\mathcal{L}_m(\Phi) = \sqrt{\left(\sum_{y_i=1}\Phi(\mathbf{x}_i)w_i^{(m)}\right)} + \sqrt{\left(\sum_{y_i=-1}\Phi(\mathbf{x}_i)w_i^{(m)}\right)}, \tag{14}$$

where $w_i^{(m)} = e^{-y_i f_{m-1}(\mathbf{x}_i)}$ can be treated as a weight of the i-th training example in the m-th iteration. This is also how the SLIPPER algorithm works [20].

In the MLRules algorithm [23], which is the next to describe, we derive rules from the maximum likelihood principle that can be seen as minimization of the logit loss (7) by the so-called gradient descent technique.

In case of the logit loss, there is no analytical solution to (12). To speed up the computations, instead of numerically solving the line search problem, we perform a single Newton-Raphson step, similarly as in [25]:

$$\alpha_m = -\left.\frac{\sum_{\Phi(\mathbf{x}_i)=1}\frac{\partial}{\partial\alpha}L_{\log}(y_i(f_{m-1}(\mathbf{x}_i)+\alpha\Phi_m(\mathbf{x}_i)))}{\sum_{\Phi(\mathbf{x}_i)=1}\frac{\partial^2}{\partial\alpha^2}L_{\log}(y_i(f_{m-1}(\mathbf{x}_i)+\alpha\Phi_m(\mathbf{x}_i)))}\right|_{\alpha=0}. \tag{15}$$

From (5) we derive the form of $\mathcal{L}_m(\Phi)$, in such a way that $\mathcal{L}_m(\Phi)$ does not depend on α. This can be done by approximating (5) up to the first order with respect to α:

$$r_m(\mathbf{x}) \simeq \arg\min_{\Phi,\alpha}\sum_{i=1}^{N}\left(L(y_i, f_{m-1}(\mathbf{x}_i)) - \alpha\Phi(\mathbf{x}_i)\tilde{w}_i^{(m)}\right), \tag{16}$$

where

$$\tilde{w}_i^{(m)} = -\left.\frac{\partial L(y_i, f(\mathbf{x}_i))}{\partial f(\mathbf{x}_i)}\right|_{f(\mathbf{x}_i)=f_{m-1}(\mathbf{x}_i)}.$$

Since $\tilde{w}_i^{(m)}$, $i=1,\ldots,N$, is the negative gradient of the function to be minimized, this technique is called *gradient descent*. It is easy to see that the optimal solution with respect to Φ is obtained by minimizing:

$$\mathcal{L}_m(\Phi) = -\sum_{\Phi(\mathbf{x}_i)=1}\alpha\tilde{w}_i^{(m)}, \tag{17}$$

since $\sum_{i=1}^{N} L(y_i, f_{m-1}(\mathbf{x}_i))$ is constant at a given iteration, and thus it does not change the solution. Observe that for a given value of α, the solution only depends on

$\sum_{\Phi(\mathbf{x}_i)=1} \tilde{w}_i^{(m)}$, so the minimization of (17) can be finally reformulated to the minimization of the following term:

$$\mathscr{L}_m(\Phi) = -\left| \sum_{\Phi(\mathbf{x}_i)=1} \tilde{w}_i^{(m)} \right|, \tag{18}$$

because the sign and the magnitude of α may be established afterwards.

Another boosted rule induction algorithm, LRI [21], uses a specific reweighting schema (cumulative error), similar to Breiman's Arc-xf algorithm [26]. For the two-class problem, however, this method can also be explained in terms of the loss function minimization, as it was done in [27] for Arc-xf. It follows that LRI minimizes a polynomial loss function by the gradient descent technique. Let us also point out that a single rule in LRI is a bit more complex in comparison with those generated by other algorithms, because the rule is a DNF-formula, i.e., disjunction of conjunctions of elementary expressions, instead of a single conjunction.

6 Extensions

6.1 Rule Coverage

Several problems associated with a typical sequential covering approach have been pointed out in [28]. One of the problems concerns the discrimination and completeness of the rule. In other words, "how good are the examples from positive class separated from negative examples, and how many positive examples are covered?" The usual way of answering this question consists in using different purity measures and their combinations, sometimes defined ad hoc, to control the process of constructing a single decision rule. In the approach followed by ENDER, the answer is very natural. In each iteration, where we find a rule which is minimizing the empirical risk, we can slightly modify the minimization technique in order to control the trade-off between discrimination and completeness.

Let us restrict α in (5) to $\alpha \in \{-\beta, \beta\}$, where β is a parameter of the algorithm. Then, (5) becomes:

$$r_m(\mathbf{x}) = \arg\min_{\Phi, \pm\beta} \left(\sum_{\Phi(\mathbf{x}_i)=1} L(y_i, f_{m-1}(\mathbf{x}_i) \pm \beta) + \sum_{\Phi(\mathbf{x}_i)=0} L(y_i, f_{m-1}(\mathbf{x}_i)) \right). \tag{19}$$

Since β is fixed, we refer this technique to as *constant-step minimization*. In the case of the exponential loss, we can prove the following theorem.

Theorem 1. *[29] The problem of solving (19) for exponential loss (6) and step length β is equivalent to minimization of*

$$\mathscr{L}_m(\Phi) = \sum_{i \in R_-} w_i^{(m)} + \ell \sum_{\Phi(\mathbf{x}_i)=0} w_i^{(m)}, \tag{20}$$

where $R_- = \{i : y_i \alpha \Phi(\mathbf{x}_i) < 0\}$ indicates the training examples misclassified by the rule, and:

$$w_i^{(m)} = e^{-y_i f_{m-1}(\mathbf{x}_i)}, \qquad \ell = \frac{1 - e^{-\beta}}{e^{\beta} - e^{-\beta}}, \qquad \beta = \log \frac{1-\ell}{\ell}.$$

Expression (20) has a nice interpretation. The first term corresponds to examples "misclassified" by the rule, while the second term – to examples which are not classified by the rule at all. Value l plays the role of a penalty for abstaining from classification and establishes a trade-off between not classified and misclassified examples. It is easy to see that for $\beta > 0$, $\ell \in [0, 0.5)$. Increasing ℓ (or decreasing β) results in more general rules, covering more examples. For $\beta \to 0$ we get the gradient descent technique applied to the exponential loss. This means that the gradient descent produces the most general rules in the sense of coverage.

6.2 Regularization

A decision rule has the form of an n-dimensional rectangle, where n is the number of attributes. It can be shown, that the class of n-dimensional rectangles has Vapnik-Chervonenkis (VC) dimension equal to $2n$, and it does not depend on the number of elementary expressions. Theoretical results [30] suggest that an ensemble with a simple base classifier (with low VC dimension) and high prediction confidence (margin) on the data set generalizes well, regardless of the size of the ensemble. Nevertheless, computational experiments show that the performance of rule ensemble can deteriorate as the number of rules grows, especially for the problems with a high level of noise. Similar phenomenon has been observed for other boosting algorithms, in particular for AdaBoost [27, 31, 32]. Therefore, in order to decrease the influence of the noise, the boosting algorithms should be used with regularization.

As pointed out in many places (see, for example, [19]), a regularized classifier can achieve much better results. The form of regularization, which is particularly useful, is the L_1-penalty, also called *lasso*. In the case of rule ensemble, this would lead to the problem consisting in fitting α_m of all possible rules and minimizing an additional term $\sum_m |\alpha_m|$. To approximate a solution of such a regularized problem, ENDER follows a strategy that is called shrinkage [19]. It consists in shrinking a newly generated rule $r_m(\mathbf{x}) = \alpha_m \Phi_m(\mathbf{x})$ towards rules already present in the ensemble:

$$f_m(\mathbf{x}) = f_{m-1}(\mathbf{x}) + \nu \cdot r_m(\mathbf{x}),$$

where $\nu \in (0, 1]$ is a shrinkage parameter that can be regarded as controlling the learning rate. For small ν, we obtain a solution that is close to the regularized one.

Such an approach works even better, when weak learners are uncorrelated. That is why the procedure for finding Φ_m works on a subsample of original data that is a fraction ζ of the set of training examples, drawn without replacement [33]. Such an approach leads to a set of rules that are more diversified and less correlated. Moreover, finding Φ_m on a subsample reduces the computational complexity. Note, however, that a small ν requires a larger M.

Independently on the fact whether Φ_m was found using a subsample or not, we calculate the value of α_m on *all* training examples. This usually decreases $|\alpha_m|$, so it plays also the role of regularization, and avoids overfitting the rule to the training set.

These above three elements (shrinking, sampling, and calculating α_m on the entire training set) used in ENDER constitute a competitive technique to pruning often used in the rule induction algorithms. Our experiments showed that it significantly improves the accuracy of the classifier.

6.3 Regression Problem

Boosting-like formulation of the rule induction algorithm allows to cope with the problem of regression [34]. In this problem, one usually considers the *squared-error loss*:

$$L_{\mathrm{se}}(y, f(\mathbf{x})) = (y - f(\mathbf{x}))^2. \tag{21}$$

It is easy to show that for this loss function, the ENDER algorithm computes the rule response according to:

$$\alpha_m = \arg\min_{\alpha} \sum_{i=1}^{N} (y_i - f_{m-1}(\mathbf{x}_i) - \alpha \Phi_m(\mathbf{x}_i))^2 = \frac{\sum_{\Phi(\mathbf{x}_i)=1}(y_i - f_{m-1}(\mathbf{x}_i))}{\sum_{i=1}^{N} \Phi(\mathbf{x}_i)}, \tag{22}$$

which is the average over the residuals $y_i - f_{m-1}(\mathbf{x}_i)$ for examples covered by the rule. The form of $\mathscr{L}_m(\Phi)$ is, in turn, the following:

$$\mathscr{L}_m(\Phi) = -\frac{\left|\sum_{\Phi(\mathbf{x}_i)=1}(y_i - f_{m-1}(\mathbf{x}_i))\right|}{\sqrt{\sum_{i=1}^{N} \Phi(\mathbf{x}_i)}}. \tag{23}$$

This form results from putting (22) to (5), and solving the latter with respect to Φ.

As we see, the application of boosted decision rules to solving the regression problem is straightforward.

6.4 Building Single Rules

There remains a question: how to generate the condition part of the rule. This question is associated with another one [28]: how close is the generated rule to the best one. In other words, how good are the procedures building condition parts of the rules.

The procedure used in ENDER follows a typical strategy of many other rule learning algorithms. It gives approximate results, as it is based on a greedy approach, but it is very efficient therefore.

Just to mention, the AQ algorithm follows a different rule induction strategy. Its main idea is to generate the so-called *star* for a selected positive example. The star is a set of candidate condition parts. One of these condition parts is selected by using the beam search in the space of the star members.

Some other procedures could be introduced to ENDER in order to generate better rules. These procedures could aim at improving the performance and/or the interpretability of the rule ensemble.

7 Experiment

We compared five rule induction algorithms: Ripper with sequential covering procedure, SLIPPER, LRI, MLRules, and ENDER with constant step minimization. We selected the following parameters for each method:

- Ripper: we used JRipper (Weka [35] implementation of the algorithm) with default parameters.
- SLIPPER: we set the maximum number of iterations to 500, the rest of parameters was default (we kept the internal cross validation, used to choose the optimal number of rules, switched on).
- LRI: according to the experiment in [21], we set the rule length to 5, froze feature selection after 50 rounds, and chose 200 rules per class and 2 disjuncts, since some previous tests showed that those values work well in practice.
- MLRules: according to the experiment in [23], we set $\zeta = 0.5, \nu = 0.1, M = 500$; those parameters have been optimized on an artificial data set.
- ENDER: according to the experiment in [29], we took the algorithm with constant step minimization $\beta = 0.2$ of the exponential loss (CS-Exp), with other parameters set to $\nu = 0.1$, $\zeta = 0.25$, $M = 500$; those parameters have been optimized on an artificial data set.

We used 20 binary classification problems, all taken from the UCI Repository [36]. Each test was performed using 10-fold cross validation (with exactly the same train/test splits for each classifier), and an average 0-1 loss on testing folds was calculated. The results are shown in Table 1.

Table 1. Test errors and ranks (in parenthesis). In the last row, the average rank is computed for each classifier.

DATA SET	MLRULES	ENDER CS-EXP	SLIPPER	LRI	JRIPPER
HABERMAN	26.2 (2.0)	26.2(3.0)	26.8 (4.0)	27.5(5.0)	25.8(1.0)
BREAST-C	25.9 (1.0)	27.2(3.0)	27.9 (4.0)	29.3(5.0)	26.9(2.0)
DIABETES	24.7 (2.0)	24.6(1.0)	25.4 (4.0)	25.4(3.0)	26.2(5.0)
CREDIT-G	24.1 (3.0)	22.8(1.0)	27.7 (4.0)	23.9(2.0)	28.2(5.0)
CREDIT-A	13.3 (3.0)	12.3(2.0)	17.0 (5.0)	12.2(1.0)	13.9(4.0)
IONOSPHERE	6.0 (2.0)	5.7(1.0)	6.5 (3.0)	6.8(4.0)	12.0(5.0)
COLIC	13.9 (1.0)	14.4(2.0)	15.0 (4.0)	16.1(5.0)	14.4(3.0)
HEPATITIS	16.2 (1.0)	18.8(4.0)	16.7 (2.0)	18.0(3.0)	20.2(5.0)
SONAR	12.0 (1.0)	16.4(3.0)	26.4 (4.0)	14.9(2.0)	29.7(5.0)
HEART-STATLOG	16.7 (1.0)	17.4(2.0)	23.3 (5.0)	19.6(3.0)	20.4(4.0)
LIVER-DISORDERS	27.5 (3.0)	24.9(1.0)	30.7 (4.0)	26.6(2.0)	33.0(5.0)
VOTE	3.4 (1.0)	3.4(2.0)	5.0 (5.0)	3.9(3.0)	4.6(4.0)
HEART-C-2	16.5 (2.0)	15.2(1.0)	19.5 (4.0)	18.5(3.0)	20.5(5.0)
HEART-H-2	18.0 (2.0)	17.3(1.0)	20.0 (5.0)	18.3(3.0)	20.0(4.0)
BREAST-W	3.1 (1.0)	3.6(3.0)	4.3 (4.5)	3.3(2.0)	4.3(4.5)
SICK	1.6 (2.0)	1.8(4.0)	1.6 (1.0)	1.8(5.0)	1.7(3.0)
TIC-TAC-TOE	11.3 (4.0)	8.1(3.0)	2.4 (1.0)	12.2(5.0)	2.8(2.0)
SPAMBASE	4.7 (2.0)	4.6(1.0)	5.9 (4.0)	4.9(3.0)	7.2(5.0)
CYLINDER-BANDS	14.4 (1.0)	19.4(3.0)	21.7 (4.0)	16.5(2.0)	31.5(5.0)
KR-VS-KP	1.0 (3.0)	1.0(2.0)	0.6 (1.0)	3.1(5.0)	1.0(4.0)
AVG. RANK	1.9	2.15	3.63	3.3	4.03

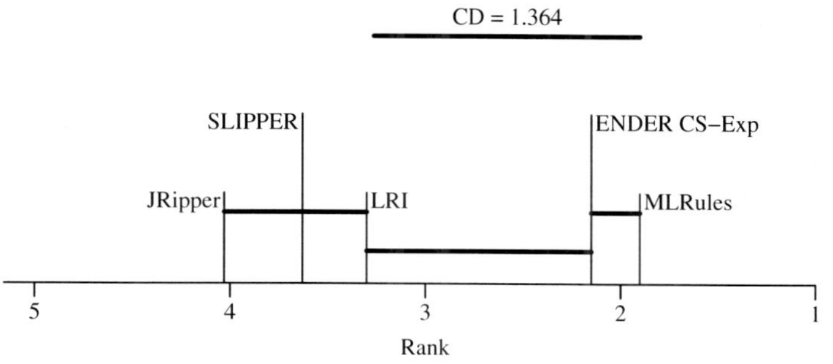

Fig. 2. Critical difference diagram

To compare multiple classifiers on multiple data sets, we follow [37], and apply the Friedman test, which uses ranks of each algorithm to check whether all the algorithms perform equally well (null hypothesis). Friedman statistics gives 27.71, which exceeds the critical value 9.488 (for confidence level 0.05), and thus we can reject the null hypothesis. Next, we proceed to a post-hoc analysis and calculate the *critical difference* (CD) according to the Nemenyi statistics. We obtain $\mathrm{CD} = 1.364$ which means that algorithms with difference in average ranks greater than 1.364 are significantly different. In Figure 2, average ranks were marked on the rank scale, and groups of the classifiers that are not significantly different were connected. This shows that MLRules outperforms LRI, SLIPPER and JRipper. ENDER CS-Exp, in turn, is significantly better than SLIPPER and JRipper. On the other hand, none of the three rule algorithms: LRI, SLIPPER, JRipper, is significantly better than any other. However, the worst results are obtained by JRipper.

The results can be interpreted in the following way. The boosting approach improves, in general, the sequential procedure. The highest improvement is achieved, however, when boosting is combined with regularization, which is applied in both, MLRules and ENDER CS-Exp. As explained above, this is justified by the recent results in theoretical analysis of boosting algorithms. The two other rule ensemble algorithms, SLIPPER and LRI, do not employ such regularization. The former uses rule pruning, but this seems to be not enough.

The main advantage of rules is their simplicity and interpretability, provided that the number of rules in the ensemble is not too high. To ease the interpretation of rules obtained by ENDER, one can apply one of techniques proposed in [22], for example, the post-processing phase in which the rules are refitted by using lasso regularization. Another possibility is to sort the rules in the ensemble using some interestingness measures, as it is done in the case of association rules. Many such measures were already introduced in the literature [38], however, recent results of comparative studies give priority to some Bayesian confirmation measures having desirable properties of symmetry and monotonicity [39, 40].

On the other hand, ENDER can be parameterized in such a way that only few short decision rules are generated – such rules can easily be interpreted while still maintaining good performance. To this aim we have to use no shrinking and no sampling. Moreover, we control the coverage of the rules by using the constant-step minimization.

Table 2. Decision rules for the CREDIT-G data set. The decision part of the rule specifies the class (sign of the rule response), the weight (absolute value of the response) of the rule (in the first parenthesis), and the numbers of training examples that are correctly classified or misclassified by the rule (in the second parenthesis, separated by ":").

#	RULE
1.	*if* CHECKING STATUS = *no checking account* *and* OTHER PAYMENT PLANS $\neq$ *bank* *and* AGE $\in [22.5, 66.5]$ *and* CREDIT AMOUNT ≤ 10924.5, *then* CUSTOMER = *good* (0.853) (302:23)
2.	*if* DURATION ≥ 15.5 *and* PURPOSE $\notin$ {*used car, retraining*} *and* SAVINGS STATUS ≤ 500 *and* AGE ≤ 63.5 *and* CHECKING STATUS ≤ 200, *then* CUSTOMER = *bad* (0.606) (161:164)
3.	*if* CREDIT HISTORY = *critical/other existing credit* *and* CREDIT AMOUNT ≤ 7806.5 *and* DURATION ≤ 37.5 *and* PURPOSE $\neq$ *education* *and* AGE ≥ 23.5 *and* OTHER PAYMENT PLANS = *none* *and* OTHER PARTIES $\in$ {*none, guarantor*}, *then* CUSTOMER = *good* (0.883) (182:14)
4.	*if* CHECKING STATUS < 0 *and* OTHER PARTIES = *none* *and* SAVINGS STATUS ≤ 500 *and* CREDIT AMOUNT ≥ 438.0 *and* FOREIGN WORKER = *no* *and* JOB $\neq$ *highqualif, selfemp, mgmt* *and* INSTALLMENT COMMITMENT ≥ 1.5 *and* PURPOSE $\neq$ *business* *and* AGE ≤ 66.0, *then* CUSTOMER = *bad* (0.636) (95:48)
5.	*if* PURPOSE = *radio, tv* *and* PROPERTY MAGNITUDE $\neq$ *no known property* *and* AGE $\in [22.5, 65.5]$ *and* CREDIT AMOUNT ≤ 4094.5 *and* CREDIT HISTORY $\neq$ *no credits, all paid* *and* DURATION ≥ 39.0 *and* OTHER PARTIES $\in$ {*none, guarantor*}, *then* CUSTOMER = *good* (0.646) (168:24)

Consider, for example, the CREDIT-G data set. We use ENDER with the exponential loss and constant-step minimization. We limit the number of rules to 5. We do not perform regularization, i.e. no shrinking and no sampling is used. To obtain good performance and interpretable rules, we tune parameter β only.

The rules are presented in Table 2. Parameter β has been set to 0.2. In 10-fold cross-validation, the algorithm obtained the misclassification error equal to 26.2, which is comparable to results obtained by other algorithms (see Table 1). Let us remark that in this data set, 700 examples belong to the class of *good* customers, and 300 examples to the class of *bad* customers. That is why the rules voting for the latter class can cover more examples from the former class (see, the second rule).

8 Conclusions

We have shown how the algorithms for rule induction evolved from the sequential covering approach to the boosted rule ensembles. This latter framework for rule induction algorithms have several important advantages.

Our results confirm the observation given in [20], that boosted rule ensembles are in fact simpler and better-understood formally than other state-of-the-art rule learners. They can be used with different loss functions. The purity measure is defined in a natural way by performing different minimization techniques. One of them allows to control the rule coverage. The proper regularization increases significantly the performance of the algorithm. It is also easy to use the boosted decision rules to regression problems.

There are still, however, several problems waiting for a better solution. One of them is the procedure for building the condition part of a single rule. Another one is the improvement of interpretability of a rule ensemble.

Acknowledgements. The authors wish to acknowledge financial support from the Ministry of Science and Higher Education (grant no. N N519 314435).

References

1. Michalski, R.S.: A Theory and Methodology of Inductive Learning. In: Michalski, R.S., Carbonell, J.G., Mitchell, T.M. (eds.) Machine Learning: An Artificial Intelligence Approach, pp. 83–129. Tioga Publishing, Palo Alto (1983)
2. Clark, P., Niblett, T.: The CN2 induction algorithm. Machine Learning 3, 261–283 (1989)
3. Cohen, W.W.: Fast Effective Rule Induction. In: Proc. of International Conference of Machine Learning, pp. 115–123 (1995)
4. Fürnkranz, J.: Separate-and-Conquer Rule Learning. Artificial Intelligence Review 13(1), 3–54 (1996)
5. Jovanoski, V., Lavrac, N.: Classification Rule Learning with APRIORI-C. In: Proc. of the 10th Portuguese Conference on Progress in Artificial Intelligence, Knowledge Extraction, Multi-agent Systems, Logic Programming and Constraint Solving, London, UK, pp. 44–51. Springer, Heidelberg (2001)
6. Stefanowski, J., Vanderpooten, D.: Induction of Decision Rules in Classification and Discovery-oriented Perspectives. International Journal on Intelligent Systems 16(1), 13–27 (2001)

7. Bazan, J.G.: Discovery of Decision Rules by Matching New Objects Against Data Tables. In: Polkowski, L., Skowron, A. (eds.) RSCTC 1998. LNCS (LNAI), vol. 1424, pp. 521–528. Springer, Heidelberg (1998)
8. Góra, G., Wojna, A.: RIONA: A New Classification System Combining Rule Induction and Instance-based Learning. Fundamenta Informaticae 54, 369–390 (2002)
9. Domingos, P.: Unifying Instance-based and Rule-based Induction. Machine Learning 24, 141–168 (1996)
10. Góra, G., Wojna, A.: Local Attribute Value Grouping for Lazy Rule Induction. In: Alpigini, J.J., Peters, J.F., Skowron, A., Zhong, N. (eds.) RSCTC 2002. LNCS (LNAI), vol. 2475, pp. 405–412. Springer, Heidelberg (2002)
11. Boros, E., Hammer, P.L., Ibaraki, T., Kogan, A., Mayoraz, E., Muchnik, I.: An Implementation of Logical Analysis of Data. IEEE Transactions on Knowledge and Data Engineering 12, 292–306 (2000)
12. Pawlak, Z.: Rough Sets. Theoretical Aspects of Reasoning about Data. Kluwer Academic Publishers, Dordrecht (1991)
13. Słowiński, R. (ed.): Intelligent Decision Support. In: Handbook of Applications and Advances of the Rough Set Theory. Kluwer Academic Publishers, Dordrecht (1992)
14. Grzymala-Busse, J.W.: LERS — A System for Learning from Examples based on Rough Sets. In: Słowiński, R. (ed.) Intelligent Decision Support, Handbook of Applications and Advances of the Rough Sets Theory, pp. 3–18. Kluwer Academic Publishers, Dordrecht (1992)
15. Skowron, A.: Extracting Laws from Decision Tables - A Rough Set Approach. Computational Intelligence 11, 371–388 (1995)
16. Stefanowski, J.: On Rough Set based Approach to Induction of Decision Rules. In: Skowron, A., Polkowski, L. (eds.) Rough Set in Knowledge Discovering, pp. 500–529. Physica Verlag, Heidelberg (1998)
17. Greco, S., Matarazzo, B., Słowiński, R., Stefanowski, J.: An algorithm for induction of decision rules consistent with the dominance principle. In: Ziarko, W.P., Yao, Y. (eds.) RSCTC 2000. LNCS (LNAI), vol. 2005, pp. 304–313. Springer, Heidelberg (2001)
18. Freund, Y., Schapire, R.E.: A Decision-theoretic Generalization of On-line Learning and an Application to Boosting. Journal of Computer and System Sciences 55(1), 119–139 (1997)
19. Hastie, T., Tibshirani, R., Friedman, J.H.: Elements of Statistical Learning: Data Mining, Inference, and Prediction. Springer, Heidelberg (2003)
20. Cohen, W.W., Singer, Y.: A Simple, Fast, and Effective Rule Learner. In: Proc. of National Conference on Artificial Intelligence, pp. 335–342 (1999)
21. Weiss, S.M., Indurkhya, N.: Lightweight Rule Induction. In: Proc. of International Conference on Machine Learning, pp. 1135–1142 (2000)
22. Friedman, J.H., Popescu, B.E.: Predictive Learning via Rule Ensembles. Annals of Applied Statistics 2(3), 916–954 (2008)
23. Dembczyński, K., Kotłowski, W., Słowiński, R.: Maximum Likelihood Rule Ensembles. In: Proc. of International Conference on Machine Learning, pp. 224–231 (2008)
24. Błaszczyński, J., Dembczyński, K., Kotłowski, W., Słowiński, R., Szelag, M.: Ensembles of Decision Rules. Foundations of Computing and Decision Sciences 31(3-4), 221–232 (2006)
25. Friedman, J.H.: Greedy Function Approximation: A Gradient Boosting Machine. Annals of Statistics 29(5), 1189–1232 (2001)
26. Breiman, L.: Bagging Predictors. Machine Learning 24(2), 123–140 (1996)
27. Mason, L., Baxter, J., Bartlett, P., Frean, M.: Functional Gradient Techniques for Combining Hypotheses. In: Bartlett, P., Schölkopf, B., Schuurmans, D., Smola, A.J. (eds.) Advances in Large Margin Classifiers, pp. 33–58. MIT Press, Cambridge (1999)
28. Fürnkranz, J.: Rule-based Classification. In: From Local Patterns to Global Models ECML/PKDD 2008 Workshop (2008)

29. Dembczyński, K., Kotłowski, W., Słowiński, R.: A General Framework for Learning an Ensemble of Decision Rules. In: Fürnkranz, J., Knobbe, A. (eds.) From Local Patterns to Global Models ECML/PKDD 2008 Workshop (2008)
30. Schapire, R.E., Freund, Y., Bartlett, P., Lee, W.S.: Boosting the Margin: A New Explanation for the Effectiveness of Voting Methods. Annals of Statistics 26(5), 1651–1686 (1998)
31. Friedman, J.H., Hastie, T., Tibshirani, R.: Additive Logistic Regression: A Statistical View of Boosting. Annals of Statistics (with discussion) 28(2), 337–407 (2000)
32. Dietterich, T.G.: An Experimental Comparison of Three Methods for Constructing Ensembles of Decision Trees: Bagging, Boosting, and Randomization. Machine Learning 40(2), 139–158 (2000)
33. Friedman, J.H., Popescu, B.E.: Importance Sampled Learning Ensembles. Research report, Dept. of Statistics, Stanford University (2003)
34. Dembczyński, K., Kotłowski, W., Słowiński, R.: Solving Regression by Learning an Ensemble of Decision Rules. In: Rutkowski, L., Tadeusiewicz, R., Zadeh, L.A., Zurada, J.M. (eds.) ICAISC 2008. LNCS (LNAI), vol. 5097, pp. 533–544. Springer, Heidelberg (2008)
35. Witten, I.H., Frank, E.: Data Mining: Practical machine learning tools and techniques, 2nd edn. Morgan Kaufmann, San Francisco (2005)
36. Asuncion, A., Newman, D.J.: UCI Machine Learning Repository (2007)
37. Demšar, J.: Statistical Comparisons of Classifiers over Multiple Data Sets. Journal of Machine Learning Research 7, 1–30 (2006)
38. Hilderman, R.J., Hamilton, H.J.: Knowledge Discovery and Measures of Interest. Kluwer Academic Publishers, Boston (2001)
39. Greco, S., Pawlak, Z., Słowiński, R.: Can Bayesian confirmation measures be useful for rough set decision rules? Engineering Applications of Artificial Intelligence 17, 345–361 (2004)
40. Brzezińska, I., Greco, S., Słowiński, R.: Mining Pareto-optimal Rules with Respect to Support and Anti-support. Engineering Applications of Artificial Intelligence 20, 587–600 (2007)

An Analysis of Relevance Vector Machine Regression

Matti Saarela[1], Tapio Elomaa[1], and Keijo Ruohonen[2]

[1] Department of Software Systems, Tampere University of Technology
[2] Department of Mathematics, Tampere University of Technology
P.O. Box 553 (Korkeakoulunkatu 1), FI-33101 Tampere, Finland
matti.saarela@iki.fi, elomaa@cs.tut.fi, keijo.ruohonen@tut.fi

Abstract. The relevance vector machine (RVM) is a Bayesian framework for learning sparse regression models and classifiers. Despite of its popularity and practical success, no thorough analysis of its functionality exists. In this paper we consider the RVM in the case of regression models and present two kinds of analysis results: we derive a full characterization of the behavior of the RVM analytically when the columns of the regression matrix are orthogonal and give some results concerning scale and rotation invariance of the RVM. We also consider the practical implications of our results and present a scenario in which our results can be used to detect potential weakness in the RVM framework.

Keywords: Relevance vector machine, regression, sparse Bayesian learning.

1 Introduction

The *relevance vector machine* (RVM) [7, 8] is a Bayesian framework for learning sparse regression models and classifiers in the supervised setting. Recently, it has been applied successfully in several practical problems (for example, see [1, 3, 5, 12, 6, 14]). The previous success, however, cannot guarantee applicability in all forthcoming problems. In order to be able to predict applicability in new situations, we need a detailed analysis of the functionality of the RVM. Even though some analytical considerations are taken into account in the derivation of the RVM [7, 8] and some analyzes have been put forward [4, 13, 16, 15], no thorough analysis can be found in the literature. For example, it is not known what kind of trade-off is induced between sparsity and fitness of the produced regression models and classifiers.

In this paper we present two kinds of analysis results for the RVM. Firstly, we analyze the RVM in the case of regression models, which consist of orthogonal basis vectors. Our analysis reveals that RVM regression with orthogonal basis vectors can be divided into three distinct subcases in which the behavior of the RVM can be completely characterized. For example, we are able to explicitly describe the trade-off between sparsity and fitness of the produced models. Furthermore, we give an example on how regression models with orthogonal basis vectors can arise in practical situations and apply our results to this example. Secondly, we examine scale and rotation invariance properties of RVM regression.

The rest of this paper is organized as follows. In Sect. 2 we briefly recapitulate the RVM framework. Our results concerning orthogonal columns of regression matrix and

J. Koronacki et al. (Eds.): Advances in Machine Learning I, SCI 262, pp. 227–246.
springerlink.com © Springer-Verlag Berlin Heidelberg 2010

invariance properties are presented in Sect. 3 and Sect. 6, respectively. In Sect. 4 we illustrate the three subcases of the analysis empirically and describe in Sect. 5 a practical scenario, in which our results are applicable. Finally, Sect. 7 contains our conclusions and suggestions for future research directions. All proofs of lemmas are presented in the appendix.

2 The RVM Briefly

In supervised learning input vectors $\boldsymbol{x}_1, \ldots, \boldsymbol{x}_n$ come associated with their respective realization y_i, $i = 1, \ldots, n$, of the dependent variable Y. In this work we are concerned with *regression*; i.e., $y_1, \ldots, y_n \in \mathbb{R}$. The RVM, introduced by Tipping [7, 8], is a framework for obtaining estimates for the weight vector $\boldsymbol{w} = [w_1, \ldots, w_m]^T$ and variance β estimates in the generalized linear model

$$y(\boldsymbol{x}) = \sum_{i=1}^{m} w_i \tau_i(\boldsymbol{x}) + \epsilon,$$

where $\tau_i : \mathbb{R}^d \to \mathbb{R}$ are so-called *basis functions* and ϵ is normally distributed with mean 0 and variance β, which we denote by $\epsilon \sim \mathcal{N}(0, \beta)$. To obtain the estimates, the RVM uses training data $\{ (y_1, \boldsymbol{x}_1), \ldots, (y_n, \boldsymbol{x}_n) \}$ to build a regression model

$$\boldsymbol{y} = \boldsymbol{\Phi w} + \boldsymbol{\epsilon}, \tag{1}$$

where

$$\begin{aligned} \boldsymbol{y} &= [y_1, \ldots, y_n]^T, \\ \boldsymbol{\Phi} &= [\boldsymbol{\phi}_1, \ldots, \boldsymbol{\phi}_m], \\ \boldsymbol{\phi}_i &= [\tau_i(\boldsymbol{x}_1), \ldots, \tau_i(\boldsymbol{x}_n)]^T, \end{aligned}$$

and $\boldsymbol{\epsilon}$ has a n-dimensional normal distribution $\mathcal{N}(\mathbf{0}, \beta I_n)$, in which $I_n \in \mathbb{R}^{n \times n}$ denotes the $n \times n$ identity matrix. Each vector $\boldsymbol{\phi}_i$, consisting of the values of basis function τ_i for the input vectors, is a *basis vector*. Further, the RVM framework considers regression model (1) from the Bayesian point of view. The approach includes fixing probability distributions for random variables in the model, performing analytical integrations and, finally, finding point estimates for model parameters.

To make the RVM favor sparse regression models, Tipping [8] adds a new parameter $\alpha_i \in (0, \infty)$ for each weight parameter w_i. The parameter α_i affects to w_i trough a conditional distribution $p(w_i \mid \alpha_i)$, which is given by

$$p(w_i \mid \alpha_i) = \mathcal{N}(0, \alpha_i^{-1}).$$

The prior distribution $p(\alpha_i)$ for the parameters α_i is fixed differently in different sources. Originally Tipping [7] considers $p(\alpha_i)$ as a uniform improper distribution on the interval $(0, \infty)$, but later Tipping [8] fixes $p(\alpha_i)$ to be a gamma distribution, whose parameters are set to zero. A similar treatment is also given for the prior distribution of

β^{-1}. Both ways of treating priors, however, lead to the same relevance vector machine construction.

In his construction Tipping [8] ends up computing maximum likelihood estimates for model parameters α_i and β from the conditional distribution

$$p(\boldsymbol{y} \mid \boldsymbol{\alpha}, \beta) = \mathcal{N}(\boldsymbol{y} \mid \boldsymbol{0}, \boldsymbol{C}). \tag{2}$$

Here the matrix $\boldsymbol{C} \in \mathbb{R}^{n \times n}$ is defined as

$$\boldsymbol{C} = \beta I_n + \boldsymbol{\Phi} \boldsymbol{A}^{-1} \boldsymbol{\Phi}^T,$$

where $\boldsymbol{A}^{-1} = \lceil \alpha_1, \ldots, \alpha_m \rfloor$ and $\lceil \alpha_1, \ldots, \alpha_m \rfloor$ stands for the diagonal matrix with elements $\alpha_1, \ldots, \alpha_m$ in the diagonal. After finding the optimal point estimates for parameters α_i and β, the point estimate $\hat{\boldsymbol{w}}$ for weight vector $\boldsymbol{w}$ can be obtained from the equation

$$\hat{\boldsymbol{w}} = \beta \left(\beta \boldsymbol{\Phi}^T \boldsymbol{\Phi} + \boldsymbol{A} \right)^{-1} \boldsymbol{\Phi}^T \boldsymbol{y}. \tag{3}$$

The actual derivation of (2) and (3) is given by Tipping [8].

Tipping [8] observes that during the optimization process some components α_i of vector $\boldsymbol{\alpha}$ approach infinity. Consequently, he deduces that the corresponding weight vector components w_i can be set to exactly zero. In other words, corresponding basis vectors $\boldsymbol{\phi}_i$ and basis functions τ_i are pruned from the model. According to empirical observations, usually a majority of basis functions is pruned from the model, which leads to sparse regression models.

3 The RVM and Orthogonal Basis Vectors

Let us now assume that the basis vectors $\boldsymbol{\phi}_i$ are orthogonal. In other words, $\boldsymbol{\phi}_i^T \boldsymbol{\phi}_k = 0$ for all $i \neq k$. Then the inequality $m \leq n$ must hold, since there can be at most n orthogonal vectors in $\mathbb{R}^n$. Clearly, finding maximum likelihood estimates for parameters $\boldsymbol{\alpha}$ and β is equivalent to finding maximum of log likelihood function [8]

$$L(\boldsymbol{\alpha}, \beta) = -\ln(|\boldsymbol{C}|) - \boldsymbol{y}^T \boldsymbol{C}^{-1} \boldsymbol{y}, \tag{4}$$

where by $|\boldsymbol{C}|$ we denote the determinant of matrix $\boldsymbol{C}$ and by $\boldsymbol{C}^{-1}$ its inverse.

Recall the Matrix determinant lemma

$$\left| A^{-1} + BC \right| = |I_m + CAB| \left| A^{-1} \right|, \tag{5}$$

where $A \in \mathbb{R}^{n \times n}$ is non-singular, $B \in \mathbb{R}^{n \times m}$, and $C \in \mathbb{R}^{m \times n}$. Recall also the so-called Sherman-Morrison-Woodbury identity

$$(A + UBV)^{-1} = A^{-1} - A^{-1} U \left(B^{-1} + V A^{-1} U \right)^{-1} V A^{-1}, \tag{6}$$

where $A \in \mathbb{R}^{n \times n}$, $B \in \mathbb{R}^{m \times m}$, $U \in \mathbb{R}^{n \times m}$, $V \in \mathbb{R}^{m \times n}$ and, in addition, the matrices A, B, and $A + UBV$ are non-singular. Proofs for both these equalities can be found, e.g., from [2].

By orthogonality of vectors $\boldsymbol{\phi}_i$ and by (5), we can express the determinant $|\boldsymbol{C}|$ in (4) as

$$
\begin{aligned}
|\boldsymbol{C}| &= \left|\beta I_n + \boldsymbol{\Phi}\boldsymbol{A}^{-1}\boldsymbol{\Phi}^T\right| \\
&= \beta^n \left|I_m + \boldsymbol{\Phi}^T\boldsymbol{\Phi}(\beta\boldsymbol{A})^{-1}\right| \\
&= \beta^{n-m}\prod_{i=1}^{m}\left(\beta + \alpha_i^{-1}\|\boldsymbol{\phi}_i\|^2\right).
\end{aligned} \tag{7}
$$

Similarly, by orthogonality and (6), we can write the term $\boldsymbol{y}^T\boldsymbol{C}^{-1}\boldsymbol{y}$ in (4) in the form

$$
\begin{aligned}
\boldsymbol{y}^T\boldsymbol{C}^{-1}\boldsymbol{y} &= \boldsymbol{y}^T\left(\beta^{-1}I_n - \beta^{-2}\boldsymbol{\Phi}\left(\boldsymbol{A} + \beta^{-1}\boldsymbol{\Phi}^T\boldsymbol{\Phi}\right)^{-1}\boldsymbol{\Phi}^T\right)\boldsymbol{y} \\
&= \beta^{-1}\|\boldsymbol{y}\|^2 - \sum_{i=1}^{m}\frac{\langle\boldsymbol{\phi}_i,\boldsymbol{y}\rangle^2}{\beta^2\alpha_i + \beta\|\boldsymbol{\phi}_i\|^2}.
\end{aligned} \tag{8}
$$

Now by (7) and (8), the target function $L(\boldsymbol{\alpha},\beta)$ can be expressed as

$$
\begin{aligned}
L(\boldsymbol{\alpha},\beta) = &\left(\sum_{i=1}^{m}\frac{\langle\boldsymbol{\phi}_i,\boldsymbol{y}\rangle^2}{\beta^2\alpha_i + \beta\|\boldsymbol{\phi}_i\|^2}\right) - \beta^{-1}\|\boldsymbol{y}\|^2 \\
&- \left(\sum_{i=1}^{m}\ln\left(\beta + \alpha_i^{-1}\|\boldsymbol{\phi}_i\|^2\right)\right) - (n-m)\ln(\beta).
\end{aligned} \tag{9}
$$

Based on observations of Tipping [8], we expect that at least in some situations some components α_i approach infinity when $L(\boldsymbol{\alpha},\beta)$ is being maximized. In these situation, however, it is somewhat misleading to refer to optimal point of $L(\boldsymbol{\alpha},\beta)$, since all α_i components are defined only in the interval $(0,\infty)$. To overcome this ambiguity, let us define the concept of configuration.

Definition 1. *Let* $\underline{\boldsymbol{\alpha}} = [\underline{\alpha}_1,\ldots,\underline{\alpha}_m]^T \in \mathbb{R}_\infty^m$, *where* $\mathbb{R}_\infty = \mathbb{R}\cup\{\,\infty\,\}$. *We say that* $\underline{\boldsymbol{\alpha}}$ *is a* configuration *of the RVM, if* $\underline{\alpha}_i = \infty$ *implies that* $\boldsymbol{\phi}_i$ *is pruned from the base and* $\underline{\alpha}_i < \infty$ *implies that* $\boldsymbol{\phi}_i$ *is not pruned from the base and* $\alpha_i = \underline{\alpha}_i$ *holds. Further, we define* $K(\underline{\boldsymbol{\alpha}}) = \{\, i \mid \underline{\alpha}_i < \infty\,\}$.

It is clear from this definition that the target function (9) value corresponding a given configuration $\underline{\boldsymbol{\alpha}}$ is of the form

$$
\begin{aligned}
\underline{L}(\underline{\boldsymbol{\alpha}},\beta) = &\left(\sum_{i\in K(\underline{\boldsymbol{\alpha}})}\frac{\langle\boldsymbol{\phi}_i,\boldsymbol{y}\rangle^2}{\beta^2\underline{\alpha}_i + \beta\|\boldsymbol{\phi}_i\|^2}\right) - \frac{\|\boldsymbol{y}\|^2}{\beta} \\
&- \left(\sum_{i\in K(\underline{\boldsymbol{\alpha}})}\ln\left(\beta + \underline{\alpha}_i^{-1}\|\boldsymbol{\phi}_i\|^2\right)\right) - (n - |K(\underline{\boldsymbol{\alpha}})|)\ln(\beta).
\end{aligned}
$$

This form can further be expressed as

$$\underline{L}(\boldsymbol{\alpha}, \beta) = \left(\sum_{i=1}^{m} l_i(\underline{\alpha}_i, \beta) \right) - \frac{\|\boldsymbol{y}\|^2}{\beta} - n \ln(\beta), \tag{10}$$

where the function $l_i(\underline{\alpha}_i, \beta)$ is defined to be

$$l_i\left(\underline{\alpha}_i, \beta\right) = \frac{\langle \boldsymbol{\phi}_i, \boldsymbol{y} \rangle^2}{\beta^2 \underline{\alpha}_i + \beta \|\boldsymbol{\phi}_i\|^2} - \ln\left(\beta + \underline{\alpha}_i^{-1} \|\boldsymbol{\phi}_i\|^2\right) + \ln(\beta),$$

if $\underline{\alpha}_i < \infty$, and $l_i\left(\underline{\alpha}_i, \beta\right) = 0$, when $\underline{\alpha}_i = \infty$.

For the target function $\underline{L}(\boldsymbol{\alpha}, \beta)$ it can be shown that for every $\beta > 0$ there exists a unique optimal configuration $\underline{\boldsymbol{\alpha}}$. We give further details on this result as follows.

Lemma 1. *Let $\beta > 0$ be given. Then there exists a unique optimal configuration $\underline{\boldsymbol{\alpha}}$ that maximizes $\underline{L}(\boldsymbol{\alpha}, \beta)$. For $\underline{\boldsymbol{\alpha}}$, the following equation holds:*

$$\underline{\alpha}_i = \begin{cases} \dfrac{\|\boldsymbol{\phi}_i\|^2}{\langle \boldsymbol{\phi}_i, \boldsymbol{y} \rangle^2 \|\boldsymbol{\phi}_i\|^{-2} - \beta}, & \langle \boldsymbol{\phi}_i, \boldsymbol{y} \rangle^2 / \|\boldsymbol{\phi}_i\|^2 > \beta, \\ \infty, & \langle \boldsymbol{\phi}_i, \boldsymbol{y} \rangle^2 / \|\boldsymbol{\phi}_i\|^2 \leq \beta. \end{cases}$$

Let $\hat{\underline{\boldsymbol{\alpha}}}(\beta)$ denote the optimal configuration corresponding to a given β. Now, in order to optimize $\underline{L}(\boldsymbol{\alpha}, \beta)$ we need to select β that maximizes $L(\hat{\underline{\boldsymbol{\alpha}}}(\beta), \beta)$. So basically we are maximizing the function $L(\beta) = \underline{L}(\hat{\underline{\boldsymbol{\alpha}}}(\beta), \beta)$. To ease the maximization, we introduce some further notation.

We denote by $I(\beta)$ the set of indices of those basis vectors that are not pruned from the base in the optimal configuration $\hat{\underline{\boldsymbol{\alpha}}}(\beta)$. More formally,

$$I(\beta) = \left\{ i \,\middle|\, \langle \boldsymbol{\phi}_i, \boldsymbol{y} \rangle^2 / \|\boldsymbol{\phi}_i\|^2 > \beta \right\}.$$

Using this notation the target function $L(\beta)$ can be expressed as

$$\begin{aligned} L(\beta) = \beta^{-1} & \left[\left(\sum_{i \in I(\beta)} \langle \boldsymbol{\phi}_i, \boldsymbol{y} \rangle^2 / \|\boldsymbol{\phi}_i\|^2 \right) - \|\boldsymbol{y}\|^2 \right] \\ & - \left(\sum_{i \in I(\beta)} \ln\left(\langle \boldsymbol{\phi}_i, \boldsymbol{y} \rangle^2 / \|\boldsymbol{\phi}_i\|^2 \right) \right) - |I(\beta)| - (n - |I(\beta)|) \ln(\beta). \end{aligned}$$

For the upcoming analysis we assume that $\langle \boldsymbol{\phi}_i, \boldsymbol{y} \rangle^2 > 0$ for every i. To see the effect of this assumption, consider the case $\langle \boldsymbol{\phi}_i, \boldsymbol{y} \rangle^2 = 0$ for some i. Then, clearly, i does not belong to $I(\beta)$ for any $\beta > 0$. Thus, we consider only those indices i, which can belong to at least some $I(\beta)$.

Next we divide the vectors $\boldsymbol{\phi}_i$ into equivalence classes. They are constructed so that the class containing $\boldsymbol{\phi}_i$ is

$$[[\boldsymbol{\phi}_i]] = \left\{ \boldsymbol{\phi}_k \,\middle|\, \langle \boldsymbol{\phi}_i, \boldsymbol{y} \rangle^2 / \|\boldsymbol{\phi}_i\|^2 = \langle \boldsymbol{\phi}_k, \boldsymbol{y} \rangle^2 / \|\boldsymbol{\phi}_k\|^2 \right\}.$$

We denote these equivalence classes by $B_1, \ldots, B_p$ and assume that they are ordered so that conditions $i < j$, $\boldsymbol{\phi} \in B_i$, and $\boldsymbol{\phi}_* \in B_j$ imply

$$\langle \boldsymbol{\phi}, \boldsymbol{y} \rangle^2 / \|\boldsymbol{\phi}\|^2 > \langle \boldsymbol{\phi}_*, \boldsymbol{y} \rangle^2 / \|\boldsymbol{\phi}_*\|^2 .$$

We also introduce constants $c_i = |B_i|$ and $\beta_i = \langle \boldsymbol{\phi}, \boldsymbol{y} \rangle^2 / \|\boldsymbol{y}\|^2$ with $\boldsymbol{\phi} \in B_i$ for every $i \in \{1, \ldots, p\}$. Additionally, we define $c_{p+1} = 0$, $\beta_0 = \infty$, and $\beta_{p+1} = 0$.

Now we observe that the set $I(\beta)$ remains constant in each distinct interval $[\beta_{i+1}, \beta_i)$. Thus, we can express $L(\beta)$ as

$$L(\beta) = \sum_{j=0}^{p} L_j(\beta) \chi(\beta \mid [\beta_{j+1}, \beta_j)),$$

where $\chi(\cdot \mid [\beta_{j+1}, \beta_j))$ is a characteristic function of the interval $[\beta_{j+1}, \beta_j)$ and $L_j(\beta)$ is defined as

$$L_j(\beta) = \beta^{-1} \left[\left(\sum_{i=1}^{j} c_i \beta_j \right) - \|\boldsymbol{y}\|^2 \right] - \left(\sum_{i=1}^{j} c_i \ln(\beta_i) \right) - \left(\sum_{i=1}^{j} c_i \right) \\ - \left(n - \sum_{i=1}^{j} c_i \right) \ln(\beta).$$

By using previous observations, we can prove for $L(\beta)$ some results that make its maximization simple. We present these results in the following three lemmas.

Lemma 2. *The target function $L(\beta)$ is differentiable in the interval $(0, \infty)$ and its derivative is*

$$\frac{d}{d\beta} L(\beta) = \sum_{j=0}^{p} \left(\frac{d}{d\beta} L_j(\beta) \right) \chi(\beta \mid [\beta_{j+1}, \beta_j)),$$

where

$$\frac{d}{d\beta} L_j(\beta) = \beta^{-2} \left[\|\boldsymbol{y}\|^2 - \left(\sum_{i=1}^{j} c_i \beta_i \right) - \left(n - \sum_{i=1}^{j} c_i \right) \beta \right].$$

Lemma 3. *The derivative function $\frac{d}{d\beta} L(\beta)$ is continuous in the interval $(0, \infty)$.*

Lemma 4. *If $m < n$ holds, then the derivative function $\frac{d}{d\beta} L(\beta)$ has at most one zero. If the zero exists, it can be expressed as*

$$\hat{\beta}_j = \left(n - \left(\sum_{i=1}^{j} c_i \right) \right)^{-1} \left(\|\boldsymbol{y}\|^2 - \left(\sum_{i=1}^{j} c_i \beta_i \right) \right),$$

where j is chosen from $\{0, \ldots, p\}$ so that the equation $g(j+1) \le \|\boldsymbol{y}\|^2 < g(j)$ holds, where

$$g(j) = \begin{cases} \infty, & j = 0, \\ \left(\sum_{i=1}^{j} c_i \beta_i \right) + \left(n - \sum_{i=1}^{j} c_i \right) \beta_j, & 1 \le j \le p, \\ \left(\sum_{i=1}^{p} c_i \beta_i \right), & j = p+1. \end{cases}$$

Furthermore, if $m = n$, *the equation* $\frac{d}{d\beta}L(\beta) = 0$ *holds only in the whole interval* $(0, \beta_p]$.

By using Lemma 4, we can finally find zeros of the target function $L(\beta)$. First we notice that the inequality

$$\sum_{i=1}^{p} c_i\beta_i = \sum_{i=1}^{m} \langle \boldsymbol{\phi}_i, \boldsymbol{y}\rangle^2 / \|\boldsymbol{\phi}_i\|^2 \le \|\boldsymbol{y}\|^2$$

holds by orthogonality of vectors $\boldsymbol{\phi}_1, \dots, \boldsymbol{\phi}_m$. By this observation, we can divide the task of finding the maxima of $L(\beta)$ into three distinct cases:

1. Case $m < n$ and $\sum_{i=1}^{p} c_i\beta_i = \|\boldsymbol{y}\|^2$:
   Consider the limit value

$$\begin{aligned}\lim_{\beta\to 0+} L(\beta) &= \lim_{\beta\to 0+} L_j(\beta) \\ &= -m - \left(\sum_{i=1}^{p} c_i \ln(\beta_i)\right) - \lim_{\beta\to 0+} (n-m)\ln(\beta) \\ &= \infty.\end{aligned}$$

   Clearly $L(\beta)$ has no global maximum, since $L(\beta)$ approaches infinity when β approaches zero.
2. Case $m < n$ and $\sum_{i=1}^{p} c_i\beta_i < \|\boldsymbol{y}\|^2$:
   We can write $L_p(\beta)$ in the form

$$\begin{aligned}L_p(\beta) &= -\frac{d_1}{\beta} - d_2\ln(\beta) + d_3 \\ &= \frac{-d_1 - d_2\beta\ln(\beta)}{\beta} + d_3,\end{aligned}$$

   where $d_1, d_2 > 0$ and $d_3 \in \mathbb{R}$. By l'Hospital's rule we get

$$\lim_{\beta\to 0+} \beta\ln(\beta) = 0.$$

   Thus, the equation

$$\begin{aligned}\lim_{\beta\to 0+} L(\beta) &= \lim_{\beta\to 0+} L_p(\beta) \\ &= \lim_{\beta\to 0+} \frac{-d_1 - d_2\beta\ln(\beta)}{\beta} + d_3 \\ &= -\infty\end{aligned} \tag{11}$$

   holds. We also get

$$\begin{aligned}\lim_{\beta\to \infty} L(\beta) &= \lim_{\beta\to \infty} L_0(\beta) \\ &= \lim_{\beta\to \infty} \left(\beta^{-1}\|\boldsymbol{y}\|^2 - n\ln(\beta)\right) \\ &= -\infty.\end{aligned} \tag{12}$$

From (11), (12), and continuity of $\frac{\mathrm{d}}{\mathrm{d}\beta}L(\beta)$ we conclude that $\frac{\mathrm{d}}{\mathrm{d}\beta}L(\beta)$ has at least one zero in the interval $(0, \infty)$. On the other hand, Lemma 4 tells us that there can be only one such zero. Hence, we conclude that $\frac{\mathrm{d}}{\mathrm{d}\beta}L(\beta)$ has exactly one zero in the interval and the zero must be a global maximum of $L(\beta)$. Moreover, the zero can be found from the location described in Lemma 4.

3. Case $m = n$:
   By Lemma 4, the derivative $\frac{\mathrm{d}}{\mathrm{d}\beta}L(\beta)$ is zero only in the interval $(0, \beta_p]$. Since $L(\beta) > -\infty$ for every $\beta \in (0, \beta]$, the equation

$$\lim_{\beta\to\infty} L(\beta) = \lim_{\beta\to\infty} L_0(\beta) = -\infty$$

   holds and $\frac{\mathrm{d}}{\mathrm{d}\beta}L(\beta)$ is continuous, the equation $\frac{\mathrm{d}}{\mathrm{d}\beta}L(\beta) < 0$ must hold for every $\beta \in (\beta_p, \infty)$. Thus, every point in the interval $(0, \beta_p]$ is a global maximum of $L(\beta)$ and there are no other global maxima.

The three cases above tell us everything that we want to know about the global maxima of $L(\beta)$. However, our original intention was to find the maxima of $\underline{L}(\underline{\boldsymbol{\alpha}}, \beta)$. By definition of $L(\beta)$, there is a correspondence between the maxima of $L(\beta)$ and $\underline{L}(\underline{\boldsymbol{\alpha}}, \beta)$, since $L(\beta) = \underline{L}(\hat{\underline{\boldsymbol{\alpha}}}(\beta), \beta)$, where $\hat{\underline{\boldsymbol{\alpha}}}(\beta)$ is a unique optimal configuration corresponding the given β. Thus, we have proved the following result.

Theorem 1. *Let function g be defined as in Lemma 4. Then the following three claims hold for the maxima of the target function $\underline{L}(\underline{\boldsymbol{\alpha}}, \beta)$.*

1. *If $m < n$ and $\sum_{i=1}^{p} c_i\beta_i = \|\boldsymbol{y}\|^2$ hold, then the target function $\underline{L}(\underline{\boldsymbol{\alpha}}, \beta)$ has no global maximum. Further, the value of $\underline{L}(\underline{\boldsymbol{\alpha}}, \beta)$ can be made infinitely large by letting β approach zero and considering pairs $(\hat{\underline{\boldsymbol{\alpha}}}(\beta), \beta)$, where $\hat{\underline{\boldsymbol{\alpha}}}(\beta)$ is an optimal configuration corresponding to the given β.*
2. *If $m < n$ and $\sum_{i=1}^{p} c_i\beta_i < \|\boldsymbol{y}\|^2$ hold, then the target function $\underline{L}(\underline{\boldsymbol{\alpha}}, \beta)$ has exactly one global maximum $(\hat{\underline{\boldsymbol{\alpha}}}(\hat{\beta}), \hat{\beta})$, where*

$$\hat{\beta} = \left(n - \left(\sum_{i=1}^{j} c_i\right)\right)^{-1} \left(\|\boldsymbol{y}\|^2 - \left(\sum_{i=1}^{j} c_i\beta_i\right)\right),$$

   the index j is chosen so that $g(j+1) \leq \|\boldsymbol{y}\|^2 < g(j)$ and $\hat{\underline{\boldsymbol{\alpha}}}(\hat{\beta})$ is a unique optimal configuration corresponding to the value $\hat{\beta}$.
3. *If $m = n$ holds, then the target function $\underline{L}(\underline{\boldsymbol{\alpha}}, \beta)$ has maximal points $(\hat{\underline{\boldsymbol{\alpha}}}(\hat{\beta}), \hat{\beta})$, where $\hat{\beta}$ belongs to the interval $(0, \beta_p]$ and $\hat{\underline{\boldsymbol{\alpha}}}(\hat{\beta})$ is a unique optimal configuration corresponding to the value $\hat{\beta}$.*

Theorem 1 finally characterizes locations of the maxima of the target function $\underline{L}(\underline{\boldsymbol{\alpha}}, \beta)$. Its three separate cases provide three unexpectedly different statements on the existence of global maxima.

- In the first case there are no global maxima at all. Rather, the target function value can be made infinitely large by pruning none of the basis vectors and letting β approach zero.

- In the second case, on the other hand, there is a single global maximum. In this case the tradeoff between sparsity and fitness of the produced regression model can be explicitly described. Informally speaking, the tradeoff is realized by pruning "a suitable amount" of basis vectors $\boldsymbol{\phi}_i$, for which the value of $\langle \boldsymbol{\phi}_i, \boldsymbol{y} \rangle^2$ is small.
- In the third case of Theorem 1, none of the basis vectors are pruned or the basis vector with the smallest value of $\langle \boldsymbol{\phi}_i, \boldsymbol{y} \rangle^2$ is pruned to achieve the optimal value of the target function.

The resulting optimal configuration and regression model can hardly be called sparse in all three cases. In the following sections we consider implications of Theorem 1 in practical situations, in which orthogonality arises.

4 Illustration of the Analysis

Let us briefly illustrate the three distinct cases of Theorem 1 empirically before actually considering how they can arise in practical situations. The RVM implementation used in this illustration is the Matlab program SparseBayes V1.0 by Tipping [9].

As the example function we use the one-dimensional $\text{sinc}(x) = \sin(x)/x$ often used to illustrate support vector regression [11, 10] and also used by Tipping [8] to illustrate the RVM.

In the first experiment illustrated in Fig. 1 thirty uniformly-spaced training examples of the sinc function (the dotted line) were generated from $[-10, \ldots, 10]$ (the dots in

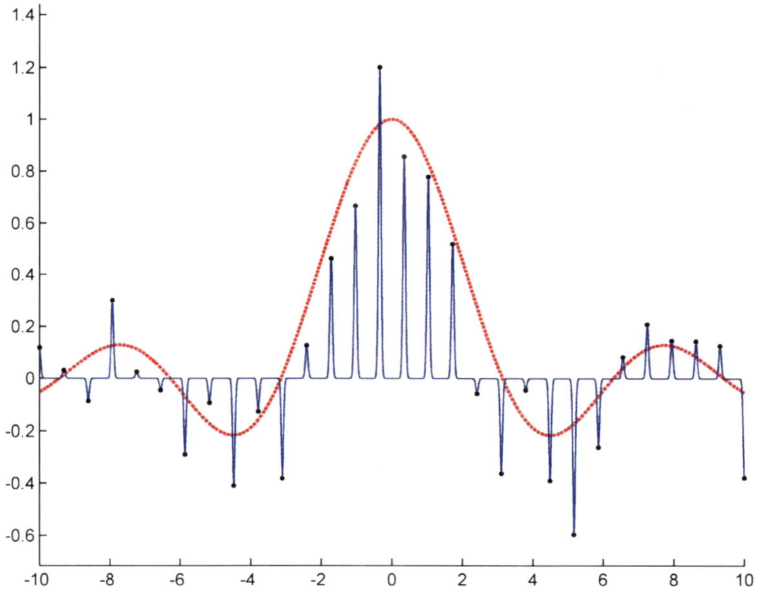

Fig. 1. The sinc function (dotted line) with as many basis vectors as there are training examples (the dots). The regression estimate based on thirty training examples is denoted by the solid line.

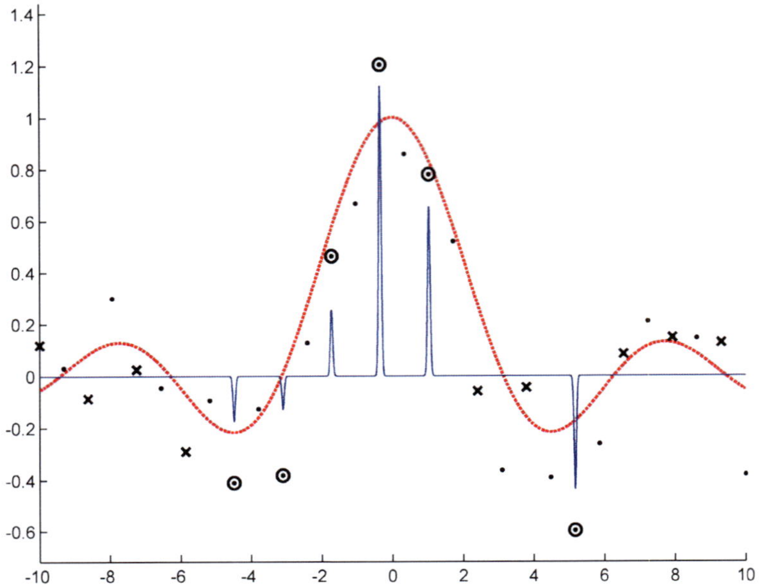

Fig. 2. The number of basis vectors is lower than that of training examples. The dots are the basis vectors of Figure 1 left out from this case. The circled dots were kept by the RVM, while the crossed ones were pruned from the regression estimate.

Fig. 1). Zero-centered normally distributed error with variance of $\sqrt{0.2}$ was used to perturb the training examples.

For each data point we choose a Gaussian radial basis function (RBF) with width parameter $r = 0.05$ to the regression model of the RVM. Because the value of the width parameter is so small, the basis vectors of the regression model that we get are almost orthogonal. In fact, within numerical accuracy the basis vectors are orthonormal, since by experimentation it holds for all $\phi_i \neq \phi_k$ that

$$\frac{|\langle \boldsymbol{\phi}_i, \boldsymbol{\phi}_k \rangle|}{\|\boldsymbol{\phi}_i\| \, \|\boldsymbol{\phi}_k\|} \leq 5 \cdot 10^{-83} \text{ and } \|\boldsymbol{\phi}_i\| \approx 1. \tag{13}$$

Also by experimentation, we observe that $\langle \boldsymbol{\phi}_i, \boldsymbol{y} \rangle^2 \, \|\boldsymbol{\phi}_i\|^{-2}$ has a different value for all basis vectors. Now, because the basis vectors are practically orthogonal and there are as many of them as training examples, the above analysis suggests that all basis vectors should remain in the regression model, or at most one of them gets pruned. The solid line in Fig. 1 is the regression estimate of RVM. Obviously it contains all the thirty basis vectors as predicted by the analysis.

In the second interesting case there are fewer orthogonal basis vectors than training examples. Moreover, the projection of vector $\boldsymbol{y}$ to the subspace spanned by the basis vectors is shorter than the vector itself. Recall that $\boldsymbol{y}$ is a vector collecting the values of the dependent variable for the training set.

To generate this situation, we remove every second basis vector from the previous illustration. According to the analysis, the RVM should now include to the regression

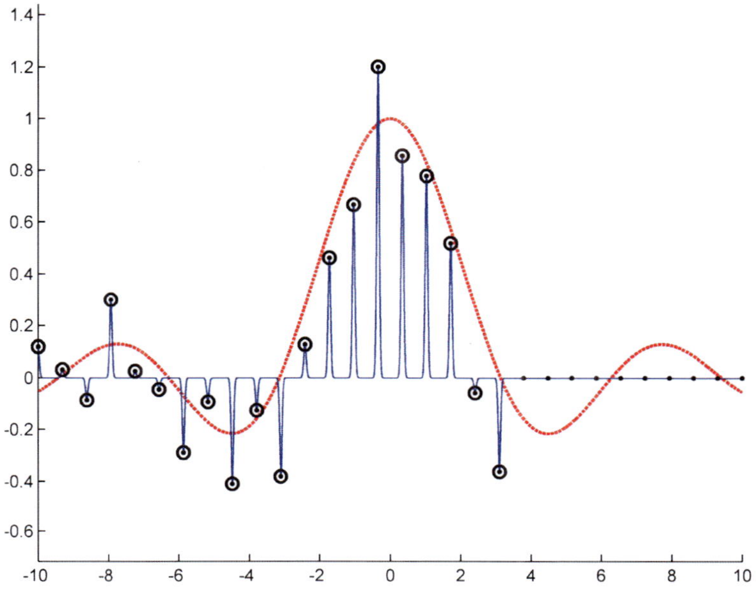

Fig. 3. The value of the dependent variable in the ten last examples of the sinc function has been set to zero. Gaussian RBFs are generated only for the twenty first examples. All basis vectors get chosen to the regression model.

model only a part of the basis vectors, those that best explain y. These are, of course, the ones whose Gaussian RBF spike coincides with a realization of the dependent variable with a large absolute value.

Figure 2 demonstrates this case. The dots stand for the basis vectors that were left out from the current setting. Crosses are the basis vectors pruned by the RVM and the circled dots are those that were kept. We see that nine out of the fifteen basis vectors were pruned by the RVM. Again the empirical behavior of the RVM conforms with the predictions of the analysis.

In the last case concerning orthogonal basis vectors there are less basis vectors than training examples, but now the projection of y to the subspace spanned by the basis vectors is of equal length to the original vector. In other words, y belongs to the subspace spanned by the basis vectors.

The analysis cannot tell us much about the behavior of the RVM in this case: The target function does not have a global maximum. Furthermore, the value of the target function can be made arbitrarily large by letting the variance parameter β approach zero. When the values of β are close enough to zero, all basis vectors belong to the model.

In order to realize this situation we modify the setting of Fig. 1: We set the value of the dependent variable of the last ten training examples to zero. The regression model contains Gaussian RBFs only for those points whose value of the dependent variable differs from zero. Hence, there are less basis vectors than training points and the vector y belongs to the subspace spanned by the basis vectors.

Figure 3 illustrates this setting. All twenty basis vectors get chosen to the regression model produced by the RVM as proposed by the analysis when β is close enough to zero.

The value of variance parameter β was approximately $3 \cdot 10^{-7}$ in our first setting, in the second one $\beta \approx 9 \cdot 10^{-9}$, and now only $\beta \approx 1 \cdot 10^{-21}$. Thus, it seems that the RVM now lets the variance parameter approach zero much more than previously. It is also worth observing that the value of the target function corresponding to the final configuration is now approximately 260, while it previously was circa 33 and 16, respectively. Hence, the value of the target function has grown significantly as the variance parameter approaches zero. Also this is in accordance with the analysis above.

5 Orthogonality in Practice

There are situations in which orthogonality can arise in practice. For example, in kernel regression it is possible to choose such kernels that the resulting regression matrix has orthogonal columns. Thus, our results from Sect. 3 are directly applicable to these situations. However, here we want to emphasize a situation in which optimization of parameters of kernel functions in the model can lead to orthogonal basis functions.

Let us consider a regression model in which the basis functions τ_i are Gaussian kernels. Furthermore, let us assume that $m = n$ and there is a one Gaussian kernel function located at each training data point $\boldsymbol{x}_i$. Thus, the linear model under consideration can be expressed as

$$y(\boldsymbol{x}) = \sum_{i=1}^{m} w_i \tau_i(\boldsymbol{x}) + \epsilon,$$

where the functions τ_i are defined to be

$$\tau_i(\boldsymbol{x}) = \exp\left(-\frac{\|\boldsymbol{x} - \boldsymbol{x}_i\|^2}{r}\right)$$

and $r > 0$ is a width parameter of the kernel function. Usually r is unknown beforehand and, thus, it must be estimated together with the other parameters. Probably the most common estimation method is cross-validation. However, Tipping [8] has proposed a method for optimizing the parameter r in the RVM regression without the need for cross-validation.

His optimization method tends to select such values of $\boldsymbol{\alpha}$, β, and r that optimize the marginal likelihood

$$p(\boldsymbol{y} \mid \boldsymbol{\alpha}, \beta, r) = \mathcal{N}(\boldsymbol{y} \mid \boldsymbol{0}, \boldsymbol{C}_r),$$

where $\boldsymbol{C}_r$ denotes the matrix $\boldsymbol{C}$, which is constructed using Gaussian kernel functions τ_i with the width parameter r. Further, Tipping deduces from illustrative figures that his method will not select inappropriately small parameter value r, meaning that the RVM will not produce a regression estimate that is severely overfitting and spiked at every $\boldsymbol{x}_i$. However, by Theorem 1 this does not hold in every situation.

Consider the case in which one of the values for the dependent variable y_i — for instance, the value y_n — is exactly zero. Even if this situation has probability zero when working with continuous random variables, the actual computations are carried out with computer and, thus, with discrete values. So it is possible to have one of the variables y_i exactly zero. If we now shrink r to be small enough, we end up with situation, in which the columns of matrix $\boldsymbol{\Phi}$ are orthogonal (if numerical accuracy is once again considered). Because of orthogonality, we can apply the results stated in Theorem 1. First we notice that vector ϕ_n will be pruned from the base in every optimal configuration, because $\langle \phi_n, \boldsymbol{y} \rangle = 0$ holds. Thus, we can leave ϕ_n out from the upcoming reasoning and we are left with $m = n - 1$ base vectors $\phi_1, \ldots, \phi_{n-1}$. For these vectors we find that they span a subspace in which $\boldsymbol{y}$ belongs to. But this means that we can apply the statement in the case 1 of Theorem 1. Again, this leads to the conclusion that likelihood value of the target function can be made infinitely large by keeping all vectors $\phi_1, \ldots, \phi_{n-1}$ in the base and shrinking β infinitely close to zero. Hence, it is possible that the RVM selects an inappropriately small value r when r is optimized by using any method that tends to optimize the likelihood given in (5).

6 Invariance Results

In practical modeling situations, we are interested in invariance properties of the RVM. We say that the RVM is invariant with respect to some change in regression model, if following two conditions hold: the predicted value of the variable y for any new data point $\boldsymbol{x}_*$ and estimate of variance parameter β remain unchanged regardless of the change made in the model. Here we consider three types of invariance properties: invariance with respect to scaling of the variable $\boldsymbol{y}$, scaling of columns of the regression matrix $\boldsymbol{\Phi}$, and rotation of columns of the regression matrix $\boldsymbol{\Phi}$. If we assume that the RVM converges to a unique maximum of the target function (4) always when such a maximum exists, we can get a couple of invariance results. We give our invariance results in following three theorems. We omit their quite straightforward proofs here.

Theorem 2. *If the target function corresponding to a configuration of the regression problem* (1) *has a unique maximum, then the RVM is invariant with respect to scaling of the variable* $\boldsymbol{y}$*. That is, the regression model can be scaled with any* $a \neq 0$*, as in the equation*

$$\boldsymbol{y} = a\left(\boldsymbol{\Phi}\boldsymbol{w} + \boldsymbol{\epsilon}\right),$$

without a change in predictions of variable y *and estimate of* β*.*

Theorem 3. *If the target function corresponding to a configuration of the regression problem* (1) *has a unique maximum, then the RVM is invariant with respect to scaling of columns of the regression matrix* $\boldsymbol{\Phi}$*. That is, the regression model can be scaled with any diagonal matrix* $\boldsymbol{D} \in \mathbb{R}^{m \times m}$*, with diagonal elements* $(\boldsymbol{D})_{ii} \neq 0$*, as in the equation*

$$\boldsymbol{y} = \boldsymbol{\Phi}\boldsymbol{D}\boldsymbol{w} + \boldsymbol{\epsilon},$$

without a change in predictions of variable y *and estimate of* β*.*

Theorem 4. *The RVM is not invariant with respect to rotation of columns of the regression matrix* $\boldsymbol{\Phi}$*. That is, applying a rotation matrix* $\boldsymbol{R} \in \mathbb{R}^{m \times m}$ *to the regression model, as in the equation*

$$\boldsymbol{y} = \boldsymbol{\Phi}\boldsymbol{R}\boldsymbol{w} + \boldsymbol{\epsilon},$$

may a induce change in predictions of the variable y *and the estimate of* β*.*

The practical significance of Theorem 2 and Theorem 3 may be limited, because we do not know how strong is the assumption of the existence of the unique maximum and the convergence of the RVM. In the case of orthogonal regression matrix columns we found that the existence of a unique maximum is possible. We do not know, however, if this applies to more general setting. The negative results of Theorem 4, on the other hand, do not suffer from the same limitation.

7 Conclusions

This work provided an analytical consideration of the RVM in the case of regression matrix that constitutes of orthogonal columns. We found that in this case it is possible to explicitly describe the optimal points of the target function of the RVM. Our results lead to the conclusion that the RVM can sometimes produce full and highly overfitting regression models instead of sparse models with good generalization capability. We also discussed, in light of our results, how problems might arise when the RVM is used for kernel regression and kernel parameters are optimized jointly with the target function of the RVM. Moreover, we presented three results concerning scale and rotation invariance properties of the RVM.

Clearly, the RVM is not fully understood, even though it is widely used. In future, it would be interesting to generalize the results put forward in this work.

Acknowledgments

This work has been supported by Academy of Finland projects INTENTS (206280), ALEA (210795), and “Machine learning and online data structures” (119699).

References

1. Agarwal, A., Triggs, B.: 3D human pose from silhouettes by relevance vector regression. In: Proceedings of the IEEE Computer Society Conference on Computer Vision and Pattern Recognition, pp. 882–888. IEEE Computer Society, Los Alamitos (2004)
2. Brookes, M.: The Matrix Reference Manual (2005), `http://www.ee.ic.ac.uk/hp/staff/dmb/matrix/intro.html`
3. Chen, S., Gunn, S.R., Harris, C.J.: The relevance vector machine technique for channel equalization application. IEEE Transactions on Neural Networks 12(6), 1529–1532 (2004)
4. Faul, A.C., Tipping, M.E.: Analysis of sparse Bayesian learning. In: Dietterich, T.G., Becker, S., Ghahramani, Z. (eds.) Advances in Neural Information Processing Systems, vol. 14, pp. 383–389. MIT Press, Cambridge (2001)
5. Li, Y., Campbell, C., Tipping, M.: Bayesian automatic relevance determination algorithms for classifying gene expression data. Bioinformatics 18(10), 1332–1339 (2002)
6. Candela, J.Q., Hansen, L.K.: Time series prediction based on the relevance vector machine with adaptive kernels. In: Proceedings of the IEEE International Conference on Acoustics, Speech, and Signal Processing, Piscataway, NJ, pp. 985–988. IEEE Signal Processing Society, Los Alamitos (2002)

7. Tipping, M.: The relevance vector machine. In: Leen, T.K., Dietterich, T.G., Tresp, V. (eds.) Advances in Neural Information Processing Systems, vol. 12, pp. 652–658. MIT Press, Cambridge (2000)
8. Tipping, M.: Sparse Bayesian learning and the relevance vector machine. Journal of Machine Learning Research 1, 211–244 (2001)
9. Tipping, M.E.: Sparse Bayesian Learning and the Relevance Vector Machine, `http://research.microsoft.com/mlp/RVM`
10. Vapnik, V.N.: Statistical Learning Theory. John Wiley & Sons, New York (1998)
11. Vapnik, V.N., Golowich, S.E., Smola, A.J.: Support vector method for function approximation, regression estimation and signal processing. In: Mozer, M., Jordan, M.I., Petsche, T. (eds.) Advances in Neural Information Processing Systems, vol. 9, pp. 281–287. MIT Press, Cambridge (1997)
12. Wei, L., Yang, Y., Nishikawa, R.M., Wernick, M.N., Edwards, A.: Relevance vector machine for automatic detection of clustered microcalcifications. IEEE Transactions on Medical Imaging 24(10), 1278–1285 (2005)
13. Weston, J., Elisseeff, A., Schölkopf, B., Tipping, M.E.: Use of the zero-norm with linear models and kernel methods. Journal of Machine Learning Research 3, 1439–1461 (2003)
14. Williams, O., Blake, A., Cipolla, R.: A sparse probabilistic learning algorithm for real-time tracking. In: Proceedings of the Ninth IEEE International Conference on Computer Vision, pp. 353–361. IEEE Computer Society, Los Alamitos (2003)
15. Wipf, D.P., Nagarajan, S.: A new view of automatic relevance determination. In: McCallum, A. (ed.) Advances in Neural Information Processing Systems, vol. 20. MIT Press, Cambridge (2008)
16. Wipf, D.P., Palmer, J.A., Rao, B.D.: Perspectives on sparse Bayesian learning. In: Thrun, S., Saul, L.K., Schölkopf, B. (eds.) Advances in Neural Information Processing Systems, vol. 16. MIT Press, Cambridge (2004)

A Proofs of Lemmas

A.1 The Proof of Lemma 1

Let us assume that $\beta > 0$ is given. From (10) we see that the target function depends on variables $\underline{\alpha}_i$ only through the sum $\sum_{i=1}^m l_i(\underline{\alpha}_i, \beta)$. Thus, $\underline{\boldsymbol{\alpha}}$ is an optimal configuration if and only if $\underline{\alpha}_i$ maximizes the function $l_i(\underline{\alpha}_i, \beta)$ for every index i. Let us consider the function $l_i(\underline{\alpha}_i, \beta)$ on the interval $(0, \infty)$. Since the equation

$$\ln\left(\beta + \underline{\alpha}_i^{-1} \|\boldsymbol{\phi}_i\|^2\right) = -\ln(\underline{\alpha}_i) + \ln\left(\beta\underline{\alpha}_i + \|\boldsymbol{\phi}_i\|^2\right)$$

holds, the derivative $\frac{\partial}{\partial \underline{\alpha}_i} l_i(\alpha_i, \beta)$ can be written in the form

$$\begin{aligned} \frac{\partial}{\partial \underline{\alpha}_i} l_i(\underline{\alpha}_i, \beta) &= \frac{\partial}{\partial \underline{\alpha}_i} \left(\frac{\beta^{-1} \langle \boldsymbol{\phi}_i, \boldsymbol{y} \rangle^2}{\beta \underline{\alpha}_i + \|\boldsymbol{\phi}_i\|^2} + \ln(\underline{\alpha}_i) - \ln\left(\underline{\alpha}_i \beta + \|\boldsymbol{\phi}_i\|^2\right) \right) \\ &= \frac{-\left(\langle \boldsymbol{\phi}_i, \boldsymbol{y} \rangle^2 \|\boldsymbol{\phi}_i\|^{-2} - \beta\right) \underline{\alpha}_i + \|\boldsymbol{\phi}_i\|^2}{\|\boldsymbol{\phi}_i\|^2 \underline{\alpha}_i \left(\beta \underline{\alpha}_i + \|\boldsymbol{\phi}_i\|^2\right)}. \end{aligned} \quad (14)$$

Let us consider first the case $\langle \boldsymbol{\phi}_i, \boldsymbol{y} \rangle^2 / \|\boldsymbol{\phi}_i\|^2 > \beta$. Now the derivative has zero at

$$\hat{\underline{\alpha}}_i = \frac{\|\boldsymbol{\phi}_i\|^2}{\langle \boldsymbol{\phi}_i, \boldsymbol{y} \rangle^2 / \|\boldsymbol{\phi}_i\|^2 - \beta}. \tag{15}$$

From (14) it can be seen, that the inequality

$$\frac{\partial}{\partial \underline{\alpha}_i} l_i(\underline{\alpha}_i, \beta) > 0$$

holds when $\underline{\alpha}_i < \hat{\underline{\alpha}}_i$ and the inequality

$$\frac{\partial}{\partial \underline{\alpha}_i} l_i(\underline{\alpha}_i, \beta) < 0$$

holds when $\underline{\alpha}_i > \hat{\underline{\alpha}}_i$. Further, we get

$$\lim_{\underline{\alpha}_i \to \infty} l_i(\alpha_i, \beta) = 0 = l_i(\infty, \beta).$$

Thus, the zero given in (15) is a unique maximum for the function $l_i(\underline{\alpha}_i, \beta)$.

Let us now consider the case $\langle \boldsymbol{\phi}_i, \boldsymbol{y} \rangle^2 / \|\boldsymbol{\phi}_i\|^2 \leq \beta$. Clearly the derivative (14) has no zeros in the interval $(0, \infty)$. In addition, we notice that the derivative is positive in the same interval. Thus, for every $\underline{\alpha}_i \in (0, \infty)$ the equation

$$l_i(\underline{\alpha}_i, \beta) < \lim_{\underline{\alpha} \to \infty} l_i(\underline{\alpha}, \beta) = 0 = l_i(\infty, 0)$$

holds. But this means that $\underline{\alpha} = \infty$ is a unique maximum for the function $l_i(\underline{\alpha}_i, \beta)$. □

A.2 The Proof of Lemma 2

Clearly, all functions $L_j(\beta)$ are differentiable in the interval $(0, \infty)$ and their derivatives can be written as

$$\frac{\mathrm{d}}{\mathrm{d}\beta} L_j(\beta) = -\beta^{-2} \left(\|\boldsymbol{y}\|^2 - \left(\sum_{i=1}^{j} c_i \beta_i \right) - \left(n - \sum_{i=1}^{j} c_i \right) \beta \right).$$

Because every function $L_j(\beta)$ is differentiable in the interval $(0, \infty)$, the target function $L(\beta)$ is differentiable in all intervals (β_{j+1}, β). Hence, we only need to examine behavior of $L_j(\beta)$ at points $\beta_1, \ldots, \beta_p$.

Let us choose a point β_j from the set $\{\beta_1, \ldots, \beta_p\}$. At this point we get the equation

$$\begin{aligned}\lim_{h \to 0} \frac{L(\beta_j + h) - L(\beta_j)}{h} &= \lim_{h \to 0} \frac{L_{j-1}(\beta_j + h) - L_{j-1}(\beta_j)}{h} \\ &= \left(\frac{\mathrm{d}}{\mathrm{d}\beta} L_{j-1} \right) (\beta_j)\end{aligned}$$

directly from definition of $L(\beta)$ and differentiability of $L_{j-1}(\beta)$. On the other hand, straightforward algebraic manipulation yields

$$\begin{aligned}
\lim_{h\to 0-}\frac{L(\beta_j+h)-L(\beta_j)}{h} &= \lim_{h\to 0-}\frac{L_j(\beta_j+h)-L_{j-1}(\beta_j)}{h}\\
&= \left[\lim_{h\to 0-}\frac{\frac{1}{\beta_j+h}-\frac{1}{\beta_j}}{h}\right]\left[\left(\sum_{i=1}^{j-1}c_i\beta_j\right)-\|y\|^2\right]\\
&\quad-\left(\lim_{h\to 0-}\frac{c_j}{\beta_j+h}\right)\\
&\quad-\left(n-\sum_{i=1}^{j-1}c_i\right)\left(\lim_{h\to 0-}\frac{\ln(\beta_j+h)-\ln(\beta_j)}{h}\right)\\
&\quad+c_j\left(\lim_{h\to 0-}\frac{\ln(\beta_j+h)-\ln(\beta_j)}{h}\right)\\
&= \beta_j^{-2}\left[\|y\|^2-\left(\sum_{i=1}^{j-1}c_i\beta_i\right)-\left(n-\sum_{i=1}^{j-1}c_i\right)\beta_j\right]\\
&= \left(\frac{\mathrm{d}}{\mathrm{d}\beta}L_{j-1}\right)(\beta_j).
\end{aligned}$$

directly from definition of $L(\beta)$ and differentiability of $L_j(\beta)$. The left limit, on the other hand, we can express in the form

$$\begin{aligned}
\lim_{h\to 0-}\frac{L(\beta_j+h)-L(\beta_j)}{h} &= \lim_{h\to 0-}\frac{L_j(\beta_j+h)-L_{j-1}(\beta_j)}{h}\\
&= \lim_{h\to 0-}\frac{1}{h}\left[\frac{1}{\beta_j+h}\left[\left(\sum_{i=1}^{j}c_i\beta_i\right)-\|\boldsymbol{y}\|^2\right]\right]\\
&\quad-\frac{1}{h}\left(\sum_{i=1}^{j}c_i\ln(\beta_i)\right)-\frac{1}{h}\left(\sum_{i=1}^{j}c_i\right)\\
&\quad-\left(n-\sum_{i=1}^{j}c_i\right)\frac{\ln(\beta_j+h)}{h}\\
&\quad-\frac{1}{h}\left[\frac{1}{\beta_j}\left[\left(\sum_{i=1}^{j-1}c_i\beta_i\right)-\|\boldsymbol{y}\|^2\right]\right]+\frac{1}{h}\left(\sum_{i=1}^{j-1}c_i\ln(\beta_i)\right)\\
&\quad+\frac{1}{h}\left(\sum_{i=1}^{j-1}c_i\right)+\left(n-\sum_{i=1}^{j-1}c_i\right)\frac{\ln(\beta_j)}{h},
\end{aligned}$$

which can be simplified to

$$\begin{aligned}
&\lim_{h\to 0-} \frac{L(\beta_j+h)-L(\beta_j)}{h} \\
&= \left[\lim_{h\to 0-} \frac{\frac{1}{\beta_j+h}-\frac{1}{\beta_j}}{h}\right]\left[\left(\sum_{i=1}^{j-1} c_i\beta_j\right)-\|y\|^2\right]-\left(\lim_{h\to 0-}\frac{c_j}{\beta_j+h}\right) \\
&\quad -\left(n-\sum_{i=1}^{j-1} c_i\right)\left(\lim_{h\to 0-}\frac{\ln(\beta_j+h)-\ln(\beta_j)}{h}\right) \\
&\quad + c_j\left(\lim_{h\to 0-}\frac{\ln(\beta_j+h)-\ln(\beta_j)}{h}\right) \qquad (16)\\
&= -\frac{1}{\beta_j^2}\left[\left(\sum_{i=1}^{j-1} c_i\beta_i\right)-\|y\|^2\right]-\frac{c_j}{\beta_j}-\left(n-\sum_{i=1}^{j-1} c_i\right)\frac{1}{\beta_j}+\frac{c_j}{\beta_j} \\
&= \beta_j^{-2}\left[\|y\|^2-\left(\sum_{i=1}^{j-1} c_i\beta_i\right)-\left(n-\sum_{i=1}^{j-1} c_i\right)\beta_j\right] \\
&= \left(\frac{\mathrm{d}}{\mathrm{d}\beta}L_{j-1}\right)(\beta_j)\,.
\end{aligned}$$

Thus, right and left limits in two previous equations are equal, so the actual limit exists. Hence, we have

$$\left(\frac{\mathrm{d}}{\mathrm{d}\beta}L\right)(\beta_j) = \lim_{h\to 0}\frac{L(\beta_j+h)-L(\beta_j)}{h} = \left(\frac{\mathrm{d}}{\mathrm{d}\beta}L_{j-1}\right)(\beta_j)\,. \qquad \square$$

A.3 The Proof of Lemma 3

Clearly, derivative functions $\frac{\mathrm{d}}{\mathrm{d}\beta}L_j(\beta)$ are continuous in the whole interval $(0,\infty)$. Thus, the derivative function $\frac{\mathrm{d}}{\mathrm{d}\beta}L(\beta)$ must be continuous in all intervals (β_{j+1},β_j). Again, this means that we need to examine $\frac{\mathrm{d}}{\mathrm{d}\beta}L_j(\beta)$ only at the points $\beta_1,\ldots,\beta_p$.

Let us choose a point β_j from the set $\{\beta_1,\ldots,\beta_p\}$. At this point we get

$$\lim_{\beta\to\beta_j+}\left(\frac{\mathrm{d}}{\mathrm{d}\beta}L\right)(\beta) = \left(\frac{\mathrm{d}}{\mathrm{d}\beta}L_{j-1}\right)(\beta_j) = \left(\frac{\mathrm{d}}{\mathrm{d}\beta}L\right)(\beta_j)$$

by Lemma 2. On the other hand, for the left limit at point β_j we get

$$\begin{aligned}
\lim_{\beta\to\beta_j-}\left(\frac{\mathrm{d}}{\mathrm{d}\beta}L\right)(\beta) &= \lim_{\beta\to\beta_j-}\left(\frac{\mathrm{d}}{\mathrm{d}\beta}L_j\right)(\beta) \\
&= \lim_{\beta\to\beta_j-}\beta^{-2}\left(\|\boldsymbol{y}\|^2-\left(\sum_{i=1}^{j} c_i\beta_i\right)-\left(n-\sum_{i=1}^{j} c_i\right)\beta\right) \\
&= \beta_j^{-2}\left(\|\boldsymbol{y}\|^2-\left(\sum_{i=1}^{j-1} c_i\beta_i\right)-\left(n-\sum_{i=1}^{j-1} c_i\right)\beta_j\right) \\
&= \left(\frac{\mathrm{d}}{\mathrm{d}\beta}L_{j-1}\right)(\beta_j)\,.
\end{aligned}$$

From the two previous equations we conclude that left and right limit of $\frac{\mathrm{d}}{\mathrm{d}\beta}L$ equal the value of $\frac{\mathrm{d}}{\mathrm{d}\beta}L$ at point β_j. Thus, the derivative function $\frac{\mathrm{d}}{\mathrm{d}\beta}L(\beta)$ is continuous at the point β_j. □

A.4 The Proof of Lemma 4

Let us consider first the case $m < n$. It is easy to see that for every $j \in \{1, \ldots, p-1\}$ the inequality

$$\begin{aligned} g(j) &= \left(\sum_{i=1}^{j} c_i\beta_i\right) + \left(n - \sum_{i=1}^{j} c_i\right)\beta_j \\ &> \left(\sum_{i=1}^{j+1} c_i\beta_i\right) + \left(n - \sum_{i=1}^{j+1} c_i\right)\beta_{j+1} \\ &= g(j+1) \end{aligned}$$

holds. Also we get trivially the inequality $g(0) > g(1)$, and $g(p) > g(p+1)$ holds, because we have $\sum_{i=1}^{p} c_i = m < n$. Thus, the function g is decreasing.

By Lemma 2, the zeros of the derivative function $\frac{\mathrm{d}}{\mathrm{d}\beta}L_j(\beta)$ are all points $\hat{\beta}_j$ that satisfy conditions

$$\left(\frac{\mathrm{d}}{\mathrm{d}\beta}L\right)\left(\hat{\beta}_j\right) = 0 \text{ and } \hat{\beta}_j \in [\beta_{j+1}, \beta_j).$$

The first condition can be written as

$$\hat{\beta}_j = \left(n - \sum_{i=1}^{j} c_i\right)^{-1} \left(\|\boldsymbol{y}\|^2 - \left(\sum_{i=1}^{j} c_i\beta_i\right)\right). \tag{17}$$

When we combine this with the second condition, we get the double inequality

$$\beta_{j+1} \le \left(n - \sum_{i=1}^{j} c_i\right)^{-1} \left(\|\boldsymbol{y}\|^2 - \left(\sum_{i=1}^{j} c_i\beta_i\right)\right) < \beta_j.$$

This inequality pair can further be expressed as

$$\left(\sum_{i=1}^{j} c_i\beta_i\right) + \left(n - \sum_{i=1}^{j} c_i\right)\beta_{j+1} \le \|\boldsymbol{y}\|^2 < \left(\sum_{i=1}^{j} c_i\beta_i\right) + \left(n - \sum_{i=1}^{j} c_i\right)\beta_j,$$

which can be shown to be equal to

$$g(j+1) \le \|y\|^2 < g(j). \tag{18}$$

Since the function g is decreasing, there can be only one index j that satisfies the inequality pair (18). Thus, the derivative function $\frac{\mathrm{d}}{\mathrm{d}\beta}L(\beta)$ can have at most one zero. If the zero exists, it is of the form given in (17). This proves Lemma 4 in the case $m < n$.

Consider next the remaining case $m = n$. In this case the equation $\sum_{i=1}^{p} c_i = m = n$ holds and the vectors $\boldsymbol{\phi}_1, \ldots, \boldsymbol{\phi}_n$ form an orthogonal base in $\mathbb{R}^n$. The latter fact implies the equation

$$\|y\|^2 = \sum_{i=1}^{n} \langle \boldsymbol{\phi}_i, \boldsymbol{y} \rangle^2 / \|\boldsymbol{\phi}_i\|^2 = \sum_{i=1}^{p} c_i \beta_i. \tag{19}$$

Now, in the interval $(0, \beta_p)$, the derivative function $\frac{\mathrm{d}}{\mathrm{d}\beta} L(\beta)$ can be expressed as

$$\frac{\mathrm{d}}{\mathrm{d}\beta} L(\beta) = \frac{\mathrm{d}}{\mathrm{d}\beta} L_n(\beta) = \beta^{-2} \left[\|\boldsymbol{y}\|^2 - \left(\sum_{i=1}^{p} c_i \beta_i \right) - \left(n - \sum_{i=1}^{p} c_i \right) \beta \right] = 0.$$

Further, it can be shown (in the same manner as in case $m < n$) that the derivative function $\frac{\mathrm{d}}{\mathrm{d}\beta} L(\beta)$ can have only one zero in the intervals $[\beta_{j+1}, \beta_j)$, where $j \in \{1, \ldots, p-1\}$. In addition, (19) implies that for zero candidate $\hat{\beta}_p$ the equation

$$\beta = \left(n - \sum_{i=1}^{p-1} c_i \right)^{-1} \left(\|y\|^2 - \sum_{i=1}^{p-1} c_i \beta_i \right) = c_p^{-1} c_p \beta_p = \beta_p \in [\beta_p, \beta_{p-1})$$

holds. Thus, $\hat{\beta}_p$ is also a zero of the derivative function $\frac{\mathrm{d}}{\mathrm{d}\beta} L(\beta)$. But this means that $\frac{\mathrm{d}}{\mathrm{d}\beta} L(\beta)$ is zero exactly in those points that belong to the interval $(0, \beta_p]$. □

Cascade Classifiers for Hierarchical Decision Systems

Zbigniew W. Raś[1], Agnieszka Dardzińska[2], and Wenxin Jiang[3]

[1] Univ. of North Carolina, Dept. of Comp. Science
Charlotte, NC 28223, USA
and Polish Academy of Sciences, Inst. of Comp. Science
01-237 Warsaw, Poland
ras@uncc.edu

[2] Bialystok Technical Univ., Dept. of Computer Science,
15-351 Bialystok, Poland
adardzin@uncc.edu

[3] Univ. of North Carolina, Dept. of Comp. Science
Charlotte, NC 28223, USA
wjiang3@uncc.edu

Abstract. Hierarchical classifiers are usually defined as methods of classifying inputs into defined output categories. The classification occurs first on a low-level with highly specific pieces of input data. The classifications of the individual pieces of data are then combined systematically and classified on a higher level iteratively until one output is produced. This final output is the overall classification of the data. In this paper we follow a controlled devise type of approach. The initial group of classifiers is trained using all objects in an information system S partitioned by values of the decision attribute d at its all granularity levels (one classifier per level). Only values of the highest granularity level (corresponding granules are the largest) are used to split S into information sub-systems where each one is built by selecting objects in S of the same decision value. These sub-systems are used for training new classifiers at all granularity levels of its decision attribute. Next, we split each sub-system further by sub-values of its decision value. The obtained tree-structure with groups of classifiers assigned to each of its nodes is called a cascade classifier. Given an incomplete information system with a hierarchical decision attribute d, we consider the problem of training classifiers describing values of d at its lowest granularity level. Taking *MIRAI* database of music instrument sounds [16], as an example, we show that the confidence of such classifiers can be lower than the confidence of cascade classifiers.

1 Introduction

One of the main goals in data mining area is to describe knowledge hidden in data sets by means of classifiers. Clearly there is a need for classifiers which are easy and quick to build, accurate and suitable for different type of data. If data are incomplete, then null-value imputation techniques, including Chase [10], [2], [3], [7] can be used before the knowledge extraction algorithms are applied.

A hierarchical classifier is usually defined as aglomerative method of classifying inputs into defined output categories [5], [11]. The classification occurs first on a low-level

J. Koronacki et al. (Eds.): Advances in Machine Learning I, SCI 262, pp. 247–256.
springerlink.com © Springer-Verlag Berlin Heidelberg 2010

with highly specific pieces of input data. The classifications of the individual pieces of data are then combined systematically and classified on a higher level iteratively until one output is produced. This final output is the overall classification of the data.

Automatic indexing of music by instruments and their types is taken as the application and testing area in our research. In [18], a multi-hierarchical decision system S with a large number of descriptors built for describing music sound objects was presented. The decision attributes in S are hierarchical and they include Hornbostel-Sachs classification and classification of instruments with respect to a playing method. The information richness hidden in these descriptors has strong impact on the confidence of classifiers built from S and used by the content-based Automatic Indexing Systems (*AIS*). Since decision attributes are hierarchical, then the indexing can be done by music instrument classes of different granularity levels (for instance aerophone, aero_double-reed, or simply an instrument can be seen as a value of a decision attribute). This way, if we fail to identify instruments playing in a given music piece, then still we may succeed to identify classes of instruments. The quality of *AIS* was verified in [18] using precision and recall based on two interpretations: user and system-based [15]. *AIS* engine follows system-based interpretation.

In this paper we propose a methodology of building cascade classifiers for regularly incomplete data sets following divisive approach. A data set is called regularly incomplete if it can be partitioned into non-singular complete data subsets. For instance the data set represented by Table 1 is regularly incomplete because it can be split into: $(\{x_1, x_2, x_7, x_8\}, \{b, c, d\})$, $(\{x_3, x_4, x_{11}, x_{12}\}, \{a, b, s, d\})$, $(\{x_5, x_6, x_9, x_{10}\}, \{a, c, d\})$.

Now, assuming that information system S is regularly incomplete, we build the initial group of classifiers based on all objects in S partitioned by values of the decision attribute d at its all granularity levels (one classifier per level). Only values of the highest granularity level (corresponding granules are the largest) are used to split S into information sub-systems where each one is build by selecting objects in S sharing the same decision value. Attributes with no-values assigned to all objects in a sub-system are removed from it. This process is recursively repeated for each sub-system by splitting it further by sub-values of its decision value. The obtained tree-type structure with groups of classifiers assigned to each of its nodes is called a cascade classifier.

Our cascade classifier has been tested on two classes of musical instrument sounds: one labelled as 4F and the other as 3B. The analysis shows a significant improvement in the precision of *AIS*, if cascade classifiers are used instead of single classifiers.

2 Multi-hierarchical Decision System and Query Language

In this section we introduce the notion of a multi-hierarchical decision system S and the query language associated with S, built from values of decision attributes. Classifier-based semantics and standard semantics of queries in S was given in [18]. The set of objects X in S forms the interpretation domain for both semantics. Standard semantics identifies all correct objects in X which should be retrieved by a query. Classifier-based semantics gives weighted set of objects which actually are retrieved be a query. The notion of precision and recall of a query answering system (*QAS*) in the proposed setting was also introduced in [18]. By improving the confidence and support of classifiers trained by S, we also improve the precision and recall of *QAS*.

Multi-hierarchical decision systems [18] can be seen as a subclass of decision systems [14] and they are mostly used for representing data which are incomplete. If a multi-hierarchical decision system contains only one decision attribute, then it is called a hierarchical decision system.

By a decision system we mean a pair $S = (U, A \cup \{d\}, V)$, where:

- U is a nonempty, finite set of objects,
- $A \cup \{d\}$ is a nonempty, finite set of attributes i.e. $a : U \longrightarrow V_a$ is a partial function for any $a \in A$, where V_a is the domain of a,
- elements in A are called classification attributes and d is a distinguished attribute called the decision. We assume that the decision is a total function,
- $V = \bigcup\{V_a : a \in A \cup \{d\}\}$.

As an example of a decision table we take $S = (\{x_1, x_2, x_3, x_4, x_5, x_6, x_7, x_8, x_9, x_{10}, x_{11}, x_{12}\}, \{a, b, c\} \cup \{d\}, V)$ represented by Table 1. Decision attribute d is hierarchical with $V_d = \{d[1], d[2], d[1,1], d[1,2], d[2,1], d[2,2]\}$. Value $d[i,j]$ should be seen as a child of value $d[i]$, for $i, j \in \{1, 2\}$.

By a multi-hierarchical decision system we mean a triple $S = (X, A \cup \{d[1], d[2], .., d[k]\}, V)$, where X is a nonempty, finite set of objects, A is a nonempty finite set of classification attributes, $\{d[1], d[2], .., d[k]\}$ is a set of hierarchical decision attributes and $V = \cup\{V_a : a \in A \cup \{d[1], d[2], .., d[k]\}\}$ is a set of their values. We assume that:

- V_a, V_b are disjoint for any $a, b \in A \cup \{d[1], d[2], .., d[k]\}$, such that $a \neq b$,
- $a : X \rightarrow V_a$ is a partial function for every $a \in A \cup \{d[1], d[2], .., d[k]\}$.

By a set of decision queries (d-queries) for S we mean a least set T_D such that:

- $0, 1 \in T_D$,
- if $w \in \bigcup\{V_a : a \in \{d[1], d[2], .., d[k]\}\}$, then $w, \sim w \in T_D$,
- if $t_1, t_2 \in T_D$, then $(t_1 + t_2), (t_1 * t_2) \in T_D$.

Table 1. Decision System

	a	b	c	d
x_1		1	2	$d[1,1]$
x_2		1	3	$d[1,1]$
x_3	1	1	0	$d[1,2]$
x_4	1	1	3	$d[1,2]$
x_5	2		2	$d[2,1]$
x_6	2		3	$d[2,1]$
x_7		1	1	$d[1,1]$
x_8		1	1	$d[1,1]$
x_9	2		1	$d[2,1]$
x_{10}	2		0	$d[2,1]$
x_{11}	1	1	2	$d[2,2]$
x_{12}	1	1	1	$d[2,2]$

Decision query t is called simple if $t = t_1 * t_2 * ... * t_n$ and $(\forall j \in \{1, 2, , n\})[(t_j \in \bigcup\{V_a : a \in \{d[1], d[2], .., d[k]\}\}) \vee (tj =\sim w \wedge w \in \bigcup\{V_a : a \in \{d[1], d[2], .., d[k]\}\})]$.

By a set of classification terms (c-terms) for S we mean a least set T_C such that:

- $0, 1 \in T_C$,
- if $w \in \bigcup\{V_a : a \in A\}$, then $w, \sim w \in T_C$,
- if $t_1, t_2 \in T_C$, then $(t_1 + t_2), (t_1 * t_2) \in T_C$.

Classification term t is called simple if $t = t_1 * t_2 * ... * t_n$ and $(\forall j \in \{1, 2, ..., n\})[(t_j \in \bigcup\{V_a : a \in A\}) \vee (t_j =\sim w \wedge w \in \bigcup\{V_a : a \in A\})]$.

By a classification rule we mean any expression of the form $[t_1 \longrightarrow t_2]$, where t_1 is a simple classification term and t_2 is a simple decision query.

Semantics M_S of c-terms in $S = (X, A \cup \{d[1], d[2], .., d[k]\}, V)$, is defined in a standard way as follows:

- $M_S(0) = 0$, $M_S(1) = X$,
- $M_S(w) = \{x \in X : w = a(x)\}$ for any $w \in V_a$, $a \in A$,
- $M_S(\sim w) = \{x \in X : (\exists v \in V_a)[(v = a(x)) \wedge (v \neq w)]\}$ for any $w \in V_a$, $a \in A$,
- if t_1, t_2 are terms, then $M_S(t_1 + t_2) = M_S(t_1) \cup M_S(t_2)$, $M_S(t_1 * t_2) = M_S(t_1) \cap M_S(t_2)$.

Now, we introduce the notation for values of decision attributes. Assume that $d[i]$ is a hierarchical decision attribute which is also interpreted as its first or highest granularity level. The set $\{d[i, 1], d[i, 2], d[i, 3], ...\}$ represents the values of attribute $d[i]$ at its second granularity level. The set $\{d[i, 1, 1], d[i, 1, 2], ..., d[i, 1, n_i]\}$ represents the values of attribute d at its third granularity level, right below the node $d[i, 1]$. We assume here that the value $d[i, 1]$ can be refined to any value from $\{d[i, 1, 1], d[i, 1, 2], ..., d[i, 1, n_i]\}$, if necessary. Similarly, the set $\{d[i, 3, 1, 3, 1], d[i, 3, 1, 3, 2], d[i, 3, 1, 3, 3], d[i, 3, 1, 3, 4]\}$ represents the values of attribute d at its forth granularity level which are finer than the value $d[i, 3, 1, 3]$.

Now, let us assume that a rule-based classifier (for instance one of the modules in systems *RSES* or *WEKA*) was used to extract rules describing simple decision queries in S. We denote this classifier by RC. The definition of semantics of c-terms is independent from a classifier but the definition of semantics M_S of d-queries is a classifier dependent.

Classifier-based semantics M_S of d-queries in $S = (X, A \cup \{d[1], d[2], .., d[k]\}, V)$, is defined as follows:

if t is a simple d-query in S and $\{r_j = [t_j \longrightarrow t] : j \in J_t\}$ is a set of all rules defining t which are extracted from S by classifier RC, then $M_S(t) = \{(x, p_x) : (\exists j \in J_t)(x \in M_S(t_j)[p_x = \Sigma\{conf(j) \cdot sup(j) : x \in M_S(t_j) \wedge j \in J_t\}/\Sigma\{sup(j) : x \in M_S(t_j) \wedge j \in J_t\}]\}$, where $conf(j)$, $sup(j)$ denote the confidence and the support of $[t_j \longrightarrow t]$, correspondingly.

Attribute value $d[j_1, j_2, ..., j_n]$ in $S = (X, A \cup \{d[1], d[2], .., d[k]\}, V)$ is dependent on $d[i_1, i_2, , i_k]$ in S, if one of the following conditions hold: (1) $n \leq k \wedge (\forall m \leq n)[i_m = j_m]$, (2) $n > k \wedge (\forall m \leq k)[i_m = j_m]$. Otherwise, $d[j_1, j_2, ..., j_n]$ is called independent from $d[i_1, i_2, ..., i_k]$ in S.

Example. The attribute value $d[2,3,1,2]$ is dependent on the attribute value $d[2,3,1,2,5,3]$. Also, $d[2,3,1,2,5,3]$ is dependent on $d[2,3,1,2]$.

Let $S = (X, A \cup \{d[1], d[2], .., d[k]\}, V)$, $w \in V_{d[i]}$, and $IV_{d[i]}$ be the set of all attribute values in $V_{d[i]}$ which are independent from w. Standard semantics N_S of d-queries in S is defined as follows:

- $N_S(0) = 0$, $N_S(1) = X$,
- if $w \in V_{d[i]}$, then $N_S(w) = \{x \in X : d[i](x) = w\}$, for any $1 \le i \le k$
- if $w \in V_{d[i]}$, then $N_S(\sim w) = \{x \in X : (\exists v \in IV_{d[i]})[d[i](x) = v]\}$, for any $1 \le i \le k$
- if t_1, t_2 are terms, then $N_S(t_1 + t_2) = N_S(t_1) \cup N_S(t_2)$, $N_S(t_1 * t_2) = N_S(t_1) \cap N_S(t_2)$.

Let $S = (X, A \cup \{d[1], d[2], .., d[k]\}, V)$, t is a d-query in S, $N_S(t)$ is its meaning under standard semantics, and $M_S(t)$ is its meaning under classifier-based semantics. Assume that $N_S(t) = X_1 \cup Y_1$, where $X_1 = \{x_i, i \in I_1\}$, $Y_1 = \{y_i, i \in I_2\}$. Assume also that $M_S(t) = \{(x_i, p_i) : i \in I_1\} \cup \{(z_i, q_i) : i \in I_3\}$ and $\{y_i, i \in I_2\} \cap \{z_i, i \in I_3\} = \emptyset$.

By precision of a classifier-based semantics M_S on a d-query t, we mean $Prec(M_S, t) = [\sum\{p_i : i \in I_1\} + \sum\{(1 - q_i) : i \in I_3\}]/[card(I_1) + card(I_3)]$.

By recall of a classifier-based semantics M_S on a d-query t, we mean $Rec(M_S, t) = [\sum\{p_i : i \in I_1\}]/[card(I_1) + card(I_2)]$.

Clearly, the precision and recall of a classifier-based semantics can be improved by using classifiers of higher confidence. In the remaining part of the paper, we show how to construct the cascade λ-representation of a multi-hierarchical decision system and how it is used to construct cascade classifiers which confidence is usually higher than the confidence of standard classifiers.

3 Cascade λ-Representation of Hierarchical Decision Systems

For simplicity reason, we only consider multi-hierarchical decision systems with one decision attribute. Such systems are called hierarchical decision systems.

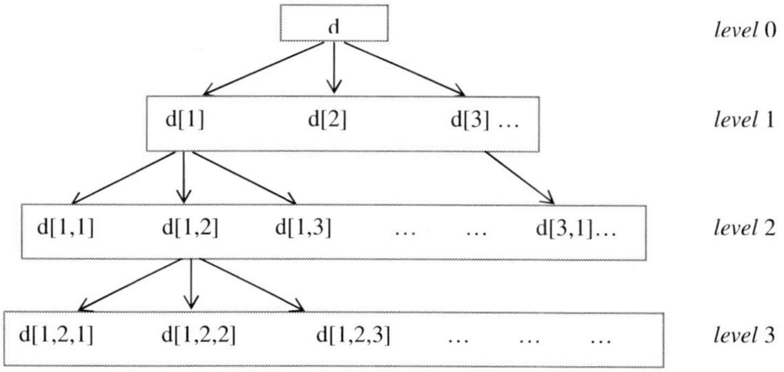

Fig. 1. Example of a decision attribute

Let $S(d) = (X, A \cup \{d\}, V)$ is a regularly incomplete decision system, where d is a hierarchical attribute. Example of a hierarchical decision attribute is given in Figure 1.

Let $\{d[1], d[2], ..., d[k]\}$ is a set of all values of the attribute d at the level 1 of its tree representation. Let $X_i = \{x \in X : d(x) = d[i]\}$ and $S_i[d_i] = (X_i, A \cup \{d[i]\}, V)$, for any $1 \leq i \leq n$. Now, assume that $CR(S)$ denotes a tree of hight one. System S is its root and $S_i(d[i])$, $(1 \leq i \leq n)$, are its children. The outgoing edge from S to $S_i(d[i])$ is labelled by $d[i]$, for any $1 \leq i \leq n$.

Cascade λ-representation of $S(d)$, where $\lambda \in [0, 1]$, is a tree with $S(d)$ defined as its root and all its descendants built by executing the instruction [if $card(V_d) > 1$, then replace $S(d)$ by $CR(S(d))$] recursively, starting from the root and then repeating for all leaves of a constructed tree. The final step of this construction process is the removal of all λ-incomplete attributes in all decision subsystems of $S(d)$ representing descendants of $S(d)$. We say that an attribute a is λ-incomplete in S, if $a(x)$ is defined for $[1 - \lambda] * 100$ percent of objects in S.

Let us go back to the example of a decision system $S(d)$ represented as Table 1. Its attributes $\{a, b\}$ are $4/12$-incomplete. To build a cascade λ-representation of $S(d)$, where $\lambda = 0$, we take its subsystems: $S^*(d) = (\{x_i : 1 \leq i \leq 12\}, \{c, d\}, V)$, $S_{[1]}(d[1]) = (\{x_i : i = 1, 2, 3, 4, 7, 8\}, \{b, c, d\}, V)$, $S_{[2]}(d[2]) = (\{x_i : i = 5, 6, 9, 10, 11, 12\}, \{a, c, d\}, V)$, $S_{[1,1]}(d[1, 1]) = (\{x_i : i = 1, 2, 7, 8\}, \{b, c, d\}, V)$, $S_{[1,2]}(d[1, 2]) = (\{x_i : i = 3, 4\}, \{a, b, c, d\}, V)$, $S_{[2,1]}(d[2, 1]) = (\{x_i : i = 5, 6, 9, 10\}, \{a, c, d\}, V)$, $S_{[2,2]}(d[2, 2]) = (\{x_i : i = 11, 12\}, \{a, b, c, d\}, V)$.

Now, the corresponding cascade λ-representation of $S(d)$, denoted as $(\{S^*(d)\} \cup \{S_k(d) : k \in J\}, \prec)$, where $J = \{[1], [2], [1, 1], [1, 2], [2, 1], [2, 2]\}$ and " $\prec$ " means *parent-child* relation, is represented in Figure 2.

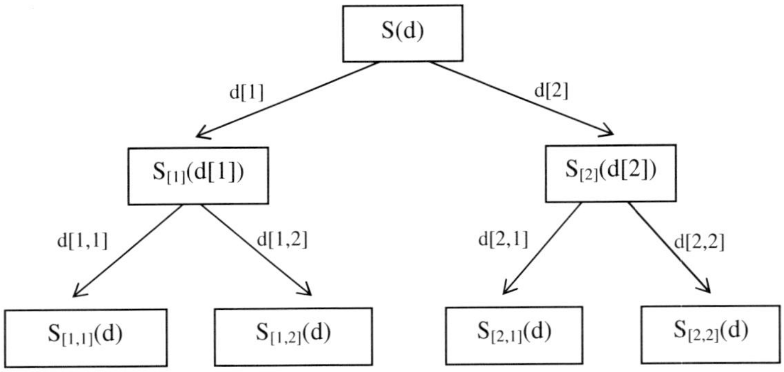

Fig. 2. Cascade λ-representation of $S(d)$

The partition of objects in $S(d)$ can be driven by an optimization function or it can be predefined, as it is done in $MIRAI$ [16], by following either Hornbostel-Sachs classification or classification of instruments with respect to a playing method.

4 Cascade Classifiers

In this section, we show how to use the cascade λ-representation of $S(d)$ to build cascade classifiers for $S(d)$.

Let $S(d) = (X, A \cup \{d\}, V)$ is a regularly incomplete decision system, where d is a hierarchical attribute. We follow the notation proposed in the previous section to represent its values, with $d[i]$ referring to a child of d and $d[i, j]$ to its grandchild. Also, we assume that $(Casc(S(d)), \prec)$, where $Casc(S(d)) = \{S_k(d) : k \in J\}$ is a cascade λ-representation of $S(d)$. Let $p_1 = [i_1, i_2, ..., i_k]$ and $p_2 = [j_1, j_2, ..., j_n]$. Relation $\prec$ is defined as: $S_{p_1}(d) \prec S_{p_2}(d)$ iff $(k \leq n)$ and $(\forall m \leq k)[i_m = j_m]$. In all other cases $\prec$ is undefined. Clearly, if d_1 is a descendent of d, then $(Casc(S(d_1)), \prec)$ is a cascade λ-representation of $S(d_1)$.

Let us assume that the height of $(Casc(S(d)), \prec)$ is n and $S_{[i_1,i_2,...,i_k]}(d) \in Casc(S(d))$. Clearly, $d[i_1, i_2, ..., i_k]$ is the root of $S_{[i_1,i_2,...,i_k]}(d)$.

By $(S, d[i_1, i_2, ..., i_k], m)$, where $k + 1 \leq m \leq n$, we denote a subtree of S with $d[i_1, i_2, ..., i_k]$ as its root and all descendent of $d[i_1, i_2, ..., i_k]$ at all levels between $k+1$ and m.

Assume now that $class(S, d[i_1, i_2, ..., i_k], m)$ denotes a classifier trained by $S(d[i_1, i_2, ..., i_k])$ with the decision attribute $d[i_1, i_2, ..., i_k]$ and its values restricted to level m of its tree representation. For example, $\{d[1, 1, 1], d[1, 1, 2], ..., d[1, 3, 3]\}$ is the set of values for $S(d[1])$ at level 3 (see Figure 1).

By a cascade classifier of type λ for $S(d)$ we mean $(class(Casc(S(d))), \prec)$, where $class(Casc(S(d))) = \{class((S, d[i_1, i_2, ..., i_k], m)) : [k + 1 \leq m \leq n] \wedge [[i_1, i_2, ..., i_k] \in J]\}$ and $Casc(S(d)) = \{S(d[i_1, i_2, ..., i_k]) : [i_1, i_2, ..., i_k] \in J\}$ is a cascade λ-representation of $S(d)$. A sample representation structure for a cascade classifier is given in Figure 3.

So, three classifiers are associated with the root level of the tree represented by Figure 3. The first one (with i=1) is trained by S with values of the decision attribute defined as the largest granules. The last one (with i=3) is based on attribute values defined as the smallest granules.

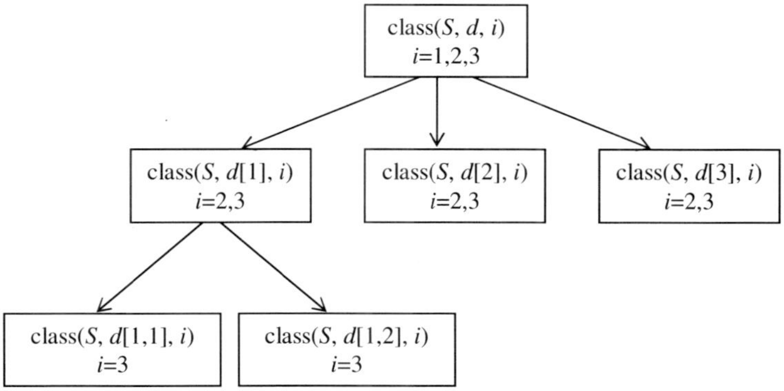

Fig. 3. Cascade classifier for $S(d)$

5 Application Domain and Testing Results

Music instrument identification [8],[9],[12],[17] is one of the important subtasks of a content-based automatic indexing, for which authors built a multi-hierarchical decision system S with all the low-level MPEG7 descriptors as well as other popular descriptors for describing music sound objects. The decision attributes in S are hierarchical and they include Hornbostel-Sachs classification (see Figure 4) and classification of instruments with respect to playing method (see Figure 5).

The information richness hidden in descriptors has strong implication on the confidence of classifiers built from S. Rule-based classifiers give us approximate definitions of values of decision attributes and they are used as a tool by content-based Automatic Indexing Systems (*AIS*) [16]. Hierarchical decision attributes allow us to have the indexing done on different granularity levels of classes of music instruments. We can identify not only the instruments playing in a given music piece but also classes of instruments if the instrument level identification fails. The quality of *AIS* can be verified using precision and recall based on two interpretations: user and system-based [15]. *AIS* engine follows system-based interpretation.

In this section we show that cascade classifiers outperform standard classifiers. The first step in the process of recognizing a dominating musical instrument in a musical piece is the identification of its pitch. If the pitch is found, then a pitch-dedicated classifier is used to identify this instrument. The testing was done for music instrument sounds of pitch 3B. The results are shown in Table 3 and Table 4. The confidence of a standard classifier $class(S, d, 3)$ for Hombostel-Sachs classification of instruments is 91.50%. However, we can get much better results by following the

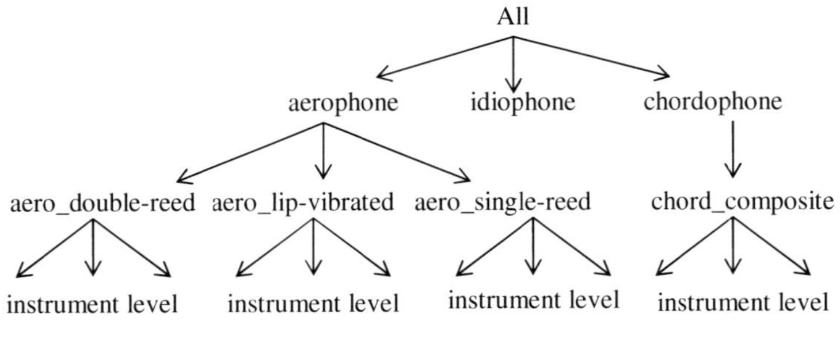

Fig. 4. Hombostel-Sachs classification of instruments

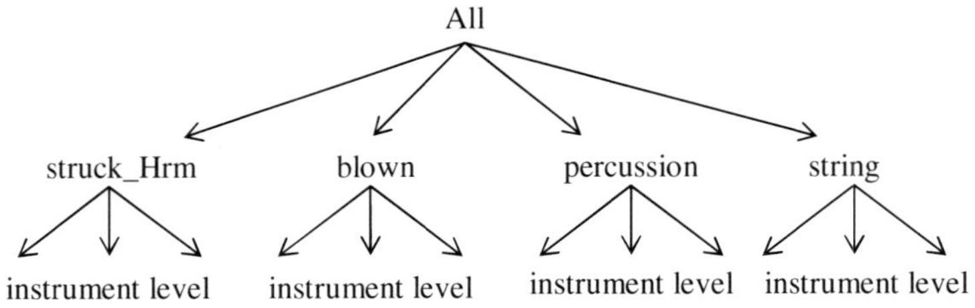

Fig. 5. Classification of instruments with respect to playing method

Table 2. Modules of cascade classifier for classification of instruments with respect to playing method and their confidence

$root$	$classname$	$classifier$	$pos\_support$	$confidence$
d	$all\_instruments$	$class(S, d, 2)$	730	91.80%
d	$all\_instruments$	$class(S, d, 1)$	750	94.26%
$d[1]$	$blown$	$class(S, d[1], 2)$	265	96.84%
$d[2]$	$string$	$class(S, d[2], 2)$	390	94.68%
$d[3]$	$struck\_Hrn$	$class(S, d[3], 2)$	69	98.84%

Table 3. Modules of cascade classifier for Hombostel-Sachs classification of instruments and their confidence

$root$	$classname$	$classifier$	$pos\_support$	$confidence$
d	$all\_instruments$	$class(S, d, 1)$	771	96.97%
d	$all\_instruments$	$class(S, d, 2)$	764	96.02%
d	$all\_instruments$	$class(S, d, 3)$	730	91.80%
$d[1]$	$aerophone$	$class(S, d[1], 2)$	269	98.26%
$d[1]$	$aerophone$	$class(S, d[1], 3)$	265	96.84%
$d[2]$	$chordophone$	$class(S, d[2], 2)$	497	98.83%
$d[2]$	$chordophone$	$class(S, d[2], 3)$	466	92.75%
$d[3]$	$idiophone$	$class(S, d[3], 2)$	19	95.95%
$d[3]$	$idiophone$	$class(S, d[3], 3)$	19	95.95%
$d[1, 1]$	$aero\_double\_reed$	$class(S, d[1, 1], 3)$	70	98.94%
$d[1, 2]$	$aero\_lip\_vibrated$	$class(S, d[1, 2], 3)$	113	95.66%
$d[1, 3]$	$aero\_side$	$class(S, d[1, 3], 3)$	10	90.91%
$d[1, 4]$	$aero\_single\_reed$	$class(S, d[1, 4], 3)$	72	99.54%
$d[2, 1]$	$chrd\_composite$	$class(S, d[2, 1], 3)$	410	93.18%

cascade approach. For instance, if we use the classifier $class(S, d, 2)$ followed by the classifier $class(S, d[1, 1], 3)$, then its precision in recognizing musical instruments in $aero\_double\_reed$ class is equal to $96.02\% * 98.94\% = 95.00\%$. Also, its precision in recognizing instruments in $aero\_single\_reed$ class is equal to $96.02\% * 99.54\% = 95.57\%$. It has to be noted that this improvement in confidence is obtained without increasing the number of attributes in the subsystems of S used to build the cascade classifier replacing S. If new attributes are added to these subsystems, then the new resulting classifiers forming the cascade classifier may have even higher confidence and the same the confidence of the cascade classifier will also get increased.

Acknowledgement

This research was partially supported by the National Science Foundation under grant IIS-0414815. Any opinions, findings, and conclusions or recommendations expressed in this material are those of the author(s) and do not necessarily reflect the views of the National Science Foundation.

References

1. Bruzzone, L., Cossu, R.: A Multiple Cascade-Classifier System for a Robust and Partially Unsupervised Updating of Land-Cover Maps, Technical Report DIT-02-026, Informatica e Telecomunicazioni, University of Trento, Italy (2002)
2. Dardzińska, A., Raś, Z.W.: Rule-Based Chase Algorithm for Partially Incomplete Information Systems. In: Tsumoto, S., Yamaguchi, T., Numao, M., Motoda, H. (eds.) AM 2003. LNCS (LNAI), vol. 3430, pp. 255–267. Springer, Heidelberg (2005)
3. Dardzińska, A., Raś, Z.W.: Chasing Unknown Values in Incomplete Information Systems. In: Lin, T.Y., Hu, X., Ohsuga, S., Liau, C. (eds.) Proceedings of ICDM 2003 Workshop on Foundations and New Directions of Data Mining, Melbourne, Florida, pp. 24–30. IEEE Computer Society, Los Alamitos (2003)
4. Duong, J., Emptoz, H.: Cascade Classifier: Design and Application to Digit Recognition. In: Proceedings of the Eighth International Conference on Document Analysis and Recognition, pp. 1065–1069. IEEE Computer Society, Los Alamitos (2005)
5. Haskell, R.-E.: Design of hierarchical classifiers. In: Sherwani, N.A., Kapenga, J.A., de Doncker, E. (eds.) Great Lakes CS Conference 1989. LNCS, vol. 507, pp. 118–124. Springer, Heidelberg (1989)
6. Huang, X., Li, S.Z., Wang, Y.: Learning with cascade for classification of non-convex manifolds. In: Proc. of CVPR Workshop on Face Processing in Video, FPIV 2004, Washington, DC (2004)
7. Im, S.: Privacy Aware Data Management and Chase. Fundamenta Informaticae 78(4), 507–524 (2007)
8. Kostek, B., Czyzewski, A.: Representing Musical Instrument Sounds for Their Automatic Classification. J. Audio Eng. Soc. 49(9), 768–785 (2001)
9. Kostek, B., Wieczorkowska, A.: Parametric Representation of Musical Sounds. Archive of Acoustics 22(1), 3–26 (1997)
10. Levene, M., Loizou, G.: Semantics for null extended nested relations. ACM Transactions on Database Systems (TODS) 18(3), 414–459 (1993)
11. Lu, C., Drew, M.S.: Construction of a hierarchical classifier schema using a combination of text-based and image-based approaches. In: SIGIR 2001 Proceedings, pp. 331–336. ACM Publications, New York (2001)
12. Martin, K.D., Kim, Y.E.: Musical instrument identification: a pattern-recognition approach. In: Proceedings of 136th Meeting of the Acoustical Society of America, Norfolk, VA (October 1998)
13. Michalski, R.S.: Attributional Ruletrees: A New Representation for AQ Learning, Reports of the Machine Learning and Inference Laboratory, MLI 02-1, George Mason University, Fairfax, VA (2002)
14. Pawlak, Z.: Information systems - theoretical foundations. Information Systems Journal 6, 205–218 (1991)
15. Raś, Z.W., Dardzińska, A., Zhang, X.: Cooperative Answering of Queries based on Hierarchical Decision Attributes. CAMES Journal, Polish Academy of Sciences, Institute of Fundamental Technological Research 14(4), 729–736 (2007)
16. Raś, Z.W., Zhang, X., Lewis, R.: MIRAI: Multi-hierarchical, FS-tree based music information retrieval system. In: Kryszkiewicz, M., Peters, J.F., Rybiński, H., Skowron, A. (eds.) RSEISP 2007. LNCS (LNAI), vol. 4585, pp. 80–89. Springer, Heidelberg (2007)
17. Yang, C.: The MACSIS Acoustic Indexing Framework for Music Retrieval: An Experimental Study. In: Proceedings of ISMIR 2002, pp. 53–62 (2002)
18. Zhang, X., Raś, Z.W., Dardzińska, A.: Discriminant Feature Analysis for Music Timbre Recognition and Automatic Indexing. In: Raś, Z.W., Tsumoto, S., Zighed, D.A. (eds.) MCD 2007. LNCS (LNAI), vol. 4944, pp. 104–115. Springer, Heidelberg (2008)

Creating Rule Ensembles from Automatically-Evolved Rule Induction Algorithms

Gisele L. Pappa[1] and Alex A. Freitas[2]

[1] Federal University of Minas Gerais, Computer Science Department, Av. Antônio Carlos, 6627, 31270-010, Belo Horizonte, MG, Brazil
glpappa@dcc.ufmg.br

[2] University of Kent, Computing Laboratory, CT2 7NF, Canterbury, Kent, UK
A.A.Freitas@kent.ac.uk

Abstract. Ensembles are a set of classification models that, when combined, produce better predictions than when used by themselves. This chapter proposes a new evolutionary algorithm-based method for creating an ensemble of rule sets consisting of two stages. First, an evolutionary algorithm (more precisely, a genetic programming algorithm) is used to automatically create complete rule induction algorithms. Secondly, the automatically-evolved rule induction algorithms are used to produce rule sets that are then combined into an ensemble. Concerning this second stage, we investigate the effectiveness of two different approaches for combining the votes of all rule sets in the ensemble and two different approaches for selecting which subset of evolved rule induction algorithms (out of all evolved algorithms) should be used to produce the rule sets that will be combined into an ensemble.

Keywords: ensembles, evolutionary algorithms, rule induction algorithms, grammar-based genetic programming.

1 Introduction

After many decades of research, classification is still one of the most studied data mining tasks. Classification problems can be solved by using a variety of types of algorithms, involving a really diverse set of representations to express classification models (the outputs of classification algorithms) [1]. Among these, rule induction algorithms produce classification models in the form of a set of IF-THEN classification rules (sometimes called decision rules), which have the advantage of being a knowledge representation intuitively comprehensible to the user.

Another type of algorithms that can be used to solve the classification task are evolutionary algorithms. Evolutionary algorithms (EAs) [2] are stochastic methods based on Darwin's theory of evolution and survival of the fittest. Since they are a very generic and flexible type of search algorithm, evolutionary algorithms can be used to build many different types of classification models. In particular, evolutionary algorithms became a common type of method to evolve decision trees [3], classification rules [4, 5] and neural networks [6], and are also often used as parameter optimizers for many other types of classifiers [7]. More recently, these methods have been often applied to build ensembles of classification models [8, 9, 10, 11, 12].

J. Koronacki et al. (Eds.): Advances in Machine Learning I, SCI 262, pp. 257–273.
springerlink.com © Springer-Verlag Berlin Heidelberg 2010

From a search perspective, a motivation for using evolutionary algorithms in data mining/machine learning is that they perform a population-based global search in the space of candidate solutions (normally candidate classification models, when the target problem is classification) [4]. This is in contrast with several other types of methods, such as the greedy (local search-based) methods typically used by more conventional types of rule induction algorithms. The global search performed by evolutionary algorithms makes them less likely to get trapped in local optima in the search space, by comparison with greedy methods.

Whatever the type of algorithm(s) used to solve a classification problem, in general an effective approach for increasing the predictive accuracy of that(ose) algorithm(s) consists of creating an ensemble of many classification models produced by that(ose) algorithm(s). An ensemble basically combines the predictions of many different classification models into a single prediction for each test example (unseen during training), and follows the principle that a "shared" decision will usually generate better results than an individual one. Actually, building an ensemble of classification models became a popular research area in classification and other supervised learning tasks (e.g. regression), since such ensembles have been shown to outperform – in terms of predictive accuracy – a single classification model [13].

In this chapter we are interested in evolutionary algorithms as a type of method for supporting the creation of an ensemble of classification models – in particular, an ensemble of rule sets. More precisely, this chapter proposes a new method for creating an ensemble of rule sets consisting of two stages. First, an evolutionary algorithm (a genetic programming algorithm) is used to *automatically* create *complete rule induction algorithms*. Secondly, the automatically-evolved rule induction algorithms are used to produce rule sets that are then combined into an ensemble. Concerning this second stage, we investigate the effectiveness of two different approaches for combining the votes of all rule sets in the ensemble (at prediction time) and two different approaches for selecting which subset of evolved rule induction algorithms (out of all evolved algorithms) should be used to produce the rule sets that will be combined into an ensemble.

The basic idea of using evolutionary algorithms – more specifically, genetic programming – for automatically creating complete rule induction algorithms (rather than manually designing rule induction algorithms as usual) was proposed in [14]. Note that by complete rule induction algorithm we mean algorithms with approximately the same level of complexity as well-known algorithms such as CN2 [15] or Ripper [16]. That work has shown that genetic programming can successfully take the automation of data mining/machine learning tasks one step further, by automatically creating rule induction algorithms competitive with manually-designed rule induction algorithms.

However, in [14] the evolved rule induction algorithms were not used to create an ensemble. That is, a set of 100 rule induction algorithms (individuals in the genetic programming algorithm's population) was evolved, but only the best one at the end of the evolution was returned as a solution to the user. As argued by Gagne et al. [8], this selection of the best individual out of many evolved individuals actually represents a missed opportunity for creating an ensemble "for free", since evolutionary algorithms naturally work with a (usually large) population of candidate solutions. At a high level of abstraction, this is the basic idea explored in this work, where we use the classification models (rule sets) produced by the rule induction algorithms automatically evolved by the genetic programming algorithm to produce an ensemble of rule sets.

Hence, the main contribution of this work is to propose a new evolutionary algorithm-based ensemble method where a genetic programming algorithm is used to evolve a population of complete rule induction algorithms, where the latter are then used to produce rule sets composing an ensemble. This is a new way of producing a diverse set of classification models. That is, instead of creating diverse classification models by injecting randomness into some *component* of a classification algorithm (e.g., injecting randomness in the attribute selection procedure in random forests – see Section 2), we inject a certain degree of randomness into the generation of the *complete* rule induction algorithms that will be used to produce the classification models that will form an ensemble. This degree of randomness is due to the stochastic nature of an evolutionary algorithm's search, which works with a diverse population of candidate solutions (rule induction algorithms in our case) spread across the search space. In any case, of course the evolutionary search for rule induction algorithms is not completely random, since it is actually a heuristic search guided by the fitness function (which favours rule induction algorithms with higher classification accuracy). Hence, it is hoped that the evolutionary search will do a good job of *automatically* creating a set of rule induction algorithms which are both accurate and diverse, which would tend to produce rule sets having those corresponding properties, which are desirable properties for an ensemble of classification models in general.

The chapter is organized as follows. Section 2 introduces some background knowledge on ensembles, and Section 3 introduces concepts of evolutionary algorithms and genetic programming. Section 4 shows how ensembles and evolutionary algorithms can be combined, and presents some related work. Section 5 gives an overview of the grammar-based genetic programming method used to evolve the rule induction algorithms, and describes how the ensemble is built. Finally, Section 6 reports some experimental results, and Section 7 draws some conclusions and presents some ideas of future work.

2 Ensembles of Classification Models

The concept of ensembles is based on the idea that combining a set of predictions coming from different classification models often generates better results than a single prediction produced by one of the models participating in the ensemble. Given a set of n classification models C_i, where i varies from 1 to n, an ensemble combines all the predictions of those classification models to predict a final class to an example.

According to [17], ensembles usually obtain better accuracy rates than single classification models because they solve three important problems that the single classification models have difficulties in dealing with: the statistical, the computational and the representational problems. The statistical problem refers to the size of the model's search space *versus* the number of training examples available. Many classification models' search spaces are simply too large to be explored considering the limited number of training examples available. Hence, many classification models with similar accuracy rates are found, and choosing one of them is a difficult task. The single classification model chosen might be one that does not generalize (on the test set) very well, by comparision with other models with similar classification accuracies in the training set. Ensembles avoid this problem, once a set of classification models is considered, and the chances of picking only bad models is reduced.

The computational problem relates to the fact that, as classification model search spaces are usually huge, one cannot guarantee the best model will be found in a feasible time. In order to cope with this problem, heuristics are used, and they can get trapped in local optima in the search space. Selecting a set of classification models instead of just one model reduces the chances of only getting classification models found around local optima.

The third problem, named representational, concerns classification model spaces that do not contain any good approximations to the "ideal" model. In this case, a set of weak classification models might guarantee better results than a single one.

Considering these problems, ensembles are usually built following one of two approaches. In the first approach, different classification models are built separately [18], and their predictions combined through some sort of voting scheme. In this kind of approach, the main difficult lies on how to choose how many and which classification models will be considered as part of the ensemble [11], as it is known that the success of an ensemble is highly dependent on how accurate and diverse its individual classification models are [19, 20]. In the second approach, this problem is avoided, as the classification models that compose the model are built incrementally, in a cooperative fashion [12].

Regardless of the approach used to build the ensembles, two important decisions have to be made: which method will be used to build the set of classification models, and which method will be used to to combine the predictions of each model into a single prediction. Concerning the methods used to build the ensembles, most of them work by manipulating the training set [21], the set of attributes [22] or the output labels of the training instances [23] before giving them to a classification algorithm. Some other methods work by injecting some randomness into the classification algorithms [24].

Among the most known methods for ensemble construction are bagging [21], boosting [25] and random forests [24]. Bagging works by creating multiple data samples, where each data sample is produced by sampling with replacement the original training set. These data samples typically have the same size the original training set has. Therefore, one data sample may have many copies of a specific example and no copies of others. The classification algorithm is then run on each of those data samples, producing a set of classification models constituting the ensemble.

Boosting, in contrast, works by setting weights to the training examples in the original training set. Initially, all the examples have the same weight. A sample of examples (also with the same size of the original training set) is probabilistically drawn from the training set. The probability of an example being chosen to be part of the sample is proportional to its weight. The classification algorithm is run multiple times, sequentially, in iteratively sampled data sets, and at each iteration the weights of the examples are updated according to the performance of the classification model in that particular iteration. In this way, examples difficult to be classified have their weights increased more frequently than other instances, which improves their probabilities of being selected for the next sample which will be used to train the next iteration of the classification algorithm.

Random forests are methods that combine bagging with randomly built decision trees. They produce a set of decision trees, each of them using a different sampled data set. In a popular approach to build a random forest, in order to build each decision tree, a number m of attributes are randomly selected from the total set of n attributes available in the training set – where m is typically much smaller than n. Out of the set of m randomly-selected attributes, the attribute that creates the best data split is used to label the current node of the decision tree, during the tree building process. Each decision tree is built to the largest possible extent, and there is no pruning. When classifying new examples, each tree gives its vote, and the votes of all decision trees are combined to classify a new example. Random forests are well-known for not overfitting.

After classification models are built using one of the methods described above or a related method, another method has to be applied to combine the predictions coming from different classification models. The most common of these combination methods is the majority voting, where the class assigned to a new example is the one predicted by the greatest number of classifiers, or some related weighted voting strategy.

At this point, it is important to note that, although most of the research in ensembles is concentrated on building ensembles using the same classification algorithm, ensembles can also be built using classification models produced by different classification algorithms (using different model representations). In this case, more sophisticated techniques to combine the results coming from different classifiers is needed. The most common of this type of methods is stacking [26].

Stacking works by creating a meta-model that learns to predict the final class to be assigned to a new example based on the classes predicted by the classification models that compose the ensemble. In order to learn a meta-model, a training set is built using a leave-one-out procedure. The most important decisions concerning stacking methods are the attributes to be used in the meta data set and the algorithm that will learn the meta model.

Ting and Witten [27], for instance, proposed a stacking method that predicts the probability distributions over the set of class values, rather than single class values. In this case, the meta-level attributes are the probabilities of each of the class values returned by each of the base level classification models in the ensemble. In this same work, the authors show there are several learning algorithms that are not suitable for learning meta-models, and they recommend a multi-response linear regression method for this task.

So far, we described some of the most used methods to create ensembles of classification models. Recently, AdaBoost [28] and random forests have been shown to be particularly powerful. At the same time, a lot of approaches using evolutionary algorithms for ensemble building were proposed [8, 11, 12]. In this chapter, we are particularly interested in what we call evolutionary ensembles. One of the main motivations to combine evolutionary algorithms and ensembles is to take advantage of the population of individuals (classification models) which is naturally evolved by the evolutionary algorithm. Section 3 gives an overview of evolutionary algorithms, and Section 4 describes methods using them to build ensembles of classification models.

3 Evolutionary Algorithms and Genetic Programming

Evolutionary algorithms [2] are stochastic methods based on Darwin's concepts of evolution and survival of the fittest, and they work by evolving a population of candidate solutions to a given target problem. Genetic Programming (GP) [29] is an area of evolutionary computation which aims to automatically evolve computer programs. Together with other types of evolutionary algorithms, its application is being successful because of its generality (applicability to potentially any problem domain), global search and associated implicit parallelism and noise tolerance [30, 31].

Essentially, a GP algorithm evolves a population of individuals, where each individual represents a "program" or "executable structure" that is a candidate solution to the target problem. These individuals are evaluated using a fitness function, which measures the goodness of the candidate solution (program) represented by the individual. The fittest individuals are usually selected to undergo reproduction, crossover and mutation operations. The new individuals produced by these operations are used to create a new population, which replaces the old one. This evolutionary process is carried out until an optimum solution or satisfactory solution is found, or a pre-defined number of generations (iterations) is reached.

The design of a GP algorithm has to consider five essential components:

1. A set of functions and terminals, used to create the first GP population.
2. The representation of the individuals.
3. The fitness function used to measure the quality of the individuals (candidate solutions).
4. A selection method.
5. Crossover and mutation operators, which produce new children individuals out of the selected parent individuals in the current generation (iteration).

The next subsections explain the main concepts involved in the design of these elements.

3.1 Functions and Terminal Sets

In a GP algorithm, the first population is randomly generated using a set of functions and terminals. The terminals are usually constants, variables and/or zero-argument functions. The most common functions are the boolean and arithmetical ones, but conditional and/or loop statements can also be used.

Although the designer of a GP algorithm has a lot of freedom to choose the function set, it should not have too many functions, since the more functions the greater the search space.

One constraint to be considered when choosing the sets of functions and terminals is that the closure property must be respected. This property states that every function in the function set has to be able to handle all the values it receives as input. Thus a division operator, for example, has to be modified to cope with division by zero. This is often implemented by making the operator return a given value, rather than an error, in case of division by zero.

3.2 Individual Representation

Most GP algorithms work with either of two standard individual representations: a tree or a linear structure [31]. A linear representation is simply a sequence of commands that are executed from left to right. In a tree representation, the execution of the commands is usually made in postfix order (reading the leftmost node of the tree first). The majority of GP algorithms use the tree representation.

3.3 Fitness Function and Selection Methods

GP algorithms evolve a population of individuals using the concepts of selection and survival of the fittest. Hence, after the initial GP population is initialized (usually randomly), individuals are evaluated using a fitness function. This function measures how well an individual solves the target problem. It is used to determine which individuals will have parts of their genetic material (i.e., parts of their candidate solution) passed onto the next generation via the action of some genetic operator (reproduction, crossover or mutation). The better the fitness of an individual, the higher the probability of that individual being selected for reproduction, crossover or mutation.

There are many selection methods, such as fitness-proportional selection, ranking selection and tournament selection. Tournament selection, for example, randomly gets a pre-defined number of individuals from the population and simulates a tournament among them. Typically, the individual with the best fitness is declared the winner of the tournament and is therefore selected for reproduction, crossover or mutation.

3.4 Crossover and Mutation Operators

Crossover swaps genetic material (parts of candidate solutions) between two individuals, whereas mutation replaces some part of the genetic material of an individual with a new randomly-generated genetic material. These two operations are applied with user-specified probabilities. The basic idea of these operators is as follows.

Crossover re-combines the genetic material of two parent individuals, in order to produce two new children. If the individuals are represented by trees, randomly-selected subtrees are swapped between the two parents. In the case of linear genomes, randomly-selected linear segments of code are swapped.

Unlike crossover, mutation acts on a single parent individual of the population. It randomly selects a subtree of the tree-based genome or a segment of code in linear genomes and replaces it by a new randomly-generated subtree or code segment. Both crossover and mutation operations can be implemented in many ways - see [31] for a detailed review of these operators.

It should be noted that, although the mechanisms of crossover and mutation are essentially a form of random moves in the search space (which makes them problem independent), a GP algorithm as a whole is a heuristic search method, because crossover and mutation are applied to parent individuals which are selected based on their (problem dependent) fitness, i.e. a measure of their ability in solving the target problem. Hence, by iteratively applying crossover and mutation operators to the best programs of the current gereration (iteration), the population of programs gradually evolves to better and better programs.

4 Evolutionary Methods for Creating Ensembles

Evolutionary algorithms and ensembles can be used together in two different broad approaches. In the first one, the evolutionary algorithm has the same role as a traditional classification algorithm: it builds a classification model from data. In this case, each individual in the EA represents a classification model, and a subset of them is selected and then combined using one of the traditional ensemble techniques, such as bagging or boosting [18]. The selection of the models (individuals) that will compose the final ensemble is the most difficult problem to be tackled in this approach.

In the second approach, evolutionary algorithms are used to optimize some components – or (broadly speaking) "parameters" – of an ensemble. These components or parameters might vary from the best set of attributes [10] to the number of classification models that should compose an ensemble [32]. Combinations of both strategies are also valid [9].

The evolutionary algorithm-based ensemble method proposed in this chapter can be considered as a new method belonging to the above second approach where, instead of optimizing a simple parameter - such as ensemble size, we "optimize" (evolve) a fundamental "macro-parameter": the classification algorithms that will produce the classification models that will compose the ensemble. In other words, a genetic programming algorithm is used to automatically create an entire set of classification (more precisely, rule induction) algorithms. After the evolved rule induction algorithms are applied to the data being mined, they create a set of classification models (rule sets), from which an ensemble of rule sets is produced. This method will be explained in detail in Section 5. Before moving to that section, however, this section reviews related work on evolutionary algorithm-based ensemble methods.

Let us now describe some methods following the first approach described above. One of the first ideas of combining ensembles and genetic programming (GP) was presented by Iba [18], who introduced the BagGP and BoostGP algorithms. In both algorithms, each individual in the population represents a classification model, and the population of a GP algorithm is divided into S sub-populations. In the case of BagGP, each subpopulation is trained with a data set of size N, sampled with replacement from the original training set (which also has size N). The best individuals of each population have the right to vote when classifying new examples in the test set.

BoostGP follows the same basic algorithmic procedure that AdaBoost does. At the beginning of the evolutionary process, each example of the training set is associated with the same weight w. The probability of an example being chosen to be part of the sampled training set, which has the same size as the original training set, is proportional to the weight it is associated with. Each sub-population is evolved sequentially with a different training set, and the weights are updated according to a loss function. After all the sub-populations are evolved, the best individual of each population votes to classify a new example.

In this same line of work, Folino *et al* [33] proposed a boosting system based on cellular genetic programming that they named ClustBoostCGPC. ClustBoostCGPC is a parallel GP algorithm, where each processing unit evolves a different population (where each individual represents a decision tree), which is locally evaluated. However, in contrast with the method proposed in [18], at the end of the evolution the well-known k-means clustering algorithm [34] is used to generate clusters of individuals for each population. The best individuals of each cluster in each population are chosen to build the ensemble. One drawback of this method is that it can generate really large ensembles, so the authors use some pruning strategies to remove individuals from a previously built ensemble.

Chen *et al* [11] proposed a multi-objetive evolutionary algorithm that, combined with a Bayesian automatic relevance determination (ARD) method [35], designed and trained ensembles of neural networks using attribute selection. The neural nets were initially built using a subset of attributes from the training set, and their parameters optimized using the Bayesian ARD method. The evolutionary algorithm was used to evolve neural networks with few attributes and low error rate. At the end of the evolutionary process, a logistic regression method was used to choose the networks that would be part of the ensemble.

Gagne *et al* [8] proposed and compared two evolutionary algorithms for ensemble learning. The first, named Off-EEL (an "off-line" ensemble construction method), used an evolutionary algorithm to evolve neural networks and built the ensemble after evolution using a greedy approach. The second approach, named On-EEL (an "on-line" ensemble construction method), used the same type of evolutionary algorithm but selected the final classification models that would form the ensemble during the evolution process. The results showed that the first approach obtained better results than the second one. They claim this result can be explained by the On-EEL getting often trapped in local optima in the search space.

Kim *et al* [9] also proposed a local selection algorithm to evolve ensembles of neural nets combined with attribute selection. In this case, each neural net was trained with a different subset of attributes, but individual neural networks and ensembles were evolved simultaneously. Each individual was represented by a binary string that encoded the attributes that should be used to train the neural network and the list of ensembles the individual's classification model belonged to. In this way, the ensembles being evolved competed directly to each other, and the individual neural networks also competed among themselves, and were allowed to move to the fittest ensembles.

In contrast to the methods described until now, Sirlantzis *et al* [32] used an evolutionary algorithm to select the best combination of a set of pre-defined classification algorithms and voting strategies, following the second approach previously described. Depending on the problem being tackled, they used 6 or 12 classification algorithms together with 4 voting strategies.

This was also the approach followed by Kim and Cho [12], where an evolutionary algorithm was used to find ensembles of pairs of attribute selection method and classification algorithm. In this case, they pre-defined 7 attribute selection methods and

6 classification algorithms. This generated 42 pairs of attribute selection method and classification algorithm, which were represented in an individual as a vector with 42 positions ("genes"). Each individual vector's gene could be represented by a binary digit or a real-valued number. In the first case, a 1 indicated that the pair of attribute selection method and classification algorithm associated with that gene is part of the ensemble being evolved. In the second case, a number different from 0 meant the pair associated with the gene was being used in the ensemble, and its vote when classifying unseen examples was weighted by the value present in that gene of the individual. Hence, each individual represented an ensemble of classification models, which would be generated using the pairs of attribute selection method and classification algorithms indicated by its gene vector.

All the evolutionary algorithms described above aim to create very accurate and diverse ensembles. In order to achieve that, beside the ideas already discussed, other mechanisms were also used to improve the effectiveness of the evolutionary algorithm-based ensemble systems. For instance, fitness based on accuracy was combined with niching mechanisms or modified according to co-evolution principles [36], and speciation was applied [37]. Moreover, multi-objective EAs (MOEAs) became commonly used as a way to optimize both accuracy and diversity simultaneously [38].

5 Building an Ensemble of Rule Sets from Rule Induction Algorithms Automatically Created by Genetic Programming

As pointed out before, the main difference among the evolutionary method proposed in this paper and other evolutionary algorithm-based ensemble methods proposed in the literature is that the evolutionary method proposed here does not evolve classification models. Instead, it evolves complete rule induction algorithms. The evolved rule induction algorithms are then used to produce rule sets composing an ensemble.

The method proposed uses a grammar-based genetic programming (GGP) algorithm to evolve a population of rule induction algorithms [39]. GGP [40] is a special type of genetic programming (GP) algorithm where the individuals are generated following the production rules of a grammar (instead of sets of terminals and functions), which enforces the generation of syntactically correct individuals only. Besides, the grammar can incorporate background knowledge about how the problem being tackled is solved. In the case of the system used in this paper, the grammar incorporates knowledge about how humans design rule induction algorithms.

As explained before, there are some elements in GP that need to be defined according to the problem being tackled. In the case of GGP, the sets of terminals and functions of a standard GP is replaced by a grammar, which is described in detail in [14, 39]. Each individual represents a complete rule induction algorithm, such as CN2 or Ripper. Fig. 1 shows an example of an individual generated by following the production rules of the grammar.

In order to extract from an individual's tree the pseudo-code of the corresponding rule induction algorithm, we read all the terminals (leaf-nodes) in the tree from left to right. The tree in Figure 1, for example, represents the pseudo-code of the CN2 algorithm [15] producing an ordered list of rules, with the beam-width (or star size, using

Fig. 1. Example of a GGP individual (candidate rule induction algorithm)

the CN2 terminology) parameter set to 5 and the statistical significant test threshold set to 0.01.

Figure 2 shows the scheme of the proposed method. As observed, it represents a standard evolutionary algorithm execution, combined with some interesting points. First, as already explained, the individuals of the first population are generated by following the production rules of a grammar. Second, the individuals are evaluated using a "meta-training set", which is composed of a set of n complete data sets (i.e. a training set and a validation set for each data set, with no overlapping of examples between these two sets). After executing the candidate rule induction algorithm in each data set i of the meta-training set, a fitness function based on the average of the value of fit_i (given by Eq. 1) over all data sets is calculated. In Eq. 1, Acc_i represents the accuracy (on the validation set) obtained by the rules discovered by the rule induction algorithm in the training set of data set i. $DefAcc_i$ represents the default accuracy (the accuracy obtained when using the most frequent class in the training examples to classify all examples in the validation set).

$$fit_i = \begin{cases} \frac{Acc_i - DefAcc_i}{1 - DefAcc_i}, \text{ if } \mathrm{Acc}_i > \mathrm{DefAcc}_i \\ \frac{Acc_i - DefAcc_i}{DefAcc_i}, \text{ otherwise} \end{cases} \quad (1)$$

Note that, for each data set in the meta-training set, the validation set has the same role as the test set in conventional classification, consisting of examples unseen during training. However, we prefer to use the term validation set (rather than test set) in this context because the fitness values of individuals are iteratively computed by accessing the validation set many times during evolution, and so the term test set would be misleading.

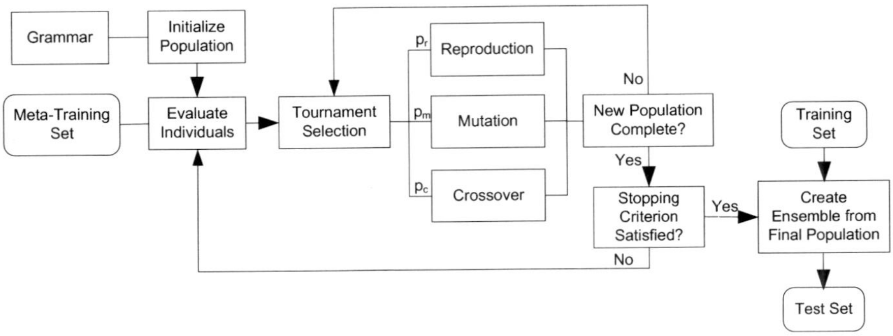

Fig. 2. Scheme of the GGP algorithm used to create the ensemble

By executing the candidate rule induction algorithms in a set of different classification problems, we aim at automatically evolving robust rule induction algorithms which can generalize well on new data sets (application domains), different from the data sets (application domains) used in the meta-training set. During the evolution, individuals created by crossover and mutation operations also have to be valid according to the grammar. A detailed description of this system can be found in [14, 39].

At the end of the evolution, all the individuals in the last population (or a subset of them, depending on the version of the system used – see next Section) are used to create an ensemble of rule sets. That is, for each rule induction algorithm in the last population, we run it in each new target data set (unseen during the GGP algorithm's evolution), producing a corresponding rule set, and then we create an ensemble consisting of the rule sets produced by those rule induction algorithms.

Once such an ensemble of rule sets has been produced, two different voting strategies can be used to combine the rule sets' predictions, namely majority voting and fitness-weighted voting. The majority voting strategy, as its name suggests, classifies an example as belonging to the class predicted by the highest number of rule sets in the ensemble. The fitness-weighted voting strategy weighs the vote of each rule set according to the fitness value attributed to its corresponding individual (rule induction algorithm) in the GGP algorithm's last generation (see Eq. 1).

As pointed out before, the success of an ensemble is highly dependent on the diversity and accuracy of the individuals belonging to it. In terms of accuracy, it is expected that, at the end of the evolutionary process, all the individuals will present accuracies at least better than the ones obtained by the baseline classification model (which always predicts the most frequent class in the training set for all examples in the test set). In terms of diversity, an analysis of the components of the grammar present in the individuals of the last generation showed that usually a diverse set of rule induction algorithms is evolved.

6 Computational Experiments

This section describes experiments performed to create ensembles of rule sets from the last population of individuals (rule induction algorithms) evolved by the GGP algorithm

Table 1. Data sets used to create ensembles

Data set	Examples	Attributes	Classes	Def. Acc.(%)
crx	690	5	2	67.7
heart-c	303	13	2	54.5
ionosphere	351	34	2	64
sonar	208	60	2	53
segment	2310	19	7	14.3

(with a population size of 100 individuals). Experiments were performed in 5 UCI data sets [41], listed in Table 1. The figures in the column *Examples* indicate the number of examples present in the data set, followed by the number of attributes and classes. The last column shows the default accuracy (the accuracy obtained when using the most frequent class in the training set to classify new examples in the test set). It is important to emphasize that these data sets are different from the ones used in the GGP algorithm's meta-training set during the evolutionary process. In the experiments reported in this work, the data sets used in the meta-training set were *monks-2*, *monks-3*, *balance-scale*, *lymph* and *zoo*.

The accuracies obtained by the ensemble of rule sets produced by the evolved rule induction algorithms (from now on referred to as Rule-Ens for short) using the two voting strategies explained in Section 5 were compared to the accuracies obtained by the rule set produced by the single best individual returned to the user as the best evolved rule induction algorithm (GGP-RI). Table 2 shows the results obtained by a 5-fold cross validation procedure (with numbers after the symbol $\pm$ representing standard deviations). The results were compared using a paired two tailed Student's t-test with significance level 0.05. Cells in dark gray represent significant wins of Rule-Ens against the GGP-RI. In Table 2 the column names "Rule-Ens Maj." and "Rule-Ens Fit." refer to Rule-Ens using the majority voting and fitness-weighted voting strategies, respectively.

The results in Table 2 show that, for the data set *segment*, both the ensemble using a majority voting and the ensemble using a fitness-weighted voting obtained better predictive accuracy rates than the GGP-RI. All the other results obtained are considered to be statistically the same as the ones obtained by the GGP-RIs.

The results in Table 2 were produced by using all individuals (rule induction algorithms) of the last population to produce rule sets and combining all those rule sets into an ensemble. In a second set of experiments, we tried to reduce the number of rule sets

Table 2. Predictive accuracy rates (%) obtained by the ensemble of rule sets built from the last population of evolved rule induction algorithms using two different voting strategies

Data Set	Rule-Ens Maj.	Rule-Ens Fit.	GGP-RIs
crx	83.2 ± 0.02	82.01 ± 0.02	79.4 ± 0.01
heart-c	80.54 ± 0.02	81.86 ± 0.02	76.6 ± 0.036
ionosphere	86.3 ± 0.01	89.91 ± 0.01	86.9 ± 0.024
segment	97.01 ± 0.002	96.45 ± 0.003	94.5 ± 0.005
sonar	74.38 ± 0.05	74.87 ± 0.04	73.1 ± 0.036

Table 3. Predictive accuracy rates (%) obtained by an ensemble of rule sets built from a set of 10 rule induction algorithms selected from the last population of evolved rule induction algorithms

Data Set	Ens-Top	Ens-Mix	GGP-RIs
crx	80.99 ± 0.02	84.2 ± 0.02	79.4 ± 0.01
heart-c	80.55 ± 0.02	82.18 ± 0.01	76.6 ± 0.036
ionosphere	85.76 ± 0.012	85.76 ± 0.01	86.9 ± 0.024
segment	97.01 ± 0.002	96.71 ± 0.003	94.5 ± 0.005
sonar	74.38 ± 0.05	74.39 ± 0.05	73.1 ± 0.036

in the ensemble. We selected a subset of rule induction algorithms to be used to produce the rule sets for the ensemble according to the following two methods.

In the first method we selected the top 10 individuals of the last population according to a fitness-based ranking (where fitness is given by Eq. 1) and created the corresponding ensemble of rule sets produced by those individuals (denoted Ens-Top). In the second method we selected the top 5 and the bottom 5 individuals of the last population according to a fitness-based ranking and combined the rule sets produced by both groups of individuals in a single ensemble (denoted Ens-Mix). This last method intuitively tends to increase the diversity of the selected rule induction algorithms (which in turn tends to increase the diversity of their produced rule sets), once similar rule induction algorithms are likely to have similar fitness values. The results obtained are presented in Table 3. Again, a paired two-tailed Student's t-test with significance level of 0.05 was used to evaluate the statistical significance of the differences in the accuracies obtained by each of the two proposed ensemble methods and the GGP-RI baseline, and significant wins of one of the proposed ensemble methods are highlighted in dark grey in the table.

Table 3 shows that both types of ensembles obtain significantly better predictive accuracies than the GGP-RI baseline in the *segment* data set, and that the ensemble produced by using the mixed strategy (which in theory has more diverse rule sets) also obtains significantly better predictive accuracies than the GGP-RI baseline in the *crx* data set. Note that in the *crx* data set only the accuracy of Ens-Mix (and not the accuracy of Ens-Top) is significantly better than the accuracy of GGP-RI. This result shows that the weaker individuals in the last population actually contribute to generate a better ensemble (since the weaker individuals tend to increase the diversity of the rule sets in the ensemble) in the *crx* data set, leading to the generation of a better ensemble by comparison with the ensemble consisting only of rule sets produced by the strongest individuals in the last population. It is important to point out that the difference in fitness values from the best to the worst individual in the last population was significant in the case of the *crx* data set, but even the worst individual of the population could not be considered a bad individual, after being evolved for 100 generations.

In conclusion, overall the ensembles built with 10 individuals selected from the last population produced better results than the ensembles produced by taking all the 100 individuals of the last population into account. In addition, mixing the rule sets produced by the 5 best and the 5 worst individuals of the last population according to fitness values produced ensembles presenting good results (significantly better than the GGP-RI baseline) in 2 out of the 5 data sets used in the experiment.

7 Conclusions and Future Research Directions

This paper proposed a new evolutionary algorithm-based method to produce an ensemble of rule sets. More precisely, the proposed method consists of two basic stages. First, it exploits the population-based search of a genetic programming algorithm to automatically evolve a diverse set of rule induction algorithms. We emphasize that what is being evolved by the genetic programming algorithm is a set of *complete rule induction algorithms*, with the same level of complexity as well-known manually-designed algorithms like CN2 or Ripper, rather than classification models as in many other evolutionary algorithms. Secondly, the set of evolved rule induction algorithms in the last population (or a selected subset of those algorithms) is used to produce a set of rule sets (one rule set for each rule induction algorithm), which are then combined into an ensemble of rule sets.

Experiments in 5 public-domain data sets showed that by using a simple majority voting scheme the ensembles of rule sets produced by using all the individuals or only 10% of the individuals in the last population were capable of obtaining predictive accuracy rates higher than the ones obtained by a single rule set (produced by the best individual in the last population) in 2 out of the 5 data sets.

Concerning future research directions, we believe these "proof of concept" results can be improved by adding some diversity measure to the genetic programming algorithm's fitness function, or by using a more sophisticated technique to select the evolved individuals (rule induction algorithms) that will be used to produce the rule sets composing the ensemble.

In addition, other voting schemes, and even a meta-classification model, could also be used to compute the ensemble prediction. An interesting experiment would involve to actually leave up to the genetic programming algorithm the task of selecting how many and which individuals should be used and which voting scheme should be used to maximize the predictive accuracy of the resulting ensemble.

Yet another research direction to be explored is to use a multi-objective version of the genetic programming algorithm (based on the concept of Pareto dominance) which simultaneously optimizes both the diversity and the accuracy of the individuals being evolved.

Acknowledgments

The first author is financially supported by FAPEMIG, Brazil.

References

1. Witten, I.H., Frank, E.: Data Mining: Practical Machine Learning Tools and Techniques with Java Implementations, 2nd edn. Morgan Kaufmann, San Francisco (2005)
2. Eiben, A.E., Smith, J.E.: Introduction to Evolutionary Computation. Springer, Heidelberg (2003)
3. Wong, M.L., Leung, K.S.: Data Mining Using Grammar-Based Genetic Programming and Applications. Kluwer, Norwell (2000)

4. Freitas, A.A.: Data Mining and Knowledge Discovery with Evolutionary Algorithms. Springer, Heidelberg (2002)
5. Tsakonas, A., Dounias, G., Jantzen, J., Axer, H., Bjerregaard, B., von Keyserlingk, D.G.: Evolving rule-based systems in two medical domains using genetic programming. Artificial Intelligence in Medicine 32(3), 195–216 (2004)
6. Yao, X.: Evolving artificial neural networks. Proceedings of the IEEE 87(9), 1423–1447 (1999)
7. Rozsypal, A., Kubat, M.: Selecting representative examples and attributes by a genetic algorithm. Intelligent Data Analysis 7(4), 291–304 (2003)
8. Gagné, C., Sebag, M., Schoenauer, M., Tomassini, M.: Ensemble learning for free with evolutionary algorithms? In: GECCO 2007: Proceedings of the 9th annual conference on Genetic and evolutionary computation, pp. 1782–1789. ACM, New York (2007)
9. Kim, Y., Street, W.N., Menczer, F.: Optimal ensemble construction via meta-evolutionary ensembles. Expert Systems With Applications 30(4) (2006)
10. Oliveira, L.S., Morita, M., Sabourin, R., Bortolozzi, F.: Multi-objective genetic algorithms to create ensemble of classifiers. In: Coello Coello, C.A., Hernández Aguirre, A., Zitzler, E. (eds.) EMO 2005. LNCS, vol. 3410, pp. 592–606. Springer, Heidelberg (2005)
11. Chen, H., Yao, X.: Evolutionary multiobjective ensemble learning based on bayesian feature selection. In: CEC 2006. IEEE Congress on Evolutionary Computation, pp. 267–274 (2006)
12. Kim, K.J., Cho, S.B.: An evolutionary algorithm approach to optimal ensemble classifiers for dna microarray data analysis. IEEE Trans. Evolutionary Computation 12(3), 377–388 (2008)
13. Brown, G., Wyatt, J.L., Harris, R., Yao, X.: Diversity creation methods: a survey and categorisation. Information Fusion 6(1), 5–20 (2005)
14. Pappa, G.L., Freitas, A.A.: Automatically evolving rule induction algorithms. In: Fürnkranz, J., Scheffer, T., Spiliopoulou, M. (eds.) ECML 2006. LNCS (LNAI), vol. 4212, pp. 341–352. Springer, Heidelberg (2006)
15. Clark, P., Boswell, R.: Rule induction with cn2: some recent improvements. In: Kodratoff, Y. (ed.) EWSL 1991. LNCS, vol. 482, pp. 151–163. Springer, Heidelberg (1991)
16. Cohen, W.W.: Fast effective rule induction. In: Proc. of the 12th International Conference on Machine Learning, pp. 115–123. Morgan Kaufmann, San Francisco (1995)
17. Dietterich, T.G.: Ensemble learning. In: Arbib, M.A. (ed.) Handbook of Brain Theory and Neural Networks, pp. 405–409. MIT Press, Cambridge (2002)
18. Iba, H.: Bagging, boosting, and bloating in genetic programming. In: Banzhaf, W., Daida, J., Eiben, A.E., Garzon, M.H., Honavar, V., Jakiela, M., Smith, R.E. (eds.) Proceedings of the Genetic and Evolutionary Computation Conference, Orlando, Florida, USA, vol. 2, pp. 1053–1060. Morgan Kaufmann, San Francisco (1999)
19. Dietterich, T.G.: Machine-learning research: Four current directions. The AI Magazine 18(4), 97–136 (1998)
20. Kuncheva, L.I., Whitaker, C.J.: Measures of diversity in classifier ensembles and their relationship with the ensemble accuracy. Machine Learning 51(2), 181–207 (2003)
21. Breiman, L.: Bagging predictors. Mach. Learn. 24(2), 123–140 (1996)
22. Stefano, C.D., Marcelli, A.: Exploiting reliability for dynamic selection of classifiers by means of genetic algorithms. In: Proc. ICDAR 2003, pp. 671–675 (2003)
23. Dietterich, T.G., Bakiri, G.: Solving multiclass learning problems via error-correcting output codes. Journal of Artificial Intelligence Research 2, 263–286 (1995)
24. Breiman, L.: Random forests. Machine Learning 45(1), 5–32 (2001)
25. Freund, Y., Schapire, R.E.: A decision-theoretic generalization of on-line learning and an application to boosting. In: Vitányi, P.M.B. (ed.) EuroCOLT 1995. LNCS, vol. 904, pp. 23–37. Springer, Heidelberg (1995)

26. Bauer, E., Kohavi, R.: An empirical comparison of voting classification algorithms: Bagging, boosting, and variants. Machine Learning 36(1-2), 105–139 (1999)
27. Ting, K.M., Witten, I.H.: Issues in stacked generalization. Journal of Artificial Intelligence Research 10, 271–289 (1999)
28. Freund, Y., Schapire, R.E.: A decision-theoretic generalization of on-line learning and an application to boosting. In: Vitányi, P.M.B. (ed.) EuroCOLT 1995. LNCS, vol. 904, pp. 23–37. Springer, Heidelberg (1995)
29. Koza, J.R.: Genetic Programming: On the Programming of Computers by the means of natural selection. The MIT Press, Cambridge (1992)
30. Baeck, T., Fogel, D.B., Michalewicz, Z.: Evolutionary Computation 1 Basic Algorithms and Operators. Institute of Physics Publishing (2000)
31. Banzhaf, W., Nordin, P., Keller, R.E., Francone, F.D.: Genetic Programming – An Introduction; On the Automatic Evolution of Computer Programs and its Applications. Morgan Kaufmann, San Francisco (1998)
32. Sirlantzis, K., Fairhurst, M., Guest, R.: An evolutionary algorithm for classifier and combination rule selection in multiple classifier systems. In: 16th International Conference on Pattern Recognition, 2002. Proceedings, vol. 2, pp. 771–774 (2002)
33. Folino, G., Pizzuti, C., Spezzano, G.: Training distributed gp ensemble with a selective algorithm based on clustering and pruning for pattern classification. IEEE Trans. Evolutionary Computation 12(4), 458–468 (2008)
34. Tan, P., Steinbach, M., Kumar, V.: An Introduction to Data Mining. Addison-Wesley, Reading (2006)
35. Neal, R.: Bayesian Learning for Neural Networks. PhD thesis, University of Toronto (1994)
36. Liu, Y., Yao, X., Higuchi, T.: Evolutionary ensembles with negative correlation learning. IEEE-EC 4(4), 380 (2000)
37. Ando, S.: Heuristic speciation for evolving neural network ensemble. In: GECCO 2007: Proc. of the 9th Conf. on Genetic and Evolutionary Computation, pp. 1766–1773. ACM, New York (2007)
38. Chandra, A., Yao, X.: Ensemble learning using multi-objective evolutionary algorithms. Journal of Mathematical Modeling and Algorithms 5(4), 417–445 (2006)
39. Pappa, G.L.: Automatically Evolving Rule Induction Algorithms with Grammar-based Genetic Programming. PhD thesis, Computing Laboratory, University of Kent, Canterbury, UK (2007)
40. Whigham, P.A.: Grammatical Bias for Evolutionary Learning. PhD thesis, School of Computer Science, University College, University of New South Wales, Australian Defence Force Academy, Canberra, Australia (1996)
41. Newman, D.J., Hettich, S., Blake, C.L., Merz, C.J.: UCI Repository of machine learning databases. University of California, Irvine (1998), `http://www.ics.uci.edu/~mlearn/MLRepository.html`

Structured Hidden Markov Model versus String Kernel Machines for Symbolic Sequence Classification

Ugo Galassi[2], Marco Botta[1], and Lorenza Saitta[2]

[1] Dipartimento di Informatica, Università di Torino
C.so Svizzera 185, 10149 - Torino, Italy
[2] Dipartimento di Informatica, Università Amedeo Avogadro
Via Bellini 25G, 15100 - Alessandria, Italy

Abstract. The problem of learning a classifier from examples is a fundamental task in Machine Learnig and is nowadays actively studied. Most approaches follow a *Discrimination* based paradigm, where the aim is to find the best way to separate the examples of one class from those of other classes. A less popular approach follows a *Characterization* based paradigm, where the aim is to build a description (model) of every class from examples, and use this model to recognizes instances of that class. This paper focuses on the specific task of classifying and tagging symbolic sequences, by introducing a Characterization approach, based on Structured Hidden Markov Models, and compares its perfomances against a widely used discriminative approach, i.e. kernel machines. This task is particulary relevant to several applications in many fields, such as molecular biology, web log analysis, network traffic analysis, user profiling, and so on. In order to assess the validity of the proposed approach, an artificial benchmark has been designed in such a way that the regularities to be discovered are well known and allow for a controlled evaluation of the real capabilities the learning algorithms investigated. The obtained results allow to point out the major advantages and weaknesses of the investigated approaches in the specific classification task addressed.

1 Introduction

The problem of inducing a classifier from examples is the first task that has been faced in Machine Learning, and is still actively investigated because of its key role in many important applications. Since the origins, two alternative approaches have been followed: *Discrimination* based versus *Characterization* based, which, depending on the cases, have been considered either as competitor or as complementary ones [4, 6].

Goal of this paper is to compare the two approaches with respect to the task of symbolic sequence classification. This task is particularly relevant because it finds application to important real world problems such as molecular biology [7, 13].

The characterization based approach tries to generate a complete model of the items belonging to the classes to discriminate accounting for all the features, which can be observed with a significant frequency. In the case of a set of sequences a characterizing model tries to reconstruct the generative process beyond the observed sequences. For this reason those models are usually referred as *Generative*. Classification is done by

J. Koronacki et al. (Eds.): Advances in Machine Learning I, SCI 262, pp. 275–295.
springerlink.com © Springer-Verlag Berlin Heidelberg 2010

comparing the likelihood of each model producing the observed sequence and assigning it to the most likely model. Instead, the discrimination based approach does not made any attempt to reconstruct the generative model. The focus here is on modeling class boundaries or learning a direct map from inputs to the class labels. The first approach has the advantage that each class is considered and trained, separately from the others. So these models are relatively easy to train. The second approach can take into account interdipendencies between classes and negative examples. This usually means that Discriminative models are less expensive in amount of *examples* required in order to obtain a statistically significant model.

In many practical cases the discriminative approach may be sufficient. Nevertheless, there are two cases in which modeling the generative process provides better results. The first is when the features which differentiate one class from another involve large portions of the sequence where the order of a set of events is important. The second is when there is a large number of classes, or one of the classes is an open class, i.e. a class where the sequences exhibit a large variability, impossible to capture with a small sample of observations.

A clarifying instance of this task, deeply investigated in pattern recognition, which clarifies the potentialities of the two alternative approaches is the recognition of isolated spoken words. When the vocabulary is small (few tens of words), gross features, like the *total energy* in the different regions of the signals, may be sufficient to solve this classification task without requiring to model the acoustic and phonetic events actually producing the observed signal. Therefore, in this case, the discrimination based approach is dominant.

Nevertheless, when the vocabulary becomes very large (several thousands of words), a generative model accounting for the single actions producing the phonemes composing the spoken word (i.e., a generative model) becomes necessary [21].

In order to compare the two approaches on sequence classification tasks, we decided to construct a benchmark, where the sequences have a structure resembling to the one of genomic sequences, while the regularities, which must be discovered, are well known. Therefore, it should be possible to check the real capabilities of the different algorithms, which frequently are in competitions in the real world. More specifically, we run a benchmark between the two top level techniques, representative of the two alternative approaches, which are currently competing in the domain od molecular biology: String kernels [16, 17, 23] and Hidden Markov Models [7].

String kernels are a discriminative method, which extend the approach of Support Vector machines introduced by V. Vapnik [25]. SVMs are well founded in terms of computational learning theory and very open to theorethical understanding and analysis. SVMs are well suited for sequences classification task because they can cope with high dimensional input spaces.

In text categorization tasks, for instance, it is necessary to cope with huge features spaces, often exceding the number of available training examples. In order to use conventional discriminative method many authors evidentiate how there is the need to make large use of feature selection methods [18, 27]. Nevertheless it is demonstrated that even features with low rank still contain considerable amount of information and are somewhat relevant [15].

For such reasons, among discriminative approaches, SVM offered, usually, the best results. But discriminative models like SVMs only account for differences emerging from features and model the differences observed between classes. They don't account for the evolution of a specific process. In real world, most systems are dynamic and are naturally described by features, whose values change significantly during an observation period. In presence of this kind of complex tasks, generative models can lead to better results.

Hidden Markov Model [21] is probably the most frequently used approach in DNA analysis [7] and belongs to the class of graphical model methods. Following this approach, a statistical approximation of a process is built; then, the process's behavior is described as a path in a *state space* governed by a set of probability distributions. The advantages of this approach are several. Firstly, it is not necessary to reconstruct the control algorithm in details, because those details can be hidden into probability distributions. Secondly, at least in principle, the statistical distributions governing the transitions between states can be estimated from a *learning set* of sequences collected from the past history. Finally, increasing the size of the *state space*, the accuracy of the model can be arbitrarily increased.

In the comparison presented in this paper, we use Structured Hidden Markov Models [11], a variant of Hidden Markov Model, which inherit its major advantages but exhibits a tractable complexity even when both the sequences and the model structure become very complex. A S-HMM is a graph built up, according to precise composition rules, with several independent sub-graphs (that are still S-HMMs). The compositional property of S-HMMs allow us to design a global model as a composition of different kinds of blocks individually learned. The major feature of S-HMMs is that they could be locally trained, considering only a sub-set of the whole model. For this reason they can be constructed and trained incrementally taming the complexity of the learning procedures.

The paper is organized as follows. In Section 2 and 3, we will introduce S-HMMs together with an algorithm called EDY, which is capable of automatically synthesizing S-HMMs from a set of learning sequences. In Section 4, we will briefly review the string kernel approach for sequence classification. In Section 5, we will present the benchmark used for an extensive comparison among the two methods. It consists of several sets of sequences artificially generated, which strongly resemble the ones obtained from the DNA databases. Section 6 presents the results of the experiment. Some conclusive remarks are reported in Section 7. The used datasets are available at *http://www.edygroup.di.unipmn.it.*

2 The Structured Hidden Markov Model

The basic assumption underlying an S-HMM (see Bouchaffra and Tan [3]) is that a sequence $O = \{O_1, O_2, O_3, ..., O_t\}$ of observations could be segmented into a set of subsequences $O^{(1)}, O^{(2)}, ..., O^{(N)}$, each one generated by a sub-process with only weak interactions with its neighbors. This assumption is realistic in many practical applications, such as, for instance, speech recognition [20, 21], and DNA analysis [7]. In speech recognition, regions are phonetic segments, like syllables, corresponding to recurrent structures in the language. In DNA, they may be biologically significant segments (*motifs*) interleaved with non-coding segments (such as *junk-DNA*). S-HMMs aim exactly at modeling such kind of processes, and, hence, they are represented as directed graphs, structured into sub-graphs (*blocks*), each one modeling a specific kind of sub-sequences.

Informally, a block consists of a set of states, only two of which (the *initial* and the *end* state) are allowed to be connected to other blocks. As an S-HMM is itself a block, a nesting mechanism is immediatly defined.

2.1 Structure of a Block

In this section, a formal definition of S-HMM will be provided. Let O denote a sequence of observations $\{O_1, O_2, \ldots, O_t\}$, where every observation O_t is a symbol V_k chosen from an alphabet of possible events V. An HMM is a stochastic automaton characterized by a set of states S, an alphabet V, and a triple $\lambda = \langle A, B, \Pi \rangle$, being:

- $A : S \times S \rightarrow [0, 1]$ a probability distribution, a_{ij}, governing the transition from state S_i to state S_j;
- $B : S \times V \rightarrow [0, 1]$ a probability distribution, $b_i(V_k)$, governing the emission of symbols in each state $S_i \in S$;
- $\Pi : S \rightarrow [0, 1]$ a distribution assigning to each state $S_i \in S$ the probability of being the start state.

A state S_i will be said a *silent* state if $\forall V_k \in V : b_i(V_k) = 0$, *i.e.*, S_i does not emit any observable symbol. When entering a silent state, the time counter must not be incremented.

In the framework provided by the classical HMM, we will now formally introduce the notion *block*, which is the basic construct of the structure of a S-HMM.

Definition 1. *A* basic block *of an S-HMM is a 4-tuple* $\lambda = \langle A, B, I, E \rangle$, *where* $I, E \in Q$ *are silent states such that:* $\pi(I) = 1$, $\forall S_i \in S : a_{iI} = 0$, *and* $\forall S_i \in S : a_{Ei} = 0$.

In other words, I and E are the input and the output states, respectively. Therefore, a composite block can be defined by connecting, through a transition network, the input and output states of a set of blocks.

Definition 2. *Given an ordered set of blocks* $\Lambda = \{\lambda_i | 1 \leq i \leq N\}$, *a* composite block *is a 4-tuple* $\lambda = \langle A_I, A_E, I, E \rangle$, *where:*

- $A_I : \mathbf{E} \times \mathbf{I} \rightarrow [0, 1]$, $A_E : \mathbf{I} \times \mathbf{E} \rightarrow [0, 1]$ *are probability distributions governing the transitions from the output states* **E** *to the input states* **I**, *and from the input states* **I** *to the output states* **E** *of the component blocks* Λ, *respectively.*
- *For all pairs* $\langle E_i, I_j \rangle$ *the transition probability* $a_{E_i I_j} = 0$ *if* $j \leq i$.
- $I \equiv I_1$ *and* $E \equiv E_N$ *are the input and output states of the composite block, respectively.*

According to Definition 2 the components of a composite block can be either basic blocks or, in turn, composite blocks. In other words, composite blocks can be arbitrarily nested. Moreover, we will keep the notation S-HMM to designate non-basic blocks only.

As a special case, a block can degenerate to the *null block*, which consists of the start and end states only, connected by an edge with probability $a_{IE} = 1$. The *null block* is useful to provide a dummy input state I or a dummy output state E, when no one of the component block is suited to this purpose.

An example of S-HMM structured into three blocks λ_1, λ_2, λ_3, and two *null blocks* λ_0, λ_4, providing the start and the end states, is depicted in Figure 1.

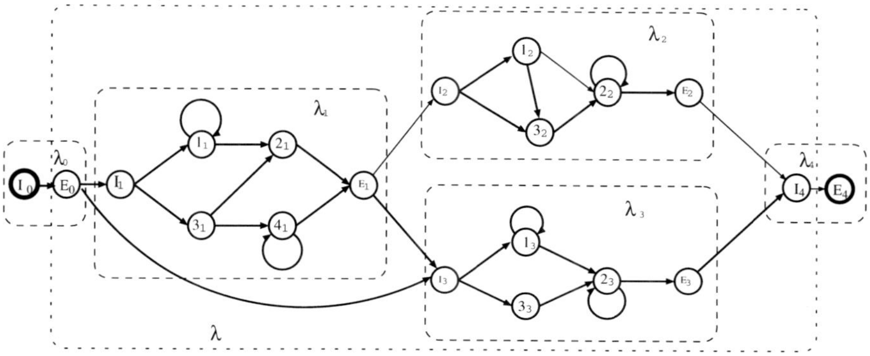

Fig. 1. Example of Structured Hidden Markov Model composed of three interconnected blocks, plus two *null blocks*, λ_0 and λ_4, providing the start and end states. Distribution A is non-null only for explicitly represented arcs.

2.2 Estimating Probabilities in S-HMM

As formally stated in [21], three problems are associated with the HMM approach:

1. given a model λ and a sequence of observations O, compute the probability $P(O|\lambda)$;
2. given a model λ and a sequence of observations O, assumed to be generated by λ, compute the most likely sequence of states in λ;
3. given a model λ and a sequence of observations O (or a set of sequences [20]), estimate the parameters in λ in order to maximize $P(O|\lambda)$.

The classical solution to Problem 1 and 3 relies on two functions α and β, plus other auxiliary functions γ and ξ, defined on α and β. The classical solution to Problem 2 relies on Viterbi algorithm [8], which implements a function computationally analogous to α. In the following we will extend α and β to S-HMMs in order to prove some properties exploited by an incremental learning algorithm.

Given a sequence of observations $O = \{O_1, O_2, ..., O_t, .., O_T\}$ and a model λ, the function $\alpha_i(t)$ computes, for every time t $(1 \leq t \leq T)$, the joint probability of being in state S_i and observing the symbols from O_1 to O_t.

Let us consider an S-HMM λ containing N blocks. We want to define the recursive equations allowing α to be computed. In order to do this, we have to extend the standard definition in order to include silent states: when leaving a silent state, the time counter is not incremented. When entering the block $\lambda_k = \langle A_k, B_k, I_k, E_k \rangle$ with N_k states, at time $r(1 \leq r < t \leq T)$, the following equations are to be used:

$$\alpha_{I_k}(r) = P(O_1, ..., O_r, S_i(r) = I_k) \tag{1a}$$

$$\alpha_j(t) = \alpha_{I_k}(t-1)a^{(k)}_{I_kj}b^{(k)}_j(O_t) + \sum_{i=1}^{N_k} \alpha_i(t-1)a^{(k)}_{ij}b^{(k)}_j(O_t)$$
$$(r+1 \leqslant t \leqslant T, 1 \leqslant j \leqslant N_k, S_j \neq I_k, S_j \neq E_k) \tag{1b}$$

$$\alpha_{E_k}(t) = \alpha_{I_k}(t) + \sum_{i=1}^{N_k} \alpha_i(t)a^{(k)}_{jE_k} \tag{1c}$$

Notice that the above equations only depend upon external states through the values of $\alpha_{I_k}(r)$ $(1 \leqslant r \leqslant T)$ computed for the input state; moreover, the block propagates $\alpha_{E_k}(t)$ $(1 \leqslant t \leqslant T)$ to the following blocks only through the output state. Finally, $\alpha_{I_1}(1) = 1$ and $\alpha_{E_N}(T) = P(O|\lambda)$.

Function $\beta_i(t)$ is complementary to $\alpha_i(t)$, and computes the probability of observing the symbols o_{t+1}, o_{t+2}, ... , o_T, given that S_i is the state at time t. For β a backward recursive definition can be given:

$$\beta_{E_k}(r) = P(o_{r+1}, ..., o_T | S_i(r) = E_k) \tag{2a}$$

$$\beta_i(t) = \beta_{E_k}(t+1)a^{(k)}_{iE_k} + \sum_{j=1}^{N_k} \beta_j(t+1)b^{(k)}_j(o_{t+1})a^{(k)}_{ij}$$
$$(1 \leqslant t \leqslant r-1, 1 \leqslant i \leqslant N_k, S_i \neq E_k, S_i \neq I_k) \tag{2b}$$

$$\beta_{I_k}(t) = \beta_{E_k}(t) + \sum_{j=1}^{N_k} \beta_j(t)a^{(k)}_{I_kj} \tag{2c}$$

From equations (2), it follows that $P(O|\lambda) = \beta_{I_1}(1)$.

Definition 3. *An S-HMM is said a forward S-HMM when for all non-basic blocks the matrices A_I and A_E define a directed acyclic graph.*

For a forward S-HMM it is easy to prove the following theorem.

Theorem 1. *In a forward S-HMM, the complexity of computing functions α and β is:*

$$C \leq T(\textstyle\sum_{h=1}^{N_C} N_h^2 + M \sum_{k=1}^{N} N_k^2)$$

being N_h the dimension of matrix $A_I^{(h)}$ of the hth block, M the cardinality of the alphabet, N_C the number of composite blocks, and N the number of basic blocks.

Proof. Notice that the second summation in the right-hand side of the formula corresponds to the computation of α and β inside the basic blocks, whereas the first summation is due to the block interconnections. Following the block recursive nesting and starting from the basic blocks, we observe that, in absence of any hypothesis on distribution A, each basic block is an HMM, whose complexity for computing α and β is upperbounded by N_k^2MT [21]. As the global network interconnecting the basic blocks is a directed forward graph, every basic block needs to be evaluated only once.

Let us consider now a composite block; the interconnecting structure is an oriented forward graph, by definition, and, then, equations (1) and (2) must be evaluated only once on the input (output) of every internal block S-HMM$_h$. As a conclusion, the complexity for this step is upperbounded by TN_h^2.

This is an upper bound on the calculation of functions α and β on a generic S-HMM. We can constrain the structure of the basic blocks adopting models with a constant branching factor, so the complexity will become quasi-linear in the number of states.

2.3 S-HMMs Are Locally Trainable

The classical algorithm for estimating the probability distributions governing state transitions and observations are estimated by means of the Baum-Welch algorithm [1, 21], which relies on functions α and β defined in the previous section. In the following we will briefly review the algorithm in order to adapt it to S-HMMs. The algorithm uses two functions, ξ and γ, defined through α and β. Function $\xi_{i,j}(t)$ computes the probability of the transition between states S_i (at time t) and S_j (at time $t+1$), assuming that the observation O has been generated by model λ:

$$\xi_{i,j}(t) = \frac{\alpha_i(t)a_{ij}b_j(O_{t+1})\beta_j(t+1)}{P(O|\lambda)} \tag{3}$$

Function $\gamma_i(t)$ computes the probability of being in state S_i at time t, assuming that observation O has been generated by model λ, and can be written as:

$$\gamma_i(t) = \frac{\alpha_i(t)\beta_i(t)}{P(O|\lambda)} \tag{4}$$

The sum of $\xi_{i,j}(t)$ over t estimates the number of times transition $S_i \rightarrow S_j$ occurs when λ generates the sequence O. In an analogous way, by summing $\gamma_i(t)$ over t, an estimate of the number of times state S_i has been visited is obtained. Then, a_{ij} can be re-estimated (a-posteriori, after seeing O) as the ratio of the sum over time of $\xi_{i,j}(t)$ and $\gamma_i(t)$:

$$\bar{a}_{ij} = \frac{\sum_{t=1}^{T-1} \alpha_i(t)a_{ij}b_j(O_{t+1})\beta_j(t+1)}{\sum_{t=1}^{T-1} \alpha_i(t)\beta_i(t)} \tag{5}$$

With a similar reasoning it is possible to obtain an a-posteriori estimate of the probability of observing $o = v_k$ when the model is in state q_j. The estimate is provided by the ratio between the number of times state q_j has been visited and symbol v_k has been observed, and the total number of times q_j has been visited:

$$\bar{b}_j(k) = \frac{\sum_{\substack{t=1 \\ o_t=v_k}}^{T-1} \alpha_j(t)\beta_j(t)}{\sum_{t=1}^{T-1} \alpha_j(t)\beta_j(t)} \tag{6}$$

From (1) and (2) it appears that, inside basic block λ_k, equations (5) and (6) are immediately applicable. Then, the Baum-Welch algorithm can be used without any change to learn the probability distributions inside basic blocks.

On the contrary, equation (5) must be modified in order to adapt it to re-estimate transition probabilities between output and input states of the blocks, which are silent states. As there is no emission, α and β propagate through transitions without time change; then, equation (5) must be modified as in the following:

$$\bar{a}_{E_i I_j} = \frac{\sum_{t=1}^{T-1} \alpha_{E_i}(t) a_{E_i I_j} \beta_{I_j}(t)}{\sum_{t=1}^{T-1} \alpha_{E_i}(t) \beta_{I_j}(t)} \tag{7}$$

It is worth noticing that functions α and β depend upon the states in other blocks only through the value of $\alpha_{I_k}(t)$ and $\beta_{E_k}(t)$, respectively. This means that, in block λ_k, given the vectors

$$\alpha_{I_k}(1), \alpha_{I_k}(2), ...\alpha_{I_k}(T)$$

and

$$\beta_{E_k}(1), \beta_{E_k}(2), ...\beta_{E_k}(T)$$

Baum-Welch algorithm can be iterated inside a block without the need of recomputing α and β in the external blocks. We will call this a *locality property*. The practical implication of the locality property is that a block can be modified and trained without any impact on the other components of an S-HMM.

3 EDY: An Algorithm for Learning S-HMMs

We will briefly describe Edy, an algorithm capable of inducing a S-Hmm from a set of examples, which will be used for the benchmark presented in the following.

Classical HMM applications rely upon the Baum-Welch algorithm to tune the model parameters, trying to maximize the probability $P(O \mid \lambda)$ when the sequence O belongs to the target class, while $P(O \mid \lambda)$ remains as low as possible when O does not belong to the target class.

The typical assumption is that, potentially, transitions are possible from any state to any other state. Althought attractive, the strategy of starting with a fully connected model, i.e. one in which all transitions are allowed, and letting the Baum-Welch algorithm refining the parameters of model, it is unfeasible in practice. In fact, for problems of realistic size, it will usually lead to very bad models, even when the training set is very large. This is due not to overfitting but to the large number of local maxima potentially entrapping the training algorithm [7]. The strategy adopted in many applications is to adapt the model topology by adding or removing transitions and states according to some iterative procedures [9, 10, 24]. However, in practice, successful HMMs are constructed by carefully deciding which transitions are to be allowed in the model, based on the knowledge about the problem under investigation.

In our approach, the S-HMM structure can change from one block to another. Constraining the structure of sub-models is helpful in avoiding local maxima. Besides, reducing the number of possible transitions in a basic S-HMM block, will also reduce the computational cost of the learning algorithm. Since the amount of free parameters and the amount of computation are directly dependent on the number of non-zero transitions probabilities, keeping low the number of possible transitions implicitly reduces the computational cost.

3.1 The Event Discovery Algorithm

The EDY (*E*vent *D*iscover*Y*) algorithm exploits co-occurrence of regularities in bounded regions of a sequence in order to detect short motifs. The S-HMM λ of a sparse pattern is constructed incrementally, starting from a learning set *LS* of sequences, going through a learning cycle in which a model is progressively extended and refined, by repeatedly incorporating new motifs and gaps. The rationale behind this architecture is that regularities due to the presence of motifs may be difficult (or impossible) to distinguish from randomness when considered in isolation, but may become evident in the context established by a partial model. Therefore the algorithm tries to discover first the motifs, which are evident in absence of any a priori information. Then using such motifs it builds up a first model which is augmented cycle after cycle by adding new motifs as long as they become detectable.

The cycle may initiate with an *empty model* or with a model supplied by an expert of the domain, and terminates when there is no more evidence of new motifs to incorporate. The basic steps for extending λ encompasses the following actions:

1. Find a not yet analyzed basic block λ_k encoding a gap.
2. For each sequence $O \in LS$ determine the most likely subsequence encoded by λ_k, using the Viterbi algorithm [21]. If such a subsequence exists, insert the observed subsequence into a temporary learning set LS_k.
3. Search for regularities (motifs) occurring in the sequences accumulated in LS_k. If no new motifs are discovered, then exit, otherwise build a composite block λ_k^1 containing the new motifs and replace λ_k with λ_k^1.
4. Locally train λ_k^1 using the Baum-Welch algorithm, and validate the final model according to some test. If the test is successful, then continue, otherwise restore model λ_k and exit.

The above procedure is iterated until no *gap* blocks remain unexplored. At every step the algorithm can use a set of refinement operators in order to refine *motif* and *gap* blocks. An overview of these operators is presented in the next subsection.

3.2 Model Refinement

As the mechanism exploited by the model extension procedure is rather primitive, at every step only a single motif model is added to the S-HMM. Then, the model refinement procedure reconstructs motif and gap models , until convergence on stable models is achieved.

The starting point is the tagged dataset $\mathcal{LS}(\lambda_t)$ constructed before calling the refinement operator. All sequence segments corresponding to motif and gap instances, which have been detected by the Viterbi algorithm, are collected into a two level hierarchical clustering. The clusters associated to the leaves correspond to states at the level γ of the model. Each leaf contains all the subsequences which have been emitted by the model μ (motif or gap) when the S-HMM was in the corresponding state. However, emissions in different states can be generated by the same motif/gap model. Then, the clusters at the second level group together the leaves whose elements are generated by the same model μ, but in different states. The root node of the tree is a dummy node,

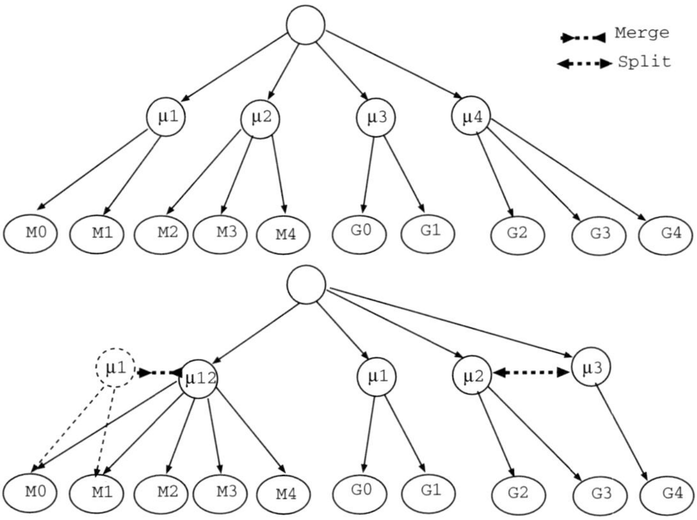

Fig. 2. Example of cluster hierarchy. Leaves corresponds to the states of the level γ, whereas second level nodes correspond to models μ of motifs and gaps.

representing the whole set of segments. During the refinement process, second level clusters can be split or merged (see right part of Figure 2), thus increasing or decreasing the set of existing motif/gap models. Given a distance measure between instances (the edit distance in the specific case), two clusters of motif/gap instances are merged if the distance between their centers is not greater than their average intra-cluster distance. Alternatively, a cluster, whose children have an intra-cluster distance much smaller than the inter-cluster distance, may be split.

The specific operators, which are applied in a refinement step, are described in the following.

Boundary refinement - This operator is meant to correct possible segmentation errors performed during the initial learning phase. Before trying to refine a motif model, the algorithm for searching local alignments is run on the new set of instances, but allowing the alignments to possibly extend into the adjoining gap regions for one or two positions. Instances of the motif can thus be extended (or reduced) if the original segmentation is found inaccurate. However, this operator is only applied a few times when a new motif is constructed, because, in the long term, it can cause instability.

Model diversification - If μ is a model associated to two different states M_j, M_k of level γ, and the two associated instance clusters C_j and C_k significantly differ, then μ is split into μ_j and μ_k, which are trained on C_j and C_k, respectively.

Model unification - When two models μ_j and μ_k have children that cannot be distinguished among themselves according to the distance criterion, the models can be merged into a single one, μ, whose parameters can be estimated from the cluster obtained as union of μ_j and μ_k's children. The procedure for merging gap models is analogous, but based on a different criterion. More precisely, considering two clusters

C_j and C_k of gap instances, the histograms h_j and h_k of the corresponding gap lengths are constructed. Histograms are compared among each other, and "similar" ones are merged. This operator is only activated optionally, as it may slow down convergence to a stable hierarchy.

Parameter refinement - As the instances of a model may be currently different from those used to initially learn it, the model's parameters are re-estimated from the new set of instances.

Gap model refinement - This operator is similar to the preceding one, except that the parameters to be estimated are those appearing in the distribution of the gap lengths.

Hierarchy Revision. The algorithm for the construction/reconstruction of the level γ of the S-HMM is very similar to the one that constructs the motif models. The difference is that it works on the abstracted sequences belonging to $\mathcal{LS}'$.

As the above algorithm is computationally inexpensive, it is repeated at every refinement step, in order to propagate to the upper level changes in the structure at the lower level.

4 Support Vector Machines

This section reviews the main ideas behind Support Vector Machines (SVMs) and kernel functions. SVMs are a class of algorithms that combine the principles of statistical learning theory with optimisation techniques and the idea of a kernel mapping. The idea of maximum margin has been introduced by Boser et al. [2], and in their simplest version they learn a separating hyperplane between two sets of points so as to maximise the distance between the separation plane and closest points of the classes to discriminate. The idea has been then elaborated by Cortes and Vapnik [5], which introduced the extension of kernel mapping. This solution has several interesting statistical properties, that make it a good candidate for valid generalisation. One of the main statistical properties of the maximal margin solution is that its performance does not depend on the dimensionality of the space where the separation takes place. In this way, it is possible to work in very high dimensional spaces, such as those induced by kernels, without overfitting.

In the classification case, SVMs work by mapping the data points into a high dimensional feature space, where a linear learning machine is used to find a maximal margin separation. In the case of kernels defined over a space, this hyperplane in the feature space can correspond to a nonlinear decision boundary in the input space.

In the case of kernels defined over sets, this hyperplane simply corresponds to a dichotomy of the input set. We now briefly describe a kernel function. A function that calculates the inner product between mapped examples in a feature space is a kernel function, that is for any mapping

$$\phi : D \to F, K(d_i, d_j) =< \phi(d_i), \phi(d_j) >$$

is a kernel function. Note that the kernel computes this inner product by implicitly mapping the examples to the feature space. The mapping ϕ transforms an n dimensional example into an N dimensional feature vector.

$$\phi(d) = (\phi_1(d), ..., \phi_N(d)) = (\phi_i(d)) \; for \; i = 1, ..., N$$

The explicit extraction of features in a feature space generally has very high computational cost but a kernel function provides a way to handle this problem. The mathematical foundation of such a function was established during the first decade of twentieth century (Mercer, 1909). A kernel function is a symmetric function,

$$K(d_i, d_j) = K(d_j, d_i),\ for\ i, j = 1, ..., n.$$

The n × n matrix with entries of the form $K_{ij} = K(d_i, d_j)$ is known as the kernel matrix. A kernel matrix is a symmetric, positive definite matrix. It is interesting to note that this matrix is the main source of information for KMs and these methods use only this information to learn a classifier. There are ways of combining simple kernels to obtain more complex ones. For example given a kernel K and a set of n vectors the polynomial construction is given by

$$K_{poly}(d_i, d_j) = (K(d_i, d_j) + c)^p$$

where p is a positive integer and c is a nonnegative constant. Clearly, we incur a small computational cost, to define a new feature space. The feature space corresponding to a degree p polynomial kernel includes all products of at most p input features. Hence polynomial kernels create images of the examples in feature spaces having huge numbers of dimensions. Furthermore, Gaussian kernels define feature space with infinite number of dimension and it is given by

$$K_{gauss}(d_i, d_j) = e^{(\frac{-\|d_i - d_j\|^2}{2\sigma^2})}$$

A Gaussian kernel allows an algorithm to learn a linear classifier in an infinite dimensional feature space.

Lodhi et al. ([16]) introduced a new kernel function to map sequences to feature vector, named as string subsequence kernel (SSK), that is based on work ([14, 26]) mostly motivated by bioinformatics applications. It maps strings to a feature vector indexed by all k tuples of characters. A k-tuple will have a non-zero entry if it occurs as a subsequence anywhere (not necessarily contiguously) in the string. The weighting of the feature will be the sum over the occurrences of the k-tuple of a decaying factor of the length of the occurrence. More formally, we have the following defintion:

Definition 4 ***String subsequence kernel(SSK).*** *Let Σ be a finite alphabet. A string is a finite sequence of characters from Σ, including the empty sequence. For strings s, t, let us denote by $|s|$ the length of the string $s = s_1 \dots s_{|s|}$, and by st the string obtained by concatenating the strings s and t. The string $s[i : j]$ is the substring $s_i \dots s_j$ of s. We say that u is a subsequence of s, if there exist indices $i = (i_1, \dots, i_{|u|})$, with $1 \le i_1 < \dots < i_{|u|} \le |s|$, such that $u_j = s_{i_j}$, for $j = 1, \dots, |u|$, or $u = s[i]$ for short. The length $l(i)$ of the subsequence in s is $i_{|u|} - i_1 + 1$. We denote by Σ^n the set of all finite strings of length n, and by Σ^* the set of all strings*

$$\Sigma^* = \bigcup_{n=0}^{\infty} \Sigma^n$$

We now define feature spaces $F_n = \Re^{\Sigma^n}$. *The feature mapping* ϕ *for a string* s *is given by defining the* u *coordinate* $\phi_u(s)$ *for each* $u \in \Sigma^n$. *We define*

$$\phi_u(s) = \sum_{i:u=s[i]} \lambda^{l(i)}$$

for some $\lambda \leq 1$. *These features measure the number of occurrences of subsequences in the string* s *weighting them according to their lengths. Hence, the inner product of the feature vectors for two strings* s *and* t *give a sum over all common subsequences weighted according to their frequency of occurrence and lengths*

$$K_n(s,t) = \sum_{u \in \Sigma^n} \phi(s) \cdot \phi(t) = \sum_{u \in \Sigma^n} \sum_{i:u=s[i]} \sum_{j:u=t[j]} \lambda^{l(i)+l(j)}$$

5 Evaluating S-HMM against String Kernels

In this section, an extensive empirical comparison between S-HMM and String Kernel approaches is presented.

5.1 Experimental Enviroment

In order to compare String Kernels versus the Structured Hidden Markov Models we designed different groups of artificial benchmarks of growing difficulty. The goal was to compare the classification power of the two methods on symbolic sequences, having a structure similar to the ones observed in molecular biology tasks. This kind of sequences can be very long and can be seen as a structured sequence of motifs, i.e. short sequences that correspond to recurring characteristic patterns, interleaved with gaps. Gaps could be seen as sub-strings corresponding to noise or to erratic behaviour.

In these kinds of tasks different classes can share some or all of the motifs. The discrimination is often based only on the order in which the motifs occur and on the intra-motifs distance, i.e. the length of gaps occurring between the same motifs.

The two benchmarks are referred in the follow as (1) the *cities* datasets and (2) the *pseudo-genomic* datasets. The first group investigates how the behaviour of the algorithms is affected by the length of sequences and by the intra-motifs distance with motifs of fixed length. The second group aims at checking how the classification precision of the two approaches is affected by the length of the motifs to discover, and by the cardinality of alphabet encoding the sequences.

These datasets are available, for the Machine Learning community, on the web[1].

5.2 Comparison on "Cities" Datasets

This group of datasets presents two classes that share the identical two motifs, corresponding to name of towns 6 characters long. The only difference between the positive and negative classes is the order in which motifs occur.

[1] http://www.edygroup.di.unipmn.it

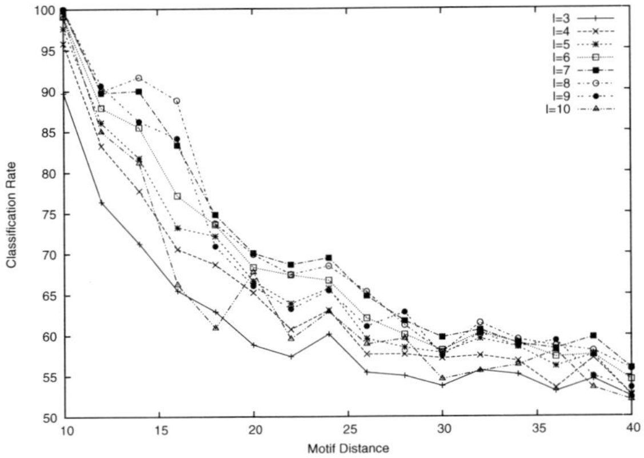

($\lambda = 0.5$)

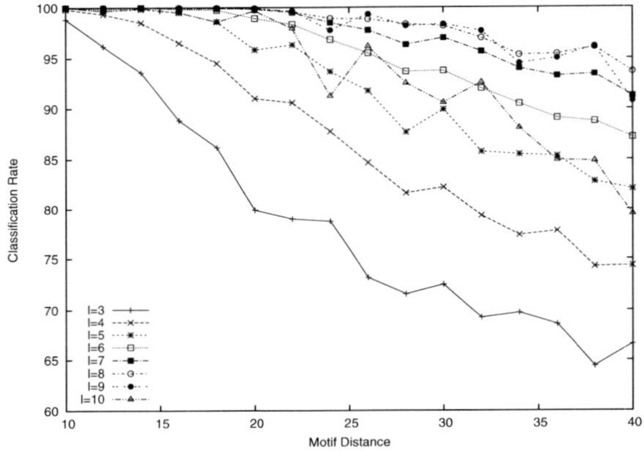

($\lambda = 0.8$)

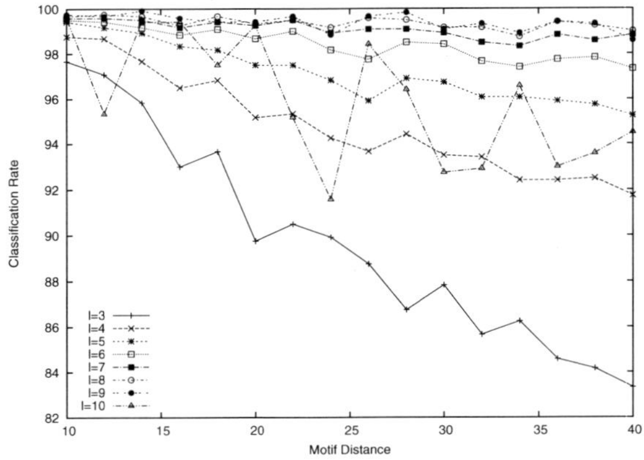

($\lambda = 0.9$)

Fig. 3. Classification rates on *Cities* benchmark obtained considering different values for λ

Globally, 96 different datasets have been generated by varying the intra-motifs distance ($0 \leq I \leq N-6$), where N ranges from 10 to 40 at steps of 2, and the sequences length M that ranges from 50 to 100 at steps of 10.

Each dataset is constituted by a learning set and a test set of 200 sequences equally distributed on the two classes.

A crucial part in Kernel approach, in order to obtain good results, is the correct tuning of algorithm parameters. This means finding the right value for the sub-string length l and for λ according to Definition 4. The right values are usually task dependent and finding them is not always a trivial task.

Comparing String Kernels versus S-HMM on *Cities* datasets we tried different values for l ($3 \leq l \leq 10$) and λ ($\lambda \in \{0.5, 0.8, 0.9\}$). As we can immediately observe from the graphs presented in Figure 3 the average better results have been obtained with $\lambda = 0.9$ and $l = 8$.

We do not show any graph about the results obtained with S-HMMs because the classification rates were in all cases 100% correct classifications.

A first difference that arise from this analysis is that S-HMM approach does not need particular tuning in order to reach good results. Nevertheless a wrong choice in learning parameter will lead String Kernel approach toward very awful results. For those reasons it is evident that S-HMM approach outperforms String Kernel on *Cities* datasets.

5.3 Comparison on "Pseudo-genomic" Datasets

This benchmark includes 96 classification problems subdivided in two groups (namely A2 and B3). For each dataset we have three classes characterized by a chain of motifs separated by gaps of varying length plus an initial and final random gap. Motifs are different for each class. Sequences belonging to A2 are generated according to a structure of three motifs appearing always in the same order. Sequences in B3 are generated according to a chain structure with forward jumps that produces sequences with a varying number of motifs (from two to six).

Using a semi-automated procedure, 32 template models (16 for each group) have been constructed; they differ in the nominal length of the motifs (5, 8, 11, 15 symbols) and in the cardinality of the alphabet (4, 7, 14, 25 symbols). From each template three different generative models (Hidden Markov Models) have been obtained, with an increasing level of perturbation on the probability distributions governing the transitions from state to state and the observations generated inside states. Considering the average edit distance δ_E between the maximum likelihood sequences generated by the model without perturbations, and the maximum likelihood sequences generated by

Table 1. Average value of parameter η and of the sequence length l in datasets A2, and B3

	N1		N2		N3	
	$\overline{\eta}$	$\overline{l}$	$\overline{\eta}$	$\overline{l}$	$\overline{\eta}$	$\overline{l}$
A2	0.186	414	0.235	388	0.256	385
B3	0.156	330	0.194	318	0.209	318

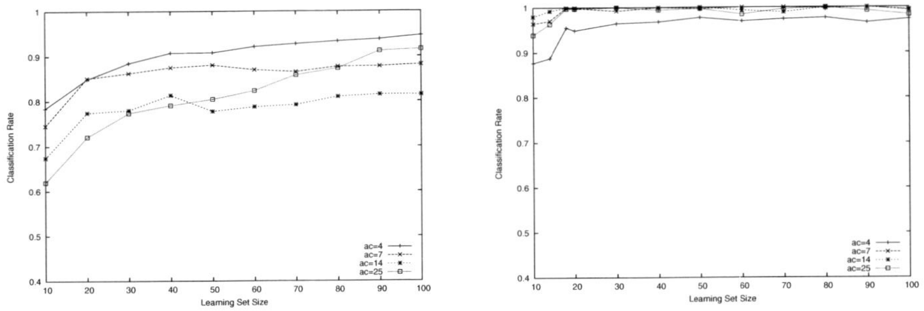

String Kernel S-HMM

AC

Fig. 4. Performances on *A2* datasets. The plot reports the classification rates on the test set versus the size of the learning set. 4 curves are plotted, one for each value of alphabet cardinality.

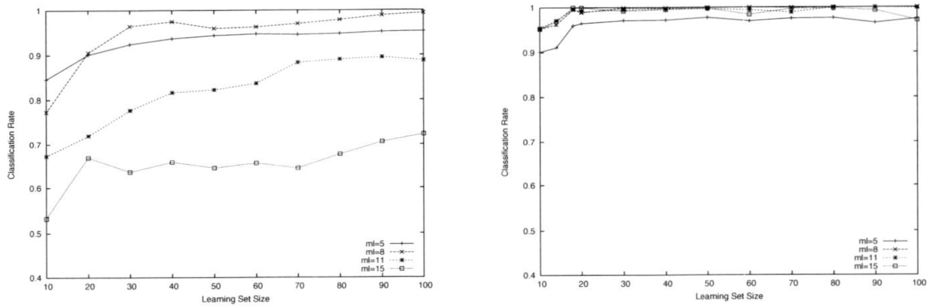

String Kernel S-HMM

ML

Fig. 5. Performances on *A2* datasets. The plot reports the classification rates on the test set versus the size of the learning set. 4 curves are plotted, one for each value of motif length.

the perturbed models the following average values have been obtained for the four class of distributions:

Class:	N1	N2	N3
δ_E:	0.11	0.19	0.28

Notice that also the gap length spread is strongly affected by the increase in the distribution spread, even if it is not accounted for in the measures reported above. For every setting of the above parameters we have generated three different models λ_n^{Ξ} (where n ranges from 1 to 3 and Ξ correspond to a specific combination of parameters) corresponding to the three classes that need to be discriminated. Finally, for every model λ_n^{Ξ}, a learning set $LS_{\lambda_n^{\Xi}}$ and a test set $TS_{\lambda_n^{\Xi}}$, each containing 100 sequences per class, have been generated.

The length of each sequence ranges from 150 to 500 symbols depending on the models. It is worth noticing that, considering the quite short motif length, the coding

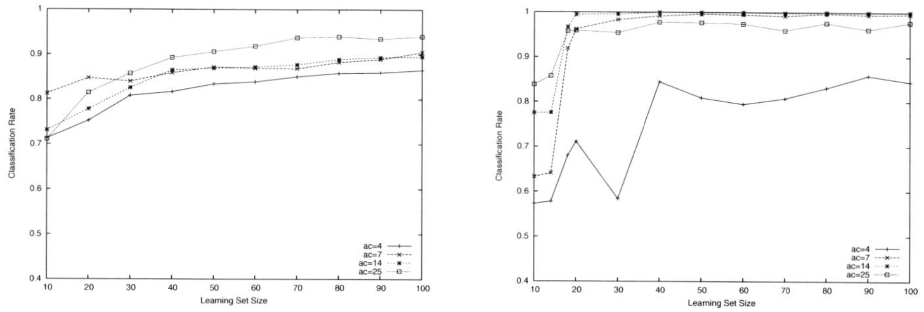

String Kernel S-HMM

AC

Fig. 6. Performances on *B3* datasets. The plot reports the classification rates on the test set versus the size of the learning set. 4 curves are plotted, one for each value of alphabet cardinality.

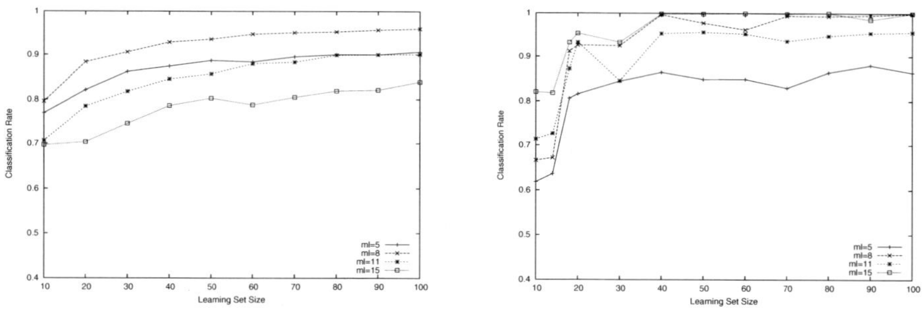

String Kernel S-HMM

ML

Fig. 7. Performances on *B3* datasets. The plot reports the classification rates on the test set versus the size of the learning set. 4 curves are plotted, one for each value of motif length.

part is much smaller than the non coding part making the task of discovering motifs quite difficult. Table 1 reports the average percentage η of the coding part over the total length of the sequences for all the datasets. Globally 96 different datasets, 48 for each kind of structure have been generated.

The analysis on pseudo-genomic datasets was focused on the dependence of the performances upon the size the learning set. For both approaches we varied the learning set size of each class, from 10 to 100 sequences at steps of 10 whilst the test set size remained fixed at 100 sequences. For the String Kernels, we considered only those values for λ and l that have performed the best results, according to analysis presented in Section 5.2 ($\lambda = 0.9$ and $l = 8$).

The results presented in Figures 4 and 5 refer to the benchmark on A2 datasets. More specifically, Figure 4 shows the dependence of the accuracy on the Alphabet Cardinality (AC); the values have been averaged on the motif length. Instead, Figure 5 shows

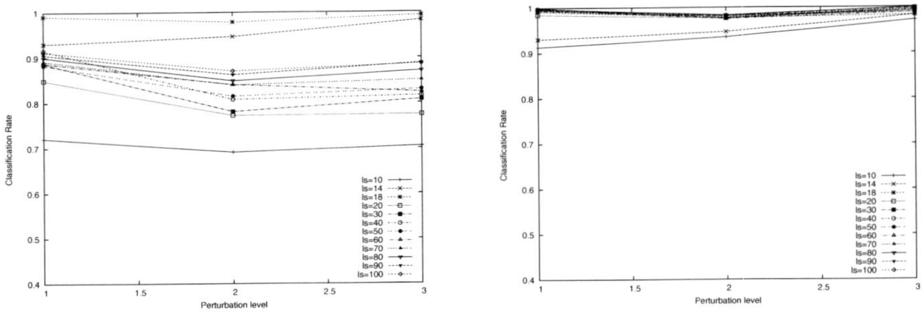

String Kernel S-HMM

ML

Fig. 8. Performances on *A2* datasets. The plot reports the classification rates on the test set versus the perturbation level. The graph presents a different plot for any learning set size.

the dependence of the accuracy on the motive length (ML); the values have been averaged on the alphabet cardinality.

In analogous way, Figures 6 and 7 report the results obtained on B3 datasets.

We do not show a detailed analysis of performance depending on the perturbation rate. In fact there is no evidence of a strong correlation between performances and perturbation rates as we can see from Figure 8. For the sake of simplicity we present only the graph for A2, nevertheless the results obtained with datasets in B3 are very similar. This fact is not surprising and only point out that both approaches cope well with the typical insertion, deletion and substitution errors that we can observe in biological symbolic sequences.

On the other hand, as one may expect, we observe that for both approaches increasing the size of the learning set leads to better results. However, the String kernel approach shows a strong decrease in performance when the motif length grows up. This is due to the bias imposed by the initial choice for the value of l. One may try to tune the value of l on the basis of the expected motif length but this might be difficult to guess and, moreover, may overfit the data. In fact, in most biological tasks, we can only have a very rough apriori estimate of this value.

Moreover, the alphabet cardinality affects the performances of the S-HMM approach. With very small alphabet cardinalities, especially in conjunction with short motifs, the learning algorithm fails at discovering all the motifs describing the events.

This decrease in performance of S-HMM is particularly evident in datasets belonging to B3. The reason is easy to understand. In B3, the number of motifs appearing in each sequence is not fixed. If the learning algorithm fails in discovering all the motifs we can have sequences where not all motifs necessary for the classification have been detected. Despite these observations S-HMM outperformed String Kernel in most cases also with very small learning sets. String Kernel approach shows slightly better results only with really small learning set (10 or less sequences per class). This is not surprising because discriminative models only have to learn the *differences* between classes, making use of positive and negative examples whilst generative models need to infer the process that generates the sequences.

6 Conclusion

We presented a comparison between a generative and a discriminative approach for symbolic sequences classification. As stated in Section 1 we focused on the most promising metodologies, commonly used in bioinformatics, i.e. String Kernel and Hidden Markov Models [7, 16, 17, 23].

We considered a variant of Hidden Markov Model called Structured Hidden Markov Model[12] which inherits the major advantages of HMM but exhibits a tractable complexity, scaling well also on very complex tasks.

A first observation is that, on the benchmarks we constructed, the S-HMMs outperformed String Kernels. The results show a strict superiority of the presented generative approach, both in terms of accuracy and in terms of computing times, when the sequences grow up in complexity and length.

We do not claim that the presented results demonstrate the superiority of generative models versus discriminative models. Nevertheless, there is a clear evidence that in the considered types of sequences the formers proved to be more effective. Moreover, it appears that, contrary to a widely-held belief, discriminative training may not always lead to the best classifiers. The question of whether generative or discriminative algorithms are more efficient has been analysed in [19, 22]. It seems that sometimes using both the positive and negative samples can be useful especially in presence of small learning sets. However, generative learning can sometimes be more efficient than discriminative learning; at least when the number of features is large compared to the number of samples and when the model considered fits well the data.

Another consideration is that String Kernels do not scale up when length the sequences and the size of the learning set grows up. This is not surprising because discriminative approaches try to model the class boundaries or the class membership probabilities directly. This requires to process simultaneously the sequences of all the classes making this kind of models harder to train. Instead, the generative approach used with S-HMMs only requires to learn one model at a time for each one of the classes. Then, the classification is done by comparing the likelihood of each class model producing the observed sequence and assigning this to the most likely one. Consequently models are learned independently from each others making the training step easier. It is important to consider that this mean also that adding new classes only require to build models for the new added classes. Then, it is not necessary to re-train the models that have been previously learned.

As a conclusion of this empirical analysis we could assert that on complex symbolic sequences, such as DNA and protein sequences, generative approaches seem to be preferrable. Especially those based on stochastic modeling like Hidden Markov Models.

References

1. Baum, L.E., Petrie, T., Soules, G., Weiss, N.: A maximisation techniques occurring in the statistical analysis of probabilistic functions of markov chains. The Annals of Mathematical Statistics 41(1), 164–171 (1970)
2. Boser, B.E.: A training algorithm for optimal margin classifiers. In: Proceedings of the 5th Annual ACM Workshop on Computational Learning Theory, pp. 144–152. ACM Press, New York (1992)

3. Bouchaffra, D., Tan, J.: Structural hidden markov models using a relation of equivalence: Application to automotive designs. Data Mining and Knowledge Discovery 12, 79–96 (2006)
4. Carbonell, J.G., Michalski, R.S., Mitchell, T.: An overview of machine learning. In: Carbonell, J.G., Michalski, R.S., Mitchell, T. (eds.) Machine Learning, an Artificial Intelligence Approach. Morgan Kaufmann, San Francisco (1983)
5. Cortes, C., Vapnik, V.: Support vector networks. Machine Learning, 273–297 (1995)
6. Davidson, P.: Integrating models of discrimination and characterization. Data Analysis 3(2), 95–109 (1999)
7. Durbin, R., Eddy, S., Krogh, A., Mitchison, G.: Biological sequence analysis. Cambridge University Press, Cambridge (1998)
8. Forney, G.D.: The viterbi algorithm. Proceedings of IEEE 61, 268–278 (1973)
9. Freitag, D., McCallum, A.: Information extraction with HMM structures learned by stochastic optimization. In: Proceedings of the Seventeenth National Conference on Artificial Intelligence, Austin, TX. AAAI Press, Menlo Park (2000)
10. Fujiwara, Y., Asogawa, M., Konagaya, A.: Stochastic motif extraction using hidden markov model. In: Altman, R.B., Brutlag, D.L., Karp, P.D., Lathrop, R.H., Searls, D.B. (eds.) Proceedings of the Second International Conference on Intelligent Systems for Molecular Biology, ISMB, pp. 121–129. AAAI, Menlo Park (1994)
11. Galassi, U.: Structured Hidden Markov Models: A General Tool for Modeling Process Behavior. PhD thesis, Università degli Studi di Torino, Dottorato di ricerca in Informatica (April 2008)
12. Galassi, U., Giordana, A., Saitta, L.: Structured hidden markov models: A general tool for modeling agent behaviors. In: Soft Computing Applications in Business. Studies in Fuzziness and Soft Computing, vol. 230, pp. 273–292. Springer, Heidelberg (2008)
13. Gussfield, D.: Algorithms on Strings, Trees, and Sequences. Cambridge University Press, Cambridge (1997)
14. Haussler, D.: Convolution kernels on discrete structures. Technical Report SC-CRL-99-10, University of California in Santa Cruz, Computer Science Department (1999)
15. Joachims, T.: Text categorization with support vector machines: Learning with many relevant features. In: Nédellec, C., Rouveirol, C. (eds.) ECML 1998. LNCS, vol. 1398, pp. 137–142. Springer, Heidelberg (1998)
16. Lodhi, H., Shawe-Taylor, J., Cristianini, N., Watkins, C.: Text classification using string kernels. Journal of Machine Learning Research 2, 419–444 (2002)
17. El Manzalawi, Y., Dobbs, D., Honavar, V.: Predicting linear b-call epitopes using string kernels. Journal of Molecular Recognition (2008), doi:10.1002/jmr.893
18. Moulinier, I., Raskinis, G., Ganascia, J.: Text categorization: a symbolic approach. In: Proceedings of the Fifth Annual Symposium on Document Analysis and Information Retrieval (1996)
19. Ng, A.Y., Jordan, M.I.: On discriminative vs. generative classifiers: A comparison of logistic regression and naive bayes. In: Dietterich, T.G., Becker, S., Ghahramani, Z. (eds.) NIPS, pp. 841–848. MIT Press, Cambridge (2001)
20. Rabiner, L., Juang, B.: Fundamentals of Speech Recognition. Prentice Hall, Englewood Cliffs (1993)
21. Rabiner, L.R.: A tutorial on hidden markov models and selected applications in speech recognition. Proceedings of IEEE 77(2), 257–286 (1989)
22. Rubinstein, Y.D., Hastie, T.: Discriminative vs informative learning. In: Proc. Third Int. Conf. on Knowledge Discovery and Data Mining, pp. 49–53. AAAI Press, Menlo Park (1997)

23. Schölkopf, B., Burgess, C., Smola, A.: Advances in Kernel Methods. MIP Press (1998)
24. Stolcke, A., Omohundro, S.: Hidden markov model induction by bayesian model merging. In: Advances in Neural Information Processing Systems, vol. 5, pp. 11–18 (1993)
25. Vapnik, V.N.: The Nature of Statistical Learning. Springer, Heidelberg (1995)
26. Watkins, C.: Dynamic alignment kernels. In: Smola, A.J., Bartlett, P.L., Schölkopf, B., Schuurmans, D. (eds.) Advances in Large Margin Classifiers, pp. 39–50. MIT Press, Cambridge (2000)
27. Yang, Y., Pedersen, J.O.: A comparative study on feature selection in text categorization. In: Fisher, D.H. (ed.) Proceedings of ICML 1997, 14th International Conference on Machine Learning, Nashville, US, pp. 412–420. Morgan Kaufmann Publishers, San Francisco (1997)

Part IV
Soft Computing

Partition Measures for Data Mining

Ronald R. Yager

Machine Intelligence Institute
Iona College
New Rochelle, NY 10801
yager@panix.com

Absract. We investigate a number of measures associated with partitions. The first of these is congruence measures, which are used to calculate the similarity between two partitions. We provide a number of examples of this type of measure. Another class of measures we investigate are prognostication measures. This measure, closely related to a concept of containment between partitions, are useful in indicating how well knowledge of an objects class in one partition predicts its class in a second partitioning. Finally we introduce a measure of the non-specificity of a partition. This measures a feature of a partition related to the generality of the constituent classes of the partition. A common task in machine learning is developing rules that allow us to predict the class of an object based upon the value of some features of the object. The more narrowly we categorize the features in the rules the better we can predict an objects classification. However counterbalancing this is the fact that to many narrow feature categories are difficult for human experts to cognitively manage, this introduces a fundamental issue in data mining. We shown how the combined use of our measures prognostication and non-specificity allow us navigate this issue.

Keywords: partitioning, clustering, granular computing, data mining, concept hierarchy.

1 Introduction

As noted by Michalski [1, 2] clustering and categorization play an important role in machine learning. Fundamental to the task of clustering and categorization is the concept of partitioning. Here we investigate some measures useful in relating partitions. One class of measures investigated here are congruence measures, which are used to calculate the similarity between two partitions. For example if we partition a set of people by education and openness to new technology we are interested in calculating how similar these two partitions are. Another application is comparing the results of two clustering algorithms. Another type of measures we investigate are prognostication measures. This type of measure, closely related to a concept of containment between partitions, is useful in indicating how well knowledge of an objects class in one partitioning indicates its class in a second partitioning. We also introduce a measure

J. Koronacki et al. (Eds.): Advances in Machine Learning I, SCI 262, pp. 299–319.
springerlink.com © Springer-Verlag Berlin Heidelberg 2010

of the non-specificity of a partition. This measures a feature related to the generality of the constituent classes of a partition.

2 Comparing Partitions

Assume X is a finite set of objects. A common task is partitioning these objects, this involves obtaining a collection of subsets of X, $A_1, ..., A_q$, called classes such that $A_i \cap A_j = \varnothing$ and $\bigcup_{j=1}^{q} A_j = X$. In this case each element $x_k \in X$ lies in exactly one of these classes. The data mining task of clustering [3] involves a partitioning of the objects. The field of rough sets is based on a partitioning [4]. The process of partitioning is not unique. If $P_1 = \langle A_1, ...A_q\rangle$ and $P_2 = \langle B_1, ..., B_p\rangle$ are two partitions of X a natural question is how similar or congruent are these two partitions. An example of this occurs when we cluster objects by different features and are interested whether the resulting partitions are the same.

Assume $\mathcal{P}$ is the set of all partitions over the space X we are interested in obtaining a mapping Cong: $\mathcal{P} \times \mathcal{P} \to [0, 1]$ indicating the degree of congruence or similarity of two partitions such that if P_1 and P_2 are two partitions we want this mapping to satisfy: **1.** $\text{Cong}(P_1, P_2) = \text{Cong}(P_2, P_1)$ and **2.** $\text{Cong}(P_1, P_1) = 1$. That is we require that Cong is a similarity relationship [5].

We can associate with any set X two special partitions. The first is the universal partition where we just have one set we denote this $P^* = \langle X\rangle$. The other special partition is the one in which each element in X is in a distinct class we denote this $P_* = \langle A_1, ..., A_n\rangle$ where $A_i = \{x_i\}$. These are the very opposite, they are the least similar of any pair of partitions. Thus another property we desire of Cong is **3.** $\text{Cong}(P^*, P_*) = \text{Min}[\text{Cong}(P_1, P_2)]$.

Here we shall consider the formulation of the congruence mapping using equivalence relations. We recall an equivalence relation E on X is a mapping $E: X \times X \to \{0, 1\}$ such that: **1.** $E(x, x) = 1$ (Identity), **2.** $E(x, y) = E(y, x)$ (Symmetry) and **3.** $E(x, z) \geq \text{Min}_y[E(x, y), E(y, z)]$ (Transitivity).

It is well known that any partitioning of X can be uniquely associated with an equivalence relation on X. If $A_1, ..., A_q$ is a partitioning of X then we can obtain an equivalence relation E such that $E(x, y) = 1$ if x and y are in the same class and $E(x, y) = 0$ if they are in different classes. Alternatively if E is an equivalence relation on X we can obtain a partitioning by assigning element x and y to the same class if $E(x, y) = 1$ and to different classes if $E(x, y) = 0$. From this perspective we can associate with each $x \in X$ an equivalence class A_x such that $y \in A_x$ if $E(x, y) = 1$. We note some special equivalence relations and the partitioning they induce. If E is such that $E(x, y) = 1$ for all x and y then E partitions X into one set $A_1 = X$. This is the partitioning we denoted as P^*. We denote this equivalence relation as E^*. At the other extreme is the case E_* where $E_*(x, y) = 0$ for $x \neq y$. In this case our classes are $A_i = \{x_i\}$, the *n* singleton sets. This is the partitioning we denoted as P_*.

Assume E_1 and E_2 are two equivalence relations. We shall say $E_1 \subseteq E_2$ if $E_1(x, y) \leq E_2(x, y)$ for all x and y. It can be shown that for any equivalence relation E we have $E_* \subseteq E \subseteq E^*$. We note that if $E_1 \subseteq E_2$ then if x and y are in the same equivalence class is under E_1 then are also in the same class under E_2. Furthermore if E_1 induces the partitioning $<A_1, ..., A_q>$ and E_2 induces the partitioning $<B_1, ..., B_p>$ then $q \geq p$.

We note that if for $x \neq y$ we indicate by $<x, y>$ an unordered pair, $<x, y> = <y, x>$, then if X has n elements we have $\binom{n}{2} = \frac{(n)(n-1)}{2}$ unordered pairs. In the following we shall use U to indicate the set of all distinct unordered pairs from X. These can also be viewed as subsets of X of cardinality 2.

In the following we shall find it convenient at times to express E as a subset of the space U of unordered pairs. Here we shall use the notation U.E to indicate the subset of U related to E. In this case $U.E(<x,y>) = E(x, y) = E(y, x)$.

We now suggest a measure of congruence between partitions that we can express in terms of their underlying equivalence relations. Assume P_1: $<A_1, ..., A_q>$ and P_2: $<B_1, ..., B_q>$ are two partitions of X based on the equivalence relations E_1 and E_2 respectively. We define

$$\mathrm{Cong}_1(P_1, P_2) = \mathrm{Cong}_1(E_1, E_2) = 1 - \frac{1}{\binom{n}{2}} \sum_U |U.E_1(<x, y>) - U.E_2(<x, y>)|$$

Here $\sum_U |U.E_1(<x, y>) - U.E_2(<x, y>)|$ is the number of pairs that have different values in E_1 and E_2.

We easily observe that $0 \leq \mathrm{Cong}_1(E_1, E_2) \leq 1$. It is also clear that $\mathrm{Cong}_1(E_1, E_2) = \mathrm{Cong}_1(E_2, E_1)$. Consider two relations E_1 and E_2 such that $E_1 = E_2$, that is $E_1(x, y) = E_2(x, y)$ for all x and y. Here they induce the same partitions. In this case $U.E_1(<x, y>) - U.E_2(<x, y>) = 0$ for all $<x, y>$and therefore $\mathrm{Cong}_1(E_1, E_1) = 1$. Thus if the classes of two partitions are the same then their congruence is one. We further note that if $E_1 \neq E_2$, there exists some x and y such that $E_1(x, y) \neq E_2(x, y)$ then $\sum_U |U.E_1(<x, y>) - U.E_2(<x, y>)| \neq 0$ and hence $\mathrm{Cong}_1(E_1, E_2) < 1$. Thus $\mathrm{Cong}_1(E_1, E_2) = 1$ if and only if $E_1 = E_2$.

Consider the equivalence relation E^*, here all $E^*(x, y) = 1$, this induces a partition with one class $A_1 = X$. If E is any other equivalence relation then $\mathrm{Cong}_1(E^*, E) = 1 - \frac{1}{\binom{n}{2}} (\#U.E = 0)$ where $\#U.E = 0$ is number of elements in U that have zero membership in U.E. From this it easily follows that if E_1 and E_2 are two equivalence relations such that $E_1 \subseteq E_2$ then $\mathrm{Cong}_1(E^*, E_2) \geq \mathrm{Cong}_1(E^*, E_1)$. Analogously we see that if E_* is such that $E_*(x, y) = 0$ for $x \neq y$ then $\text{Cong}(E_*, E) \geq 1 - \frac{1}{\binom{n}{2}}$ (# E = 1).

From this it follows that if E_1 and E_2 are such that $E_1 \subseteq E_2$ then $\text{Cong}(E_*, E_2) \leq \text{Cong}(E_*, E_1)$.

Consider the partitions E^* and E_*. In this case $|U.E^*(<x, y>) - U.E_*(<x, y>) = 1$ for all <x, y> and hence $\text{Cong}_1(E^*, E_*) = 0$ We can further see that for any other two pair of partitions $\text{Cong}_1(E_1, E_2) > 0$. Thus $\text{Cong}_1(E_1, E_2) = 1$ iff $E_1 = E_2$ and $\text{Cong}_1(E_1, E_2) = 0$ iff $E_1 = E^*$ and $E_2 = E_*$. We also note here that for any partition E always has less equivalence classes then E_* and more than E^*.

Consider two equivalence relations E_1 and E_2 where $E_1 \subseteq E_2$, $E_1(x, y) \leq E_2(x, y)$. Let $P_1: < A_1, \ldots, A_q>$ and $P_2: < B_1, \ldots, B_p>$ be their respective partitions. We observer that for any A_j there exists a B_k such that $A_j \subseteq B_k$.

We can provide an alternative expression for $\sum_U |U.E_1(<x, y>) - U.E_2(<x, y>)|$.

Let G be a subset of U defined such that

$$G = (U.E_1 \cap \overline{U.E_2}) \cup (\overline{U.E_1} \cap U.E_2)$$

Thus G is the set of elements in U which have $U.E_1(<x, y>) = 1$ and $U.E_2(<x, y>) = 0$ or $U.E_1(<x, y>) = 0$ and $U.E_2(<x, y>) = 1$. It is easy to see that

$$\text{Card}(G) = \sum_U |U.E_1(<x, y>) - U.E_2(<x, y>)|$$

hence we can express $\text{Cong}_1(E_1, E_2) = 1 - \frac{1}{\binom{n}{2}}\text{Card}(G)$. We further observe that if we let $G_a = U.E_1 \cap \overline{U.E_2}$ and $G_b = \overline{U.E_1} \cap U.E_2$ then $G = G_a \cup G_b$ and $G_a \cap G_b = \varnothing$, G_a and G_b are disjoint. From this it follows that

$$\text{Card}(G) = \text{Card}(G_a) + \text{Card}(G_b)$$

Using this formulation we can prove the following property about the measure Cong_1.

Theorem: Assume E_1, E_2 and E_3 are three equivalence relations such that $E_1 \subset E_2 \subset E_3$ then

$$\text{Cong}_1(E_3, E_2) \geq \text{Cong}_1(E_3, E_1).$$

Consider the space U of unordered pairs <x, y>. Let Z be an $\binom{n}{2}$ dimensional space its coordinates $(z_1, \ldots, z_m)$ are such that $m = \binom{n}{2}$. Assume a bijective mapping

M associating with each element <x, y> a dimension in the space Z, M: U $\rightarrow$ Z. Using this we can associate with any equivalence relation E a vector V.E in this space Z such that the ith component of the vector $V.E(z_i) = U.E(<x, y>)$ where <x, y> is the pair corresponding to the dimension z_i. More formally $V.E(z_1) = U.E(M^{-1}(<x, y>)$. Thus the z_i coordinate in V.E takes the value one if the unordered pair <x. y> associated with z_i is in E otherwise it takes the value zero.

Consider now two equivalence relations E_1 and E_2 defined on X. We can associate with each of these a vector as defined above, we denote these $V.E_1$ and $V.E_2$. Consider now the term $\sum_{<x, y> \in U} |U.E_1(<x, y>) - U.E_2(<x, y>)|$. With m = $\binom{n}{2}$ we see that this can be alternatively expressed as $\sum_{j=1}^{m} |V.E_1(z_j) - V.E_2(z_j)|$. This alternative expression is a well-known measure of distance, a metric, between the two vectors. It is an example of the more general Minkowski metric

$$L_p[(a_1, ..., a_m), (b_1, ..., b_m) = (\sum_{i=1}^{m} |a_i - b_i|^p)^{1/p}$$

for $p > 0$. In the case where $p = 1$ this is called city block metric. We note that if the arguments are restricted to the unit interval then for any p this assumes its maximal value for the case when $a_i = 0$ and $b_i = 1$ for all i.

Thus we can view $\sum_{j=1}^{m} |V.E_1(z_j) - V.E_2(z_j)|$ as a distance between two vectors. We denote this $L_1(V.E_1, V.E_2))$. It is well known that $L_1(V.E_1, V.E_2)$ satisfies the properties of a metric.

Let **E** be the set of all equivalence relations on X. Let **V.E** be the set of all associated vectors as defined above. Then Lp is a mapping

$$Lp: \mathbf{V.E} \times \mathbf{V.E} \rightarrow R$$

such that for any $V.E_1$, $V.E_2$ and $V.E_3$ associated with the equivalence relations E_1, E_2 and E_3 we have

1) $Lp(V.E_1, V.E_2) = Lp(V.E_2, V.E_1)$
2) $Lp(V.E_1, V.E_2) = 0 \quad$ if $V.E_1 = V.E_2$
3) $Lp(V.E_1, V.E_3) \leq Lp(V.E_1, V.E_2) + Lp(V.E_2, V.E_3)$

Furthermore since $Lp(V.E_1, V.E_2)$ is a metric then

$$d(E_1, E_2) = \frac{1}{\binom{n}{2}} Lp(V.E_1, V.E_2)$$

is also a metric, distance measure, however it maps into the unit interval. From this we see that for any p the term $1 - d(E_1, E_2)$ is a similarity relation and can be used as a measure of congruence between E_1 and E_2. Furthermore we note for the special equivalence relation E^* all the components in the vector $V.E^*$ are ones while for he special equivalence relation E_* all the components in the vector $V.E_*$ are zeros. From this it follows that $1 - d(E_*, E^*)$ will have the smallest congruence for any two equivalence relations.

3 An Alternative Measure of Congruence

In the preceding we have introduced a measure of similarity, congruence, between two partitions using the underlying equivalence relations. Here we now consider formulating a congruence measure from the perspective of the partitions themselves.

Assume we have two partitions of the space X, $P_1 = \langle A_1, \ldots, A_q \rangle$ and $P_2 = \langle B_1, \ldots, B_p \rangle$ here $X = \bigcup_{j=1}^{q} A_j$ and $A_i \cap A_j = \varnothing$ for $i \neq j$ and $X = \bigcup_{j=1}^{p} B_j$ and $B_i \cap B_j = \varnothing$ for $i \neq j$. Without loss of generality we shall assume $q = p$. If $q > p$ we can augment the partition P_2 by adding $q - p$ subsets, $B_{p+1} = B_{p+2} = \ldots = B_q = \varnothing$. Thus in the following we assume the two partitions have the same number of classes, q.

We now introduce an operation called a pairing of P_1 and P_2, denoted $g(P_1, P_2)$, which associates with each subset A_i of P_1 a unique partner B_j from P_2. Formally if $Q = \{1, 2, \ldots, q\}$ then a pairing is a mapping $g: Q \to Q$ that is bijective, one to one and onto. Essentially g is a permutation of Q. We then have that a pairing $g(P_1, P_2)$ is a collection of q pairs, $(A_j, B_{g(j)})$. We shall now associate with each pairing a score, $\text{Score}(g(P_1, P_2))$, defined as follows. Denoting

$$D_{g.j} = A_j \cap B_{g(j)} \qquad \text{for } j = 1 \text{ to } q$$

we obtain

$$\text{Score}(g(P_1, P_2)) = \frac{\sum_{j=1}^{q} \text{Card}(D_{g.j})}{\text{Card}(X)}$$

In the following we illustrate the process of determining $\text{Score}(g(P_1, P_2))$.

Example: Let $X = \{x_1, x_2, x_3, x_4, x_5, x_6, x_7\}$.

Assume P_1 is $A_1 = \{x_1, x_2, x_3\}$, $A_2 = \{x_4, x_5, x_6\}$ and $A_3 = \{x_7\}$

Assume P_2 is $B_1 = \{x_2, x_4, x_6\}$ and $B_2 = \{x_1, x_3, x_5, x_7\}$.

We first augment P_2 with $B_3 = \varnothing$

One Pairing is $g(j) = j$ in this case we get the pairs (A_1, B_1), (A_2, B_2), (A_3, B_3). From this

$$D_{g.1} = A_1 \cap B_1 = \{x_2\}$$
$$D_{g.2} = A_2 \cap B_2 = \{x_5\}$$
$$D_{g.3} = A_3 \cap A_3 = \varnothing$$

In this case $Score(g(P_1, P_2)) = \frac{2}{7}$.

Another pairing is g(1) = 3, g(2) = 1, g(3) = 2 here our pairs are

$$(A_1, B_3), (A_2, B_1), (A_3, B_2)$$

In this case

$$D_{g.1} = A_1 \cap B_3 = \varnothing$$
$$D_{g.2} = A_2 \cap B_1 = \{x_4, x_6\}$$
$$D_{g.3} = A_3 \cap B_2 = \{x_7\}$$

In this case $Score(g(P_1, P_2)) = \frac{3}{7}$

We now shall use this to obtain a measure of congruence, $Cong_2(P_1, P_2)$. Let G be the set of all pairings, $g \in G$. We define

$$Cong_2(P_1, P_2) = \underset{g \in G}{Max} [Score(g(P_1, P_2))]$$

Thus this measure of congruence is the score of the largest pairing.

We see that for any pairing g, $0 \le \sum_{j=1}^{q} Card(D_{g.j}) \le Card(X)$. From this it follows that

$$0 \le Cong_2(P_1, P_2) \le 1.$$

More precisely since for any two partitions we can always find a pairing g in which $\sum_{j=1}^{q} Card(D_{g.j}) \geq 1 \ge 1$ we see that

$$\frac{1}{Card(X)} \le Cong_2(P_1, P_2) \le 1$$

Let us look at some special cases of the measure of congruence. Here we assume Card(X) = n. First consider the case when $P_1 = P_2$ that is

$$P_1 = A_1, A_2, ..., A_q$$

$$P_2 = B_1, B_2, ... , B_q$$

where $A_j = B_j$. We see the pairing g such that g(j) = j has the pairs (A_j, B_j) for j = 1 to q. In this case $D_{g.j} = A_j \cap B_j = A_j$. We get that $\sum_{j=1}^{q} Card(D_{g.j}) = \sum_{j=1}^{q} Card(A_j) = n$.

For this g we have $g(P_1, P_2) = \frac{n}{n} = 1$. From this we get that $Cong_2(P_1, P_2) = \underset{g \in G}{Max}\,[Score(g(P_1, P_2))] = 1$. Thus if the partitions are the same the congruence is 1. It can be shown that $Cong_2(P_1, P_2) = 1$ if and only if $P_1 = P_2$.

Consider now the situation in which one of the partitions P^* is simply X the whole space, it is the grand partition and the other partition $P = A_1, ..., A_q$. We see here that there are only q distinct pairing mappings possible. Each one corresponds to an association of X with one of the A_j and the other classes in P are associated with ∅. For the pairing g_j that associates X with A_j we have $X \cap A_j = A_j$. In this case we get $Score(g_j) = \frac{Card(A_j)}{n}$. Thus here then $Cong_2(P^*, P) = \underset{j}{Max}\,[\frac{Card(A_j)}{n}]$. Thus for the grand partition P^* its congruence with any other partition is simply $\frac{1}{n} Max_j[Card(A_j)]$, its the proportion of elements in the largest class of P.

Consider now the case in which one partition is the partition consisting of classes which are singletons, $P_* = A_1, ..., A_n$, where $A_i = \{x_i\}$. Let $P = B_1, ... B_q$ be any arbitrary partitioning. In this case $Cong_2(P_* P) = q/n$ where q is the number of classes in P. In the special case where P is P^* we get that $Cong_2(P_*, P^*) = 1/n$.

4 Obtaining the Optimal Pairing

In the general case the calculation of $Cong_2(P_1, P_2)$ requires the solution of an optimization problem, we must determine the pairing g that maximize the score. In the following we shall provide a view of this optimization problem that allows for a solution.

We have two partitions P_1: $A_1, ..., A_q$ and P_2: $B_1, ..., B_q$, we have assumed the inclusion of null sets if necessary. We can associate with these two partitions the matrix N show below. In this matrix $n_{ij} = Card(A_i \cap B_j)$, it is the number of elements that A_i and B_j have in common. We note that the sum of the elements in this i^{th} row is equal to $Card(A_i)$ while the sum of the element is the jth column is $Card(B_j)$. The sum of all the elements in the matrix is n, Card(X).

$$\begin{array}{c|ccc|} & B_1 & B_j & B_q \\ \hline A_1 & n_{11} & & n_{1q} \\ A_i & & n_{ij} & \\ A_q & n_{q1} & & n_{qq} \\ \hline \end{array}$$

The problem of calculating $Cong_2(P_1, P_2)$ can be viewed in the perspective of this matrix. A pairing g is an association of each row A_i with a unique column B_j, each pairing results in the selection of q elements consisting of one from each row and one from each column. Furthermore each pair (A_i, B_j) in a paring g has associated value n_{ij}. The score of the pairing g is the sum of these q associated values. The problem of determining $Cong_2(P_1, P_2)$ is then to find the selection of these q values that provide the largest total. This problem is a well known optimization problem in operation research, it is called the assignment problem [6]. Techniques have been developed for solving this problem. We note that if q is small then we can solve this problem by enumeration. We note that if g is an assignment function, g(i) assigns $B_{g(i)}$ to A_i, then the score of g equals $\sum_{i=1}^{q} n_{ig(i)}$.

Here we shall present a view of this optimization problem that allows for easy solution by enumeration if q is small or by use of genetic algorithms [7] if q is large.

To calculate $Cong_2(P_1, P_2)$ we must calculate $Max_{g\in G}[Score(g(P_1, P_2))$. Using a mapping g the pairs are $(A_i, B_{g(i)})$ and the score is

$$Score(g(P_1, P_2)) = \sum_{i=1}^{q} Card[A_i \cap B_{g(i)}] = \sum_{i=1}^{q} n_{ig(i)}$$

Consider the simple assignment g(i) = i here then A_i and B_i are assigned as pairs. In this case

$$Score(g(P_1, P_2)) = \sum_{i=1}^{q} n_{ii}$$

it is the sum of the elements on the diagonal of the matrix N.

We note that any mapping $g: Q \to Q$ can be viewed as a permutation of the B_j. We see that for any permutation g of $<B_1, ..., B_q>$ we get an interchange of the columns of N. For this permutation, however, its score can be obtained as the sum of the elements on the main diagonal.

Thus one way to obtain the value of $Cong_2(P_1, P_2)$ is the do all the possible permutation of the columns. Total up the values of the main diagonal for each permutation. The largest of these totals is the value of $Cong_2(P_1, P_2)$. In cases where q is small this can be easily done by enumeration. In cases where q is large the total number of permutations, Card(G), becomes too large for enumeration. Here we can use genetic algorithms to get an optimal solution [7].

We can express this optimization problem in the form that can be solved using a genetic algorithm. Consider our genomes to be permutations of the [1, 2, ..., q]. In the case where q = 6 an example of this would be $\tilde{g}$ = [2 3 4 5 6 1].

Each genome $\tilde{g}$ can be seen to induce a matrix in which the ith column is $B_{\tilde{g}(i)}$. Given a particular genome $\tilde{g}$ our fitness function is

$$F = \sum_{i=1}^{q} n_{i\tilde{g}(i)}$$

We can now search that the whole space of genomes G using the established procedures for searching a permutation space with genetic algorithms.

5 A Measure of Prognostication

In the following we shall investigate a different type of measure of the relationship between two partitions. Assume P_1 and P_2 are two partitions of X,

$$P_1: A_1, ..., A_q$$
$$P_2: B_1, ..., B_r$$

Here we will try to measure how well we are to able tell what class an object is in the partition P_2 from our knowledge of the class it is in partition P_1. This can be seen as a kind of prognostication or prediction. We call denote this measure $\mathrm{Prog}(P_1, P_2)$. As we shall see this will not be symmetrical, that is $\mathrm{Prog}(P_1, P_2) \neq \mathrm{Prog}(P_2, P_1)$ but it will be such that $\mathrm{Prog}(P_j, P_j)$ will assume the highest value. As we shall subsequently see this concept is closely related to the idea of containment between the partitions. In the following we again assume Card(X) = n.

We first define what we shall call an association. Let Q = {1, 2, ..., q} and let R = {1, 2, ..., r} we define an association to be a mapping h: Q → R. Thus for each $i \in Q$ we get $h(i) \in R$. We note that h need not be bijective. This for $i \neq k$ we can have h(i) = h(k). Essentially this mapping associates with each class A_i in partition P_1, a class B_j in partition P_2. While more then one A_i can be associated with each B_j, each A_i is only associated with only one B_j. For a given h we define $\mathrm{Fam}(j) = \{A_i/h(i) = j\}$, it is the family of classes in P_1 that map into B_j in partition P_2. We let H be the set of all association mappings.

We now obtain for any association mapping h a value

$$\mathrm{Value}(h) = \frac{1}{n} \sum_{j=1}^{r} \mathrm{Card}(B_j \cap \bigcup_{i \in \mathrm{Fam}(j)} A_i)$$

Using these values we define

$$\backslash\mathrm{Prog}(P_1, P_2) = \mathrm{Max}_{h \in H}[\mathrm{Value}(h)]$$

Thus the degree of prognostication of P_1 for P_2 is the largest value of any association.

We now make some observations about the properties of Value(h) and $\mathrm{Prog}(P_1, P_2)$. First we observe that for any h, Value (h) ≥ 0 and Value(h) ≤ 1. From this we conclude that $0 \leq \mathrm{Prog}(P_1, P_2) \leq 1$.

Let us now consider some special cases of P_1 and P_2. First consider the case where $P_1 = P_2$, here

$$P_1: A_1, A_2, ..., A_q$$
$$P_2: B_1, B_2, ..., B_q$$

where $B_i = A_i$. Consider now the association mapping h(i) = i. In this case Value(h) $= \frac{1}{n} \sum_{j=1}^{q} \text{Card}(A_j \cap B_j) = 1$. From this we can conclude that $\text{Prog}(P_1, P_2) = 1$.

Another interesting special case is where P_2 is P^* it just consists of one set X,

$$P_1: A_1, ..., A_q$$
$$P_2: B_1$$

where $B_1 = X$. Here there is only one association mapping, h(i) = 1 for all i. In this case Value(h) $= \frac{1}{n} \text{Card}(X \bigcup_{i = 1 \text{ to } q} A_j) = 1$. Thus here again we get Prog(P, P*) = 1.

Consider now another special case. Here we assume P_1 is P_*, each $A_i = \{x_i\}$ it is a singleton set and P_2 is any arbitrary partition. Thus in this case

$$P_1: A_1, A_2, ..., A_n$$

$$P_2: B_1, B_2, ..., B_r$$

Consider the association mapping h(i) such that h(i) = j where $x_i \in B_j$. In this case

$$\text{Value}(h) = \frac{1}{n} \sum_{j=1}^{r} \text{Card}(B_j \cap \bigcup_{x_i \in B_j} A_i) = 1$$

Again here we get $\text{Prog}(P_*, P) = 1$.

Actually these three preceding situations can be seen as special cases of a more general relationship between P_1 and P_2. This relationship is a hierarchical or containment relationship between the two partitions. We define this in the following. Assume P_1: $A_1, A_2, ..., A_q$ and P_2: $B_1, B_2, ..., B_r$, we say $P_1 \subseteq P_2$ if for each A_i there exists a B_j such that $A_i \subseteq B_j$. It is clear that $r \leq q$.

For the situation in which $P_1 \subseteq P_2$ consider the association mapping h where h(i) = j where $A_i \subseteq B_j$. In this case $\bigcup_{i \in \text{Fam}(j)} A_i = B_j$ and hence

$$\text{Value}(h) = \frac{1}{n} \sum_{j=1}^{r} \text{Card}(B_j \cap \bigcup_{h(i) = j} A_i) = \frac{1}{n} \sum_{j=1}^{r} \text{Card}(B_j) = 1$$

Thus in this case of $P_1 \subseteq P_2$ we always get $\text{Prog}(P_1, P_2) = 1$. Here then we can view $\text{Prog}(P_1, P_2)$ essentially as a measure of containment of the cluster P_1 in P_2.

Let us look in the opposite direction. Assume are partitions are

$$P_1: A_1 = X$$

$$P_2: B_1, ..., B_r$$

In this case there exists r possible association mappings. $h_1, ..., h_r$ where for h_j we have $h_j(1) = j$. For any of these mappings we have $\text{Value}(h_j) = \frac{1}{n}\text{Card}(B_j)$. Thus here $\text{Prog}(P^*, P_2) = \frac{1}{n}\text{Max}_j[\text{Card}(B_j)]$, it is the proposition of the elements in the largest cluster of P_2.

Consider now the case where P_2 is singletons, $B_j = \{x_j\}$ and P_1 is arbitrary

$$P_1: A_1, ..., A_q$$
$$P_2: B_1, ..., B_n$$

With some thought it can be seen there the maximum of Value(h) occurs for any mapping h that associates A_i with one of the sets B_j whose element is in A_i, $x_j \in A_i$. In the case we get

$$\text{Value}(h) = \frac{1}{n}\sum_{j=1}^{n}\text{Card}(B_j \cap \bigcup_{h(i)=j} A_i) = \frac{q}{n}$$

Thus $\text{Prog}(P_1, P_*) = \frac{q}{n}$

We can provide a lower bound on Prog (P_1, P_2). Again assume $P_1: A_1, ..., A_q$ and $P_2: B_1, ..., B_r$. Consider a mapping h_j such that for each i, $h_j(i) = j$. Here every class in P_1 is associated with B_j. In this case $\text{Value}(h_j) = \text{Card}(B_j \cap \bigcup_{i=1}^{q} A_i) = \text{Card}(B_j \cap X) = \text{Card}(B_j)$. From this we can conclude that $\text{Prog}(P_1, P_2) \geq \text{Max}_j[\text{Card}(B_j)]$.

We now shall provide a view of the situation which will greatly simplify the calculation of $\text{Prog}(P_1, P_2)$. Again assume we have two partitions P_1: $\langle A_1, ..., A_q\rangle$ and P_2: $\langle B_1, ..., B_r\rangle$. Consider the matrix N shown below. In this matrix the value n_{ij} is the number of elements A_i and B_j have in common, that is $\text{Card}(A_i \cap B_j) = n_{ij}$. Here the sum of the elements in the ith row is $\text{Card}(A_i)$ and the sum of the elements in the jth column is $\text{Card}(B_j)$.

$$\begin{array}{c} \\ A_1 \\ A_i \\ A_q \end{array} \begin{array}{c} B_1 \quad B_j \quad B_r \\ \left[\begin{array}{ccc} & & \\ & n_{ij} & \\ & & \end{array}\right] \end{array}$$

We recall that any association mapping h assigns to each class A_i in P_1 a class in P_2 which we denote as $B_{h(i)}$. Consider now

$$\text{Value}(h) = \frac{1}{n} \sum_{j=1}^{r} \text{Card}(B_j \cap \bigcup_{\substack{i \text{ s.,t.} \\ h(i) = j}} A_i)$$

We observe since $A_i \cap A_j = \varnothing$, they are disjoint, we can rewrite the above as

$$\text{Value}(h) = \frac{1}{n} \sum_{j=1}^{r} \left(\sum_{\substack{i \text{ s.t.} \\ h(i) = j}} \text{Card}(B_j \cap A_i) \right)$$

Furthermore since A_i is associated with only one B_j we can reformulate this as

$$\text{Value}(h) = \frac{1}{n} \sum_{j=1}^{r} \left(\sum_{\substack{i, \text{ s.t.} \\ h(i) = j}} \text{Card}(B_{h(i)} \cap A_i) \right) = \frac{1}{n} \sum_{i=1}^{q} \text{Card}(B_{h(i)} \cap A_i)$$

$$\text{Value}(h) = \frac{1}{n} \sum_{i=1}^{q} n_{ih(i)}$$

Thus for a given h its Value(h) is the sum of the $n_{ih(i)}$. Essentially for a given h for each A_i we select one element in the matrix N from its corresponding row and then add these up. From this observation the determination of h^* such that $\text{Val}(h^*) = \text{Max}_h[\text{Value}(h)]$ becomes clear. We select from each row the largest element in N. From this we easily calculate the value of Prog

$$\text{Prog}(P_1, P_2) = \frac{1}{n} \sum_{i=1}^{q} \text{Max}_j[n_{ij}]$$

Thus the calculation of $\text{Prog}(P_1, P_2)$ simply requires us to determine the largest value in each row of the matrix N.

It the following we shall let $g: Q \rightarrow R$ be the unique association such that $n_{ig(i)} = \text{Max}_j[n_{ij}]$. We now can easily understand the situation in which $P_1 \subseteq P_2$. In this case for each A_i there exists one B_j such that $A_i \subseteq B_j$. Let us denote this g(i), thus $A_i \subseteq B_{g(i)}$. In this case $A_i \cap B_{g(i)} = A_i$ and hence $n_{ig(i)} = \text{Card}(A_i)$ and furthermore $n_{ij} = 0$ for $j \neq g(i)$. Thus each row has only one no-zero entry and the value of that is $\text{Card}(A_i)$. It is clear how we pick these elements and this gives us $\text{Prog}(P_1, P_2) = 1$.

Thus the determination of $\text{Prog}(P_1, P_2)$ can be seen as type of assignment problem. We assign each A_i to one B_j, the one with the maximal common elements. Since we can assign multiple A_i to same B_j there is no conflict and each assignment can be done independently. Thus this is easier than the previous problem.

Let us define $U_i = |A_i| - n_{ig(i)} = \sum_{j \neq g(i)} n_{ij}$. It is the number of elements in A_i not in $B_{g(i)}$. We now see that

$$\mathrm{Prog}(P_1, P_2) = \frac{1}{n} \sum_{i=1}^{q} n_{ig(i)} = \frac{1}{n} \sum_{i=1}^{q} |A_i| - U_i = 1 - \frac{1}{n} \sum_{i=1}^{q} U_i ,$$

it is one minus the proportion of the element not in the assignment set.

We see the following the following theorem holds

Theorem: $\mathrm{Prog}(P_1, P_2) = 1$ if and only if for each A_i there exists an B_j such that $A_i \subseteq B_j$.

Proof: This is this the necessary and sufficient for $\mathrm{Prog}(P_1, P_2) = 1$

Definition: Assume P_1: $<A_1, ..., A_q>$ is a partition of X and P_2: $<B_1, ..., B_p>$ is another partition of X where $q \geq p$. If for each A_i there exists a B_j such that $A_i \subseteq B_j$ then we say that P_1 is a ***refinement*** of P_2. Alternatively we say P_2 is a coarsing of P_1.

Note: Earlier we showed if P_1 is a refinement of P_2 then $\mathrm{Prog}(P_1, P_2) = 1$.

Theorem: Assume P_1 is a refinement of P_2. Let P_3 be any arbitrary partition of X then

$$\mathrm{Prog}(P_1, P_3) \geq \mathrm{Prog}(P_2, P_3).$$

Proof: Assume P_1: $<A_1, ..., A_q>$, P_2: $<B_1, ..., B_p>$and P_3: $<D_1, ..., D_m>$. Here $q \geq p$. Let g_2 be the optimal mapping between P_2 and P_3 hence $\mathrm{Prog}(P_2, P_3)$ $\sum_{j=1}^{p} |B_j \cap D_{g_2(j)}|$. Because of the relationship between P_1 and P_2 each B_j is the union of a subset of classes from P_1. We shall denote the subset of these classes making up B_j as C(j) and indicate an element in C(j) as A_{k_j}. Using this notation then $B_j = \bigcup_{k_j \in C(j)} A_{k_j}$

$$B_j = \bigcup_{K=1}^{n_j} A_{jK}$$

Using this we see that $\mathrm{Prog}(P_2, P_3) = \sum_{j=1}^{p} |B_j \cap D_{g_2(j)}| = \sum_{j=1}^{p} \left(\sum_{k_j \in C(j)} |A_{k_j} \cap D_{g_2(j)}| \right)$

Let g_1 be a mapping associating a class in P_1 with a class in P_3 such that for any A_{k_j} we have $|A_{k_j} \cap D_{g_1(k_j)}| = \mathrm{Max}_{\text{all } D_i} |A_{k_j} \cap D_i|$. From this we have $\mathrm{Prog}(P_1, P_3) =$

$\sum_{i=1}^{q} |A_i \cap D_{g_1(i)}|$ however $\sum_{i=1}^{q} |A_i \cap D_{g_1(i)}| = \sum_{j=1}^{p} (\sum_{k_j \in C(j)} |A_{k_j} \cap D_{g_1(j)}|) =$ and

therefore $Prog(P_1, P_3) = \sum_{j=1}^{p} (\sum_{k_j \in C(j)} |A_{k_j} \cap D_{g_1(j)}|)$.

Since $|A_{k_j} \cap D_{g_1(k_j)}| \geq |A_{k_j} \cap D_{g_2(k_j)}|$ the result follows.

6 Applications of the Prog Measure

One application of $Prog(P_1, P_2)$ is to provide a measure of congruence. In particular we can define

$$Cong_3(P_1, P_2) = Prog(P_1, P_2) \wedge Prog(P_2, P_1) \qquad (\wedge = Min)$$

Let us see what are the properties of this definition. First we we see that it is symmetric

$$Cong_3(P_1, P_2) = Cong_3(P_2, P_1)$$

Furthermore since $Prog(P_1, P_1) = 1$ then we have $Cong_3(P_1, P_1) = 1$. We further observe that if $P_1 \subseteq P_2$ then we showed $Prog(P_1, P_2) = 1$ and hence $Cong_3(P_1, P_2) = Prog(P_2, P_1)$.

Consider the case where P_1 is the partition into singleton, P_*, and P_2 is the whole space, P^*. Here $Prog(P_*, P^*) = 1$ while $Prog(P^*, P_*) = 1/n$ and hence $Cong_3((P_*, P^*) = 1/n$.

Another application of $Prog(P_1, P_2)$ is in helping build relations between variables. Again assume X is a collection of objects. Let U and V be two attributes associated with these objects. Let P_1: $<A_1, ..., A_r>$ be a partition of the objects in X based upon the attribute U. Let P_2: $<B_1, ..., B_q>$ be a partition of the object in X based upon the second attribute V. Here we are interested being able to determine in which class in P_2 an object falls given our knowledge of which class in the partition P_1 the object lies. Here we want relations like

If an objects U value is in A_i then its V value is in B_j

Here the quantity $Prog(P_1, P_2)$ helps us determine which is the best association to make. The function g that maximizes $Prog(P_1, P_2)$ provides the best association. In this case the value of $Prog(P_1, P_2)$ provides some measure of the usefulness or quality of the relation obtained.

An important example of this problem is the following. Assume we have some preferred partitioning, P_2: $<B_1, ..., B_r>$, of the objects in X based upon some attribute V. Let $\mathcal{F}$ be a family of partitions of X based upon another attribute U. Here $\mathcal{F} = \{P_{11}, ..., P_{1s}\}$ where each P_{1k} is a partition of X based on U. That is P_{1k}: $<A_{k1}, ..., A_{kq_k}>$. Our objective here is to find the partition in $\mathcal{F}$ which gives the best rules for the connection between the U and V attributes. Here we must first search through each P_{1k} to find its best association function, g_k, and obtain the value for Prog(P_{1k}, P_2). Using this we then select the partition P_{1k*} such that Prog(P_{1k*}, P_2) = $\text{Max}_{k \in \mathcal{F}}$[Prog($P_{1k}$, P_2)]. This type of problem naturally arises in situations in which a decision maker knows what action to perform if the object is in any one of the classes of P_2, B_j, but makes observations about the objects in terms of attribute U. A typical example of this occurs in the medical domain. In this type of problem the decision maker, doctor, would know from his training what treatment to recommend if a patient has a particular disease. However he usually doesn't directly observe a disease, instead he observes measurable physical attributes of a patient. Here U would correspond to measurable physical attributes of a patient and the partition P_2 would correspond to different diseases. In order to prescribe the treatment he must be able to determine the current disease from the symptoms U. The goal in selecting the optimal partition of U is of course to accurately determine the disease from the measurements of the physical attributes U. In order to accomplish this task we must enable the decision to comprehend the values of the attributes. Human comprehension of data is generally enhanced by using granularization to view the values of the attributes. For example blood pressure of 180 is better viewed as high blood pressure. Body temperature of 99.3 is most usefully viewed as normal. With this understanding we see that each partition of P_{1k} of U corresponds to a different granularization of the attribute. We then see that $\mathcal{F}$ provides a collection of different granularizations which are comprehensible to the decision maker. Thus by initially introducing $\mathcal{F}$ we are essentially trying to constrain the space in which we are looking for the optimal partitioning to be those which are cognitively comprehensible to the decision maker. Once having introduced $\mathcal{F}$ our objective then becomes selecting from these different comprehensible granulations the one that leads to the most accurate determination of the disease from the systems.

The following figure#1 illustrates the essential features of this paradigm

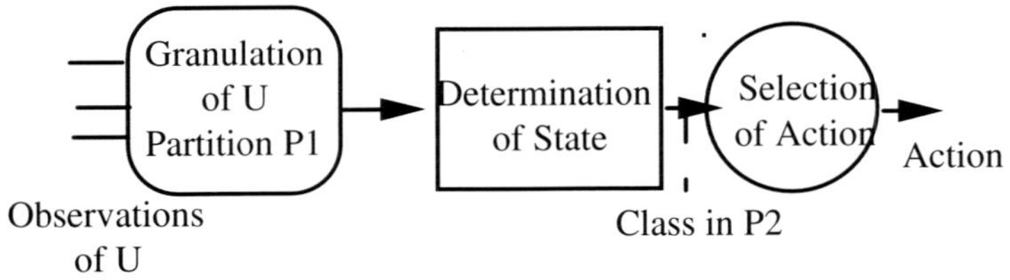

Fig. 1. Basic Paradigm

Examples of this paradigm also occur in the financial environment. Here experts have "rules of thumb" describing what action to take if they know the state of the economy, (recession, inflation etc). However they observe primary data about the economy and from these observations they must determine the state in order to know what action to take.

7 Measure of Generality (Non-specificity) of a Partition

Yager [8] and Klir [9] investigated the concepts of specificity and its complement non-specificity in considerable detail. It is closely related to the idea of generality.

Klir [9] describes one measure of non-specificity of a set, if A is a subset of X then Klir's measure of non-specificity is $\log_2(\text{Card}(A))$. In the following we shall use $|A|$ to denote Card(A). When A is a singleton, $A = \{x\}$ we get $\log_2(|A|) = 0$. For the case when $A = X$, $\log_2(|A|) = \log_2(n)$, where $|X| = n$. Since $\log_2(|A|)$ is monotonic with respect to $|A|$ we see that $0 \leq \log_2(|A|) \leq \log_2(n)$. A normalized version of Klir's measure of non-specificity is $NS_K(A) = \frac{\log_2(|A|)}{\log_2(n)}$. Here then for any subset $A \subseteq X$ we have $0 \leq NS_K(A) \leq 1$.

Based on Yager's [8] work on specificity we obtain an alternative definition for non–specificity

$$NS_Y(A) = \frac{|A| - 1}{n - 1}$$

This takes values in the unit interval. It attains its maximal value of one for the case when $A = X$ and its minimum value of zero for the case when A is a singleton.

It can be seen that these two definitions are cointensive, they are both measures of the granularity of the subset A. The larger the subset the closer the non–specificity is to one. For singletons we get a non–specificity of zero. In using these definitions we will only be considering non-empty sets. The choice between these two definitions of non–specificity depends is context dependent, it depends upon what application and what features are of interest. In the following we shall use the neutral term NS when we have no need to distinguish between these two definitions.

For our purposes we see that the measure of non-specificity can be used to provide a measure of generality. The larger the set A (the more non-specific) the more general the concept the set it is representing.

We now consider the extension of the concept of non-specificity from sets to partitions. Our interest here is in quantifying the idea of generality. We are interested in quantifying how general the classes comprising a partitioning are.

Assume P: $\langle A_1, \ldots, A_q \rangle$ is a partition of X. We are now interested in the calculation of the non-specificity of P, NS(P). In the following we shall propose such a definition. This definition will make use of the measure of non-specificity of the individual classes in P. The definition we shall use is the following is

$$NS(P) = \sum_{i=1}^{q} \frac{|A_i|}{n} NS(A_i)$$

We see it is a weighted average of the non-specificities of the component classes in P. The weights are determined by the number of elements in the class.

We see that this definition is independent of the indexing of the classes, each class is treated in the same manner. We also note that it is idempotent, if for all i, $NS(A_i) = a$ then $NS(P) = a$. This if all the classes have the same degree of non-specificity then this is the non-specificity of the partition as a whole. We also note that since each of $NS(A_i) \in [0, 1]$ then $NS(P) \in [0, 1]$.

Let us look at some notable special cases. Consider the case where $P = P^*$, that is $q = 1$ and $A_1 = X$. We see that $AS(A_1) = 1$ and $|A_1| = n$ and therefore $NS(P^*) = 1$. Thus the non–specificity of a partition consisting of just the whole space is one.

At the opposite extreme is the case when the classes are just singletons P_*, here we have n classes with $A_i = \{x_i\}$. In the case $NS(A_i) = 0$ and hence from the idempotency we get $NS(P) = 0$.

We should note that fact that $\bigcup_{i=1}^{q} A_i = X$ places some constraints on the possible manifestations of NS(P). An interesting property holds between the measures of non-specificity of two partitions P_1 and P_2 when $P_1 \subseteq P_2$, $NS(P_1) \le NS(P_2)$.

Let us look at the manifestation of NS(P) for the two cases of specificity of sets, $NS_y(A)$ and $NS_K(A)$. For the case of NS_y we get that $NS_y(P) = \sum_{i=1}^{q} \frac{a_i}{n} . \frac{a_i - 1}{n - 1}$ where $a_i = |A_i|$. After some algebraic manipulation we have $NS_y(P) = \frac{1}{n-1}(\sum_{i=1}^{q} \frac{a_i^2}{n} - \sum_{i=1}^{q} \frac{a_i}{n})$. Since $\sum_{i=1}^{q} a_i = n$ then

$$NS_y(P) = \frac{1}{n-1} (\sum \frac{a_i^2}{n} - 1) = \frac{n}{n-1} \sum_{i=1}^{q} \frac{a_i^2}{n^2} - \frac{1}{n-1}$$

Letting $w_i = \frac{a_i}{n}$, the proportion of elements is A_i, we get

$$NS_y(P) = \frac{n}{n-1} \sum_{i=} w_i^2 - \frac{1}{n-1} = \frac{1}{n-1} (n \sum_{i=}^{n} w_i^2 - 1)$$

We further observe that if n is large then $NS_y(P) \to \sum_{i=1}^{n} w_i^2$. Thus this is a kind of variance.

For the case where we use $NS_K(A)$ we have $NS_K(P) = \sum_{i=1}^{q} \frac{a_i}{n} \frac{\log_2(a_i)}{\log_2(n)}$. Again l etting $\frac{a_i}{n} = w_i$ we get

$$NS_K(P) = \sum_{i=1}^{q} w_i \frac{\log_2(n\, w_i)}{\log_2(n)} = \frac{1}{\log(n)} \sum_{i=1}^{q} (w_i \log_2(w_i) + w_i \log_2(n))$$

Since $\sum w_i = 1$ we see

$$NS_K(P) = 1 + \frac{1}{\log(n)} \sum_{i=1}^{q} w_i \log_2(w_i)$$

We see this as a kind of entropy.

8 Understanding a Fundamental Issue in Data Mining

A common task that occurs in data mining is the following. Let X be a class of objects having some feature V called the target feature. Let P: $<B_1, B_2, ..., B_r>$ be a partition of the objects based the feature V. In addition assume we have a concept hierarchy over X based upon some other features, U. We recall that a concept hierarchy consists of a group of partition of increasing generality. Let $P_1, ..., P_r$ be a collection of partitions on X forming a concept hierarchy where P_i: $<A_{i1}, A_{i2}, ..., A_{iq_i}>$. The requirement for these partitions to form a concept hierarchy is that $P_k \subset P_j$ for $k \leq j$. We refer to P_i as the ith level partition. In this case the generality is increasing as the level increases. Concept hierarchies play a fundamental role in the way humans conceptualize the world. A typical example of a concept hierarchy is the following. Let U be the attribute corresponding to an objects home address. A concept hierarchy would be the following

P_1 is a partition by the *addresses*, P_2 is a partition by *neighborhood*, P_3 is a partition by *City*, P_4 is a partition by *state* and P_5 is a partition by *Region*

Our objective here is to build rules predicting the class in P, B_j, to which an object belongs based on its value for the feature U. In order to make these rules human comprehensible we want to use the categories (concepts) associated with the concept hierarchy

In order to accomplish this task we can consider each level of the hierarchy, P_i. Using the Prog measure described earlier we can find for each level i the function g_i that makes the best assignment between the classes in the partition P_i and the target partition P, Prog(P_i, P). Let us denote the value associated with each of these as Val(g_i). The larger this the better the association. From our preceding discussion we showed that if $P_i \subset P_j$ then the respective assigning functions g_i and g_i are such that Val(g_i) $\geq$ Val(g_j). Thus here then the best rules are obtained using the categories in the lowest level of the concept hierarchy P_1.

However as we noted an important objective in formulating the rules relating U and V is human comprehension and understanding. As reported by Miller in his classic paper [10] human information processing capacity deteriorates if one has to manage to many classes. Thus in order achieve our objective we have a preference for rules using fewer more general concepts for the categories used to describe the values of the variable U. Thus here then we prefer partitions that use more general concepts. Parenthetically we note that an additional benefit of using more general classes for U in formulating our rules is that we can allow less precision in our determination of the values U when using the derived rules. The need for less precision is usually more economical on resources.

As we previously noted the generality of a partitioning can be measured by the its non–specificity. In addition we showed that if P_i and P_j are two partitions such that $P_1 \subset P_j$ then $S_p(P_i) \leq Sp(P_j)$. In particular we see that generality is increased as we go up the concept hierarchy.

We now see the fundamental conflict associated with our task of trying to build rules to help predict an objects class in the target partition P based upon an observation of its U value. Our desire of attaining correct prediction benefits by using as our partitioning for U one down the bottom of the concept hierarchy while our desire for generalization and human comprehension benefits by going up the concept hierarchy. While we shall not pursue this further here we believe that the measures we have introduced, $S_p(P_i)$ and $Prog(P_i, P)$, can play a role in helping adjudicating this conflict. We also note that in [11] we looked at issues closely related to this problem.

9 Conclusion

We investigated a number of measures relating partitions. One class of measures we considered were congruence measures. These measures are used to calculate the similarity between two partitions. We discussed a number of examples of this type of measure. Another class of measures we investigated were prognostication measures. This type of measure, closely related to a concept of containment between partitions, is useful in indicating how well knowledge of an objects class in one partitioning indicates its class in a second partitioning. We applied our measures to some data mining applications using the structure of the concept hierarchy. We also introduced a measure of the non-specificity of a partition. This measures a feature of a partition related to the generality or granularity of the constituent classes of the partition.

References

[1] Michalski, R.S., Stepp, R.E.: Learning from observation: Conceptual clustering. In: Michalski, R.S., Carbonell, J.G., Mitchell, T.M. (eds.) Machine Learning: An Artificial Intelligence Approach. Morgan Kaufmann, San Mateo (1983)

[2] Michalski, R.S., Stepp, R.E.: Automated construction of classifications: Conceptual clustering versus numerical taxonomy. IEEE Transactions on Pattern Analysis and Machine Intelligence 5, 396–410 (1983)

[3] Han, J., Kamber, M.: Data Mining: Concepts and Techniques. Morgan Kaufmann, San Francisco (2001)

[4] Pawlak, Z.: Rough Sets- Theoretical Aspects of Reasoning About Data. Kluwer, Hingham (1991)

[5] Zadeh, L.A.: Similarity relations and fuzzy orderings. Information Sciences 3, 177–200 (1971)
[6] Hillier, F.S., Lieberman, G.J.: Introduction to Operations Research. McGraw Hill, New York (2005)
[7] Goldberg, D.E.: Genetic Algorithms in Search Optimization and Machine Learning. Addison-Wesley, Reading (1989)
[8] Yager, R.R.: On measures of specificity. In: Kaynak, O., Zadeh, L.A., Turksen, B., Rudas, I.J. (eds.) Computational Intelligence: Soft Computing and Fuzzy-Neuro Integration with Applications, pp. 94–113. Springer, Berlin (1998)
[9] Klir, G.J.: Uncertainty and Information. John Wiley & Sons, New York (2006)
[10] Miller, G.A.: The magical number seven, plus or minus two: Some limitations on our capacity for processing information. Psychological Review 63, 81–97 (1956)
[11] Yager, R.R., Petry, F.E.: Evidence resolution using concept hierarchies. IEEE Transactions on Fuzzy Systems 16, 299–308 (2008)

An Analysis of the FURIA Algorithm for Fuzzy Rule Induction

Jens Christian Hühn and Eyke Hüllermeier

Department of Mathematics and Computer Science
University of Marburg
Hans-Meerwein-Straße
35039 Marburg
Germany
huehnj@informatik.uni-marburg.de, eyke@informatik.uni-marburg.de

Abstract. This paper elaborates on a novel fuzzy rule-based classification method called FURIA, which is short for "Fuzzy Unordered Rule Induction Algorithm". FURIA has recently been developed as an extension of the well-known RIPPER algorithm. It learns fuzzy rules instead of conventional rules and unordered rule sets instead of rule lists. Moreover, to deal with uncovered examples, it makes use of an efficient rule stretching method. First experimental results have shown that FURIA significantly outperforms the original RIPPER in terms of classification accuracy. Elaborating on the advantages of a fuzzy approach, this paper makes an attempt to distill and quantify the influence of rule fuzzification on the performance of the algorithm. Moreover, going beyond the conventional classification problem, we investigate the performance of FURIA in the context of bipartite ranking, in which a fuzzy approach appears to be even more appealing.

1 Introduction

The interest in learning rule-based models for classification goes far beyond the field of machine learning itself and also includes other research areas, notably fuzzy systems (Hül05). This is hardly surprising, given that rule-based models have always been a cornerstone of fuzzy systems and a central aspect of research in that field. To a large extent, the popularity of rule-based models can be attributed to their comprehensibility, a distinguishing feature and key advantage in comparison to many other (black-box) classification models. Despite the existence of many sound algorithms for rule induction, the field still enjoys great popularity and, as shown by recent publications (IY05; Cv06; JCC07; FGHd07), offers scope for further improvements.

This paper investigates some properties of a novel fuzzy rule-based classification method called *Fuzzy Unordered Rule Induction Algorithm*, or FURIA for short, which was recently introduced in (HH09). FURIA is a fuzzy rule learner that builds upon RIPPER, a state-of-the-art rule induction method developed by Cohen (Coh95). The main modifications of RIPPER include changes in the pruning procedure, the learning of an unordered instead of an ordered rule set, a novel rule stretching mechanism and, perhaps most importantly, a fuzzification strategy that turns conventional rules into fuzzy rules with soft boundaries.

J. Koronacki et al. (Eds.): Advances in Machine Learning I, SCI 262, pp. 321–344.
springerlink.com © Springer-Verlag Berlin Heidelberg 2010

First experimental studies suggest that FURIA is superior to the original RIPPER in terms of classification accuracy, and that the rule fuzzification has an important part in the improvements. In this paper, we elaborate in more detail on the influence of rule fuzzification. In particular, we try to isolate the effect of rule fuzzification from the effect of other modifications. Moreover, going beyond the conventional classification problem, we investigate the performance of FURIA in the context of bipartite ranking.

To make the paper self-contained, we recall the basics of both the FURIA and RIPPER algorithms in the next section. In Section 3, we give some general arguments in favor of using fuzzy rules. An experimental evaluation of FURIA as well as an empirical analysis of the influence of rule fuzzification are presented in Section 4. The paper ends with a summary and concluding remarks in Section 5.

2 FURIA: Fuzzy Unordered Rule Induction Algorithm

This section gives a brief introduction of FURIA; for technical details, we refer to (HH09). As FURIA builds upon the separate-and-conquer rule induction procedure of RIPPER, we start with a recapitulation of the basics of this algorithm and give an overview of the main modifications afterward.

2.1 RIPPER

RIPPER was introduced by (Coh95) as a successor of the IREP algorithm for rule induction (FW94). Even though the key principles remained unchanged, RIPPER improves IREP in many details and, moreover, is able to cope with multi-class problems.

Consider a polychotomous (multi-class) classification problem with m classes $\mathbb{L} \stackrel{\text{df}}{=} \{\lambda_1, \lambda_2, \ldots, \lambda_m\}$. Suppose instances to be represented in terms of attributes A_i, $(i = 1, \ldots, n)$, which are either numerical (real-valued) or nominal, and let $\mathbb{D}_i$ denote the corresponding domains. Thus, an instance is represented as an n-dimensional attribute vector

$$x = (x_1, x_2, \ldots, x_n) \in \mathbb{D} \stackrel{\text{df}}{=} \mathbb{D}_1 \times \ldots \times \mathbb{D}_n.$$

A single RIPPER rule is of the form $r = \langle r_A \,|\, r_C \rangle$, consisting of a premise part r_A and a consequent part r_C. The premise part r_A is a conjunction of predicates (selectors) which are of the form $(A_i = v_i)$ for nominal and $(A_i \,\theta\, v_i)$ for numerical attributes, where $\theta \in \{\leq, =, \geq\}$ and $v_i \in \mathbb{D}_i$. The consequent part r_C is a class assignment of the form (class $= \lambda$), where $\lambda \in \mathbb{L}$. A rule $r = \langle r_A \,|\, r_C \rangle$ is said to *cover* an instance $x = (x_1, x_2, \ldots, x_n)$ if the attribute values x_i satisfy all the predicates in r_A.

RIPPER learns such rules in a greedy manner, following a separate-and-conquer strategy (Für99). Prior to the learning process, the training data is sorted by class labels in ascending order according to the corresponding class frequencies. Rules are then learned for the first $m-1$ classes, starting with the smallest one. Once a rule has been created, the instances covered by that rule are removed from the training data, and this is repeated until no instances from the target class are left. The algorithm then proceeds with the next class. Finally, when RIPPER finds no more rules to learn, a default rule (with empty antecedent) is added for the last (and hence most frequent) class.

Rules for single classes are learned until either all positive instances are covered or the last rule r that has been added was "too complicated". The latter property is implemented in terms of the total description length (Qui95).

2.2 Learning Individual Rules

Each individual rule is learned in two steps. The training data, which has not yet been covered by any rule, is therefore split into a *growing* and a *pruning* set. In the first step, the rule will be specialized by adding antecedents which were learned using the growing set. Afterward, the rule will be generalized by removing antecedents using the pruning set.

When RIPPER learns a rule for a given class, the examples of that class are denoted as *positive* instances, whereas the examples from the remaining classes are denoted as *negative* instances.

A new rule is learned on the growing data, using a propositional version of the FOIL algorithm (Qui90; QCJ93). It starts with an empty conjunction and adds selectors until the rule covers no more negative instances, i.e., instances not belonging to the target class. The next selector to be added is chosen so as to maximize FOIL's information gain criterion (IG), which is a measure of improvement of the rule in comparison with the default rule for the target class:

$$\mathrm{IG}_r \stackrel{\mathrm{df}}{=} p_r \times \left(\log_2 \left(\frac{p_r}{p_r + n_r} \right) - \log_2 \left(\frac{p}{p+n} \right) \right), \tag{1}$$

where p_r and n_r denote, respectively, the number of positive and negative instances covered by the rule; likewise, p and n denote the number of positive and negative instances covered by the default rule.

The above procedure typically produces rules that overfit the training data. To remedy this effect, a rule is simplified so as to maximize its performance on the pruning data through a cut-off at the position that maximizes the rule-value metric

$$V(r) \stackrel{\mathrm{df}}{=} \frac{p_r - n_r}{p_r + n_r}$$

Therewith, all those antecedents will be pruned that were learned after the antecedent maximizing $V(r)$; shorter rules are preferred in the case of a tie.

The ruleset RS produced by the learning algorithm outlined so far, called IREP*, is taken as a starting point for a subsequent optimization process in which all rules are re-examined. For each rule r, two alternative rules – the *replacement rule* and the *revision rule* – are considered. To decide which version of r to retain, the MDL (Minimum Description Length (Qui93)) criterion is used. Afterward, the remaining positives are covered using the IREP* algorithm.

The RIPPER*k* algorithm iterates the optimization of the ruleset and the subsequent covering of the remaining positive examples with IREP* k times, hence the name RIPPER (Repeated Incremental Pruning to Produce Error Reduction).

2.3 Modifications of RIPPER

A first modification of RIPPER concerns the type of rule model that is learned and, related to this, the use of default rules: FURIA learns to separate each class from all other classes, which means that no default rule is used and the order of the classes is irrelevant.

When using an unordered rule set without default rule, two problems can occur in connection with the classification of a new query instance. First, a conflict may occur since the instance is equally well covered by rules from different classes. As will be seen in Section 2.6, this problem is rather unlikely to occur and, in case it still does, can easily be resolved. Second, it may happen that the query is not covered by any rule. To solve this problem, we propose a novel rule stretching method. The idea is to modify the rules in a local way so as to make them applicable to the query.

FURIA uses a rule stretching approach that exploits the order in which the antecedents were learned, treating them as a list $\langle \alpha_1, \alpha_2, \ldots, \alpha_m \rangle$ instead of a set $\{\alpha_1, \alpha_2, \ldots, \alpha_m\}$. The idea is that the ordering reflects the importance of the antecedents, an assumption that is clearly justified in light of the underlying rule learning algorithm. As generalizations, we then only allow lists of the form $\langle \alpha_1, \alpha_2, \ldots, \alpha_k \rangle$ with $k \leq m$. For the minimal generalization, k is simply given by $j-1$, where α_j is the first antecedent which is not satisfied by the query instance. By Laplace-correcting the relative number of remaining antecedents, k/m, preference is given to longer and, hence, more specific rules.

Computationally, the above rule stretching strategy is much more efficient than the original proposal of Eineborg and Boström (Bos04). The complexity for re-evaluating a rule r is linear in its number of antecedents. Moreover, since the evaluations of all generalizations of a rule can be calculated and stored directly in the course of the rule learning process, in which antecedents are learned in a successive way, there is no need for storing the training data.

The RIPPER algorithm can be divided into the building and the optimization phase. The rule building is done via the IREP* algorithm, which essentially consists of a propositional FOIL algorithm, the pruning strategy and the stopping conditions. Interestingly, we found that the pruning strategies in IREP* have a negative influence on the performance of FURIA. We therefore omitted the pruning step and instead learned the initial ruleset on the whole training data directly.

In the optimization phase, the pruning was retained, as its deactivation was not beneficial. This is in agreement with the goal to minimize the MDL. The coverage of the remaining positive instances, which is again accomplished by IREP*, also benefited from omitting the pruning, just like IREP* in the building phase.

FURIA still applies pruning when it comes to creating the replacement and the revision rule. Here, the original pruning strategy is applied, except in case the pruning strategy tries to remove all antecedents from a rule, thereby generating a default rule. In this case, the pruning will be aborted, and the unpruned rule will be used for the MDL comparison in the optimization phase. We found that those pruning strategies are still sufficient to avoid overfitting. Thus, the removal of the pruning in the IREP* part has no negative impact on classification accuracy.

2.4 Representation of Fuzzy Rules

A selector constraining a numerical attribute A_i (with domain $\mathbb{D}_i = \mathbb{R}$) in a RIPPER rule can obviously be expressed in the form $(A_i \in I)$, where $I \subseteq \mathbb{R}$ is an interval: $I = (-\infty, c]$ if the rule contains a selector $(A_i \leq c)$, $I = [b, \infty)$ if it contains a selector $(A_i \geq b)$, and $I = [b, c]$ if it contains both (in the last case, two selectors are combined).

Essentially, a fuzzy rule is obtained through replacing intervals by fuzzy intervals, namely fuzzy sets with trapezoidal membership function.

A fuzzy interval of that kind is specified by four parameters and will be written $I^F = [a, b, c, d]$:

$$I^F(v) \stackrel{\text{df}}{=} \begin{cases} 1 & b \leq v \leq c \\ \frac{v-a}{b-a} & a < v < b \\ \frac{d-v}{d-c} & c < v < d \\ 0 & \text{else} \end{cases}$$

b and c are, respectively, the lower and upper bound of the core of the fuzzy set; likewise, a and d are, respectively, the lower and upper bound of the support. Note that, as in the non-fuzzy case, a fuzzy interval can be open to one side ($a = b = -\infty$ or $c = d = \infty$.) In fact, as will be seen later on, the fuzzy antecendents successively learned by FURIA are fuzzy half-intervals of exactly that kind.

A fuzzy selector $(A_i \in I_i^F)$ *covers* an instance $x = (x_1 \ldots x_n)$ to the degree $I_i^F(x_i)$. A fuzzy rule r^F involving k selectors $(A_i \in I_i^F)$, $i = 1, \ldots, k$, covers x to the degree

$$\mu_{r^F}(x) = \bigotimes_{i=1\ldots k} I_i^F(x_i) \;, \tag{2}$$

where $\otimes$ is a so-called t-norm operator such as the product or the minimum.

2.5 Rule Fuzzification

To obtain fuzzy rules, the idea is to fuzzify the final rules from our modified RIPPER algorithm. More specifically, using the training set $D \subseteq \mathbb{D}$ for evaluating candidates, the idea is to search for the best fuzzy extension of each rule, where a fuzzy extension is understood as a rule of the same structure, but with intervals replaced by fuzzy intervals. Taking an interval I from an original RIPPER rule as the core $[b, c]$ of the sought fuzzy interval I^F, the problem is to find optimal bounds for the respective supports, i.e., to determine a and d.

For the fuzzification of a single interval I_i on attribute A_i it is important to consider only the relevant training data D^i, i.e., to ignore those instances that are excluded by any other fuzzy interval I_j^F, $j \neq i$:

$$D^i \stackrel{\text{df}}{=} \left\{x = (x_1, x_2, \ldots, x_k) \in D \,|\, I_j^F(x_j) > 0 \text{ for all } j \neq i\right\} \subseteq D \tag{3}$$

We partition D^i into the subset of positive instances, D_+^i, and negative instances, D_-^i. To measure the quality of a fuzzification, the purity will be used:

$$\text{pur} = \frac{p_i}{p_i + n_i}, \tag{4}$$

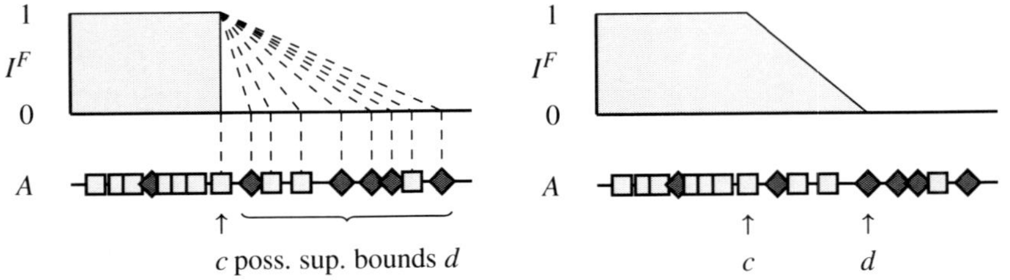

(a) Examination of possible support bounds of a crisp interval.

(b) Fuzzification that maximizes the interval's fuzzy purity.

Fig. 1. Fuzzification process

where

$$p_i \stackrel{\text{df}}{=} \sum_{x \in D^i_+} I_i(x) \quad \text{and} \quad n_i \stackrel{\text{df}}{=} \sum_{x \in D^i_-} I_i(x).$$

Rules are fuzzified in a greedy way, as shown by Algorithm 1. In each iteration, a fuzzification is computed for every antecedent, namely the best fuzzification in terms of (4). This is done by testing all values

$$\{x_i \,|\, x = (x_1, x_2, \ldots, x_k) \in D^i, x_i < b_i\}$$

as candidates for a_i and, likewise, all values

$$\{x_i \,|\, x = (x_1, x_2, \ldots, x_k) \in D^i, x_i > c_i\}$$

as candidates for d_i (see Fig. 1a). Ties are broken in favor of larger fuzzy sets, that is, larger distances from the core. The fuzzification is then realized for the antecedent with the highest purity, cf. Figure 1b. This is repeated until all antecedents have been fuzzified.

Algorithm 1. The antecedent fuzzification algorithm for a single rule r

Let A be the set of numeric antecedents of r
while $A \neq \emptyset$ **do**
  $a_{\max} \leftarrow$ null // a_{max} *denotes the antecedent with the highest purity*
  $\text{pur}_{\max} \leftarrow 0$ // pur_{max} *is the highest purity value, so far*
  for $i \leftarrow 1$ to size(A) **do**
    compute the best fuzzification of $A[i]$ in terms of purity
    $\text{pur}_{A[i]} \leftarrow$ be the purity of this best fuzzification
    if $\text{pur}_{A[i]} > \text{pur}_{\max}$ **then**
      $\text{pur}_{\max} \leftarrow \text{pur}_{A[i]}$
      $a_{\max} \leftarrow A[i]$
    end if
  end for
  $A \leftarrow A \setminus a_{\max}$
  Update r with $a_{\max}$
end while

Note that the fuzzification of a single antecedent may change the relevant training data (3), which is hence recomputed in each iteration. In fact, each fuzzification may increase the number of covered instances, which in turn may also influence the rule purity. Furthermore, note that, after the complete premise part of a rule has been fuzzified, the whole procedure could in principle be repeated until convergence is achieved (convergence is guaranteed, as purity can only increase in each iteration). We did not implement this option, however, as we observed that, except for very rare cases, convergence is already achieved after the first iteration.

2.6 Classifier Output

Suppose that fuzzy rules $r_1^{(j)}, r_2^{(j)}, \ldots, r_k^{(j)}$ have been learned for class λ_j. For a new query instance x, the support of this class is defined by

$$s_j(x) \stackrel{\text{df}}{=} \sum_{i=1,\ldots,k} \mu_{r_i^{(j)}}(x) \cdot CF\left(r_i^{(j)}\right) , \tag{5}$$

where $CF(r_i^{(j)})$ is the *certainty factor* of the rule $r_i^{(j)}$. It is defined as follows:

$$CF\left(r_i^{(j)}\right) = \frac{2\frac{|D_T^{(j)}|}{|D_T|} + \sum_{x \in D_T^{(j)}} \mu_{r_i^{(j)}}(x)}{2 + \sum_{x \in D_T} \mu_{r_i^{(j)}}(x)} , \tag{6}$$

where $D_T^{(j)}$ denotes the subset of training instances with label λ_j. The authors in (IN01, IY05) argued that weighing rules according to (5) allows for modeling more flexible decision boundaries and thereby improves classification accuracy. The certainty factor (6) is the m-estimate for $m = 2$ (PFTV92).

The class predicted by FURIA is the one with maximal score. In the case where x is not covered by any rule, which means that $s_j(x) = 0$ for all classes λ_j, a classification decision is derived using the rule stretching procedure (see above). In the case of a tie, a decision in favor of the class with highest frequency is made.

3 Advantages of Fuzzy Rules

Fuzzy rules are more general than conventional rules and have a number of potential advantages. For example, conventional (non-fuzzy) rules produce models with "sharp" decision boundaries and, correspondingly, abrupt transitions between different classes. This property is questionable and not very intuitive. Instead, one would expect the support for a class provided by a rule to decrease from "full" (inside the core of the rule) to "zero" (near the boundary) in a gradual rather than an abrupt way. As explained above, fuzzy rules have "soft" boundaries, which is one of their main characteristics. Admittedly, if a definite classification decision has to be made, soft boundaries have again to be turned into crisp boundaries. Interestingly, however, these boundaries are potentially more flexible in the fuzzy case. For example, by using suitable aggregation operators for combining fuzzy rules, they are not necessarily axis-parallel (PFTV92).

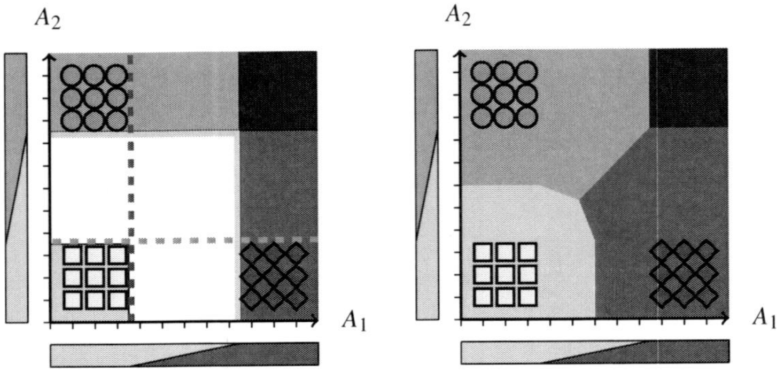

(a) Rule cores and support edge **(b)** Decision boundaries

Fig. 2. Three axis-parallel fuzzy rules (light gray region, dark gray region, and region marked by the intersection of the dotted lines) learned for a three-class problem (left) and the induced decision boundaries (right)

As an illustration, consider the three-class problem in Fig. 2a. For the top-left and bottom-right class, a simple one-dimensional rule is sufficient. The rule cores tightly fit the data, while the rule supports reach to the lower left class. To separate the latter, a rule with conditions on both attributes is necessary. Again, the rule core covers the data in a firm way, while the support decreases toward the other classes. As can be seen in Fig. 2b, the decision boundaries induced by the rules are non-linear, even though all rule conditions are axis-parallel.

In general, one may wonder to what extent rule fuzzification in FURIA can actually become effective in the sense of increasing the rule purity. To answer this question, one should first notice that a trivial fuzzification does always exist, namely the one that puts the support bound to the first instance behind the core bound. Even though this fuzzification does not change the purity on the training data, it is meaningful when it comes to classifying new instances. The reason is that it extrapolates the original rule very carefully to an area that might not have been covered by any other rule. The previous example was exactly of that kind.

The question remains, of course, whether there are non-trivial fuzzifications. Admittedly, such fuzzifications cannot exist when a rule contains only one fuzzy interval as numerical antecedent. In this case, RIPPER has learned this rule with the objective to maximize the information gain (1) which implicitly contains the purity. Consequently, the last instance – marking the core boundary – covered by that interval has to be from the class represented by the rule. A non-trivial fuzzification cannot exist, since the growing procedure would have selected the support bound as core position.

A non-trivial fuzzification can be found, however, when a rule uses at least two numeric attributes in its antecedence part. In such cases, it can happen that the myopic rule-learning strategy disables RIPPER (like any separate-and-conquer rule learner without look-ahead) to find the globally optimal solution. As an illustration, consider the following example.

Example 1. Assume examples from classes □ and ◆ lying in $A_1 \times A_2 \subseteq \mathbb{R}^2$, cf. Fig. 3a. Moreover, consider a rule learner that tries to optimize the purity. The optimal non-fuzzy rule for the latter class is

$$\text{IF } A_1 \in (-\infty, 6] \wedge A_2 \in (-\infty, 6] \text{ THEN class} = ◆$$

with a purity of $\frac{4}{5}$, cf. Fig. 3b. The myopic separate-and-conquer strategy is not able to find this rule, however, because it adds antecedents successively. Thus, it first finds the rule

$$\text{IF } A_1 \in (-\infty, 5] \text{ THEN class} = ◆$$

having a purity of $\frac{7}{13}$. The first antecedent of the optimal rule

$$\text{IF } A_1 \in (-\infty, 6] \text{ THEN class} = ◆$$

is refused since it has a purity of only $\frac{1}{2}$.

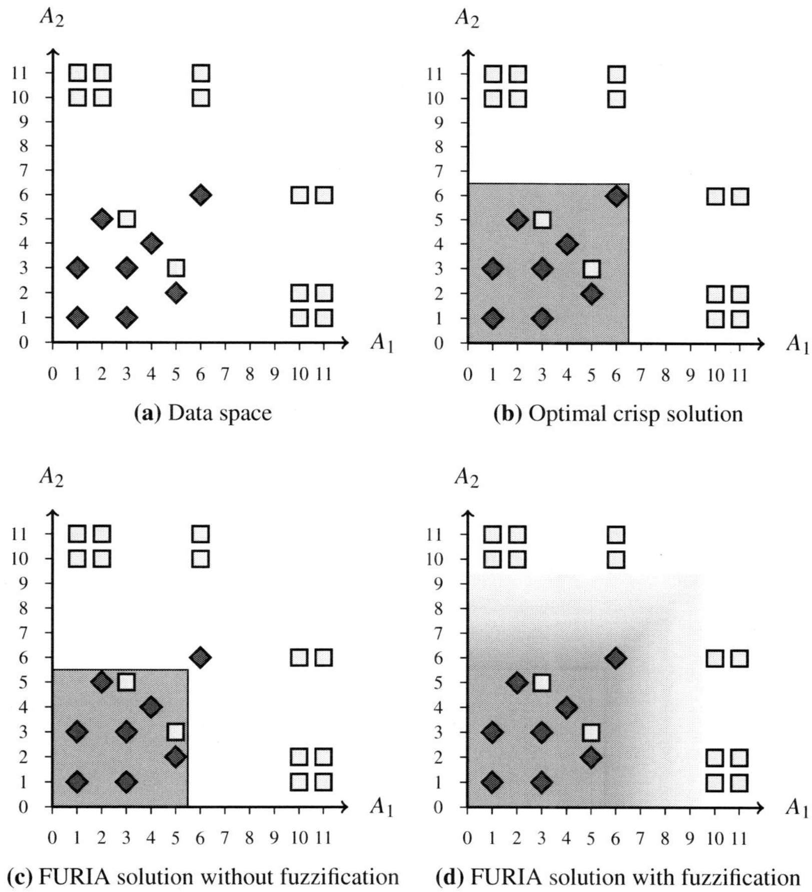

(a) Data space **(b)** Optimal crisp solution

(c) FURIA solution without fuzzification **(d)** FURIA solution with fuzzification

Fig. 3. A two-dimensional classification problem showing the failure of myopic separate-and-conquer learning and the correction through post-generalization (fuzzification)

The next step in the rule learning procedure can not undo the mistake and finds the rule

$$\text{IF } A_1 \in (-\infty, 5] \wedge A_2 \in (-\infty, 5] \text{ THEN class} = \blacklozenge$$

with a purity of $\frac{7}{9} < \frac{4}{5}$, cf. Figure 3c. Note that, due to the symmetry of the problem, the same values result when considering A_2 first.

In this situation a non-trivial fuzzification will lead to the rule

$$\text{IF } A_1 \in (-\infty, -\infty, 5, 10] \wedge A_2 \in (-\infty, -\infty, 5, 10] \text{ THEN class} = \blacklozenge \ .$$

The example at position $(6,6)$ is now covered to a degree of $\frac{10-6}{10-5} = 0.8$ by each condition, resulting in a purity of $\frac{39}{49}$ when using the minimum t-norm in (2).

This example suggests that the rule fuzzification of FURIA, which is also a kind of generalization, may help to undo mistakes that were made due to short-sighted (myopic) learning strategies. In the next section, we shall investigate this conjecture on an experimental basis.

4 Experiments

To analyze the performance of FURIA, several experimental studies were conducted. As a starting point, we used the RIPPER implementation of WEKA ("JRip"), cf. Section 2.1, for implementing the rule learning part of FURIA.

Our testbed for the experiments consists of 45 real-world classification datasets. Most of the datasets, 24 in total, were taken from the UCI repository (AN07), 13 datasets originate from the Statlib repository (MV07) and 3 come from an agricultural domain (Bul07; Bar07; Har07). The remaining five datasets were generated from meteorological station data, published by the German Weather Service (DWD). Table 1 gives an overview of the datasets and their main characteristics. We only used datasets with at least as many numeric as nominal attributes, since the fuzzy techniques presented in this work are ineffective for nominal data.

Table 1. An overview of the datasets used in the experiments

Dataset	# Inst.	# Classes	# Attributes Cont.	Nom.	Miss.	Origin
analcatdata-bankruptcy	50	2	5	1	0	Statlib
analcatdata-cyyoung8092	97	2	7	3	0	Statlib
analcatdata-cyyoung9302	92	2	6	4	0	Statlib
analcatdata-esr	32	2	2	0	0	Statlib
analcatdata-lawsuit	264	2	3	1	0	Statlib
biomed	209	2	7	1	2	Statlib
haberman	306	2	2	1	0	UCI
heart-statlog	270	2	13	0	0	UCI
ionosphere	351	2	34	0	0	UCI
liver-disorders	345	2	6	0	0	UCI

Table 1. *(continued)*

Dataset	# Inst.	# Classes	# Attributes Cont.	Nom.	Miss.	Origin
pima diabetes	768	2	8	0	0	UCI
prnn-synth	250	2	2	0	0	Statlib
schizo-	340	2	12	2	11	Statlib
sonar	208	2	60	0	0	UCI
wisconsin-breast-cancer	699	2	9	0	1	UCI
analcatdata-authorship	841	4	70	0	0	Statlib
analcatdata-halloffame	1340	3	15	2	1	Statlib
analcatdata-votesurvey	48	4	3	1	0	Statlib
cars	406	3	6	1	2	Statlib
collins	500	15	20	3	0	Statlib
ecoli	336	8	7	0	0	UCI
eucalyptus	736	5	14	5	9	agricult.
glass	214	6	9	0	0	UCI
iris	150	3	4	0	0	UCI
metStatCoordinates	4748	16	3	0	0	own
metStatRainfall	4748	16	12	0	0	own
metStatRST	336	12	3	0	0	own
metStatSunshine	422	14	12	0	0	own
metStatTemp	673	15	12	0	0	own
mfeat-factors	2000	10	216	0	0	UCI
mfeat-fourier	2000	10	76	0	0	UCI
mfeat-karhunen	2000	10	64	0	0	UCI
mfeat-morphological	2000	10	6	0	0	UCI
mfeat-zernike	2000	10	47	0	0	UCI
optdigits	5620	10	64	0	0	UCI
page-blocks	5473	5	10	0	0	UCI
pasture	36	3	21	1	0	agricult.
pendigits	10992	10	16	0	0	UCI
segment	2310	7	19	0	0	UCI
squash-unstored	52	3	20	3	8	agricult.
synthetic control	600	6	60	1	0	UCI
vehicle	846	4	18	0	0	UCI
vowel	990	11	10	2	0	UCI
waveform-5000	5000	3	40	0	0	UCI
wine	178	3	13	0	0	UCI

To evaluate a classifier in terms of its predictive performance, a dataset is randomly split into two parts, $2/3$ for training and $1/3$ for testing. To reduce random effects, the performance measure (e.g., classification rate or AUC) is averaged over 100 repetitions of this procedure.

To analyze the results, we follow the two-step procedure recommended by (Dem06): First, a Friedman test is conducted to test the null hypothesis of equal classifier performance (Fri37; Fri40). In case this hypothesis is rejected, which means that the classifiers' performance differs in a statistically significant way, a posthoc test is conducted to compare the classifiers in a pairwise way. Here, we apply the Bonferroni-Dunn test (Dun61) which, just like the Friedman test, is a rank-based statistic: On each dataset, the classifiers are ranked according to their performance, i.e., the best classifier receives rank 1, the second-best rank 2, and so on. The overall performance of a classifier is then measured in terms of its average rank, and what the aforementioned tests do is essentially looking at the differences between these averages.

To visualize the statistical test results, Demšar proposed a number line from 1 to k, which represents the average ranks of the k classifiers according to the Friedman test. Every single classifier is marked on that number line at the position of its average rank, and an interval that has twice the width of the critical distance is centered at the rank of the reference classifier. Thus, another classifier falling outside this interval differs from the reference classifier in a statistically significant way.

4.1 Classification Performance

In the first experiment, we compared FURIA to other classifiers using the classification rate as a performance measure, i.e., the percent of correct classifications. For RIPPER we used the WEKA default settings. Moreover, we added two fuzzy rule-based classifiers from the KEEL suite (AFSG$^+$09): The fuzzy grid-based CHI algorithm and the genetic fuzzy rule learner SLAVE. The CHI algorithm is based on (CWY95, CYP96) and uses rule weighing as proposed by (IY05).[1] The SLAVE algorithm makes use of genetic algorithms to learn a fuzzy classifier (GP99; GP01).[2] Finally, we also included the well-known C4.5 decision tree learner as a benchmark classifier (Qui93; Qui95).

Table 2. Average accuracies and ranks

data set	FURIA	RIPPER	C4.5	CHI	SLAVE
acd-authorship	95.67(1)	93.05(3)	93.50(2)	71.60(5)	91.87(4)
acd-bankruptcy	82.57(1)	81.97(2)	81.29(3)	74.40(5)	77.80(4)
acd-cyyoung8092	80.02(2)	80.04(1)	79.86(3)	70.72(5)	79.32(4)
acd-cyyoung9302	82.64(2)	82.01(3)	80.82(4)	80.27(5)	83.90(1)
acd-esr	80.90(2)	82.38(1)	80.36(3)	79.55(4)	77.72(5)
acd-halloffame	92.92(1)	92.87(3)	92.87(2)	92.18(5)	92.68(4)
acd-lawsuit	98.00(1)	97.54(3)	97.94(2)	94.93(4)	94.81(5)
acd-votesurvey	36.92(3)	34.40(4)	38.75(2)	40.19(1)	29.51(5)
biomed	88.31(1)	87.40(3)	87.80(2)	80.64(5)	84.74(4)
cars	79.08(2)	75.93(3)	82.15(1)	68.97(5)	70.68(4)
collins	96.35(1)	92.89(3)	96.10(2)	42.63(5)	50.87(4)

[1] We used the following parameter setting: 3 fuzzy sets, product t-norm, maximum inference, and weighting scheme number 2 from (IY05).

[2] We used the following parameter setting: 5 fuzzy sets, 500 iterations without change, mutation probability 0.01, use weights, population size 100.

Table 2. (*continued*)

data set	FURIA	RIPPER	C4.5	CHI	SLAVE
ecoli	83.12(1)	80.57(4)	81.35(2)	77.43(5)	81.03(3)
eucalyptus	60.62(1)	58.69(3)	59.98(2)	54.09(5)	58.16(4)
glass	68.22(1)	63.18(3)	66.69(2)	61.39(5)	61.83(4)
haberman	72.72(3)	72.16(4)	71.75(5)	73.08(2)	73.31(1)
heart-statlog	79.75(1)	78.44(2)	77.08(4)	68.66(5)	78.44(3)
ionosphere	89.59(2)	88.64(4)	88.72(3)	66.40(5)	89.83(1)
iris	94.76(2)	93.45(4)	94.25(3)	92.27(5)	94.92(1)
liver-disorders	67.15(1)	65.93(2)	63.40(3)	58.75(5)	59.77(4)
metStatCoord.	93.02(1)	92.04(3)	92.87(2)	46.79(5)	58.77(4)
metStatRainfall	64.51(1)	60.66(2)	59.47(3)	24.51(5)	29.35(4)
metStatRST	33.56(4)	36.08(3)	38.60(2)	25.24(5)	42.02(1)
metStatSunshine	49.05(1)	44.48(3)	46.78(2)	37.93(4)	28.83(5)
metStatTemp	50.71(2)	47.45(3)	53.18(1)	30.63(4)	22.10(5)
mfeat-factors	92.09(1)	87.05(4)	87.96(3)	89.19(2)	86.83(5)
mfeat-fourier	76.69(1)	71.37(4)	74.42(2)	69.27(5)	73.49(3)
mfeat-karhunen	86.47(1)	79.13(4)	80.20(3)	82.55(2)	78.37(5)
mfeat-morpholog.	72.09(1)	70.74(3)	71.60(2)	57.93(5)	67.08(4)
mfeat-zernike	73.67(1)	67.58(5)	69.11(3)	72.37(2)	68.26(4)
optdigits	94.78(1)	89.68(3)	89.51(4)	45.90(5)	93.45(2)
page-blocks	97.02(1)	96.79(3)	96.89(2)	91.96(5)	93.58(4)
pasture-prod.	74.67(1)	68.46(3)	73.67(2)	44.23(5)	53.63(4)
pendigits	97.77(1)	95.54(4)	95.92(3)	97.45(2)	87.26(5)
pima diabetes	74.71(1)	74.56(2)	73.43(4)	72.55(5)	73.65(3)
prnn-synth	83.57(2)	82.50(4)	83.18(3)	84.14(1)	81.51(5)
schizo-	80.52(1)	75.33(2)	74.93(3)	56.08(5)	56.29(4)
segment	96.50(1)	94.53(3)	95.95(2)	83.65(5)	88.87(4)
sonar	77.01(1)	72.41(3)	72.09(4)	74.61(2)	68.50(5)
squash-unstored	76.44(1)	71.74(3)	76.08(2)	70.56(4)	65.56(5)
synthetic control	89.75(2)	82.85(4)	90.00(1)	68.33(5)	89.23(3)
vehicle	70.10(2)	67.80(3)	71.38(1)	61.99(5)	64.08(4)
vowel	75.43(2)	64.71(3)	75.60(1)	59.49(5)	63.84(4)
waveform	82.24(1)	78.72(2)	75.05(4)	72.38(5)	75.34(3)
wine	93.25(1)	90.02(5)	91.22(4)	92.77(2)	92.46(3)
w.-breast-cancer	95.68(1)	95.58(2)	94.51(4)	90.20(5)	95.49(3)
average	1.40	3.07	2.60	4.24	3.69

The overall picture conveyed by the results, summarized in Table 2, is clearly in favor of FURIA, which outperforms the other methods on most data sets. In fact, the Friedman test rejects the null-hypothesis quite safely (at a significance level < 0.01) and, hence, indicates that there are significant differences between the classifiers' performance. The results of the Bonferroni-Dunn test are summarized graphically in Fig. 4. As can be seen, FURIA is significantly better than all other classifiers at the significance level of 0.05.

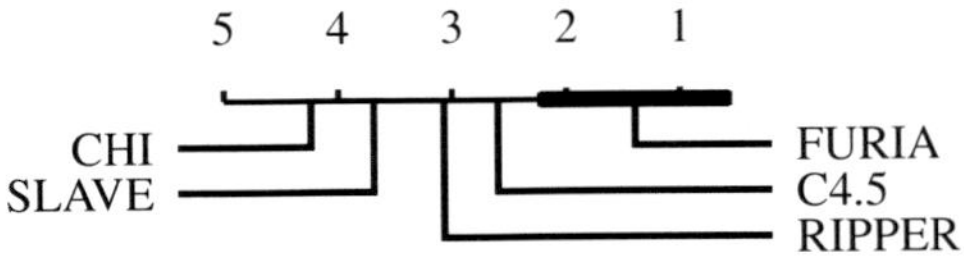

Fig. 4. Visualization of the Bonferroni-Dunn test according to Demšar (Dem06). FURIA is significantly better than its competitors in terms of classification rate (significance level 0.05).

4.2 Ranking Performance

In a second study, we analyzed the ranking performance of the classifiers, using the AUC (Area Under the ROC Curve) as an evaluation measure (PF97; PFK98). This study is motivated by the recent paper (HV09), according to which fuzzy models are especially suitable for this kind of problem. In bipartite ranking, the problem is to order the test instances from the most likely positive to the most likely negative. Typically, a ranking of this kind is established by sorting the instances according to the scores they receive from the classifier. A ranking error occurs when a negative instance is ranked higher than a positive one, and the AUC essentially counts the number of such errors. For multi-class problems, we used an extension of the AUC as proposed in (PD03).

Table 3. Average AUC and ranks

data set	FURIA	RIPPER.	C4.5	CHI	SLAVE.
acd-authorship	0.98(1)	0.96(3)	0.96(4)	0.90(5)	0.98(2)
acd-bankruptcy	0.87(1)	0.82(5)	0.83(4)	0.87(2)	0.86(3)
acd-cyyoung8092	0.74(2)	0.71(3)	0.67(4)	0.65(5)	0.77(1)
acd-cyyoung9302	0.77(2)	0.71(4)	0.70(5)	0.75(3)	0.79(1)
acd-esr	0.67(1)	0.62(3)	0.63(2)	0.60(4)	0.59(5)
acd-halloffame	0.83(3)	0.79(5)	0.81(4)	0.87(2)	0.89(1)
acd-lawsuit	0.95(2)	0.92(4)	0.93(3)	0.81(5)	0.96(1)
acd-votesurvey	0.52(3)	0.50(4)	0.54(2)	0.55(1)	0.49(5)
biomed	0.90(3)	0.86(5)	0.87(4)	0.93(1)	0.91(2)
cars	0.88(2)	0.82(4)	0.90(1)	0.86(3)	0.81(5)
collins	1.00(1)	0.97(3)	0.98(2)	0.88(4)	0.83(5)
ecoli	0.91(3)	0.89(4)	0.89(5)	0.94(1)	0.91(2)
eucalyptus	0.79(3)	0.82(2)	0.82(1)	0.79(4)	0.77(5)
glass	0.81(1)	0.76(5)	0.79(3)	0.77(4)	0.79(2)
haberman	0.62(2)	0.60(3)	0.56(5)	0.65(1)	0.60(4)
heart-statlog	0.82(1)	0.79(3)	0.78(4)	0.76(5)	0.81(2)
ionosphere	0.91(2)	0.88(4)	0.88(5)	0.89(3)	0.92(1)
iris	0.97(3)	0.96(5)	0.96(4)	1.00(1)	0.98(2)
liver-disorders	0.68(1)	0.64(2)	0.62(3)	0.57(5)	0.60(4)
metStatCoord.	0.98(2)	0.98(1)	0.98(3)	0.78(5)	0.82(4)
metStatRainfall	0.87(1)	0.85(2)	0.81(3)	0.76(4)	0.66(5)
metStatRST	0.65(3)	0.70(2)	0.71(1)	0.59(4)	0.58(5)
metStatSunshine	0.77(2)	0.75(4)	0.74(5)	0.80(1)	0.75(3)
metStatTemp	0.76(3)	0.77(2)	0.77(1)	0.66(5)	0.68(4)

Table 3. (*continued*)

data set	FURIA	RIPPER	C4.5	CHI	SLAVE
mfeat-factors	0.98(2)	0.95(4)	0.94(5)	0.98(1)	0.97(3)
mfeat-fourier	0.92(2)	0.92(4)	0.88(5)	0.92(1)	0.92(3)
mfeat-karhunen	0.96(2)	0.92(4)	0.90(5)	0.98(1)	0.95(3)
mfeat-morpholog.	0.88(5)	0.94(1)	0.92(3)	0.93(2)	0.89(4)
mfeat-zernike	0.91(2)	0.90(4)	0.87(5)	0.95(1)	0.90(3)
optdigits	0.99(1)	0.96(3)	0.95(4)	0.82(5)	0.96(2)
page-blocks	0.95(1)	0.93(3)	0.93(2)	0.82(5)	0.83(4)
pasture-prod.	0.86(1)	0.79(3)	0.83(2)	0.64(5)	0.71(4)
pendigits	1.00(2)	0.98(4)	0.98(5)	1.00(1)	0.99(3)
pima diabetes	0.73(4)	0.71(5)	0.74(3)	0.80(1)	0.76(2)
prnn-synth	0.85(3)	0.84(5)	0.84(4)	0.90(1)	0.90(2)
schizo-	0.86(1)	0.78(3)	0.81(2)	0.56(5)	0.59(4)
segment	0.99(1)	0.98(2)	0.98(3)	0.97(4)	0.96(5)
sonar	0.81(2)	0.74(4)	0.72(5)	0.82(1)	0.75(3)
squash-unstored	0.83(1)	0.77(3)	0.81(2)	0.71(5)	0.76(4)
synthetic control	0.97(3)	0.93(5)	0.95(4)	0.99(1)	0.99(2)
vehicle	0.85(1)	0.85(2)	0.85(4)	0.85(3)	0.83(5)
vowel	0.93(1)	0.88(5)	0.91(3)	0.91(2)	0.89(4)
waveform	0.91(2)	0.88(4)	0.83(5)	0.91(3)	0.91(1)
wine	0.97(3)	0.93(5)	0.93(4)	0.98(1)	0.97(2)
w.-breast-cancer	0.97(3)	0.96(4)	0.95(5)	0.98(1)	0.97(2)
average	2.02	3.53	3.51	2.84	3.09

The results of this study, summarized in Table 3 and Fig. 5, convey the same message as the previous one: FURIA is significantly better than the other classifiers. Worth mentioning is also the improved performance of the other two fuzzy classifiers, CHI and SLAVE. The fact that the fuzzy approaches outperform the non-fuzzy ones in terms of ranking performance is completely in agreement with the results in (HV09).

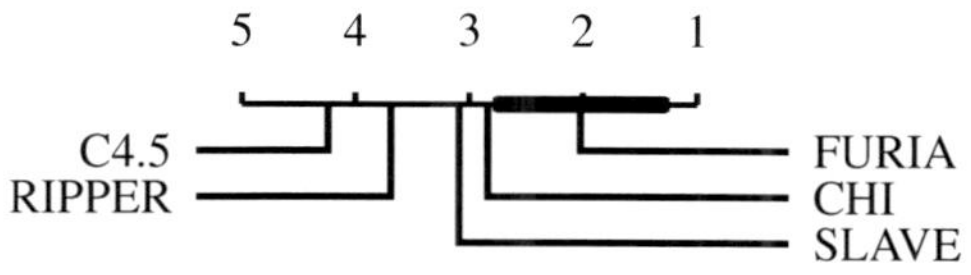

Fig. 5. Visualization of the Bonferroni-Dunn test according to Demšar (Dem06). FURIA is significantly better than its competitors in terms of AUC (significance level 0.1).

4.3 The Effect of Fuzzification

The previous results have shown that FURIA is a significant improvement in comparison to RIPPER. Since FURIA differs from RIPPER in several ways, it is interesting to investigate the influence of the different modifications. One may wonder, for example, to what extent the improvements can be attributed to the use of fuzzy instead of conventional rules or, more generally, to the post-generalization step which involves the fuzzification of rules as a key component. Needless to say, distilling and quantifying the influence of single modifications is rather difficult, if not impossible, given that all components of an algorithm interact in a complex and non-trivial way. Nevertheless, to get a rough idea, we shall at least make an attempt in this direction. To this end, we compare different variants of FURIA in terms of their predictive performance:

FURIA-prod. The original FURIA algorithm using the product t-norm in (2) to combine rule antecedents.

FURIA-min. The FURIA algorithm using the minimum t-norm instead of the product.

FURIA-crisp. This is a "crisp" version of FURIA, in which the fuzzy intervals are turned back into normal intervals. To this end, all membership degrees $I^F(x_i) > 0$ are lifted to the degree 1.

FURIA-w/o. This version does not apply any post-generalization technique; the rules remain in the form as they were learned by RIPPER.

To minimize the interference with other extensions, we disable the rule stretching technique. This, however, means that test instances may remain uncovered, i.e., that FURIA will abstain from the classification of some instances. To take this into consideration, we measure three instead of only a single performance metric: The *accuracy* is the percentage of correct classifications, the *error* the percentage of incorrect classifications, and the *abstention* the percentage of unclassified instances; obviously, these three quantities sum to 1.

The complete results of this test can be found in Tables 6–9. A synopsis can also be found in Table 4 and Table 5. To cope with the large amount of information, we try to answer a number of relevant questions step by step.

4.3.1 Benefit of Post-Generalization

In Section 3 and Example 1, we argued that post-generalizing the rules may correct an improper placement of the interval boundaries. Our first question therefore concerns the

Table 4. Wins and losses in terms of accuracy and error on the test data for variants of FURIA

	Accuracy				Error			
	prod	min	crisp	w/o	prod	min	crisp	w/o
FURIA-prod	-	22	37	45	-	23	37	0
FURIA-min	8	-	37	45	7	-	37	0
FURIA-crisp	7	7	-	45	7	7	-	0
FURIA-w/o	0	0	0	-	45	45	45	-

Table 5. Wins and losses in terms of accuracy and error on the training data for variants of FURIA

	Accuracy				Error			
	prod	min	crisp	w/o	prod	min	crisp	w/o
FURIA-prod	-	1	6	43	-	1	6	28
FURIA-min	9	-	6	43	9	-	6	28
FURIA-crisp	26	26	-	42	26	26	-	29
FURIA-w/o	0	0	1	-	10	10	9	-

contribution of the post-generalization for avoiding the myopia of separate-and-conquer rule induction. To answer this question we have to look at the performance on the training data. If the generalization process is beneficial, the performance of FURIA-w/o should be worse than the performance of the other variants.

In agreement with this expectation, we find that FURIA-w/o has the worst accuracy on virtually all data sets, cf. Table 5. Moreover, FURIA-w/o has also a higher error. The other variants have a lower classification error on at least 28 data sets and are tying on 7. A simple sign test rejects the hypothesis of equal performance with an error probability of 0.01. From this we can conjecture that the post-generalization has a beneficial impact on both accuracy and error on the training data. Especially the observation of a smaller training error gives evidence to the claim that the post-generalization mitigates the initial myopia.

4.3.2 Benefit of Fuzzy Rules

Our second question concerns the impact of the use of fuzzy instead of conventional rules on the classification performance. Making use of fuzzy intervals, a fuzzy rule can cover an instance to a certain degree: While instances in the core are definitely covered by the rule, the degree of coverage decreases linearly toward the boundary of the rule. Of course, even though this notion makes sense intuitively, it is a-priori not clear that it also leads to an improved performance. Therefore, we compare the fuzzy versions FURIA-prod and FURIA-min to the non-fuzzy variant FURIA-crisp. As can be seen in Table 4, FURIA-prod and FURIA-min win 37 and lose only 7 data sets over FURIA-crisp in terms of both accuracy and error on the test data.

A surprising observation is that, in contrast to the test data, FURIA-crisp wins 26 over both fuzzy competitors FURIA-prod and FURIA-min and ties 13 data sets in terms of accuracy and error on the training data (cf. Table 5). Interestingly, a similar observation has been made by Kuwajima et al. (KNI08). In a related analysis, they also found that a stronger fuzzification leads to worse results on the training data but not on the test data.

In summary, these results clearly suggest that a gradual extrapolation, which gives a higher support to instances in the core of a rule than to instances in the boundary region, is not only intuitively appealing but also advantageous with regard to generalization performance.

4.3.3 Choice of Fuzzy Aggregation Operator

Finally, we investigated whether the product t-norm is preferable to the minimum t-norm for fuzzy inference. Since the product has isobars with a round shape, it leads to decision boundaries that are smoother than the boundaries produced by the minimum. Moreover, it extrapolates in a more "careful" way, since the product yields membership degrees that are smaller than those produced by the minimum.[3] Finally, the product gives rise to an interesting probabilistic interpretation: Looking at the membership degrees as (cumulative) probabilities of the location of the interval boundaries, which are considered as random variables, and making a simplifying assumption of independence

[3] For an instance x covered by I_1 and I_2 with $0 < I_1(x_1) \neq I_2(x_2) < 1$, we have $\top_{\text{prod}}\big(I_1(x_1), I_2(x_2)\big) < \top_{\min}\big(I_1(x_1), I_2(x_2)\big)$.

between the boundaries of different attributes, the product combination corresponds to the probability that an instance is covered by the rule.

In terms of predictive performance, we indeed find that FURIA-prod is significantly better than FURIA-min (at a significant level of 0.05): It wins on 22 data sets and ties on 15 for accuracy, and for the error measure it wins even 23 data sets. It is worth mentioning that the results are again different on the training data: FURIA-min is better than FURIA-prod on 9 data sets and worse on only one for both classification accuracy and error. Again, it seems that a more "careful" extrapolation, as realized by the product t-norm, is advantageous in terms of generalization performance.

5 Summary and Conclusion

A major goal of this paper is to show that rule induction, an important subfield of machine learning, can strongly benefit from tools and concepts that have been developed in the field of fuzzy logic. To this end, we analyzed FURIA, a fuzzy rule-based classification method that builds on RIPPER. Our experimental results have shown that FURIA significantly outperforms the original RIPPER, as well as other rule learners used for comparison, in terms of both classification and ranking performance.

Elaborating on the advantages of a fuzzy approach, we tried to distill and quantify the influence of rule fuzzification on the strong classification performance of FURIA. Conducting experiments with different variants of the algorithm, we indeed found strong evidence for the importance of fuzzy rules. In particular, it seems that a gradual extrapolation as realized by fuzzy rules is advantageous with regard to the generalization performance of a rule-based classifier.

Going beyond the conventional classification problem, we also investigated the performance of FURIA in the context of bipartite ranking. As noted in the recent literature, fuzzy models appear to be especially suitable for this type of problem, a supposition which is fully supported by our results. In future work, we therefore plan to investigate the advantages of a fuzzy approach in the context of ranking in more detail.

Further Tables

Table 6. Average accuracy on the test data for variants of FURIA

data set	prod	min	crisp	w/o
analcatdata-authorship	92.80921	92.80921	92.55734	91.22824
analcatdata-bankruptcy	81.32557	81.32925	82.58047	76.12663
analcatdata-cyyoung8092	76.01159	76.04189	76.19168	73.30849
analcatdata-cyyoung9302	78.74153	78.70927	79.05907	76.49093
analcatdata-esr	80.09091	80.09091	80.91818	75.97121
analcatdata-halloffame	90.87807	90.87588	90.84069	89.38324
analcatdata-lawsuit	97.77013	97.77013	97.72594	97.02380
analcatdata-votesurvey	8.56054	8.56054	8.56054	7.33627
biomed	85.42540	85.41131	85.18755	82.39893

Table 6. *(continued)*

data set	prod	min	crisp	w/o
cars	74.60520	74.59071	74.51809	72.33939
collins	96.01711	96.02881	95.10308	89.67611
ecoli	80.58999	80.58138	79.96263	76.89322
eucalyptus	47.92876	47.92878	47.72092	46.76565
glass	60.66062	60.64654	59.69095	54.49754
haberman	67.24621	67.24621	67.32295	66.93906
heart-statlog	73.32883	73.32883	73.14369	71.92296
ionosphere	86.64396	86.62716	86.51006	84.33937
iris	94.00000	94.00000	93.33333	91.74510
liver-disorders	57.66717	57.66717	57.47095	56.12342
metStatCoordinates	92.18901	92.18282	91.98087	90.28669
metStatRainfall	56.11860	56.11798	55.69124	51.63397
metStatRST	22.53966	22.53966	22.23236	20.24124
metStatSunshine	41.36317	41.37031	41.12752	36.62300
metStatTemp	42.04130	42.03699	41.84980	40.39076
mfeat-factors	88.99853	88.99265	88.74559	86.35588
mfeat-fourier	70.69559	70.69265	70.27794	66.62353
mfeat-karhunen	82.40294	82.40441	81.67794	77.45000
mfeat-morphological	68.09853	68.09853	68.09706	66.80441
mfeat-zernike	68.08382	68.09265	67.57206	63.87941
optdigits	92.55994	92.55993	92.31394	90.40031
page-blocks	96.52939	96.52831	96.37623	95.90225
pasture-production	64.65293	64.65293	63.30037	58.21520
pendigits	97.07328	97.07007	96.70082	95.02126
pima diabetes	68.94244	68.94244	68.91185	67.83203
prnn-synth	81.04536	81.04536	80.99885	79.49978
schizo-	72.38598	72.37736	71.90262	69.93216
segment	95.53575	95.53829	95.13213	93.69444
sonar	70.20954	70.19525	69.63149	66.04831
squash-unstored	72.87754	72.87754	73.19556	67.90609
synthetic control	85.52941	85.51961	84.48039	80.25980
vehicle	62.84809	62.84462	62.62561	61.07522
vowel	71.10840	71.08169	67.91235	61.32328
waveform-5000	75.34261	75.34261	75.34909	70.52370
wine	90.57375	90.55709	89.61175	86.27115
wisconsin-breast-cancer	94.25726	94.25726	94.14377	93.64308

Table 7. Average error on the test data for variants of FURIA

data set	prod	min	crisp	w/o
analcatdata-authorship	2.68560	2.68560	2.93748	2.52823
analcatdata-bankruptcy	16.78513	16.78145	15.53023	12.45956
analcatdata-cyyoung8092	17.39773	17.36742	17.21764	15.79217
analcatdata-cyyoung9302	14.82836	14.86062	14.51082	13.19684

Table 7. (*continued*)

data set	prod	min	crisp	w/o
analcatdata-esr	17.35909	17.35909	16.53182	15.54545
analcatdata-halloffame	5.71984	5.72203	5.75722	5.16018
analcatdata-lawsuit	1.93935	1.93935	1.98355	1.74996
analcatdata-votesurvey	18.96895	18.96895	18.96895	16.91095
biomed	9.42371	9.43779	9.66155	8.17166
cars	14.39157	14.40606	14.47868	12.12514
collins	2.97016	2.95846	3.88419	0.12375
ecoli	13.89950	13.90812	14.52686	12.33990
eucalyptus	23.42483	23.42482	23.63268	21.91884
glass	23.12261	23.13669	24.09228	18.70914
haberman	22.99615	22.99615	22.91941	22.73690
heart-statlog	16.25442	16.25442	16.43956	15.65444
ionosphere	8.51115	8.52796	8.64505	7.45617
iris	4.70588	4.70588	5.37255	4.27451
liver-disorders	25.18079	25.18079	25.37701	24.30197
metStatCoordinates	6.13532	6.14151	6.34346	5.23401
metStatRainfall	20.59847	20.59909	21.02583	16.57657
metStatRST	27.89949	27.89949	28.20679	22.68913
metStatSunshine	30.61461	30.60746	30.85025	23.06217
metStatTemp	25.78395	25.78826	25.97545	22.89670
mfeat-factors	4.86618	4.87206	5.11912	4.13824
mfeat-fourier	16.02500	16.02794	16.44265	13.05000
mfeat-karhunen	8.48088	8.48088	9.20735	6.58676
mfeat-morphological	22.51029	22.51029	22.51176	21.50441
mfeat-zernike	15.50000	15.49853	16.15735	12.97059
optdigits	3.39419	3.39419	3.64018	2.84412
page-blocks	2.65105	2.65213	2.80421	2.32270
pasture-production	15.12546	15.12546	16.47802	12.53571
pendigits	1.69079	1.69400	2.06325	1.44917
pima diabetes	20.77898	20.77898	20.80957	20.08571
prnn-synth	14.57498	14.57498	14.62149	13.65491
schizo-	13.29685	13.30547	13.78021	12.19070
segment	2.41159	2.40904	2.81521	1.85139
sonar	18.64786	18.66214	19.22590	16.59549
squash-unstored	20.36859	20.36859	20.05057	17.38321
synthetic control	6.91176	6.92157	7.96078	5.85784
vehicle	16.90668	16.91015	17.12915	15.50935
vowel	16.82586	16.85257	20.02191	11.63290
waveform-5000	14.07581	14.07581	14.06934	12.21165
wine	4.35848	4.37514	5.32048	3.10506
wisconsin-breast-cancer	3.66027	3.66027	3.77375	3.50447

Table 8. Average accuracy on the training data for variants of FURIA

data set	prod	min	crisp	w/o
analcatdata-authorship	99.58021	99.58021	99.58021	99.53697
analcatdata-bankruptcy	97.06350	97.06350	97.06350	96.91199
analcatdata-cyyoung8092	90.53704	90.53704	90.53704	90.27212
analcatdata-cyyoung9302	92.34319	92.34319	92.34319	92.21176
analcatdata-esr	88.24697	88.24697	88.24697	88.05866
analcatdata-halloffame	95.40479	95.40479	95.37877	95.18996
analcatdata-lawsuit	99.10482	99.10482	99.10482	99.09907
analcatdata-votesurvey	24.05953	24.05953	24.05953	24.05953
biomed	96.04186	96.04186	96.05630	95.85345
cars	87.73836	87.73836	87.65253	87.48447
collins	99.73038	99.73038	99.73038	99.42730
ecoli	90.76378	90.76378	90.78166	90.50300
eucalyptus	61.47577	61.47577	61.51902	61.14015
glass	81.52163	81.52163	81.54296	80.81389
haberman	72.38894	72.38894	72.40874	72.25012
heart-statlog	85.61704	85.61704	85.63392	85.44872
ionosphere	96.34704	96.34704	96.34704	96.19581
iris	97.86869	97.86869	97.93939	97.76768
liver-disorders	72.34301	72.34301	72.38699	71.82926
metStatCoordinates	96.76641	96.76704	96.76321	96.45910
metStatRainfall	73.55085	73.55245	73.61245	72.42661
metStatRST	36.89565	36.89565	36.91824	36.50843
metStatSunshine	69.69173	69.69530	69.71691	69.11754
metStatTemp	63.47378	63.47378	63.47820	63.21942
mfeat-factors	98.97727	98.97727	98.97727	98.93636
mfeat-fourier	89.99167	89.99167	90.00758	89.55682
mfeat-karhunen	97.18485	97.18561	97.18939	96.93182
mfeat-morphological	74.55000	74.55000	74.57955	74.28182
mfeat-zernike	82.13636	82.13712	82.16591	81.69091
optdigits	99.35111	99.35111	99.35084	99.26996
page-blocks	97.86254	97.86032	97.71748	97.74737
pasture-production	89.06472	89.06472	89.06472	89.06472
pendigits	99.64230	99.64244	99.64285	99.53106
pima diabetes	75.01928	75.01928	75.06860	74.48856
prnn-synth	87.12079	87.12079	87.26008	86.89067
schizo-	88.35659	88.35659	88.39222	88.05802
segment	98.83123	98.83188	98.83713	98.64758
sonar	95.69622	95.69622	95.69622	95.50670
squash-unstored	93.04192	93.04192	93.04192	93.01250
synthetic control	99.13384	99.13384	99.13636	98.97980
vehicle	74.76969	74.76969	74.79478	74.42219
vowel	93.58258	93.58411	93.60706	92.87406
waveform-5000	91.31896	91.31956	91.33168	90.22170
wine	98.97811	98.97811	98.97811	98.77371
wisconsin-breast-cancer	98.29834	98.29834	98.28316	98.23550

Table 9. Average error on the training data for variants of FURIA

data set	prod	min	crisp	w/o
analcatdata-authorship	0.00540	0.00540	0.00540	0.00540
analcatdata-bankruptcy	1.60974	1.60974	1.60974	1.76125
analcatdata-cyyoung8092	4.57378	4.57378	4.57378	4.55791
analcatdata-cyyoung9302	2.98185	2.98185	2.98185	3.01464
analcatdata-esr	5.06558	5.06558	5.06558	5.06558
analcatdata-halloffame	1.16123	1.16123	1.18725	0.98257
analcatdata-lawsuit	0.59672	0.59672	0.59672	0.59672
analcatdata-votesurvey	2.82503	2.82503	2.82503	2.82503
biomed	1.13803	1.13803	1.12359	1.18860
cars	3.78576	3.78576	3.87160	3.57683
collins	0.00000	0.00000	0.00000	0.00000
ecoli	4.42350	4.42350	4.40562	4.45045
eucalyptus	10.25930	10.25930	10.21605	10.20780
glass	4.53830	4.53830	4.51697	4.51702
haberman	17.35651	17.35651	17.33671	17.27228
heart-statlog	5.15872	5.15872	5.14183	5.17557
ionosphere	1.39920	1.39920	1.39920	1.44680
iris	1.24242	1.24242	1.17172	1.27273
liver-disorders	10.95915	10.95915	10.91518	11.01625
metStatCoordinates	1.89078	1.89014	1.89397	1.92398
metStatRainfall	4.29721	4.29562	4.23562	4.33423
metStatRST	11.50757	11.50757	11.48498	11.49858
metStatSunshine	4.61678	4.61321	4.59160	4.67405
metStatTemp	7.76461	7.76461	7.76018	7.76904
mfeat-factors	0.05455	0.05455	0.05455	0.06136
mfeat-fourier	1.20985	1.20985	1.19394	1.23258
mfeat-karhunen	0.29091	0.29015	0.28636	0.31742
mfeat-morphological	15.98258	15.98258	15.95303	15.84545
mfeat-zernike	4.17803	4.17727	4.15379	4.19773
optdigits	0.01752	0.01752	0.01779	0.01806
page-blocks	1.39415	1.39637	1.53921	1.14417
pasture-production	1.12862	1.12862	1.12862	1.12862
pendigits	0.03722	0.03708	0.03667	0.04218
pima diabetes	14.12000	14.12000	14.07068	13.76685
prnn-synth	7.99655	7.99655	7.85726	8.02090
schizo-	2.40335	2.40335	2.36772	2.42565
segment	0.22497	0.22431	0.21907	0.23021
sonar	0.79367	0.79367	0.79367	0.81557
squash-unstored	2.75223	2.75223	2.75223	2.75223
synthetic control	0.06313	0.06313	0.06061	0.07323
vehicle	5.75072	5.75072	5.72563	5.76862
vowel	0.80510	0.80357	0.78063	0.85253
waveform-5000	0.33240	0.33180	0.31968	0.34210
wine	0.05939	0.05939	0.05939	0.08496
wisconsin-breast-cancer	0.42705	0.42705	0.44223	0.43569

References

[AFSG+09] Alcalá-Fernandez, J., Sánchez, L., García, S., del Jesus, M.J., Ventura, S., Garrell, J.M., Otero, J., Romero, C., Bacardit, J., Rivas, V.M., Fernández, J.C., Herrera, F.: KEEL: a software tool to assess evolutionary algorithms for data mining problems. Soft Computing 13(3), 307–318 (2009)

[AN07] Asuncion, A., Newman, D.J.: UCI machine learning repository (2007), http://archive.ics.uci.edu/ml/index.html (Obtained on 22nd of August 2007)

[Bar07] Barker, D.: Dataset: Pasture production (2007), http://weka.sourceforge.net/wiki/index.php/Datasets (Obtained on 20th of October 2007)

[Bos04] Boström, H.: Pruning and exclusion criteria for unordered incremental reduced error pruning. In: Proceedings of the Workshop on Advances in Rule Learning, ECML, pp. 17–29 (2004)

[Bul07] Bulloch, B.: Dataset: Eucalyptus soil conservation (2007), http://weka.sourceforge.net/wiki/index.php/Datasets (Obtained on 20th of October 2007)

[Coh95] Cohen, W.W.: Fast effective rule induction. In: Prieditis, A., Russell, S. (eds.) Proceedings of the 12th International Conference on Machine Learning, ICML, Tahoe City, CA, USA, July 9-12, pp. 115–123. Morgan Kaufmann, San Francisco (1995)

[Cv06] Cloete, I., van Zyl, J.: Fuzzy rule induction in a set covering framework. IEEE Transactions on Fuzzy Systems 14(1), 93–110 (2006)

[CWY95] Chi, Z., Wu, J., Yan, H.: Handwritten numeral recognition using self-organizing maps and fuzzy rules. Pattern Recognition 28(1), 59–66 (1995)

[CYP96] Chi, Z., Yan, H., Pham, T.: Fuzzy Algorithms: With Applications to Image Processing and Pattern Recognition. World Scientific Publishing Co., Inc., River Edge (1996)

[Dem06] Demšar, J.: Statistical comparisons of classifiers over multiple data sets. Journal of Machine Learning Research 7, 1–30 (2006)

[Dun61] Dunn, O.J.: Multiple comparisons among means. Journal of the American Statistical Association 56, 52–64 (1961)

[FGHd07] Fernández, A., García, S., Herrera, F., del Jesús, M.J.: An analysis of the rule weights and fuzzy reasoning methods for linguistic rule based classification systems applied to problems with highly imbalanced data sets. In: Masulli, F., Mitra, S., Pasi, G. (eds.) WILF 2007. LNCS (LNAI), vol. 4578, pp. 170–178. Springer, Heidelberg (2007)

[Fri37] Friedman, M.: The use of ranks to avoid the assumption of normality implicit in the analysis of variance. Journal of the American Statistical Association 32(200), 675–701 (1937)

[Fri40] Friedman, M.: A comparison of alternative tests of significance for the problem of m rankings. The Annals of Mathematical Statistics 11(1), 86–92 (1940)

[Für99] Fürnkranz, J.: Separate-and-Conquer rule learning. Artificial Intelligence Review 13(1), 3–54 (1999)

[FW94] Fürnkranz, J., Widmer, G.: Incremental reduced error pruning. In: Cohen, W.W., Hirsh, H. (eds.) Proceedings of the 11th International Conference on Machine Learning, ICML, New Brunswick, NJ, USA, pp. 70–77. Morgan Kaufmann, San Francisco (1994)

[GP99] González, A., Perez, R.: Slave: a genetic learning system based on an iterative approach. IEEE Transactions on Fuzzy Systems 7(2), 176–191 (1999)

[GP01] González, A., Perez, R.: Selection of relevant features in a fuzzy genetic learning algorithm. IEEE Transactions on Systems, Man, and Cybernetics, Part B 31(3), 417–425 (2001)

[Har07] Harvey, W.: Dataset: Squash harvest stored / unstored (2007), `http://weka.sourceforge.net/wiki/index.php/Datasets` (Obtained on 20th of October 2007)

[HH09] Hühn, J.C., Hüllermeier, E.: Furia: an algorithm for unordered fuzzy rule induction. In: Data Mining and Knowledge Discovery (2009)

[Hül05] Hüllermeier, E.: Fuzzy sets in machine learning and data mining: Status and prospects. Fuzzy Sets and Systems 156(3), 387–406 (2005)

[HV09] Hüllermeier, E., Vanderlooy, S.: Why fuzzy decision trees are good rankers. IEEE Transactions on Fuzzy Systems (2009)

[IN01] Ishibuchi, H., Nakashima, T.: Effect of rule weights in fuzzy rule-based classification systems. IEEE Transactions on Fuzzy Systems 9(4), 506–515 (2001)

[IY05] Ishibuchi, H., Yamamoto, T.: Rule weight specification in fuzzy rule-based classification systems. IEEE Transactions on Fuzzy Systems 13(4), 428–436 (2005)

[JCC07] Juang, C., Chiu, S., Chang, S.: A self-organizing TS-Type fuzzy network with support vector learning and its application to classification problems. IEEE Transactions on Fuzzy Systems 15(5), 998–1008 (2007)

[KNI08] Kuwajima, I., Nojima, Y., Ishibuchi, H.: Effects of constructing fuzzy discretization from crisp discretization for rule-based classifiers. Artificial Life and Robotics 13(1), 294–297 (2008)

[MV07] Meyer, M., Vlachos, P.: Statlib (2007), `http://lib.stat.cmu.edu/`

[PD03] Provost, F., Domingos, P.: Tree induction for probability-based ranking. Machine Learning 52(3), 199–215 (2003)

[PF97] Provost, F.J., Fawcett, T.: Analysis and visualization of classifier performance: Comparison under imprecise class and cost distributions. In: Heckerman, D., Mannila, H., Pregibon, D. (eds.) Proceedings of the Third International Conference on Knowledge Discovery and Data Mining, KDD, Newport Beach, CA, USA, pp. 43–48. AAAI Press, Menlo Park (1997)

[PFK98] Provost, F.J., Fawcett, T., Kohavi, R.: The case against accuracy estimation for comparing induction algorithms. In: Shavlik, J.W. (ed.) Proceedings of the Fifteenth International Conference on Machine Learning, ICML, Madison, WI, USA, pp. 445–453. Morgan Kaufmann, San Francisco (1998)

[PFTV92] Press, W.H., Flannery, B.P., Teukolsky, S.A., Vetterling, W.T.: Numerical Recipes in FORTRAN: The Art of Scientific Computing, 2nd edn. Cambridge University Press, Cambridge (1992)

[QCJ93] Quinlan, J.R., Cameron-Jones, R.M.: FOIL: A midterm report. In: Brazdil, P.B. (ed.) ECML 1993. LNCS, vol. 667, pp. 3–20. Springer, Heidelberg (1993)

[Qui90] Quinlan, J.R.: Learning logical definitions from relations. Machine Learning 5(3), 239–266 (1990)

[Qui93] Quinlan, J.R.: C4.5: programs for machine learning. Morgan Kaufmann, San Francisco (1993)

[Qui95] Quinlan, J.R.: MDL and categorial theories (continued). In: Prieditis, A., Russell, S.J. (eds.) Proceedings of the 12th International Conference on Machine Learning, ICML, Lake Tahoe, CA, USA, pp. 464–470. Morgan Kaufmann, San Francisco (1995)

Increasing Incompleteness of Data Sets— A Strategy for Inducing Better Rule Sets

Jerzy W. Grzymala-Busse[1,2] and Witold J. Grzymala-Busse[3]

[1] Department of Electrical Engineering and Computer Science, University of Kansas, Lawrence, KS 66045, USA
[2] Institute of Computer Science, Polish Academy of Sciences, 01–237 Warsaw, Poland
[3] Touchnet Information Systems, Inc., Lenexa, KS 66219, USA

Abstract. Our main objective was to verify the following hypothesis: for some complete (i.e., without missing attribute vales) data sets it is possible to induce better rule sets (in terms of an error rate) by increasing incompleteness (i.e., removing some existing attribute values) of the original data sets. In this paper we present detailed results of experiments showing that rule sets induced from some incomplete data sets are significantly better than the rule set induced from the original (complete) data set, with the significance level of 5%, two-tailed test. Additionally, we discuss criteria for inducing better rules by increasing incompleteness.

Keywords: incomplete data sets, missing attribute values, approximations for incomplete data, LERS data mining system, MLEM2 algorithm.

1 Introduction

In this paper we show that by increasing incompleteness of a data set (i.e., by removing attribute values in the data set) we may improve quality of the rule sets induced from such a modified data set. In our experiments we replaced randomly existing attribute values in the original data sets by symbols that were recognized by the rule induction module as missing attribute values. In other words, the rule sets were induced from data sets in which some values were erased using a Monte Carlo method. The process of such replacements was done incrementally, with an increment equal to 5% of the total number of attribute values for a given data set.

We assume that missing attribute values have three different interpretations. The first type of missing attribute values will be called *lost*. Such values were originally specified, but currently they are unavailable, e.g., they were incidentally erased, forgotten to be recorded, etc. A rough set approach to incomplete data sets in which all attribute values were lost was presented for the first time in [1], where two algorithms for rule induction, modified to handle lost attribute values, were introduced. Research on lost values was continued, e.g., in works [2, 3].

The next type of missing attribute vales will be called a *"do not care" condition*. Such missing attribute values were off the point during collection of data, i.e., an expert decided that the attribute value was irrelevant for a classification or diagnosis of

J. Koronacki et al. (Eds.): Advances in Machine Learning I, SCI 262, pp. 345–365.
springerlink.com © Springer-Verlag Berlin Heidelberg 2010

the case. For example, a data set describing flu patients may contain, among other attributes, an attribute *Color of hair*. Though some scrupulous patients may fill in this value, other patients may assume that this attribute is irrelevant for the flu diagnosis and leave it unspecified. If we suspect that this attribute does matter, the best interpretation for missing attribute values is replacing them by all possible existing attribute values. A rough set approach to incomplete data sets in which all attribute values were "do not care" conditions was presented for the first time in [4], where a method for rule induction was introduced in which each missing attribute value was replaced by all values from the domain of the attribute. Later this research was continued, e.g., in [5, 6].

The third type of missing attribute value will be called an *attribute-concept value*. Its interpretation is similar to a "do not care" condition, however, it is restricted to a specific concept. A *concept* (*class*) is a set of all cases classified (or diagnosed) the same way. Using this interpretation, we will replace a missing attribute value by all values of the same attribute typical for attribute values for the concept to which this case belongs. Let us consider a patient, sick with flu, with a missing attribute value for *Color of hair*. Other patients, sick with flu, filled in values *brown* and *grey* for this attribute. On the other hand, healthy patients characterized the color of their hair as *blond* and *brown*. Using attribute-concept value interpretation, this missing attribute value is replaced by *brown* and *grey*. If we would use "do not care" condition interpretation, the same missing attribute value should be replaced by *blond*, *brown*, and *grey*. This approach to missing attribute values was introduced in [7].

In general, incomplete decision tables are described by characteristic relations, in a similar way as complete decision tables are described by indiscernibility relations [8, 9, 10].

In rough set theory, one of the basic notions is the idea of lower and upper approximations. For complete decision tables, for a given indiscernibility relation and for any subset of the set U of all cases, its lower and upper approximations are unique.

For incomplete decision tables, for a given characteristic relation and a subset of U, there are three important and different possibilities to define lower and upper approximations, called *singleton*, *subset*, and *concept* approximations [8]. Singleton lower and upper approximations were studied in [3, 5, 6, 11]. Note that similar definitions of lower and upper approximations, though not for incomplete decision tables, were studied in [12]. Some other rough-set approaches to missing attribute values were presented in [4, 13, 14, 15] as well.

A preliminary version of this paper was presented at the 6-th International Conference on Rough Sets and Current Trends in Computing, RSCTC'2008, Akron, OH, October 23–25, 2008 [16].

2 Blocks of Attribute-Value Pairs—Complete Data

Input data sets are presented in the form of a *decision table*. An example of the decision table is shown in Table 1. Rows of the decision table represent *cases*, while columns are labeled by *variables*. The set of all cases will be denoted by U. In Table 1, U = {1, 2,..., 7}. Independent variables are called *attributes* and a dependent variable is called a *decision* and is denoted by d. The set of all attributes will be denoted by A. In

Table 1, $A = \{Temperature, Headache, Cough\}$. Any decision table defines a function ρ that maps the direct product of U and A into the set of all values. For example, in Table 1, $\rho(1, Temperature) = 38$. A decision table with completely specified function ρ will be called *completely specified*, or, for the sake of simplicity, *complete*.

In real world, data are frequently affected by missing attribute values. In other words, the corresponding function ρ is incompletely specified (partial). A decision table with an incompletely specified function ρ will be called *incomplete*. Function ρ describing Table 1 is completely specified.

An important tool to analyze complete decision tables is a block of the attribute-value pair. Let a be an attribute, i.e., $a \in A$ and let v be a value of a for some case. For complete decision tables if $t = (a, v)$ is an attribute-value pair then a *block* of t, denoted by $[t]$, is a set of all cases from U that for attribute a have value v.

Rough set theory [17, 18] is based on the idea of an indiscernibility relation, defined for complete decision tables. Let B be a nonempty subset of the set A of all attributes. The *indiscernibility relation* $IND(B)$ is a relation on U defined for $x, y \in U$ as follows

$$(x, y) \in IND(B) \text{ if and only if } \rho(x, a) = \rho(y, a) \text{ for all } a \in B.$$

The indiscernibility relation $IND(B)$ is an equivalence relation. Equivalence classes of $IND(B)$ are called *elementary sets* of B and are denoted by $[x]_B$, where $x \in U$. For example, for Table 1, elementary sets of $IND(A)$ are $\{1\}$, $\{2\}$, $\{3\}$, $\{4, 5\}$, $\{6\}$, and $\{7\}$. Additionally, IND(A) = {(1, 1), (2, 2), (3, 3), (4, 4), (4, 5), (5, 4), (5, 5), (6, 6), (7, 7)}. The indiscernibility relation $IND(B)$ may be computed using the idea of blocks of attribute-value pairs. For Table 1,

[(Temperature, 36)] = {6, 7},
[(Temperature, 38)] = {1, 3, 4, 5},
[(Temperature, 40)] = {2},
[(Headache, severe)] = {1},
[(Headache, no)] = {2, 6, 7},
[(Headache, mild)] = {3, 4, 5},

Table 1. A complete decision table

		Attributes		Decision
Case	Temperature	Headache	Cough	Flu
1	38	severe	mild	yes
2	40	no	severe	yes
3	38	mild	severe	yes
4	38	mild	no	yes
5	38	mild	no	no
6	36	no	mild	no
7	36	no	no	no

[(Cough, mild)] = {1, 6},
[(Cough, severe)] = {2, 3}.
[(Cough, no)] = {4, 5, 7}.

The indiscernibility relation $IND(B)$ is known when all elementary blocks of IND(B) are known. Such elementary blocks of B are intersections of the corresponding attribute-value pairs, i.e., for any case $x \in U$,

$$[x]_B = \cap\{[(a, v)] \mid a \in B, \rho(x, a) = v\}.$$

We will illustrate the idea how to compute elementary sets of B for Table 1 and $B = A$.

$[1]_A = \{1, 3, 4, 5\} \cap \{1\} \cap \{1, 6\} = \{1\}$,
$[2]_A = \{2\} \cap \{2, 6, 7\} \cap \{2, 3\} = \{2\}$,
$[3]_A = \{1, 3, 4, 5\} \cap \{3, 4, 5\} \cap \{2, 3\} = \{3\}$,
$[4]_A = [5]_A = \{1, 3, 4, 5\} \cap \{3, 4, 5\} \cap \{4, 5, 7\} = \{4, 5\}$,
$[6]_A = \{6, 7\} \cap \{2, 6, 7\} \cap \{1, 6\} = \{6\}$,
$[7]_A = \{6, 7\} \cap \{2, 6, 7\} \cap \{4, 5, 7\} = \{7\}$,

For completely specified decision tables lower and upper approximations are defined using the indiscernibility relation. Any finite union of elementary sets, associated with B, will be called a *B-definable set*. Let X be any subset of the set U of all cases. Let us recall that the set X is called a *concept* and is usually defined as the set of all cases defined by a specific value of the decision. In general, X is not a B-definable set. However, set X may be approximated by two B-definable sets, the first one is called a *B-lower approximation* of X, denoted by $\underline{B}X$ and defined as follows

$$\cup\{[x]_B \mid x \in U, [x]_B \subseteq X\},$$

The second set is called a *B-upper approximation* of X, denoted by $\overline{B}X$ and defined as follows

$$\cup\{[x]_B \mid x \in U, [x]_B \cap X \neq \emptyset).$$

Data set presented in Table 1 is inconsistent, the lower approximations and upper approximations for both concepts are

$$\underline{A}\{1, 2, 3, 4\} = \{1, 2, 3\},$$

$$\underline{A}\{5, 6, 7\} = \{6, 7\},$$

$$\overline{A}\{1, 2, 3, 4\} = \{1, 2, 3, 4, 5\},$$

$$\overline{A}\{5, 6, 7\} = \{4, 5, 6, 7\}.$$

Rules in the LERS format (every rule is equipped with three numbers, the total number of attribute-value pairs on the left-hand side of the rule, the total number of cases correctly classified by the rule during training, and the total number of training cases matching the left-hand side of the rule) [19] are:

certain rules:
1, 2, 2
(Cough, severe) -> (Flu, yes),
1, 1, 1
(Headache, severe) -> (Flu, yes),
1, 2, 2
(Temperature, 36..37) -> (Flu, no),

and possible rules:
1, 4, 5
(Temperature, 37..40) -> (Flu, yes),
1, 2, 3
(Cough, no) -> (Flu, no),
1, 2, 2
(Temperature, 36..37) -> (Flu, no).

Note that the above rules were induced by the MLEM2 (Modified Learning from Examples Module, version 2) option of the LERS (Learning from Examples based on Rough Sets) data mining system [19, 20, 21].

3 Blocks of Attribute-Value Pairs—Incomplete Data

For the rest of the paper we will assume that all decision values are specified, i.e., they are not missing. Additionally, we will assume that lost values will be denoted by "?", "do not care" conditions by "*", and attribute-concept values by "−". Additionally, we will assume that for each case at least one attribute value is specified.

An example of an inconsistent data set is presented in Table 2. For incomplete decision tables, a block of an attribute-value pair must be modified in the following way:

- If for an attribute a there exists a case x such that $\rho(x, a) = ?$, i.e., the corresponding value is lost, then the case x should not be included in any blocks $[(a, v)]$ for all values v of attribute a,

Table 2. An incomplete decision table

	Attributes			Decision
Case	Temperature	Headache	Cough	Flu
1	*	severe	mild	yes
2	40	−	severe	yes
3	38	mild	−	yes
4	?	mild	no	yes
5	38	mild	?	no
6	36	*	mild	no
7	36	no	?	no

- If for an attribute a there exists a case x such that the corresponding value is a "do not care" condition, i.e., $\rho(x, a) = *$, then the case x should be included in blocks $[(a, v)]$ for all specified values v of attribute a.
- If for an attribute a there exists a case x such that the corresponding value is an attribute-concept value, i.e., $\rho(x, a) = -$, then the corresponding case x should be included in blocks $[(a, v)]$ for all specified values $v \in V(x, a)$ of attribute a, where

$$V(x, a) = \{\rho(y, a) \mid \rho(y, a) \text{ is specified}, y \in U, \ \rho(y, d) = \rho(x, d)\}.$$

For Table 2, $V(2, Headache) = \{severe, mild\}$ and $V(3, Cough) = \{mild, severe, no\}$, so

[(Temperature, 36)] = {1, 6, 7},
[(Temperature, 38)] = {1, 3, 5},
[(Temperature, 40)] = {1, 2},
[(Headache, severe)] = {1, 2, 6},
[(Headache, no)] = {6, 7},
[(Headache, mild)] = {2, 3, 4, 5, 6},
[(Cough, mild)] = {1, 3, 6},
[(Cough, severe)] = {2, 3},
[(Cough, no)] = {3, 4}.

For incomplete data sets the idea of the elementary block is extended to a *characteristic set*. For a case $x \in U$ the *characteristic set* $K_B(x)$ is defined as the intersection of the sets $K(x, a)$, for all $a \in B$, where the set $K(x, a)$ is defined in the following way:

- If $\rho(x, a)$ is specified, then $K(x, a)$ is the block $[(a, \rho(x, a)]$ of attribute a and its value $\rho(x, a)$,
- If $\rho(x, a) = ?$ or $\rho(x, a) = *$ then the set $K(x, a) = U$,
- If $\rho(x, a) = -$, then the corresponding set $K(x, a)$ is equal to the union of all blocks of attribute-value pairs (a, v), where $v \in V(x, a)$ if $V(x, a)$ is nonempty. If $V(x, a)$ is empty, $K(x, a) = U$.

For Table 2,

$$K_A(1) = U \cap \{1, 2, 6\} \cap \{1, 3, 6\} = \{1, 6\},$$
$$K_A(2) = \{1, 2\} \cap (\{1, 2, 6\} \cup \{2, 3, 4, 5, 6\}) \cap \{2, 3\} = \{2\},$$
$$K_A(3) = \{1, 3, 5\} \cap \{2, 3, 4, 5, 6\} \cap (\{1, 3, 6\} \cup \{2, 3\} \cup \{3, 4\}) = \{3\},$$
$$K_A(4) = U \cap \{2, 3, 4, 5, 6\} \cap \{3, 4\} = \{3, 4\},$$
$$K_A(5) = \{1, 3, 5\} \cap \{2, 3, 4, 5, 6\} \cap U = \{3, 5\},$$
$$K_A(6) = \{1, 6, 7\} \cap U \cap \{1, 3, 6\} = \{1, 6\},$$
$$K_A(7) = \{1, 6, 7\} \cap \{6, 7\} \cap U = \{6, 7\}.$$

Characteristic set $K_B(x)$ may be interpreted as the set of cases that are indistinguishable from x using all attributes from B and using a given interpretation of missing attribute values. Thus, $K_A(x)$ is the set of all cases that cannot be distinguished from x using all attributes. In [12] $K_A(x)$ was called a successor neighborhood of x, see also [12, 22, 23].

Obviously, when a data set is complete, for given $B \subseteq A$, all characteristic sets $K_B(x)$ are identical with elementary blocks $[x]_B$.

The characteristic relation $R(B)$ is a relation on U defined for $x, y \in U$ as follows

$$(x, y) \in R(B) \text{ if and only if } y \in K_B(x).$$

The characteristic relation $R(B)$ is reflexive but—in general—does not need to be symmetric or transitive. Also, the characteristic relation $R(B)$ is known if we know characteristic sets $K_B(x)$ for all $x \in U$. In our example, $R(A)$ = $\{$(1, 1), (1, 6), (2, 2), (3, 3), (4, 3), (4, 4), (5, 3), (5, 5), (6, 1), (6, 6), (7, 6), (7, 7)$\}$. The most convenient way to define the characteristic relation is through the characteristic sets.

For decision tables, in which all missing attribute values are lost, a special characteristic relation was defined in [3], see also, e.g., [2, 11].

4 Approximations

For incompletely specified decision tables lower and upper approximations may be defined in a few different ways. First, the definition of definability should be modified. A union of some intersections of attribute-value pair blocks, in any such intersection all attributes should be different and attributes are members of B, will be called B-*locally definable* sets. A union of characteristic sets $K_B(x)$, where $x \in X \subseteq U$ will be called a B-*globally definable* set. Any set X that is *B*-globally definable is *B*-locally definable, the converse is not true. In this paper we quote three different definitions of lower and upper approximations [8, 9, 10].

Let X be a concept, let B be a subset of the set A of all attributes, and let $R(B)$ be the characteristic relation of the incomplete decision table with characteristic sets $K(x)$, where $x \in U$. Our first definition uses a similar idea as in the previous articles on incompletely specified decision tables [2, 3, 5, 6, 11], i.e., lower and upper approximations are sets of singletons from the universe U satisfying some properties. Thus, lower and upper approximations are defined by constructing both sets from singletons. We will call these approximations *singleton*. Namely, a singleton B-lower approximation of X is defined as follows:

$$\underline{B}X = \{x \in U \mid K_B(x) \subseteq X\}.$$

A singleton B-upper approximation of X is

$$\overline{B}X = \{x \in U \mid K_B(x) \cap X \neq \emptyset\}.$$

For our example of the decision table presented in Table 2 let us say that $B = A$. Then the singleton A-lower and A-upper approximations of the two concepts: $\{$1, 2, 3, 4$\}$ and $\{$5, 6, 7$\}$ are:

$$\underline{A}\{1, 2, 3, 4\} = \{2, 3, 4\},$$
$$\underline{A}\{5, 6, 7\} = \emptyset,$$
$$\overline{A}\{1, 2, 3, 4\} = \{1, 2, 3, 4, 5, 6\},$$
$$\overline{A}\{5, 6, 7\} = \{1, 5, 6, 7\}.$$

Note that the set $\overline{A}\{5,6,7\}$ is not even A-locally definable. Indeed, case 5 is a member of two attribute-value blocks: [(Temperature, 38)] and [(Headache, mild)], both blocks contain case 3 as well, so any intersection of blocks containing case 5 must contain case 3. Hence, as it was previously argued in [8, 9, 10], singleton approximations should not be used for rule induction. Obviously, if a set is not B-locally definable then it cannot be expressed by rule sets using attributes from B.

We may define lower and upper approximations for incomplete decision tables by using characteristic sets instead of elementary sets. There are two ways to do this. Using the first way, a *subset* B-lower approximation of X is defined as follows:

$$\underline{B}X = \cup\{K_B(x) \mid x \in U, K_B(x) \subseteq X\}.$$

A *subset* B-upper approximation of X is

$$\overline{B}X = \cup\{K_B(x) \mid x \in U, K_B(x) \cap X \neq \emptyset\}.$$

Since any characteristic relation $R(B)$ is reflexive, for any concept X, singleton B-lower and B-upper approximations of X are subsets of the subset B-lower and B-upper approximations of X, respectively [10]. For the same decision table, presented in Table 2, the subset A-lower and A-upper approximations are

$$\underline{A}\{1,2,3,4\} = \{2,3,4\},$$

$$\underline{A}\{5,6,7\} = \emptyset,$$

$$\overline{A}\{1,2,3,4\} = \{1,2,3,4,5,6\},$$

$$\overline{A}\{5,6,7\} = \{1,3,5,6,7\}.$$

The second possibility is to modify the subset definition of lower and upper approximation by replacing the universe U from the subset definition by a concept X. A *concept* B-lower approximation of the concept X is defined as follows:

$$\underline{B}X = \cup\{K_B(x) \mid x \in X, K_B(x) \subseteq X\}.$$

Obviously, the subset B-lower approximation of X is the same set as the concept B-lower approximation of X [8]. A concept B-upper approximation of the concept X is defined as follows:

$$\overline{B}X = \cup\{K_B(x) \mid x \in X, K_B(x) \cap X \neq \emptyset\} = \cup\{K_B(x) \mid x \in X\}.$$

The concept B-upper approximation of X is a subset of the subset B-upper approximation of X [8]. For the decision table presented in Table 2, the concept A-lower and A-upper approximations are

$$\underline{A}\{1,2,3,4\} = \{2,3,4\},$$

$$\underline{A}\{5,6,7\} = \emptyset,$$

$$\overline{A}\{1, 2, 3, 4\} = \{1, 2, 3, 4, 6\},$$

$$\overline{A}\{5, 6, 7\} = \{1, 3, 5, 6, 7\}.$$

Note that for complete decision tables, all three definitions of lower approximations, singleton, subset and concept, coalesce to the same definition. Also, for complete decision tables, all three definitions of upper approximations coalesce to the same definition. This is not true for incomplete decision tables, as our example shows.

For Table 2, certain rules, induced from the *concept* lower approximations are

1, 2, 2
(Cough, severe) -> (Flu, yes),
1, 2, 2
(Cough, no) -> (Flu, yes),

and possible rules, induced from the *concept* upper approximations, are

1, 2, 2
(Cough, severe) -> (Flu, yes),
1, 2, 3
(Headache, severe) -> (Flu, yes),
1, 2, 2
(Cough, no) -> (Flu, yes),
1, 3, 5
(Temperature, 36..39) -> (Flu, no).

5 Experiments

Let us recall that we assumed that for each case at least one attribute value was specified. Such an assumption limits the percentage of missing attribute values used for experiments; for example, for the *wine* data set, starting from 70% of randomly assigned missing attribute values, this assumption was violated.

Table 3. Data sets used for experiments

Data set	Number of		
	cases	attributes	concepts
Bankruptcy	66	5	2
Breast cancer - Slovenia	277	9	2
Hepatitis	155	19	2
Image segmentation	210	19	7
Iris	150	4	3
Lymphography	148	18	4
Wine	178	12	3

Table 4. Breast cancer - Slovenia data set. Certain rule sets.

Percentage of lost values	Average error rate	Standard deviation	Z score
0	28.68	0.65	
5	28.86	0.67	−1.06
10	28.72	0.75	−0.26
15	28.53	0.73	0.81
20	28.18	0.60	**3.05**
25	28.97	0.38	**−2.10**
30	28.50	0.61	1.10
35	27.69	0.66	**5.79**
40	28.88	0.38	−1.48
45	28.77	0.36	−0.71

For rule induction from incomplete data we used the MLEM2 data mining algorithm, for details see [19]. We used rough set methodology [17, 18], i.e., for a given interpretation of missing attribute vales, *lower* and *upper approximations* were computed for all concepts and then rule sets were induced, *certain* rules from lower approximations and *possible* rules from upper approximations. Note that for incomplete data there is a few possible ways to define approximations, we used *concept* approximations [7, 8].

As follows from our experiments, some of the rule sets induced from such incomplete data are better than the rule sets induced form original, complete data sets. More precisely, the error rate, a result of ten-fold cross validation, is significantly lower, with the significance level of 5%, than the error rate for rule sets induced from the original data.

In our experiments seven typical data sets were used, see Table 3. All of these data sets are available from the UCI ML Repository, with the exception of the *bankruptcy*

Table 5. Iris data set. Certain rule sets.

Percentage of lost values	Average error rate	Standard deviation	Z score
0	6.04	1.08	
5	5.00	0.82	**4.23**
10	5.93	0.98	0.42
15	5.82	1.22	0.81
20	6.36	1.03	−1.14
25	6.56	1.04	−1.87
30	5.69	0.99	1.33
35	6.93	1.82	**−2.3**

Table 6. Wine data set. Certain rule sets.

Percentage of lost values	Average error rate	Standard deviation	Z score
0	7.66	1.32	
5	7.17	1.74	1.22
10	7.13	2.00	1.20
15	8.76	1.85	**−2.66**
20	7.06	1.38	1.72
25	7.27	1.55	1.06
30	6.20	1.39	**4.17**
35	6.55	1.16	**3.43**
40	6.8	1.28	**2.56**
45	7.73	1.48	−0.21
50	7.21	0.82	1.58
55	8.01	1.29	−1.05
60	7.30	1.00	1.00
65	8.41	0.98	0.98

Table 7. Wine data set. Possible rule sets.

Percentage of lost values	Average error rate	Standard deviation	Z score
0	7.66	1.32	
5	7.21	1.92	1.06
10	7.32	1.34	0.98
15	8.46	1.75	**−2.01**
20	7.17	1.72	1.23
25	7.64	1.63	0.05
30	6.33	1.15	**4.15**
35	6.57	1.12	**3.44**
40	6.22	1.29	**4.27**
45	7.79	1.30	−0.39
50	7.12	0.68	**2.00**
55	7.68	0.98	−0.06
60	6.89	0.78	**2.74**
65	8.31	1.17	**−2.04**

data set. These data sets were completely specified (all attribute values were completely specified), with the exception of *breast cancer - Slovenia* data set, which originally contained 11 cases (out of 286) with missing attribute values. These 11 cases were removed.

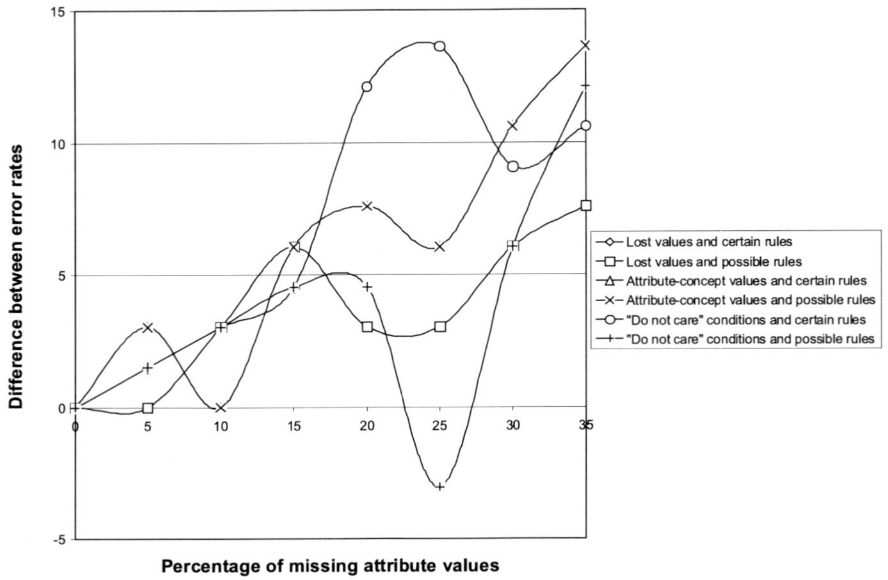

Fig. 1. Bankruptcy data set. Difference between error rates for testing with complete data sets and data sets with missing attribute values.

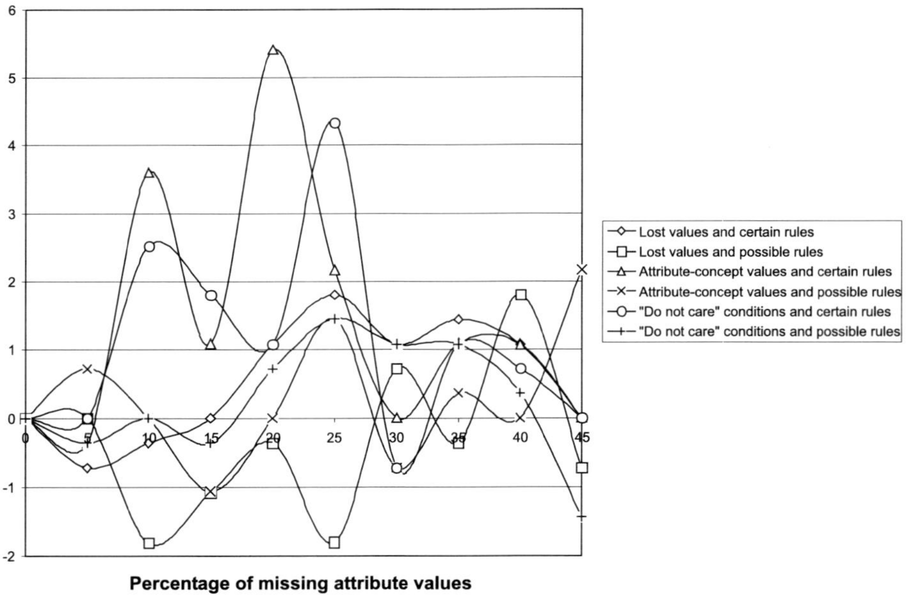

Fig. 2. Breast cancer - Slovenia data set. Difference between error rates for testing with complete data sets and data sets with missing attribute values.

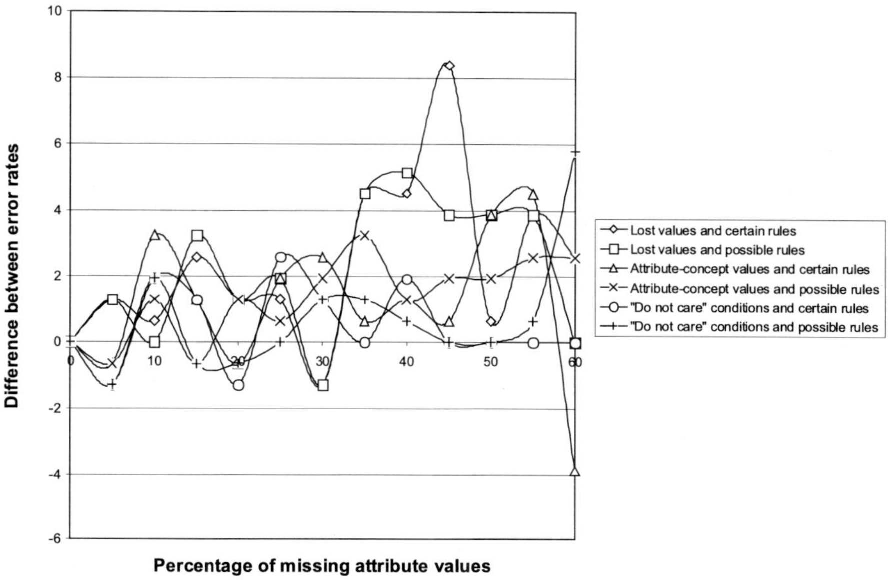

Fig. 3. Hepatitis data set. Difference between error rates for testing with complete data sets and data sets with missing attribute values.

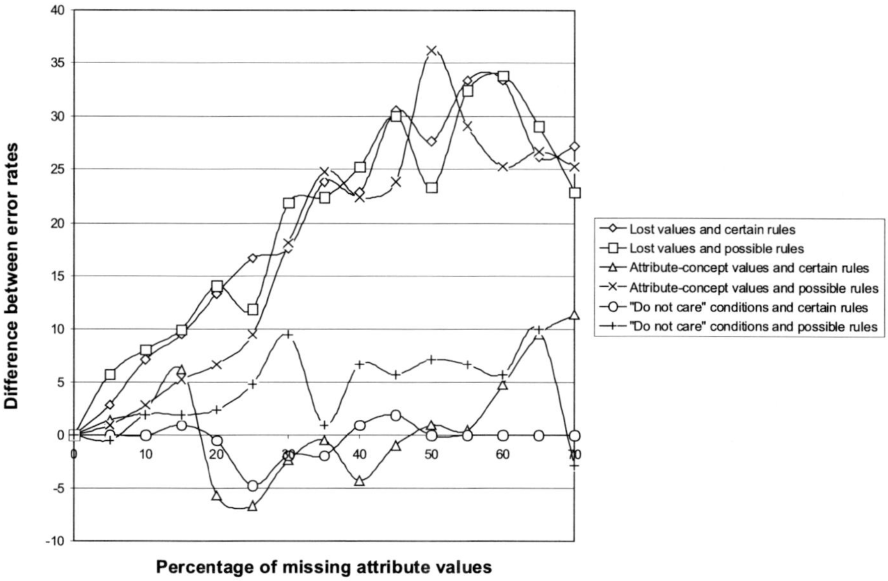

Fig. 4. Image segmentation data set. Difference between error rates for testing with complete data sets and data sets with missing attribute values.

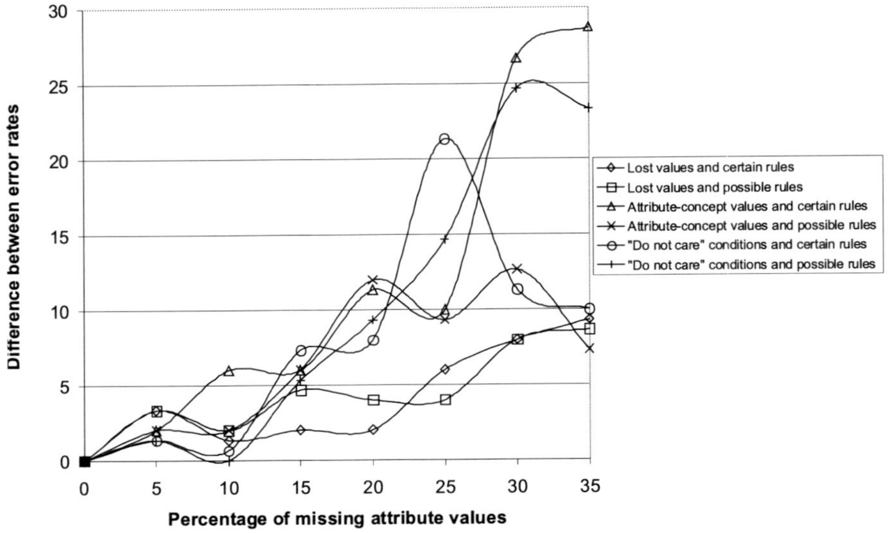

Fig. 5. Iris data set. Difference between error rates for testing with complete data sets and data sets with missing attribute values.

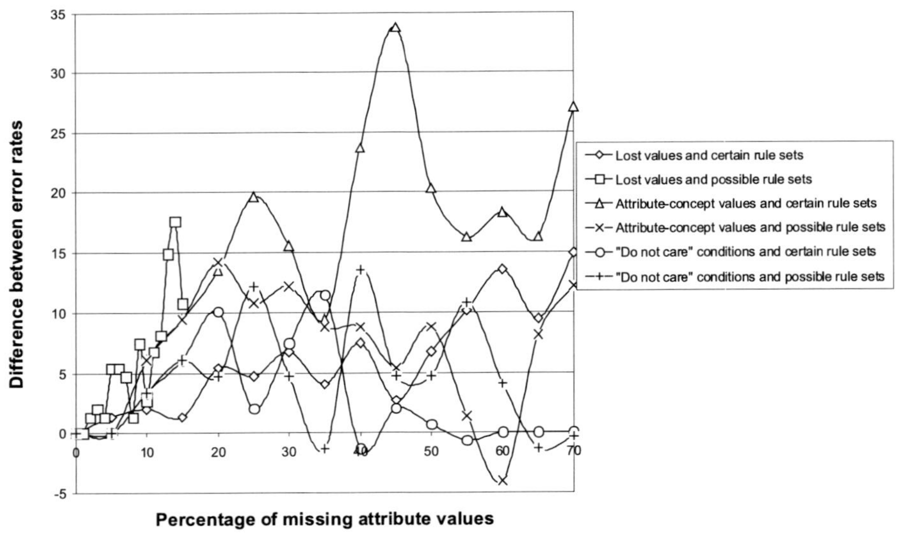

Fig. 6. Lymphography data set. Difference between error rates for testing with complete data sets and data sets with missing attribute values.

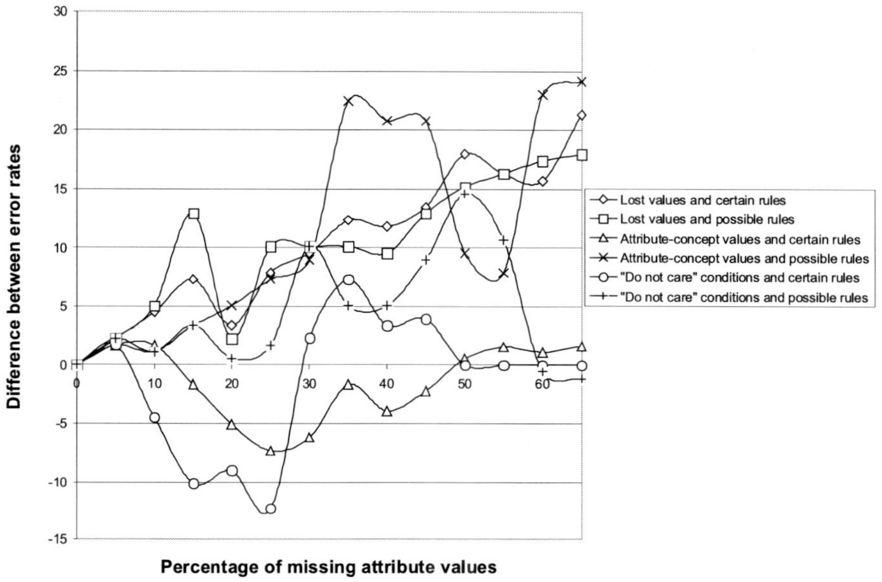

Fig. 7. Wine data set. Difference between error rates for testing with complete data sets and data sets with missing attribute values.

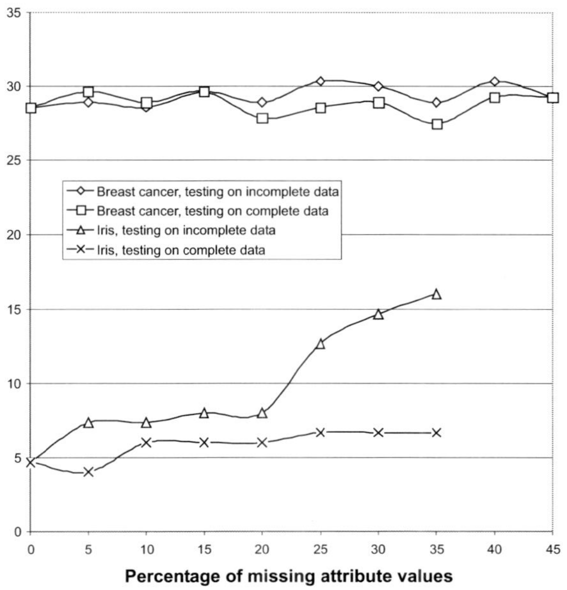

Fig. 8. Breast cancer - Slovenia and Iris data sets. Testing on complete and incomplete data sets, all missing attribute values are interpreted as *lost*, certain rule sets.

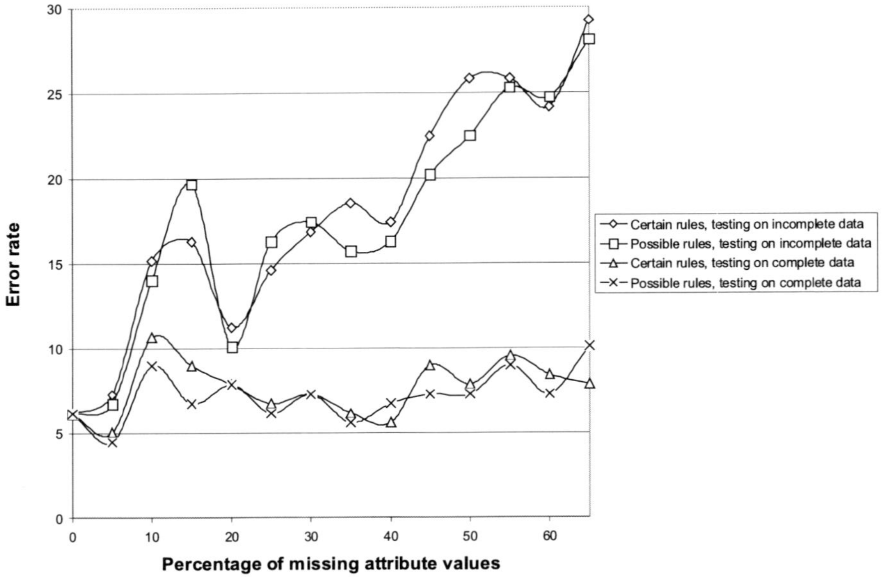

Fig. 9. Wine data set. Testing on complete and incomplete data sets, all missing attribute values are interpreted as *lost*.

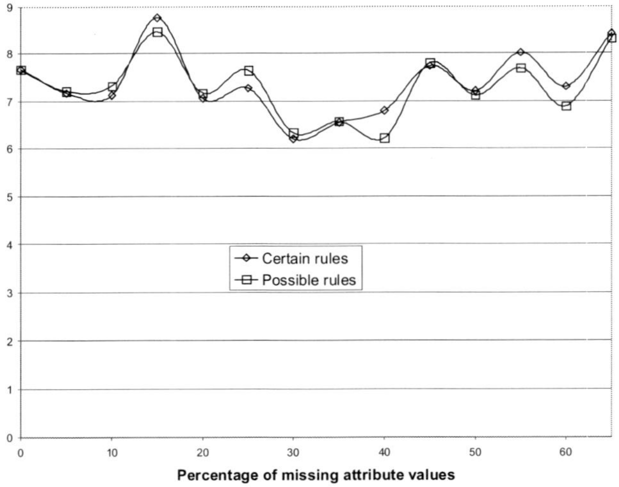

Fig. 10. Wine data set. Average error rate, a result of 30 experiments of 10-fold cross validation, rule induction from incomplete data sets, all missing attribute values are interpreted as *lost*, testing on complete data sets.

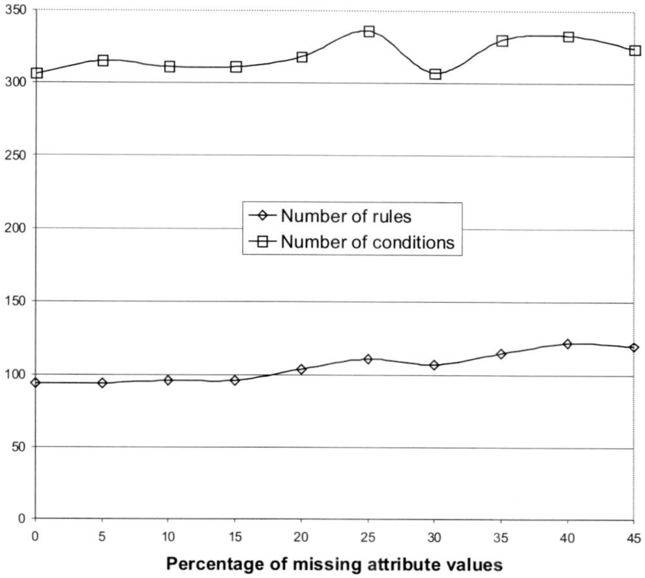

Fig. 11. Breast cancer - Slovenia data set. Size of rule sets induced from data with missing attribute values, all missing attribute values are interpreted as *lost*.

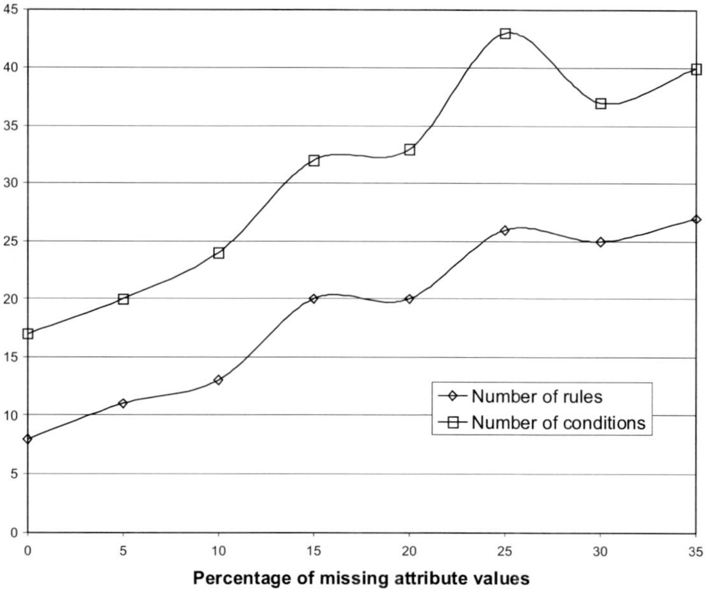

Fig. 12. Iris data set. Size of rule sets induced from data with missing attribute values, all missing attribute values are interpreted as *lost*.

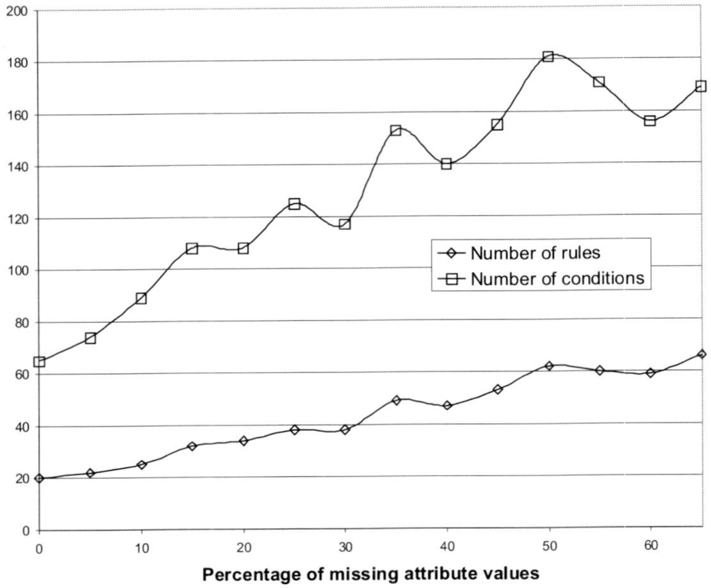

Fig. 13. Iris data set. Size of rule sets induced from data with missing attribute values, all missing attribute values are interpreted as *lost*.

In two data sets: *bankruptcy* and *iris* all attributes were numerical. These data sets were processed as numerical (i.e., discretization was done during rule induction by MLEM2). The *image segmentation* data set was converted into symbolic using a discretization method based on agglomerative cluster analysis (this method was described, e.g., in [24]).

Preliminary results [25] show that for some data sets by increasing incompleteness we may improve quality of rule sets. Therefore we decided to conduct extensive experiments on three data sets: *breast cancer - Slovenia*, *iris*, and *wine*, repeating 30 times ten-fold cross validation experiments (changing the random case ordering in data sets) for every percentage of lost values and then computing the Z score using the well-known formula

$$Z = \frac{\overline{X_1} - \overline{X_2}}{\sqrt{\frac{s_1^2 + s_2^2}{30}}},$$

where $\overline{X_1}$ is the mean of 30 ten-fold cross validation experiments for the original data set, $\overline{X_2}$ is the mean of 30 ten-fold cross validation experiments for the data set with given percentage of lost values, s_1 and s_2 are sample standard deviations for original and incomplete data sets, respectively.

Note that though rule sets were induced from incomplete data, for testing such rule sets the original, complete data were used so that the results for incomplete data are fully comparable with results for the original data sets. Obviously, if the Z score is larger than

1.96, the rule set induced from the data set with given percentage of lost values is significantly better than the corresponding rules set induced from the original data set, with the significance level of 5%, two-tailed test.

Table 4 presents Z-scores for *breast cancer - Slovenia* data set. In two cases (for 20% and 35% of lost values) new rule sets, induced from incomplete data and tested on complete data, perform significantly better than the rules induced from complete data, in one case (for 25% of lost values) significantly worse. In remaining cases there is no significant difference in performance.

Similarly, for *iris* data set (see Table 5), one rule set (for 5% of lost values) performs significantly better than the rule set induced from complete data sets, and one rule set (for 35% of lost values) performs worse. For remaining cases there is no signiciant difference in performance.

For *wine* data set (Tables 6 and 7), there are three and five rule sets performing better than the rule sets induced from the original data sets, for certain and possible rule sets, respectively. Additionally, in only one case for certain rule sets and for two cases for possible rule sets the rule sets induced from incomplete data perform worse than the rule sets induced from the original data.

The problem is how to recognize a data set that is a good candidate for improving rule sets by increasing incompleteness. One possible criterion is a large difference between two error rates: one induced from incomplete data and tested on incomplete data and the other induced from incomplete data and tested on the original data set. The corresponding differences of these error rates are presented on Figures 1–7.

Another criterion of potential usefulness of inducing rules from incomplete data is the graph of an error rate for rule sets induced from incomplete data and tested on original, complete data. Such graphs were presented in [25]. In this paper we present these graphs, restricted to the *breast cancer - Slovenia* and *iris* data sets, see Figure 8, and to *wine* data set, see Figure 9. Additionally, performance of certain and possible rules for *wine* data set is presented on Figure 10. A flat graph representing the error rate for rule sets induced from incomplete data and tested on complete data, roughly speaking, parallel to the *percentage of missing attribute values* axis represents a data set that may satisfy the criterion. It is clear that the *breast cancer - Slovenia*, *iris*, and *wine* data sets satisfy both criteria. Note that all graphs, presented in Figures 1–9, were plotted for single experiments of ten-fold cross validation.

The main objective of this paper is proven: for some data sets it is possible to improve the quality of rule sets by increasing incompleteness of data sets (or replacing existing attribute values by symbols of missing attribute values).

Additionally, the size of the rule sets induced from incomplete data, both in terms of the number of rules and the total number of conditions, is larger for incomplete data, see Figures 11−13. This fact follows from the principles of the MLEM2 algorithm: MLEM2 is less likely to induce simpler rules if the search space is smaller.

The question is why sometimes we may improve the quality of rule sets by increasing incompleteness of the original data set. A possible explanation for occasional improvement of the quality of rule sets is redundancy of information in some data sets, such as *breast cancer - Slovenia*, *iris*, and *wine* data sets, so that it is still possible to induce not only good but sometimes even better rule sets than the rule set induced from the original data set.

6 Conclusions

As follows form our experiments, there are some cases of the rule sets, induced from incomplete data sets, with an error rate (result of ten-fold cross validation) significantly smaller (with a significance level of 5%, two-tailed test) than the error rate for the rule set induced from the original data set. Thus, we proved that there exists an additional technique for improving rule sets, based on increasing incompleteness of the original data set (by replacing some existing attribute values by symbols of missing attribute values). Note that this technique is not always successful. A possible criterion for success is a large difference between the error rate for rule sets induced from incomplete data and tested on original data and on incomplete data. Another criterion is a flat graph for an error rate versus percentage of missing attribute vales for rule sets induced from incomplete data and tested on original, complete data.

References

1. Grzymala-Busse, J.W., Wang, A.Y.: Modified algorithms LEM1 and LEM2 for rule induction from data with missing attribute values. In: Proceedings of the Fifth International Workshop on Rough Sets and Soft Computing (RSSC 1997) at the Third Joint Conference on Information Sciences (JCIS 1997), pp. 69–72 (1997)
2. Stefanowski, J.: Algorithms of Decision Rule Induction in Data Mining. Poznan University of Technology Press, Poznan (2001)
3. Stefanowski, J., Tsoukiàs, A.: On the extension of rough sets under incomplete information. In: Zhong, N., Skowron, A., Ohsuga, S. (eds.) RSFDGrC 1999. LNCS (LNAI), vol. 1711, pp. 73–82. Springer, Heidelberg (1999)
4. Grzymala-Busse, J.W.: On the unknown attribute values in learning from examples. In: Raś, Z.W., Zemankova, M. (eds.) ISMIS 1991. LNCS, vol. 542, pp. 368–377. Springer, Heidelberg (1991)
5. Kryszkiewicz, M.: Rough set approach to incomplete information systems. In: Proceedings of the Second Annual Joint Conference on Information Sciences, pp. 194–197 (1995)
6. Kryszkiewicz, M.: Rules in incomplete information systems. Information Sciences 113, 271–292 (1999)
7. Grzymala-Busse, J.W.: Three approaches to missing attribute values—a rough set perspective. In: Proceedings of the Workshop on Foundation of Data Mining, in conunction with the Fourth IEEE International Conference on Data Mining, pp. 55–62 (2004)
8. Grzymala-Busse, J.W.: Rough set strategies to data with missing attribute values. In: Workshop Notes, Foundations and New Directions of Data Mining, in conjunction with the 3-rd International Conference on Data Mining, pp. 56–63 (2003)
9. Grzymala-Busse, J.W.: Data with missing attribute values: Generalization of indiscernibility relation and rule induction. Transactions on Rough Sets 1, 78–95 (2004)
10. Grzymała-Busse, J.W.: Characteristic relations for incomplete data: A generalization of the indiscernibility relation. In: Tsumoto, S., Słowiński, R., Komorowski, J., Grzymała-Busse, J.W. (eds.) RSCTC 2004. LNCS (LNAI), vol. 3066, pp. 244–253. Springer, Heidelberg (2004)

11. Stefanowski, J., Tsoukias, A.: Incomplete information tables and rough classification. Computational Intelligence 17, 545–566 (2001)
12. Yao, Y.Y.: Relational interpretations of neighborhood operators and rough set approximation operators. Information Sciences 111, 239–259 (1998)
13. Grzymała-Busse, J.W., Hu, M.: A comparison of several approaches to missing attribute values in data mining. In: Ziarko, W.P., Yao, Y. (eds.) RSCTC 2000. LNCS (LNAI), vol. 2005, p. 378. Springer, Heidelberg (2001)
14. Hong, T.P., Tseng, L.H., Chien, B.C.: Learning coverage rules from incomplete data based on rough sets. In: Proceedings of the IEEE International Conference on Systems, Man and Cybernetics, pp. 3226–3231 (2004)
15. Nakata, M., Sakai, H.: Rough sets handling missing values probabilistically interpreted. In: Ślęzak, D., Wang, G., Szczuka, M.S., Düntsch, I., Yao, Y. (eds.) RSFDGrC 2005. LNCS (LNAI), vol. 3641, pp. 325–334. Springer, Heidelberg (2005)
16. Grzymala-Busse, J.W., Grzymala-Busse, W.J.: Inducing better rule sets by adding missing attribute values. In: Chan, C.-C., Grzymala-Busse, J.W., Ziarko, W.P. (eds.) RSCTC 2008. LNCS (LNAI), vol. 5306, pp. 160–169. Springer, Heidelberg (2008)
17. Pawlak, Z.: Rough sets. International Journal of Computer and Information Sciences 11, 341–356 (1982)
18. Pawlak, Z.: Rough Sets. Theoretical Aspects of Reasoning about Data. Kluwer Academic Publishers, Dordrecht (1991)
19. Grzymala-Busse, J.W.: MLEM2: A new algorithm for rule induction from imperfect data. In: Proceedings of the 9th International Conference on Information Processing and Management of Uncertainty in Knowledge-Based Systems, pp. 243–250 (2002)
20. Chan, C.C., Grzymala-Busse, J.W.: On the attribute redundancy and the learning programs ID3, PRISM, and LEM2. Technical report, Department of Computer Science, University of Kansas (1991)
21. Grzymala-Busse, J.W.: LERS—a system for learning from examples based on rough sets. In: Slowinski, R. (ed.) Intelligent Decision Support. Handbook of Applications and Advances of the Rough Set Theory, pp. 3–18. Kluwer Academic Publishers, Dordrecht (1992)
22. Yao, Y.Y., Lin, T.Y.: Generalization of rough sets using modal logics. Intelligent Automation and Soft Computing 2, 103–119 (1996)
23. Lin, T.Y.: Topological and fuzzy rough sets. In: Slowinski, R. (ed.) Intelligent Decision Support. Handbook of Applications and Advances of the Rough Sets Theory, pp. 287–304. Kluwer Academic Publishers, Dordrecht (1992)
24. Chmielewski, M.R., Grzymala-Busse, J.W.: Global discretization of continuous attributes as preprocessing for machine learning. International Journal of Approximate Reasoning 15, 319–331 (1996)
25. Grzymala-Busse, J.W., Grzymala-Busse, W.J.: Improving quality of rule sets by increasing incompleteness of data sets. In: Proceedings of the Third International Conference on Software and Data Technologies, pp. 241–248 (2008)

Knowledge Discovery Using Rough Set Theory

Yaile Caballero[1], Rafael Bello[2], Leticia Arco[3], Maria García[2], and Enislay Ramentol[1]

[1] Department of Computer Science
University of Camagüey, Camagüey, Cuba
yailec@yahoo.com,
enislay.ramentol@reduc.edu.cu

[2] Department of Computer Science, Universidad Central de Las Villas, Santa Clara, Cuba
{rbellop,leticiaa,mmgarcia}@uclv.edu.cu

Abstract. Rough Set Theory (RST) opened a new direction in the development of incomplete information theories and is a powerful data analysis tool. In this investigation, the possibility of using this theory to generate a priori knowledge about a dataset is demonstrated. A proposal is developed for previous characterization of training sets, using RST estimation measurements. This characterization offers an assessment of the quality of data in order to use them as a training set in machine learning techniques. The proposal has been experimentally studied using international databases and some known classifiers such as MLP, C4.5 and K-NN, and satisfactory results have been obtained.

Keywords: knowledge discovery, Rough Set Theory.

1 Introduction

For scientists, data represent carefully gathered observations of certain studied phenomena; in businesses, data keep information on markets, competitors and clients; in industrial processes, they gather values on the fulfillment of objectives. The true value of data lies in their possibility of providing useful information for decision making, or the exploration and understanding of the phenomena that gave rise to them [1]. Data analysis is important in such fields as: biotechnology, medicine, economics, finance, and environmental studies, among others. In each, data preprocessing is essential.

The process of extracting knowledge from databases is known as KDD (Knowledge Discovery in Databases). This process includes various stages, from data gathering to applying acquired knowledge in decision making. Between these stages is one that could be called the nucleus of the KDD process, and consists in extracting knowledge from data. This phase, crucial in order to obtain appropriate results, depends on the Machine Learning algorithm applied, but also is influenced to a large degree by the quality of the data that arrive from the previous phase [1].

In learning systems, induction is used frequently. Among the examples of example-based learning are the Artificial Neural Nets (ANN) [2], the K-Nearest Neighbor method (k-NN) [3], and the C4.5 algorithm [4], among others. Supervised

J. Koronacki et al. (Eds.): Advances in Machine Learning I, SCI 262, pp. 367–383.
springerlink.com © Springer-Verlag Berlin Heidelberg 2010

classification can be tedious and even sometimes inapplicable when dealing with numerous training sets and upper dimensional data. These two factors determine the computational expense of the application of most existent supervised classification methods.

One of the most recently theories for use in data analysis is the Rough Set Theory (RST) [5-14]. This theory is considered with a high interest in areas such as Artificial Intelligence and Theory of incomplete information, because it constitutes a very useful tool for handling uncertainty information [15-19]. RST has been used for rule generation [20-24] and attribute selection [20, 25-28], among other applications [29-38].

The development of intelligent systems and the increase in information storage in electronic form, have led a peak of automated learning techniques. These techniques are based on the use of the information upon an application domain such as a training set (TS), from which the learning method extracts the knowledge needed to solve problems from that domain. The methods to study the quality of the TS are usually applied post-learning, so that they include the computational expense of applying the learning method and applying some method of validation. Finding methods to permit an evaluation of the TS ahead of time is highly relevant for the field of Machine Learning.

As this article will show, this search for such methods motivated the present investigation. First, important concepts about Rough Set Theory are introduced in part two, and later, in part three we present the proposal related to knowledge discovery using Rough Set Theory.

2 Rough Set Theory

Rough Set Theory (RST) was first introduced by Z. Pawlak in 1982 [5]. It is based on the idea that to approximate any concept, for example, a class in a problem of supervised classification, a pair of sets, calls lower and upper approximations can be used. With this theory it is possible to treat both quantitative and qualitative data, and it is unnecessary to eliminate inconsistencies prior to the analysis; with respect to output information. RST can be used to determine the grade of inconsistency in data, the relevance of the attributes, to generate relations among them (in form of rules), among other uses [23, 24, 39-46]. Inconsistency describes a situation in which two or more values are in dispute to be assigned to a variable [47, 48], and it is considered one type of uncertainty.

Many authors have written about Rough Sets, and generally view this theory as the best tool for modelling this kind of uncertainty, and as a new direction in the development of incomplete information theory [7, 8, 49-53] . In this section, the fundamental concepts of Rough Sets will be described, as much for the classic case as for the approach based on similarity relations.

2.1 Main Definitions of Rough Set Theory

The philosophy of Rough Sets is based on the assumption that for any object x of a universe U there exists a certain amount of associated information (data and knowledge), expressed by means of certain attributes that describe the object [6, 54].

Diverse computational models operate on data collections. In each case this collection has its own characteristics, mainly organizational, and receives a particular denomination. For example, for a database manager the collection is a data base, while for an Artificial Neuronal Network it is a training set. In the case of Rough Set Theory, the basic structure of information is the Information System.

Definition 1. Information System and Decision System.

Let be a set of attributes A={ a_1, a_2,...,a_n} and a nonempty, finite set called the universe U, with example elements (objects, organizations, situations or states) described using the attributes ai; Information System is the name given to the pair (U, A) [6]. If to each element of U is attached a new attribute d called decision, indicating the decision made in that state or situation, then a Decision System is created (U, $A \cup$ [23]), where $d \notin A$.

Definition 2. Information Function.

For each attribute a_i there is an associated domain called v_i, leading to the function f: $UxA \rightarrow V$, $V=\{v_1,v_2,\ldots,v_p\}$ such that $f(x,a_i) \in v_j$ for each $a_i \in A$, $x \in U$, called the information function [6].

The attribute of decision d induces a partition of the object universe U. Let be a set of whole numbers $\{1,\ldots,l\}$, $X_i=\{x \in U$: $d(x)=i\}$, then $\{X_1,\ldots,X_l\}$ is a collection of equivalency classes, called decision classes, where two objects belong to the same class if they have the same decision attribute value.

It is said that an attribute $a_i \in A$ separates or discerns an object x from another y, and it is written $Discerns(a_i, x, y)$, if and only if:

$$f(x,a_i) \neq f(y,a_i) \tag{1}$$

The discernibility relation is based on the comparison of the values of an attribute, for which strict equality (or inequality) has been used. Nevertheless, it is possible to use a less strictly defined condition of comparison, in this form:

$$Discern\ (a_i, x, y) \Leftrightarrow |f(x,a_i) - f(y,a_i)| > \varepsilon \tag{2}$$

Definition 3. Indiscernibility Relation

For each subgroup of attributes B of A $B \subseteq A$ there is an associated binary indiscernibility relation denoted with symbol R, which is the set of pairs of objects that are indiscernible from one another. [6].

$$R = \{(x, y) \in UxU : f(x,a_i) = f(y,a_i) \forall a_i \in B\} \tag{3}$$

An equivalence relation is a type of indiscernibility relation, it can be defined by forming subsets of elements of U which have equal value for a subset of attributes B of A, $B \subseteq A$. The basic concepts of the RST are the lower and upper approximations of a subset $X \subseteq U$. These concepts originally were introduced in reference to an indiscernibility relation R. Let R be a defined binary relation of U which represents the indiscernibility of U, it can be said that R(x) means the set of objects which are indiscernible from x. Thus, $R(x) = \{y \in U : yRx\}$. In classic Rough Set Theory, R is defined as an equivalence relation; that is, a binary relation $R \subseteq UxU$ that is reflective, symmetrical and transitive. R induces a partition of U into equivalence classes corresponding to $R(x)$, $x \in U$.

This classic approach of RST is extended by accepting that objects that are not indiscernible but sufficiently close or similar can be grouped in the same class [55]. The aim is to construct a similarity relation *R'* from the indiscernibility relation *R* by relaxing the original conditions for indiscernibility. This relaxation can be performed in many ways, thus giving many possible definitions for similarity. Several functions of comparison of attributes exist (similarity functions), which are associate to the type of the attribute that compares [56, 57]. However, this similarity relation *R'* must satisfy some minimal requirements. *R* being an indiscernibility relation (equivalence relation) defined on *U*, *R'* is a similarity relation extending *R* iff $\forall x \in U$, $R(x) \subseteq R'(x)$ and $\forall x \in U$, $\forall y \in R'(x)$, $R(y) \subseteq R'(x)$, where $R'(x)$ is the similarity class of *x*, i.e. $R'(x)=\{y \in U: yR'x\}$. *R'* is reflexive, any similarity class can be seen as a grouping of indiscernibility classes and *R'* induces a covering of *U* [58]. This shows that an object can belong simultaneously to different similarity classes, which means that the covering induced by R' on U is not necessarily a partition.

The approximation of a set $X \subseteq U$, using an indiscernibility relation R, has been induced as a pair of sets called lower-R and upper-R approximations of X. This study uses a more general definition of approximations, which refers to any reflective relation R'. The lower-R' ($R'_*(X)$) and upper-R' ($R'^*(X)$) approximations of X are defined respectively as shown in (4) and (5).

$$R'_*(X) = \{x \in X : R'(x) \subseteq X\} \tag{4}$$

$$R'^*(X) = \bigcup_{x \in X} R'(x) \tag{5}$$

Considering the expressions defined in (4) and (5), the boundary region of X is defined for the R' relation [59]:

$$BN_B(X) = R'^*(X) - R'_*(X) \tag{6}$$

If set BNB is empty then set X he is exact with respect to the R' relation. In opposite case, set X is inexact or rough with respect to R'.

The use of similarity relations offers great possibilities for the construction of the approximations; nevertheless, this greater flexibility comes at a cost, since it is more difficult from the computational perspective to look for the relevant approximations in this larger space [51].

There exist several functions for attribute comparison (similarity functions), which are related to the attribute type. Some of these similarity functions are described in [56, 57].

Using the upper and lower approximations of a concept X, three regions are defined to characterize the approximation space: the positive region that is the lower-R' approximation, the boundary region that is set BNB and the negative region

(NEG(X)) that is the difference between the universe and the upper-R' approximation. Sets R'*(X) (also denoted as POS(X)), R'*(X), BNB(X) and NEG(X) are the principle concepts of Rough Set Theory.

An important aspect of Rough Set Theory is the reduction of attributes based on the concept of reduct. A reduct is a reduced set of attributes that preserves the partition of the universe [6, 28]. The use of reducts in the selection and reduction of attributes has been studied widely [6, 12, 28, 51, 60-64].

2.2 Classic RST Measures of Inference

Rough Set Theory offers certain measures to for information system analysis [65, 66]. The principle measures are presented next. In (7) – (10) the lower-R' ($R'_*(X)$) and upper-R' ($R'^*(X)$) approximations of X are used, as defined, respectively, in (4) and (5).

Approximation Accuracy. A rough set X can be characterized numerically by a coefficient called approximation accuracy, where $|X|$ denotes the cardinality of X, $X \neq \phi$. See (7).

$$\alpha(X) = \frac{|R'_*(X)|}{|R'^*(X)|} \tag{7}$$

Obviously, $0 \leq \alpha(X) \leq 1$. If $\alpha(X)=1$, X is hard (exact), if $\alpha(X)<1$, X is rough (vague, inexact), always with respect to the set of attributes considered [65].

Approximation Quality. The following coefficient

$$\gamma(X) = \frac{|R'_*(X)|}{|X|} \tag{8}$$

expresses the proportion of objects that can correctly be classified in class X. In addition, $0 \leq \alpha(X) \leq \gamma(X) \leq 1$, and $\gamma(X)=0$ if and only if $\alpha(X)=0$, while $\gamma(X)=1$ if and if $\alpha(X)=1$ [65].

Considering that $X_1,\ldots,X_l$ are the decision system classes, the measure is defined:

Classification Quality. This coefficient describes the inexactness of the approximate classifications:

$$\Gamma(DS) = \frac{\sum_{i=1}^{l} |R'_*(X_i)|}{|U|} \tag{9}$$

The measure classification quality expresses the proportion of objects that can be correctly classified in the system. If the classification quality coefficient is equal to 1, then the decision system is consistent, in any another case it is inconsistent [65].

Rough Membership Function. This function quantifies the degree of relative overlapping between $R'(x)$ (similarity class of x) and the class to which object x belongs. It is defined as follows:

$$\mu_X(x) = \frac{|X \cap R'(x)|}{|R'(x)|} \tag{10}$$

The rough membership function can be interpreted as an estimation based on the frequencies of $Pr(x \in X \mid x, R'(x))$, that is to say, the conditional probability that object x belongs to set X [49].

3 A Priori Characterization of Training Sets Based on RST Estimation Measures

In this section, new RST measures are proposed, and a proposal is developed for using these new measures to predict ahead of time the quality of a training set, as well as selecting what type of supervised classifier will be the most advisable to use in the learning process: a neuronal network (Multi Layer Perceptron, MLP), a decision tree (C4.5) or a method of lazy learning (k-NN).

3.1 New Measures to Evaluate Decision Systems Based on RST

In this section, generalized options for the accuracy and quality of supervised classification are proposed. In many applications, experts can weigh the classes or they can be followed heuristically to define to what extent the classes are important.

Generalized classification accuracy. In this measure, for each class the possibility occurs of ascribing a weight that influences in the evaluation of the decision system.

$$\mathrm{A}_G(DS) = \frac{\sum_{i=1}^{l} \left(\alpha(X_i) \cdot w(X_i)\right)}{\sum_{i=1}^{l} w(X_i)} \tag{11}$$

where l is the number of classes of the decision system, $\alpha(X_i)$ is calculated as indicated in (7). In (10), the rough membership function is defined; the calculation of the average of this function applied to the objects of a class can be considered a measurement of the importance of the class, that is, the value for $w(X_i)$.

Generalized classification quality. In (12) as given below, weighting by classes is permitted when averaging the quality of each approximation:

$$\Gamma_G(DS) = \frac{\sum_{i=1}^{l} \left(\gamma(X_i) \cdot w(X_i)\right)}{\sum_{i=1}^{l} w(X_i)} \tag{12}$$

where l is the amount of classes of the decision system, $\gamma(X_i)$ can be calculated as shown in (8). In [67] the rough involvement function is proposed, this quantifies the degree to which the R'(x) (class of similarity of x) covers class X, and is shown in (13):

$$\nu_X(x) = \frac{|X \cap R'(x)|}{|X|} \tag{13}$$

The average rough involvement of the objects for the class can be considered as a value to be assigned to the weight of each class ($w(X_i)$), in (12).

In (11) and (12), $w(X_i)$ is the weight of class X_i and is a value between 0 and 1 defined by experts. Considering the average of the rough membership and the rough involvement by classes, respectively, constitutes a way to give to value to these $w(X_i)$.

Moreover, other measure is introduced to quantify the behavior of the dataset. It is called the coefficient of general approximation, defined by (14) is proposed. This coefficient shows a proportion between the quantity of objects that could to be well classified (those that belong to the lower approximation of the classes) and the amount of objects that may or not belong to the classes of the decision system.

Generalized coefficient of approximation. The quality of the information system without differentiating by classes is evaluated.

$$\mathrm{T}(DS) = \frac{\sum_{i=1}^{l} |R'_*(X_i)|}{\sum_{i=1}^{l} |R'^*(X_i)|} \tag{14}$$

where l is the number of classes in the decision system, $R'_*(X_i)$ is the lower approximation [described in (4)] and $R'^*(X_i)$ is the upper approximation [described in (5)].

3.2 Experimental Study of the Relation between Measures of Inference and Classifier Performance

For the experimental study the following datasets were selected: Balance-Scale, Breast-Cancer-Wisconsin, Bupa (Liver Disorders), Dermatology, E-Coli, Heart-Disease (Hungarian), Iris, Lung-Cancer, Monks-1, Pima-Indians-Diabetes, Promoter-Gene-Sequence, Tic-Tac-Toe, Wine Recognition, Yeast, Zoo, Glass, Hayes-Roth, Soybean, Ionosphere, Page-blocks, Postoperative, Waveform, Credit-Screening, Hepatitis and Lymphography. The description of these data sets appears in table 1.

Table 1. Description of the international datasets used in the experiments

Datasets	# of instances	# of attributes	# of classes	Absence of information
Balance-Scale	625	4	3	no
Ballons	20	4	2	no
Breast-Cancer-Wisconsin	699	10	2	yes
Bupa (Liver Disorders)	345	6	2	no
Credit	876	20	2	yes
Credit-Screening	125	17	2	yes
Dermatology	366	34	6	yes
E-Coli	336	7	8	no
Exactly	780	13	2	yes
Glass	214	9	7	no
Hayes-Roth	133	4	3	yes
Heart-Disease (Hungarian)	294	13	5	yes
Hepatitis	155	19	2	yes
House-Votes	435	16	2	yes
Iris	150	4	3	no
Ionosphere	351	34	2	no
LED	226	24	10	no
Lung-Cancer	32	56	3	yes
Lymphography	148	18	4	no
M-of-N	1000	13	2	yes
Monks-1	432	7	2	no
Mushroom (Agaricus-Lepiota)	8124	22	2	yes
Page-blocks	5473	10	5	no
Pima-Indians-Diabetes	768	8	2	no
Postoperative patient	90	8	3	yes
Promoter-Gene-Sequence	106	57	2	no
Soybean (large)	307	35	19	yes
Tic-Tac-Toe	958	9	2	no
Waveform	5000	40	3	no
Wine Recognition	179	13	3	no
Yeast	1484	9	10	no
Zoo	101	17	7	no

The following scheme of experimental study was applied to each data set:

1. Creation of training and control samples
The training sets and control sets were obtained following the principle of cross-validation, 75% of the objects were selected for the training set and 25% for the control set.

2. Supervised classification process
The supervised classification process was performed using the Weka tool, specifically with the supervised classifiers K-Nearest Neighbors (KNN), Multilayer Perceptron (MLP) and C4.5 (J48 in Weka).

3. Calculation of the Measures of Inference
For each training set the measures calculated: approximation precision (7), approximation quality (8), classification quality (9), generalized classification precision (11), generalized classification quality (12), and generalized approximation coefficient (14), as described in sections 2 and 4. The weights (w(Xi)) used to calculate the measures defined in (11) and (12) were the average of the rough membership by class and the average rough involvement by class, respectively.

4. Calculation of statistical correlations
With the aid of SPSS 13.0, the Pearson coefficients of correlation between measures of inference described in (7) and (8), and the performance by class of the classifiers were found; equally, the coefficients of correlation between the measures described in (9), (11), (12), and (14).,and results of the general performance of the supervised classification were evaluated. Coefficients of correlation in all the cases were close to 1, with a bilateral meaning less than 0.01, assuring that the correlation coefficient was significant ($p<0.01$).

3.3 Knowledge Generation from the Inference Measures Using Machine Learning

It would be interesting to infer knowledge from these results, that is, for a certain training set, it is possible to decide which supervised classifier (between k-NN, MLP and C4.5) is most advisable to apply, and, additionally, to be able to give a qualitative valuation of the general accuracy results that this classifier will achieve for that training set (low, average, high, very high). For this purpose, the rule generator C4.5 has been used, along with a Multilayer Perceptron, and the results achieved by both are compared.

In addition, once the supervised classifier most suitable for a training set has been selected, its accuracy can be valorized through a function estimator method, in this case the K- Nearest Neighbors (k-NN) was used.

Construction of training datasets
For this, six new datasets were created, two corresponding to each supervised classifier studied (KNN, C4.5 and MLP): a dataset with the discretized accuracy values (figure 1) and a dataset with the numerical accuracy values (figure 2). The first is used to infer the best classifier and the last one to estimate the performance that this classifier will obtain.

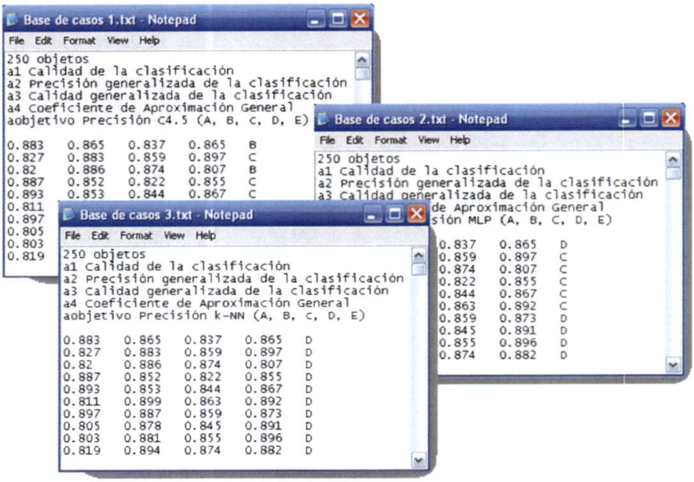

Fig. 1. Datasets constructed with the measure results and the discretized accuracy of the classifiers

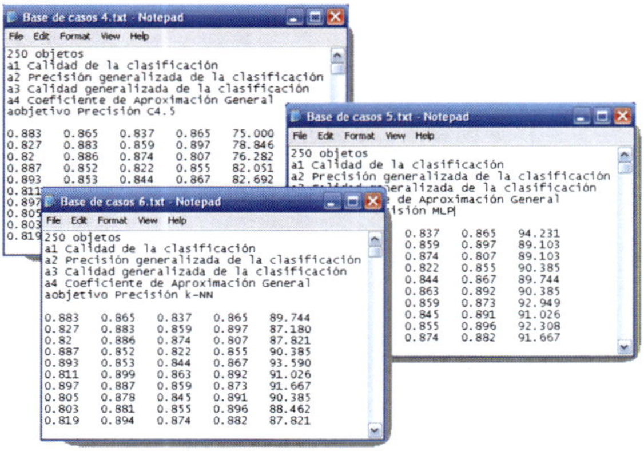

Fig. 2. Data sets constructed with the results of the inference measures and classifier accuracy

In these data sets the objects (different datasets) are described by four predictor attributes, represented by the measures: classification quality, generalized classification accuracy, generalized classification quality, and generalized approximation coefficient, all with values between zero and one. The objective attribute (class), for the three datasets with the discretized accuracy, will have the values: A→ "non-applicable, very low accuracy"; B→ "applicable, low accuracy"; C→ "applicable, average accuracy"; D→ "applicable, high accuracy" and E→ "applicable, very high accuracy". To be able to categorize classifier accuracy in A, B, C, D and E, the percentiles of the accuracy were calculated for each classifier using SPSS 13.0. In this manner, the accuracy values were associated with the different categories with the following scale: A←values below the 20 percentile; B← values between the 20 and 40 percentile; C← values between the 40 and 60 percentile; D← values between the 60 and 80 percentile; and E← the values above the 80 percentile.

Here other discretization techniques can be applied to categorize classifier accuracy in A, B, C, D, and E. Examples include entropy-based discretization methods or error-based discretization [68].

The datasets were constructed based on the results of the inference measures and the classifier accuracy. Each dataset is formed by 250 examples, since each of the 25 datasets which made up the study were randomly divided into ten training and test samples.

Automated rules generation

Using the C4.5 method, the supervised classification process was carried out in order to determine, for a new dataset (new example), a qualitative valuation of the precision that the studied classifiers would achieve. The three datasets used for the training are in figure 1.

The results of the performance of the C4.5 for estimating classifier accuracy, based on the proposed measures, are shown in table 2. These are the results of applying a cross-validation for ten test samples.

Table 2. Measures of general performance when using the C4.5 method

Measures of general performance	**for classifier accuracy**		
	k-NN	**MLP**	**C4.5**
Instances correctly classified (%)	95,906	95,313	96,491
Kappa Stadigraph	0,909	0,897	0,923
Average of absolute error	0,045	0,050	0,050
Root of average quadratic error	0,164	0,203	0,185
Relative absolute error (%)	9,811	10,976	10,943

Supervised classification using a multilayer Perceptron neuron network

A neuron network was constructed with four neurons in the entrance layer, a hidden layer and five neurons in the exit layer, one for each category. This forms a neuron network that is trained with three datasets (one for each studied classifier with the discretized objective attribute (class)), able to decide whether a given new training set is applicable or not to each classifier, and, in case of being applicable, also gives a valuation of accuracy that will be achieved (low, average, high, very high). The performance of this neuronal network for the cross-validation test (ten partitions) is shown in table 3.

Table 3. Measures of general performance when using the multilayer Perceptron

Measures of general performance	For classifier accuracy		
	k-NN	MLP	C4.5
Instances classified correctly (%)	92,949	92,308	91,667
Kappa Stadigraph	0,876	0,863	0,857
Average of absolute error	0,070	0,081	0,081
Root of average quadratic error	0,184	0,219	0,199
Relative absolute error (%)	18,58	21,355	21,303

When a new training set is introduced (a new example), in order to predict ahead of time the suitable classifier to apply, the following steps are followed:

S1. The new example is randomly portioned into the training set (75%) and the test set (25%) ten separate times in order to form ten training sets and ten test sets.

S2. To each of the ten training sets obtained in the step S1, the following measures are calculated: classification quality, generalized classification accuracy, generalized classification quality, and generalized approximation coefficient. For each measure, the average value of the ten results is obtained.

S3. The new example is now described by four attributes associated with the measures, the values of which are the averages obtained in the S2 step.

S4. For each classifier (k-NN, MLP, C4.5), the accuracy quality is determined, following the classes (A→ "non-applicable, very low accuracy"; B→ "applicable, low accuracy"; C→ "applicable, average accuracy"; D→"applicable, high accuracy" and E→ "applicable, very high accuracy"). This can be performed using C4.5 or the multilayer Perceptron, both trained with the datasets shown in figure 1.

S5. Following step S4, the classifier that was assigned the best score (E being better, A worse) is chosen as the ideal classifier for the training set.

Estimation of the accuracy value of the most appropriate classifier

After choosing the classifier with which the highest accuracy will be achieved for a given training set, the accuracy value is estimated based on the datasets in figure 2, using the k-NN method. The optimal value of k is calculated and an approximated value of the accuracy of the selected classifier is obtained. The results obtained after making a cross-validation (with ten test samples) can be observed in table 4.

Table 4. Measures of general performance when using the method k-NN

Measures of general performance	For classifier accuracy		
	k-NN	**MLP**	**C4.5**
Coefficient of correlation	0,969	0,914	0,913
Average of absolute error	0,004	0,040	0,073
Root of average quadratic error	0,200	0,198	0,344
Relative absolute error (%)	5,915	8,840	10,413

5 Conclusions

The inference measures based on Rough Set: classification quality (classic measure), generalized classification accuracy, generalized classification quality and the generalized approximation coefficient, which refer to the performance of the entire decision system, are highly correlated with the general performance of the supervised classifiers Multilayer Perceptron, C4.5 and k-NN. Similarly, the measures approximation quality and approximation accuracy are highly correlated, which particularize by classes, with the performance by classes of the classifiers previously mentioned.

The proposal to construct training datasets from the results of the inference measures of RST and of the accuracy of supervised classifiers (Multilayer Perceptron, k-NN, C4.5), was successful in knowledge generation, achieving good results in supervised classification with respect to a priori inference of the accuracy that will be achieved with new training sets.

All the above shows that Rough Set Theory permits the construction of algorithms that efficiently solve diverse tasks in the field of Machine Learning.

References

1. Ruiz, R.: Heurísticas de selección de atributos para datos de gran dimensionalidad. Departamento de Lenguajes y Sistemas Informáicos. Universidad de Sevilla, Sevilla (2006)
2. Rosemblatt, F.: Principles of Neurodynamics, New York (1962)
3. Cover, T.M., Hart, P.E.: Nearest neighbour pattern classification. Institute of Electronical and Electronics Engineers Transactions on Information Theory 13, 21–27 (1967)
4. Quinlan, J.R.: C-4.5: Programs for machine learning, San Mateo, California (1993)
5. Pawlak, Z.: Rough Sets. International Journal of Computer and Information Sciences 11, 341–356 (1982)
6. Komorowski, J., Pawlak, Z.: Rough Sets: A tutorial. Rough Fuzzy Hybridization: A new trend in decision-making, pp. 3–98. Springer, Heidelberg (1999)
7. Greco, S.: Rough sets theory for multicriteria decision analysis. European Journal of Operational Research 129, 1–47 (2001)

8. Pal, S.K.: Web mining in Soft Computing framework: Relevance, State of the art and Future Directions. IEEE Transactions on Neural Networks (2002)
9. Segovia, M.J.: Predicción de insolvencias con el método Rough Set. Universidad Complutense de Madrid, España (2003)
10. Tay, F.E., Shen, L.: Fault diagnosis based on Rough Set Theory. Engineering Applications of Artificial Intelligence 16, 39–43 (2003)
11. Yao, Y.Y.: On Generalizing Rough Set Theory. In: Wang, G., Liu, Q., Yao, Y., Skowron, A. (eds.) RSFDGrC 2003. LNCS (LNAI), vol. 2639. Springer, Heidelberg (2003)
12. Caballero, Y.: Rough Set Theory Measures to Knowledge Generation. In: Proceedings of Seven International Conference on Intelligent Systems Design and Applications, ISDA2007, Rio de Janeiro, Brazil. IEEE Computer Society, Los Alamitos (2007); Order Number P2976. Library of Congress Numbrer 2007930106, ISBN 0-7695-2976-3
13. Mitra, S.: Computational Intelligence in Bioinformatics. In: Peters, J.F., Skowron, A. (eds.) Transactions on Rough Sets III. LNCS, vol. 3400, pp. 134–152. Springer, Heidelberg (2005)
14. Bello, R., Puris, A., Nowe, A., Martínez, Y., García, M.M.: Two step ant colony system to solve the feature selection problem. In: Martínez-Trinidad, J.F., Carrasco Ochoa, J.A., Kittler, J. (eds.) CIARP 2006. LNCS, vol. 4225, pp. 588–596. Springer, Heidelberg (2006)
15. Chin, K.S., Liang, J., Dang, C.: Rough Set Data Analysis Algorithms for Incomplete Information Systems. In: Wang, G., Liu, Q., Yao, Y., Skowron, A. (eds.) RSFDGrC 2003. LNCS (LNAI), vol. 2639. Springer, Heidelberg (2003)
16. Peters, J.F.: Rough Ethology: Towards a Biologically-Inspired Study of Collective Behavior in Intelligent Systems with Approximation Spaces. In: Peters, J.F., Skowron, A. (eds.) Transactions on Rough Sets III. LNCS, vol. 3400, pp. 153–174. Springer, Heidelberg (2005)
17. Skowron, A., Świniarski, R.W., Synak, P.: Approximation Spaces and Information Granulation. In: Peters, J.F., Skowron, A. (eds.) Transactions on Rough Sets III. LNCS, vol. 3400, pp. 175–189. Springer, Heidelberg (2005)
18. Ślęzak, D.: Rough Sets and Bayes Factor. In: Peters, J.F., Skowron, A. (eds.) Transactions on Rough Sets III. LNCS, vol. 3400, pp. 202–229. Springer, Heidelberg (2005)
19. Wolski, M.: Formal Concept Analysis and Rough Set Theory from the Perspective of Finite Topological Approximations. In: Peters, J.F., Skowron, A. (eds.) Transactions on Rough Sets III. LNCS, vol. 3400, pp. 230–243. Springer, Heidelberg (2005)
20. Koczkodaj, W.W.: Myths about Rough Set Theory. ACM 41 (1998)
21. Ohrn, A., Komorowski, J., Skowron, A., Synak, P.: The Design and Implementation of a Knowledge Discovery Toolkit Based on Rough Sets. In: Pulkowski, Skorn (eds.) The ROSETTA System. Rough Sets in Knowledge discovery 1: Methodology and Applications. Studies in Fuzziness and Soft Computing, vol. 18, pp. 376–399 (1998)
22. Lee, S., Propes, N., Zhang, G., Zhao, Y., Vachtsevanos, G.: Rough Set Feature Selection and Diagnostic Rule Generation for Industrial Applications. In: Alpigini, J.J., Peters, J.F., Skowron, A., Zhong, N. (eds.) RSCTC 2002. LNCS (LNAI), vol. 2475, p. 568. Springer, Heidelberg (2002)
23. Tsumoto, S.: Automated extraction of hierarchical decision rules from clinical databases using rough set model. Expert systems with Applications 24, 189–197 (2003)
24. Grzymala-Busse, J.W., Siddhaye, S.: Rough set approaches to rule induction from incomplete data. In: 10th International Conference on Information Processing and Management of Uncertainty in Knowledge-Bases systems IPMU 2004, Perugia, Italy, vol. 2, pp. 923–930 (2004)
25. Pawlak, Z.: Rough Sets. Comm. of ACM 38 (1995)

26. Yao, Y.S., Wong, C.: Methodologies for Knowledge Discovery and Data Mining. In: Zhou, Z.a. (eds.) On Information-Theoretic Measures of attribute importance, pp. 231–238 (1999)
27. Dunstsh, I., Gunter, G.: Rough set data analysis (2000)
28. Zhong, N., Dong, J., Ohsuga, S.: Using Rough sets with heuristics for feature selection. Journal of Intelligent Information Systems 16, 199–214 (2001)
29. Kierczak, M., Rudnicki, W.R., Komorowski, J.: Construction of rough sets-based classifiers for predicting HIV resistance to nucleoside reverse transcriptase inhibitors. In: The International Symposium on Fuzzy and Rough Sets, ISFUROS 2006, Santa Clara, Cuba (2006)
30. Revett, K., Gorunesco, F., Gorunesco, M.: A Rough Sets based investigation of a Beta-Carotene/Retinol dataset. In: The International Symposium on Fuzzy and Rough Sets, ISFUROS2006, Santa Clara, Cuba (2006)
31. Caballero, Y., Bello, R., Salgado, Y., Márquez, Y., León, P., Alvarez, D., Zaldívar, J.M.: La Teoría de los Conjuntos Aproximados en el mejoramiento de los conjuntos de entrenamiento en Bioinformática. In: II Congreso Internacional de Bioinformática y Neuroinformática. Informática 2007, La Habana, Cuba (2007)
32. Pal, S.K., Mitra, P.: Rough Sets, EM Algorithm, MST and Multispectral Image Segmentation. In: Wang, G., Liu, Q., Yao, Y., Skowron, A. (eds.) RSFDGrC 2003. LNCS (LNAI), vol. 2639. Springer, Heidelberg (2003)
33. Hu, X.T., Lin, T.Y., Han, J.: A New Rough Sets Model Based on Database Systems. In: Wang, G., Liu, Q., Yao, Y., Skowron, A. (eds.) RSFDGrC 2003. LNCS (LNAI), vol. 2639. Springer, Heidelberg (2003)
34. Zheng, Z., Wang, G., Wu, Y.: A Rough Set and Rule Tree Based Incremental Knowledge Acquisition Algorithm. In: Wang, G., Liu, Q., Yao, Y., Skowron, A. (eds.) RSFDGrC 2003. LNCS (LNAI), vol. 2639. Springer, Heidelberg (2003)
35. Bazan, J.G., Szczuka, M.: The Rough Set Exploration System. In: Peters, J.F., Skowron, A. (eds.) Transactions on Rough Sets III. LNCS, vol. 3400, pp. 37–56. Springer, Heidelberg (2005)
36. Hor, C.-L., Crossley, P.A.: Knowledge Extraction from Intelligent Electronic Devices. In: Peters, J.F., Skowron, A. (eds.) Transactions on Rough Sets III. LNCS, vol. 3400, pp. 82–111. Springer, Heidelberg (2005)
37. Kostek, B., Szczuko, P., Żwan, P., Dalka, P.: Processing of Musical Data Employing Rough Sets and Artificial Neural Networks. In: Peters, J.F., Skowron, A. (eds.) Transactions on Rough Sets III. LNCS, vol. 3400, pp. 112–133. Springer, Heidelberg (2005)
38. Suraj, Z., Grochowalski, P.: The Rough Set Database System: An Overview. In: Peters, J.F., Skowron, A. (eds.) Transactions on Rough Sets III. LNCS, vol. 3400, pp. 190–201. Springer, Heidelberg (2005)
39. Choubey, S.K.: A comparison of feature selection algorithms in the context of rough classifiers. In: Fifth IEEE International Conference on Fuzzy Systems, vol. 2, pp. 1122–1128 (1996)
40. Chouchoulas, A., Shen, Q.: A rough set-based approach to text classification. LNAI, vol. 11, pp. 118–127. Springer, Heidelberg (1999)
41. Piñero, P., Arco, L., García, M.M., Caballero, Y., Yzquierdo, R., Morales, A.: Two New Metrics for Feature Selection in Pattern Recognition. In: Sanfeliu, A., Ruiz-Shulcloper, J. (eds.) CIARP 2003. LNCS, vol. 2905, pp. 488–497. Springer, Heidelberg (2003)
42. Sugihara, K., Tanaka, H.: Rough Sets approach to information systems with interval decision values in evaluation problems. In: The International Symposium on Fuzzy and Rough Sets, ISFUROS 2006, Santa Clara, Cuba (2006)

43. Midelfart, H., Komorowski, J., Ñorsett, K., Yadetie, F., Sandvik, A., Laegreid, A.: Learning rough set classifiers from gene expression and clinical data. Fundamenta Informaticae 53, 155–183 (2003)
44. Miao, D., Hou, L.: An Application of Rough Sets to Monk's Problems Solving. In: Wang, G., Liu, Q., Yao, Y., Skowron, A. (eds.) RSFDGrC 2003. LNCS (LNAI), vol. 2639. Springer, Heidelberg (2003)
45. Zhao, Y., Zhang, H., Pan, Q.: Classification Using the Variable Precision Rough Set. In: Wang, G., Liu, Q., Yao, Y., Skowron, A. (eds.) RSFDGrC 2003. LNCS (LNAI), vol. 2639. Springer, Heidelberg (2003)
46. Greco, S., Inuiguchi, M., Slowinski, R.: Rough Sets and Gradual Decision Rules. In: Wang, G., Liu, Q., Yao, Y., Skowron, A. (eds.) RSFDGrC 2003. LNCS (LNAI), vol. 2639. Springer, Heidelberg (2003)
47. Bosc, P., Prade, H.: An introduction to fuzzy set and possibility theory based approaches to the treatment of uncertainty and imprecision in database management system. In: Proc. of Second Workshop Uncertainty management, Information Systems: from Needs to Solution, California (1993)
48. Parsons, S.: Current approaches to handling imperfect information in data and knowledges bases. IEEE Trans. on Knowledge and Data Engineering 8 (1996)
49. Grabowski, A.: Basic Properties of Rough Sets and Rough Membership Function. Journal of Formalized Mathematics 15 (2003)
50. Grzymala-Busse, J.W.: Managing uncertainty in machine learning from examples. In: Proceedings of the Workshop Intelligent Information System III, Polonia (1994)
51. Pal, S.K., Skowron, A.: Rough Fuzzy Hybridization: A New Trend in Decision-Making (1999)
52. Orlowska, E. (ed.): Incomplete Information. Rough sets analysis. Physica-Verlag (1998)
53. Skowron, A., Peters, J.F.: Rough Sets: Trends and Challenges. In: Wang, G., Liu, Q., Yao, Y., Skowron, A. (eds.) RSFDGrC 2003. LNCS (LNAI), vol. 2639. Springer, Heidelberg (2003)
54. Bazan, J., Son, N.H., Skowron, A., Szczuka, M.: A View on Rough Set Concept Approximations. In: Wang, G., Liu, Q., Yao, Y., Skowron, A. (eds.) RSFDGrC 2003. LNCS (LNAI), vol. 2639. Springer, Heidelberg (2003)
55. Slowinski, R., Vanderpooten, D.: Similarity relation as a basis for rough approximations. In: Advances in Machine Intelligence & Soft-Computing, vol. IV, pp. 17–33 (1997)
56. Wilson, D.R., Martínez, T.R.: Improved Heterogeneous Distance Functions. Journal of Artificial Intelligence Research 6, 1–34 (1997)
57. García, J.M.: KNN Workshop. Suite para el Desarrollo de Clasificadores Basados en Instancias. Departamento Computación. Facultad de Matemática, Física y Computación. Universidad Central "Marta Abreu" de Las Villas (2003)
58. Skowron, A., Stepaniuk, J.: Intelligent systems based on rough set approach. In: International Workshop Rough Sets. State of the Art and Perspectives, pp. 62–64 (1992)
59. Deogun, J.S.: Exploiting upper approximations in the rough set methodology. In: Fayyad, U.Y.U. (ed.) First International Conference on Knowledge Discovery and Data Mining, Canada, pp. 69–74 (1995)
60. Kohavi, R., Frasca, B.: Useful Feature Subsets and Rough Set Reducts. In: Third International Workshop on Rough Sets and Soft Computing (1994)
61. Carlin, U.S.: Rough set analysis of medical datasets and A case of patient with suspected acute appendicitis. In: ECAI 1998 Workshop on Intelligent data analysis in medicine and pharmacology (1998)

62. Ahn, B.S.: The integrated methodology of rough set theory and artificial neural networks for business failure predictions (2000)
63. Lazo, M., Ruiz, J., Alba, E.: An overview of the evolution of the concept of testor. Pattern Recognition, 753–762 (2001)
64. Santiesteban, Y., Pons, A.: LEX: un nuevo algoritmo para el cálculo de los testores típicos. Revista Ciencias Matemáticas 21 (2003)
65. Skowron, A., et al. (eds.): RSFDGrC 1999. LNCS (LNAI), vol. 1711. Springer, Heidelberg (1999)
66. Arco, L., Bello, R., García, M.: On clustering validity measures and the Rough Set Theory. 5th Mexican International Conference on Artificial Intelligence. IEEE Computer Society Press (2006)
67. Caballero, Y., Arco, L., Bello, R., Marx, J.: New Measures for Evaluating Decision Systems using Rough Set Theory: The Application in Seasonal Weather Forecasting. In: Marx, J., Sonnenschein, M., Müller, M., Welsch, H., Rautenstrauch, C. (eds.) Third International ICSC Symposium on Information Technologies in Environmental Engineering (ITEE 2007), pp. 161–174. Springer Verlag, Carl von Ossietzky Universität Oldenburg, Heidelberg (2007)
68. Witten, I., Frank, E.: Transformation: Engineering the input and output. In: Witten, I., Frank, E. (eds.) Data Mining. Practical Machine Learning Tools and Techniques, pp. 296–304 (2005)

Machine Learning Techniques for Prostate Ultrasound Image Diagnosis

Aboul Ella Hassanien[1], Hameed Al-Qaheri[2], Václav Snášel[3], and James F. Peters[4]

[1] Information Technology Department, FCI, Cairo University
5 Ahamed Zewal Street, Orman, Giza, Egypt
aboitcairo@gmail.com, a.hassanien@fci-cu.edu.eg

[2] System Department, CBA, Kuwait University, Kuwait
alqaheri@cba.edu.kw

[3] Department of Computer Science, Faculty of Electrical Engineering and Computer Science, VŠB - Technical University of Ostrava, 17. listopadu 15, 708 33 Ostrava - Poruba, Czech Republic
vaclav.snasel@vsb.cz

[4] Computational Intelligence Laboratory, Department of Electrical & Computer Engineering, University of Manitoba, Winnipeg, Manitoba R3T 5V6 Canada
jfpeters@ee.umanitoba.ca

Abstract. Estimation of prostate location and volume is essential in determining a dose plan for ultrasound-guided brachytherapy, a common prostate cancer treatment. However, manual segmentation is difficult, time consuming and prone to variability. In this chapter, we present a machine learning scheme, employing a combination of fuzzy sets, wavelets and rough sets, for analyzing prostate ultrasound images in order diagnose prostate cancer. To address the image noise problem we first utilize an algorithm based on type-II fuzzy sets to enhance the contrast of the ultrasound image. This is followed by performing a modified fuzzy c-mean clustering algorithm in order to detect the boundary of the prostate pattern. Then, a wavelet features are extracted and normalized, followed by application of a rough set analysis for discrimination of different regions of interest to determine whether they represent cancer or not. The experimental results obtained, show that the overall classification accuracy offered by the employed rough set approach is high compared with other machine learning techniques including decision trees, discriminant analysis, rough neural networks, and neural networks.

Keywords: machine learning, rough set; fuzzy type-II, wavelet, prostate ultrasound imaging; intelligent hybrid approach; computational intelligence.

1 Introduction

Prostate cancer is one of the most frequent cancers in the male population, and is a major cause of mortality in developed countries. Detection of prostate carcinoma at an early stage is crucial for successful treatment. Different types of diagnostics, such as digital rectal examination (DRE) and prostate-specific antigen (PSA), are used to detect

J. Koronacki et al. (Eds.): Advances in Machine Learning I, SCI 262, pp. 385–403.
springerlink.com © Springer-Verlag Berlin Heidelberg 2010

prostate cancer at an early stage. However, DRE may miss small tumors, while PSA values are dependent on several factors that are not caused only by prostate cancer [25, 33]. Therefore, diagnosis often also relies on imaging methods for accurate localisation and staging of the disease [25].

Several imaging types can be used for a more reliable diagnosis. The use of transrectal ultrasound (TRUS) has become common for this purpose because of its ability to visualise the prostate gland with no injurious effects, its low cost and its real time characteristic. While TRUS is currently the most widely used imaging method, it has relatively low predictive value in detecting cancer due to the considerable overlap of benign and malignant lesion characteristics as well as the fact that its predictive ability is highly dependent on the radiologist's interpretation. To improve the performance, a computer-aided diagnosis system, which should provide useful information in real time to the radiologist, is required in order to allow for more accurate and efficient diagnosis with lower error rates. Unfortunately, no TRUS computer-aided diagnosis system for prostate cancer detection exists until now [25, 33].

Over the last few decades, machine leaning plays an essential role in the medical imaging field, including medical image analysis, computer-aided diagnosis, organ/lesion segmentation, image fusion, image-guided therapy, image annotation and image retrieval, because objects such as lesions and anatomy in medical images cannot be modeled accurately by simple equations; thus, tasks in medical imaging require learning from examples. ML provides methods, techniques, and tools that can help solving diagnostic and prognostic problems in a variety of medical domains. ML is being used for the analysis of the importance of clinical parameters and of their combinations for prognosis, e.g. prediction of disease progression, for the extraction of medical knowledge for outcomes research, for therapy planning and support, and for overall patient management. ML is also being used for data analysis, such as detection of regularities in the data by appropriately dealing with imperfect data, interpretation of continuous data used in the Intensive Care Unit, and for intelligent alarming resulting in effective and efficient monitoring[10, 16].

Machine learning systems are a well-established paradigm with current systems having many of the characteristics of biological computers and being capable of performing a variety of tasks that are difficult or impossible to do using conventional techniques. The combination or integration of more distinct intelligent methodologies can be done in any form, either by a modular integration of two or more intelligent methodologies, which maintains the identity of each methodology, or by integrating one methodology into another, or by transforming the knowledge representation in one methodology into another form of representation, characteristic to another methodology. Neural networks and rough sets are widely used for classification and rule generation [8, 12, 14, 15, 27, 28]. Instead of solving a problem using a single machine learning technique like neural network, rough sets, fuzzy image processing alone.

In this chapter we show how a combination of various machine learning techniques can be used to analyst TRUS images for the purpose of diagnosing prostate cancer. Our scheme utilisers fuzzy logic based algorithms for improving the quality of TRUS images. Then we perform segmentation utilising a modified fuzzy c-mean clustering algorithm in order to detect the boundary of the prostate pattern. Wavelet features are extracted from these regions and analysed by a rough set algorithm that defines a set reducts consisting of a minimal set of necessary attributes. Rough set classification is then performed for the actual diagnosis.

The rest of the chapter is organized as follows. Section 2 gives a brief introduction to the machine learning techniques used in our scheme. Section 3 presents our proposed prostate ultrasound image diagnosis approach in detail. Experimental results are discussed in Section 4 while Section 5 concludes.

2 Preliminaries: Machine Learning Techniques

Machine Learning is the theory of fitting models to data while simultaneously accounting for the computational complexity of the learning algorithm and the model's performance on future data. Its primary goal is to produce learning procedures that are computationally efficient and perform well (near optimal) on future data. Recently various machine learning techniques and approaches have been applied to handle the different challenges posed by image analysis. The main constituents of machine learning systems include fuzzy logic, neural networks, genetic algorithms, and rough sets. Each of them contributes a distinct methodology for addressing problems in its domain. This is done in a cooperative, rather than a competitive, manner. The result is a more intelligent and robust system providing a human-interpretable, low cost, exact enough solution, as compared to traditional techniques. This section provides a brief introduction into the type-II fuzzy sets, wavelets and rough sets.

2.1 Fuzzy Sets

Fuzzy image processing
Generally, fuzzy image processing [13, 23, 42] gathers all approaches that understand, represent and process the images, their segments and features as fuzzy sets [42].

Definition 1. *(Fuzzy image) An image I of the size $M \times N$ pixels having L gray levels g ranging from 0 to $L-1$. The image can be considered as an array of fuzzy singletons, each having a value of membership denoting its degree of brightness relative to some levels. Using the fuzzy set notation, we can represent the image I as follows:*

$$I = \bigcup_{ij} \frac{\mu(g_{ij})}{g_{ij}}, i = 0, \ldots M-1, j = 0, \ldots N-1 \tag{1}$$

where g_{ij} is intensity of the (M,N)-th pixel and μ_{ij} is its membership value.

The membership function characterizes a suitable property of image (e.g. edginess, darkness, textural properties) and can be defined globally for the whole image or locally for its segments. In recent years, some researchers have applied the concept of fuzziness to develop new algorithms for image processing tasks, for example image enhancement, segmentation, etc. Hence, more specifically, fuzzy image processing is a rule-based methodology that uses fuzzy logic to reason about image data.

The type-II fuzzy sets

Fuzzy logic systems usually employ the type-I fuzzy sets and represent uncertainty by numbers in the range [0,1] which are referred to as degrees of membership. Type-II fuzzy sets are an extension of type-I fuzzy sets with an additional dimension that represent the uncertainty about the degrees of membership. The type-II fuzzy sets are useful in circumstances where it is difficult to determine the exact membership function for a fuzzy set. The type-I membership functions are precise in the sense that once they have been chosen all the uncertainty disappears. However, type-II membership functions are fuzzy themselves. The simplest type-II fuzzy sets are interval type-II sets whose elements degree of membership are intervals with secondary membership degree of 1.0 [9].

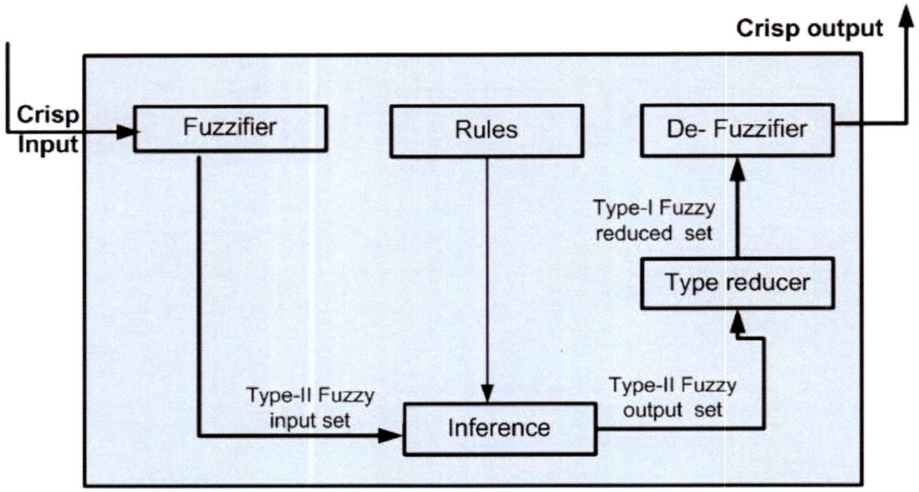

Fig. 1. Type-II fuzzy logic system

Figure (1) shows the architecture of the type-II fuzzy logic system. It works as follows: the crisp inputs are first fuzzified into either type-0 (known as singleton fuzzification) or type-I, which then activate the inference engine and the rule base to produce output type-II fuzzy sets. These the type-II fuzzy sets are then processed by a type-reducer (which combines the output sets and then performs a centroid calculation), leading to an interval type-II fuzzy sets called the type-reduced set. A defuzzifier then defuzzifies the type-reduced set to produce crisp outputs [9].

2.2 Rough Sets

Rough sets theory provides a novel approach to knowledge description and to approximation of sets. Rough theory was introduced by Pawlak during the early 1980s [39] and elaborated in [37, 41]. It is based on the concept of approximation spaces and models of the sets and concepts. In rough sets theory, the data is collected in a table, called a decision table. Rows of a decision table correspond to objects, and columns correspond to features. In the data set, set of examples with a class label to indicate the class to which each example belongs are given. We call the class label a decision feature, the rest of the features are conditional. Let $\mathcal{O}, \mathcal{F}$ denote a set of sample objects and a set of functions representing object features, respectively. Assume that $B \subseteq \mathcal{F}, x \in \mathcal{O}$. Further, let $[x]_B$ denote

$$[x]_B = \{y : x \sim_B y\}.$$

Rough sets theory defines three regions based on the equivalent classes induced by the feature values: lower approximation $\underline{B}X$, upper approximation $\overline{B}X$ and boundary $BND_B(X)$. A lower approximation of a set X contains all equivalence classes $[x]_B$ that are subsets of X, and upper approximation $\overline{B}X$ contains all equivalence classes $[x]_B$ that have objects in common with X, while the boundary $BND_B(X)$ is the set $\overline{B}X \setminus \underline{B}X$, *i.e.*, the set of all objects in $\overline{B}X$ that are not contained in $\underline{B}X$. So, we can define a rough set as any set with a non-empty boundary. The indiscernibility relation $\sim_B$ (or by Ind_B) is a mainstay of rough set theory. Informally, $\sim_B$ is a set of all objects that have matching descriptions. Based on the selection of B, $\sim_B$ is an equivalence relation that partitions a set of objects $\mathcal{O}$ into equivalence classes (also called elementary sets). The set of all classes in a partition is denoted by $\mathcal{O}/\sim_B$ (also by $\mathcal{O}/Ind_B$). The set $\mathcal{O}/Ind_B$ is called the quotient set. Affinities between objects of interest in the set $X \subseteq \mathcal{O}$ and classes in a partition can be discovered by identifying those classes that have objects in common with X. Approximation of the set X begins by determining which elementary sets $[x]_B \in \mathcal{O}/\sim_B$ are subsets of X.

2.3 Wavelet Transform

Wavelet is a well-known image analysis methodology [35]. It is known that the use of wavelets provides a richer feature space. It provides approximation with excellent time and frequency resolution. The produced wavelet coefficient yields a smaller set of more robust features, which can improve the probability of correct classification [26].

The fundamental idea behind wavelets is to analyze the signal at different scales or resolutions, which is called multiresolution [31, 32]. Wavelets are a class of functions used to localize a given signal in both space and scaling domains. A family of wavelets can be constructed from a mother wavelet. Compared to Windowed Fourier analysis, a mother wavelet is stretched or compressed to change the size of the window. In this way, big wavelets give an approximate image of the signal, while smaller and smaller wavelets zoom in on details. Therefore, wavelets automatically adapt to both the high-frequency and the low-frequency components of a signal by different sizes of windows. Any small change in the wavelet representation produces a correspondingly small change in the original signal, which means local mistakes will not influence the entire transform. The wavelet transform is suited for nonstationary signals, such as very brief signals and signals with interesting components at different scales [31]. Wavelets mean small waves that segments data into different frequency components and transfer each component with different resolution that is matched to its scale. The main idea of wavelet analysis is to see both coarse and detail data without heavy computational penalty. The goal of most modern wavelet researches is to create a set of basis functions and transform them in order to give an information.

The wavelet transform (WT) decomposes a signal $f(t)$ by performing inner products with a collection of analysis function $\psi(a,b)$, which are scaled and translated version of the wavelet ψ. The wavelet coefficient $W(a,b)$ of the function $f(t)$ is defined as follows:

$$W(a,b) = \langle f, \psi_{(a,b)} \rangle = \int_{-\infty}^{+\infty} f(t)\psi_{(a,b)}(t)dt \tag{2}$$

$$\psi_{(a,b)}(t) = a^{-1/2}\psi(\frac{t-b}{a}) \tag{3}$$

It refers to the degree of similarity between the basis functions (wavelet) and the original signal at the current scale. The amplitude of the WT therefore tends to be maximum at those scales and locations where the signal most resembles the analysis template. The continuous wavelet transform is a reversible transform, $f(t)$, which can be restored using the the following:

$$f(t) = \frac{1}{C_\psi} \int_0^{\infty} \int_{-\infty}^{+\infty} \frac{1}{\sqrt{a}} W(a,b)(\frac{x-b}{a})\frac{dbdb}{a^2} \tag{4}$$

where C_ψ is a constant that depends on the choice of the wavelet, $a > 0$ is the scale parameter and b is the position parameter.

When the scale a varies, the filter ψ is only reduced or dilated, while keeping the same pattern [31]. The reconstruction is only possible if C_ψ is defined by admissibility condition, which restricts the class of functions that can be wavelet.

3 Machine Learning Prostate Ultrasound Image Diagnosis

In this chapter, a machine learning scheme, employing a combination of fuzzy logic, wavelets and rough sets, for analyzing prostrate ultrasound images in order diagnose prostate cancer is presented. Figure (2) illustrates the architecture of the proposed scheme. It is comprised of three fundamental building phases as follows:

- **Pre-processing.** In the first phase of the investigation, a pre-processing algorithms based on type-II fuzzy sets and fuzzy c-mean are presented. It is adopted and used to improve the quality of the images and to make the feature extraction phase more reliable. It contains several sub-processes; (a) enhancement of the contrast of the whole image using fuzzy type-II algorithm;(b) detection of the boundary of the prostate region; and (c) enhancement of the edges surrounding the region of interest.
- **Feature extraction.** In the second phase, wavelet-based features are extracted, normalized and represented in a database as vector values. The third phase is rough set data analysis. It is done by computing the minimal number of necessary attributes, together with their significance, and generating the sets of rules.
- **Rough set data analysis and classification.** The last phase is the prediction of new objects. It is dependent on the type of generated rules and strategies based on the rough model. These three phases are described in detail in the following section along with the steps involved and the characteristics feature for each phase.

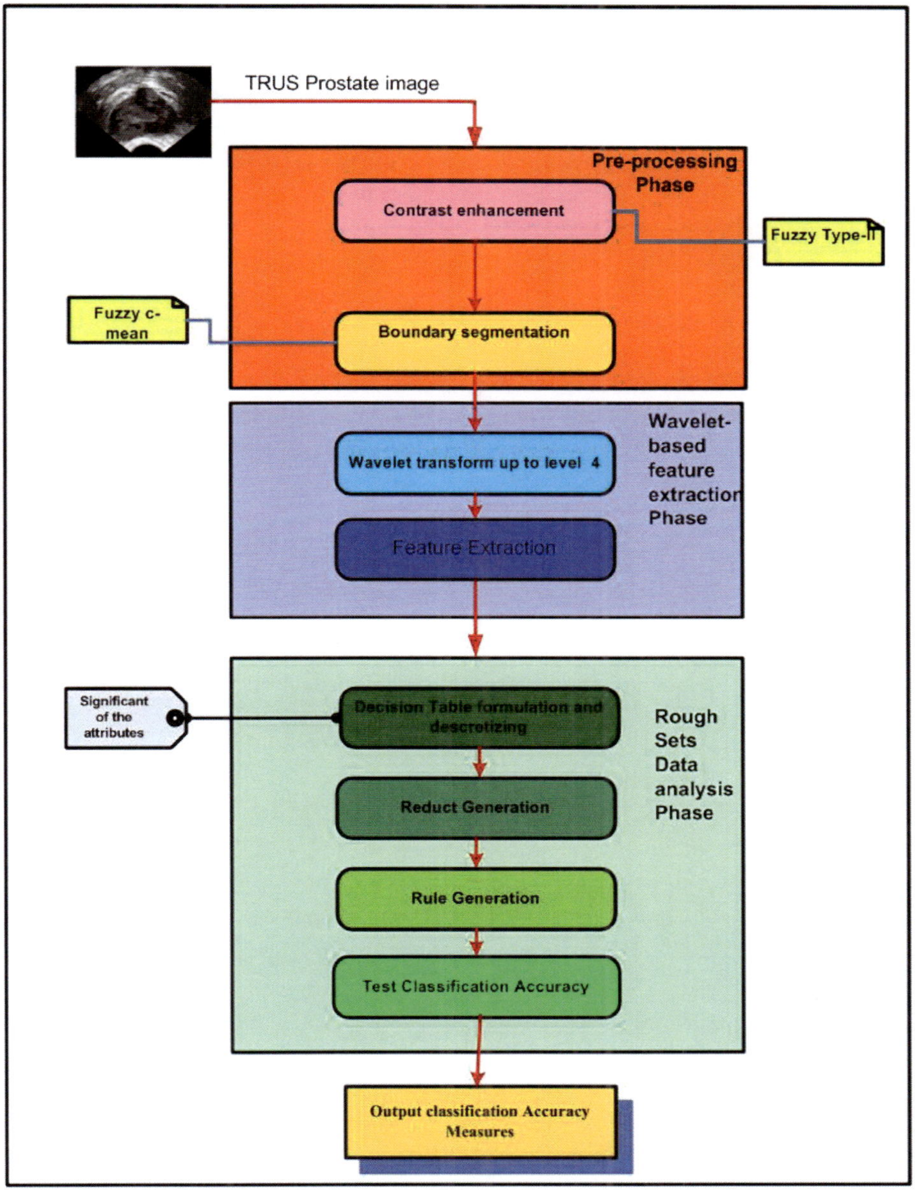

Fig. 2. Machine learning scheme for prostate Ultrasound image diagnosis

3.1 Pre-processing Phase

Ultrasound imaging techniques are widely used in medical diagnosis [25, 33]. Its non-invasive nature, low cost, portability, and real-time image formation make ultrasound imaging an attractive means for medical diagnosis. One of the limitations of ultrasound images is poor image quality affected by speckle noise. Speckle reduction remains a difficult problem due to the lack of reliable models to estimate noise. Ultrasound images are very difficult to diagnose because of the existence of speckle which hampers the perception and the extraction of fine details from the image [6]. To increase the efficiency of the classification and prediction process, a pre-processing stage is considered to enhance the quality of the input prostate images before feature extraction and classification is done. In this chapter, algorithm based on fuzzy type-II is adopted and used to enhance the contrast of the Ultrasound prostate image. Also, a modified fuzzy c-mean-based segmentation is used to achieve an accurate region of interest (ROI) identification and to detect the boundary of the prostate pattern.

Fuzzy type-II contrast enhancement algorithm

Fuzzy type-II set obtained by blurring a type-I membership function. It uses interval-based sets to construct the type-II fuzzy set by defining the upper and lower membership values using the following equations:

$$\mu_{UPPER}(x) = \mu(x)^{0.5} \tag{5}$$

$$\mu_{LOWER}(x) = \mu(x)^{2} \tag{6}$$

where $0 \leq \mu(x) \leq 1$ is the membership function for value x. Algorithm-1 shows the main steps of the introduced Type-II fuzzy enhance image process.

Standard Fuzzy c-mean clustering algorithm

Fuzzy c-means [34] is based on the idea of finding cluster centers by iteratively adjusting their positions and evaluation of an objective function which is typically defined as

$$E = \sum_{j=1}^{C} \sum_{i=1}^{N} \mu_{ij}^{k} ||x_i - c_j||^2 \tag{7}$$

where μ_{ij}^{k} is the fuzzy membership of sample (or pixel) x_i and the cluster identified by its centre c_j, and k is a constant that defines the fuzziness of the resulting partitions.

E can reach the global minimum when pixels nearby the centroid of corresponding clusters are assigned higher membership values, while lower membership values are assigned to pixels far from the centroid [11]. In here, the membership is proportional to the probability that a pixel belongs to a specific cluster where the probability is only dependent on the distance between the image pixel and each independent cluster centre. The membership functions and the cluster centres are updated by

$$\mu_{ij} = \frac{1}{\sum_{m=1}^{C} \left(\frac{||x_j - c_i||}{||x_j - c_m||)^{2/(k-1)}} \right)} \tag{11}$$

Algorithm 1. Type-2 Fuzzy Image Enhancement

Input: TRUS prostate image
Output: Fuzzy type-I and Fuzzy type-II enhanced images of the input image
Processing:

Step-1 Fuzzy hyperbolization

1: For each grey level Do
2: Compute the the fuzzy type-I membership value

$$\mu(g_{mn}) = \frac{g_{mn} - g_{min}}{g_{max} - g_{min}} \tag{8}$$

{where g_{min} and g_{max} are the image minimum and maximum grey levels}

Step-2:Fuzzy type-I enhanced image calculation

3: Compute the new grey levels of the enhanced image using the following formula:

$$\tilde{g}_{mn} = \frac{L-1}{e^{-1}-1} \times [e^{-\mu(g_{mn})^{\beta}-1}] \tag{9}$$

{where the parameter β as a fuzzifier is set to 1.7 and the number of grey levels L is set to 256.}

Step-3 Calculation of Fuzzy type-II membership function

4: Compute $\mu_{LOWER}(x)$ and $\mu_{UPPER}(x)$ using Equations (5 and 6).
{A type-II fuzzy set is obtained by blurring a type-1 membership function. For this purpose we use interval-based sets to construct the type-II fuzzy set by defining the lower membership values $\mu_{LOWER}(x)$ and the upper membership values $\mu_{UPPER}(x)$}
5: Divide the image into sub-images
6: **For** every grey level value **Do**
7: Calculate a window of size 21×21
{This is done for the enhancement to be localized according to each grey level's neighbors}.
8: Compute the fuzzy Type-II membership function using the following Equation

$$\mu_{T_{II}(g_{mn})} = (\mu_{LOWER} \times \alpha) + (\mu_{UPPER} \times (\alpha - 1)) \tag{10}$$

where $\alpha = \frac{g_{mean}}{L}$

Step-4 Fuzzy type-2 enhanced image calculation

9: Compute the new grey levels of the enhanced image as in fuzzy type-1 using equation (7)

and

$$c_i = \frac{\sum_{j=1}^{N} u_{ij}^k x_j}{\sum_{j=1}^{N} u_{ij}^k} \tag{12}$$

Modified Fuzzy c-mean clustering algorithm

The FCM objective function is minimized when high membership values are assigned to pixel whose intensities are close to the centroid of its particular class, and low membership values are assigned when the pixel data is far from the centroid. The FCM algorithm is computationally expensive and sensitive to the noise. To solve these problems, we present a modified version of the fuzzy c-mean clustering algorithm [3, 13, 34]

Algorithm 2. Modified fuzzy c-mean Clustering

Input: Enhanced prostate image I
Set the initial values of centroid c
Output: Set of clustered regions

1: **for** i=1 to c **do**
2: Update the partition matrix {Membership calculation}
3: $D_{ik} = \|y_k - \lambda_i * v_k\|^2$;
4: $\phi_i = \sum_{y_r \in N_k} \|y_r - \lambda_r * v_i\|^2$;
5: $v_{ik}^* = \frac{1}{\sum_{j=1}^{c} (\frac{D_{ik} + \frac{\alpha}{N_R}\phi_i}{D_{jk} + \frac{\alpha}{N_R}\phi\delta_j})^{\frac{1}{p-1}}}$;
6: **end for**
{N_R stands for the cardinality of the set of neighbors that exist in a window around x_k. Influence of N_R is controlled by the parameter α; higher value of the parameter α is better}
7: Update the centroid
8: Estimate $v_i^* = \frac{\sum_{k=1}^{N} u_{ik}^p \left[(y_k - \lambda_k) + \frac{\alpha}{N_R} \sum_{y_r \in N_k} (y_r - \lambda_r)\right]}{(1+\alpha) \sum_{k=1}^{N} u_{ik}^p}$;
9: Estimate a new gain filed $\lambda_j = \frac{\sum_{i=1}^{c} v_{ik}^p y_j v_i}{\sum_{i=1}^{c} v_{ik}^p}$;
10: **repeat**
11: statement 2
12: **until** $\|V_{new} - V_{old}\| < \epsilon$; {Until convergence}
{ϵ is the threshold set by the user and V is a vector of cluster centers}

3.2 Feature Extraction Phase: Wavelet Transform

In this work the features will be defined in wavelet transform domain. In the case of images, we consider the image an $N \times N$ square matrix of signal values. We decompose the image row-wise for every row using 1-D decomposition algorithm. The resulting two matrices are transposed and processed row-wise again to obtain four $\frac{N}{2} \times \frac{N}{2}$ square matrices. This procedure can be repeated any number of times to represent the resolution level we want to reach. The total number of coefficients after decomposition always equals to the initial input coefficient. This matrix will be considered as our co-occurrence matrix. Since cancer creates discontinuities in the tissue, in order for cancer to survive, it develops its own blood supply system which is different than the supply system of normal tissue [5]. Therefore, the variance of the edge is higher when the continuity breaks down, so cancerous tissue is expected to have more edges, and the time between the edges is smaller than non cancerous regions. The activity level of this area is high. The activity level can be measured as the entropy of this coefficient. The traditional entropy measure is global quantity calculated for the whole image, and then it is not appropriate for quantifying the distribution of the information at different scales of resolution.

Strack et al. [30] introduced the concept of multiresolution into an entropy measure. It considers the information contained in the image as the summation of information at different resolution levels. We introduce a new feature within a wavelet transform domain [36]. The new feature is called Multiresolution Local Contrast Entropy $MLCE$. Let w_j be the detail coefficients of observed data over an l wavelet transform. The multiresolution entropy is:

$$ME = \sum_{j=1}^{l} \sum_{k=1}^{N} \frac{w_{j(k)^2}}{2\sigma^2{}_j} \tag{13}$$

N, l and σ are the number of data, the resolution level, and the noise variance estimated in the finest scale, respectively. We claim that the more significant feature is the probability of the Multiresolution Local Contrast Entropy $(MLCE)$ for each coefficient. The $MLCE$ is calculated by:

$$MLCE(x, y) = \frac{ME(x, y)}{\sum_{i=1}^{N} \sum_{j=1}^{N} ME(i, j)} \tag{14}$$

It assigns a value to the wavelet coefficient according to its contribution weight to the local contrast entropy. It will be calculated as follows: From each subband at the fifth level of resolution, compute the average multiresolution local contrast entropy of the coefficient as:

$$v_i = \frac{1}{n_i} \sum_{i=1}^{n_i} MLCE_i, \quad i = 1, \ldots 32 \;\; \& \;\; n_i = 128 \tag{15}$$

where v_i is the ith feature in a wavelet packet feature vector. We will consider the average of this vector space as the AMLCE feature. To identify prostate cancer, texture features are needed as a discriminative measurement for the samples to measure the homogeneity, the coarseness, the periodicity and the linearity of texture. The rest of the features will be calculated using the coefficient matrix produced to the fifth level of resolution. We use the texture features used in [29, 38] to identify cancerous tissue. Then the feature is calculated by the average of each feature over all the levels. The set of selected features can be seen in Table (1).

Table 1. Selected Features

No	Name	Description
1	AMLCE	Average Multi-resolution local contrast entropy
2	RMSE	Root mean square error
3	ASM	Angular second moment
4	CON	Contrast
5	DIS	Dissimilarity
6	COR	Correlation
7	IDM	Inverse difference moment
8	SD	Standard deviation
9	ADWC	Average of Detailed Wavelet Coefficient
10	NSSDWC	Sum of Square of Detailed Wavelet Coefficient

3.3 Rough Sets Data Analysis Phase

One way to construct a simple model computed from data, easier to understand and with more predictive power, is to create a set of minimal number of rules. Some condition values may be unnecessary in a decision rule produced directly from the database. Such values can then be eliminated to create a more comprehensible minimal rule preserving essential information. The rough set data analysis contains the following four processes:

- Discretizing the image features (i.e attributes);
- Significant of the attributes;
- Generate the reducts with minimal number of attributes;
- Rule generation and rule accuracy: generate a list of rules, compute the overall accuracy of the generated rules.

These four processes are described in detail in the following section along with the steps involved and the characteristics feature for each process.

Rough Discretization process

When dealing with attributes in concept image classification, it is obvious that they may have varying importance in the problem being considered. Their importance can be pre-assumed using auxiliary knowledge about the problem and expressed by properly chosen weights. However, in the case of using the rough set approach to concept classification, it avoids any additional information aside from what is included in the information table itself. Basically, the rough set approach tries to determine from the data available in the information table whether all the attributes are of the same strength and, if not, how they differ in respect of the classifier power. Therefore, some strategies for discretization of real valued features must be used when we need to apply learning strategies for data classification (*e.g.*, equal width and equal frequency intervals). It has been shown that the quality of learning algorithm is dependent on this strategy, which has been used for real-valued data discritization [2]. A pixel grey value in an image is always a numeric value. Many classification algorithms such as rough set theory, require that training data contain only discretized feature values. Otherwise, too many equivalent classes will be produced and the algorithms will be over sensitive to noise. To use such an algorithm when there are numeric-valued features, all numeric values must first be converted into discrete values -a process called discretization. This process is performed by dividing the values of continuous attributes into a small number of intervals, where each interval is mapped to a discrete categorical, nominal, symbolic symbol. Discretization can significantly influence the effectiveness of a classification algorithm. We used the discretization algorithm introduced in our previous work in [2].

Minimal reducts and rule generation processes

Reduct is an important concept in rough set theory and data reduction is a main application of rough set theory in pattern recognition and data mining. As it has been proven that finding the minimal reduct of an information system is a NP hard problem [2].

The computation of the reducts from a decision table is a way of selecting relevant features [7]. It is a global method in the sense that the resultant reducts represent the minimal sets of features which are necessary to maintain the same classification accuracy given by the original and complete set of attributes. A straight manner for selecting relevant features is to assign a measure of relevance to each attribute and choose the attributes with higher values. Based on the reduct system, we generate the list of rules that will be used for building the classifier model for the new objects. In decision tables, there often exist conditional attributes that do not provide (almost) any additional information about the objects. So, we should remove those attributes since it reduces complexity and cost of decision process [7]. A decision table may have more than one reduct. Anyone of them can be used to replace the original table. Finding all the reducts from a decision table is NP-complete problem. Fortunately, in applications, it is usually not necessary to find all of them – one or a few of them are sufficient enough. A natural question is, which reducts are the best. The selection depends on the optimality criterion associated with the attributes. If it is possible to assign a cost function to attributes, then the selection can be naturally based on the combined minimum cost criteria. In the absence of an attribute cost function, the only source of information to select the reduct is the contents of the table. In this chapter, we adopt the criteria that the best reducts are the ones with the minimal number of attributes and – if there are more such reducts – with the least number of combinations of values of its attributes cf. [5]. We present a reduct algorithm based on the entropy information measure introduced in [24].

The generated reducts are used to generate decision rules. The decision rule, at its left side, is a combination of values of attributes such that the set of (almost) all objects matching this combination have the decision value given at the rule's right side. The rule derived from reducts can be used to classify the data. The set of rules is referred to as a classifier and can be used to classify new and unseen data. The main steps of the rule generation and classification algorithm are provided in our previous published work (cf. [4]).

4 Experimental Results and Discussions

Our dataset consists of 212 TRUS cancer and non cancer images. A number of experiments are carried out to determine if the feature extraction and the classification schemes proposed can predict the malignancy in a TRUS image. All the experiments are implemented in MATLAB. According to the introduced hybrid machine learning scheme, the experiments go through the following stages:

4.1 Contrast Enhancement and Boundary Segmentation Results

Our data set consists of 212 TRUS cancer and non cancer images. A number of experiments are carried out to determine if the feature extraction and the classification schemes proposed can predict the malignancy in a TRUS image. All the experiments were implemented on PC with a 3 GHz P4 processor using MATLAB 7.01. Due to the differences in conditions under which acquisition occurred, the images are not at the same level of contrast. A pre-processing step is then required. We apply type-II fuzzy sets to enhance the contrast of the prostate image. Figure 3, shows the fuzzy based enhancement results. The four rows are prostate TRUS images, and the first column of each row is the original prostate image, the second column shows the result of locally adaptive type-I fuzzy approach presented in [9], the third is the result of the type-II fuzzy approach. In the four results fuzzifier $\beta = 1.7$ and the minimum and maximum size of local windows are set to 21 and 21, respectively. From Figure (3) we observe that type-II fuzzy algorithm is able to perform image contrast enhancement better than its type-I.

Figure 4 shows the segmented boundary result using a modified fuzzy c-mean algorithm. Figure 4(a & b) shows the original and the type-II fuzzy enhanced image, respectively. While Figure 4(c) shows the segmented boundary result using the

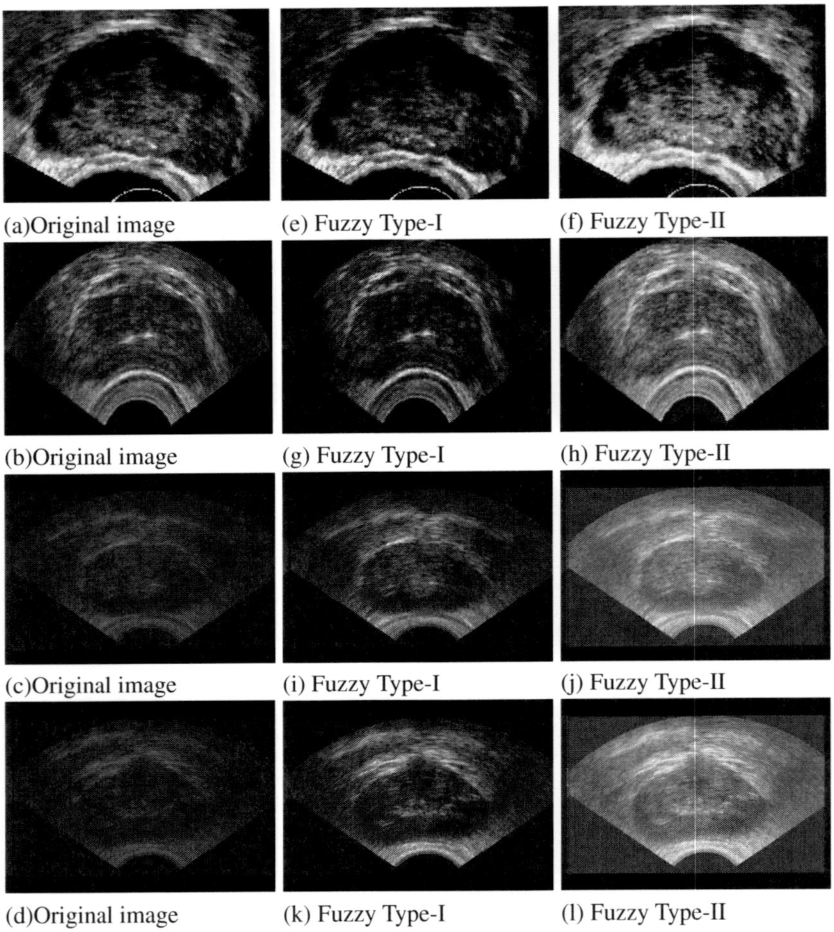

Fig. 3. Type-I and type-II fuzzy sets enactments results

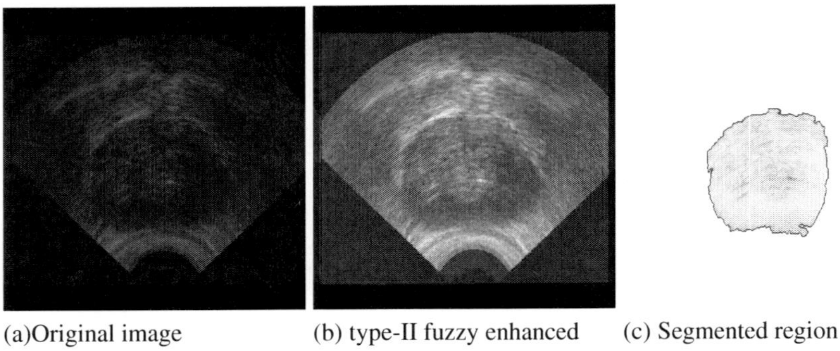

Fig. 4. Modified Fuzzy c-mean segmentation results

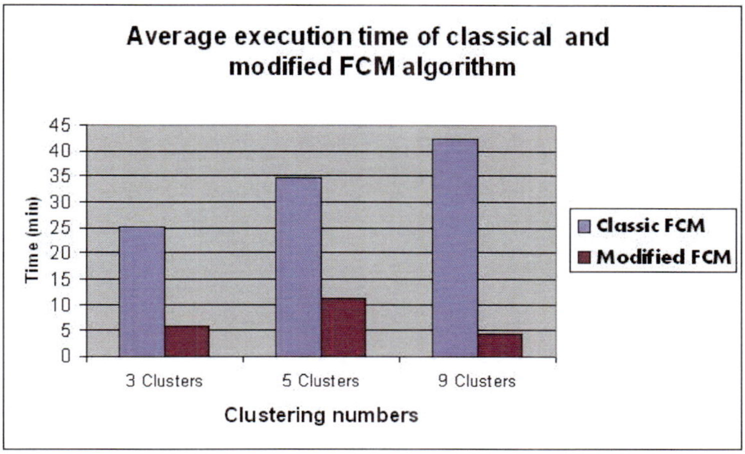

Fig. 5. Time comparison between FCM (Classical and Modified)

modified fuzzy c-mean algorithm. Figure 2 represents the modified FCM visual clustering results with different initiation parameters. The weight parameter of cost function is ranging from 0.001 to 0.0000001 and the clusters' number ranging from three to nine. From the obtained results, we observe that both algorithms get the same good results with higher number of clusters and small weight of the cost function. The average of segmentation accuracy of the segmentation algorithm is about 2.1437 $\%$ error, which means that it is robust enough. The segmentation accuracy has been done by measuring the similarity between the obtained results and the gold standard. Figure 5 shows the computation time of the modified c-mean clustering algorithm compared with the classical fuzzy c-mean. From Figure 3, we observe that the computation time using the modified fuzzy c-mean is less than the segmentation using the classical fuzzy c-mean.

4.2 Reducts and Number of Generated Rules

Table 2 shows the generated two reducts and their discrimination factors. We can observe that we have two minimal reduction sets: $R1$ and $R2$ with the same discrimination factor.

To evaluate the sets of induced rules, the rule importance measure is taken into account. It is defined as follows:

$$Importance_{rule} = \frac{\tau_r}{\rho_r}, \tag{16}$$

where τ_r is the number of times a rule appears in all reducts and ρ_r is the number of reduct sets.

Table 2. Generated reduct sets and their discrimination factors

Minimal Reduct Set	Attribute	Discrimination Factor
R1	AMLCE,SD,NSSDWC,ADWC,CON	0.401
R2	AMLCE,SD,NSSDWC,ADWC,RMSE	0.401

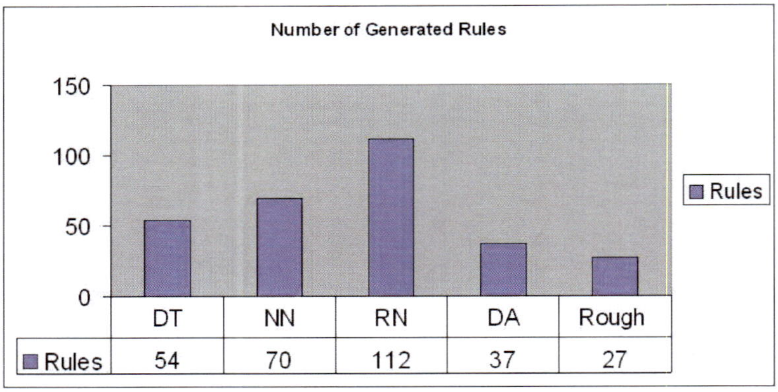

Fig. 6. Generated rules: Comparative analysis among Decision Tree (DT), Discriminant Analysis (DA) Rough Neural (RN), Neural Network (NN) and Rough Sets

Figure 6 depicts the number of generated rules. We can observe that the number of generated rules for all algorithms is larger than the rule generated using rough sets. Actually, it is less than the number of objects. It makes classification acceptably fast. Therefore, it is not necessary in this application to prune the rules during their generation. On the other hand, rule-based classification process associated with neural networks is not easy to explain as rules that are meaningful to the user. Moreover, in decision tree algorithm, more robust features are required to improve the performance of decision tree.

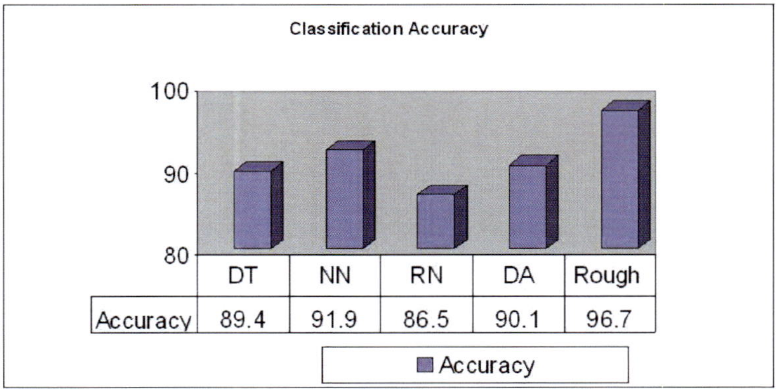

Fig. 7. Classification accuracy: Comparative analysis among Decision Tree (DT), Discriminant Analysis (DA) Rough Neural (RN), Neural Network (NN) and Rough Sets

4.3 Classification Accuracy Evaluation

In our implementation, we divide the 212 samples of data into two sets of samples such that a first set is used for testing the model that has been developed from the training set. To assess the performance of the model on the test data set, in this chapter we used the sensitivity and specificity measures to estimate the classifier accuracy for introduced approach.

Figure 7 illustrates the overall rough sets classification accuracy in terms of sensitivity and specificity compared with Decision Tree (DT), Discriminant Analysis (DA) Rough Neural (RN), and Neural Network (NN). Empirical results reveal that the proposed rough set approach performs better than the other classifiers.

5 Conclusions and Future Works

We have developed an advanced intelligent scheme based on machine learning techniques for diagnosing cancer in prostate TRUS images with a high degree of accuracy. In our scheme contrast enhancement process is applied first based on the type-II fuzzy sets. This is followed by employing a modified fuzzy c-mean segmentation algorithm to detect the boundary of the region of interest. This is followed by performing a modified fuzzy c-mean clustering algorithm in order to detect the boundary of the prostate pattern. Wavelet-based features are then extracted, followed by application of a rough set analysis for classification between cancer and non-cancer cases. Experimental results confirmed the usefulness of our approach while the employed rough set classifier was also compared to diagnosis based on decision trees, discriminant analysis, rough neural networks, and neural networks to demonstrate the efficacy of our proposed method.

Our future work is to develop near set-based methods that offer an approach to classifying perceptual objects by means of near sets that are discovered bycomparisons of descriptions of objects contained in partitions of images. It is fairly apparent that near set methods can be useful in object recognition, especially in solving medical imaging problems. The near set approach to object description, feature selection, and automatic image segmentation based on the partition of an image into equivalence classes offer a practical as well as straightforward approach to classifying images. It is in the domain of medical image segmentation that the near set approach holds the greatest promise for medical imaging. Moreover, near set approach to image feature selection is tractable, since it is always possible to find the information content of partitions of sample objects at the single feature level within a reasonable time. A more comprehensive review can be found in sources such as [1, 17, 18, 19, 20, 21, 22].

References

1. Hassanien, A.E., Abraham, A., Peters, J.F., Schaefer, G., Henry, C.: Rough Sets and Near Sets in Medical Imaging: A Review. IEEE Trans. on Information Technology in Biomedicine (in press, 2009)
2. Hassanien, A.E., Abdelhafez, M.E., Own, H.S.: Rough Sets Data Analysis in Knowledge Discovery: A Case of Kuwaiti Diabetic Children Patients. Advances in Fuzzy System (2008)

3. Hassanien, A.E.: Fuzzy-rough hybrid scheme for breast cancer detection. Image and Computer Vision Journal 25(2), 172–183 (2007)
4. Hassanien, A.E., Abraham, A.: Rough Morphology Hybrid Approach for Mammography Image Classification and Prediction. International Journal of Computational Intelligence and Applications 7(1), 17–42 (2008)
5. Hassanien, A.E., Selzak, D.: Rough Neural Intelligent Approach for Image Classification: A Case of Patients with Suspected Breast Cancer. International Journal of Hybrid Intelligent Systems 3(4), 205–218 (2006)
6. Saad, A.S.: Visual enhancement of digital ultrasound images: wavelet versus Gauss-Laplace contrast pyramid. Int. J. CARS 2, 117–125 (2007)
7. Bazan, J., Nguyen, H.S., Nguyen, S.H., Synak, P., Wróblewski, J.: Rough Set Algorithms in Classification Problem. In: Polkowski, L., Tsumoto, S., Lin, T.Y. (eds.) Rough Set Methods and Applications, pp. 49–88. Physica Verlag (2000)
8. Pous, C., Caballero, D., Lopez, B.: Diagnosing patients with a combination of principal component analysis and case based reasoning. International Journal of Hybrid Intelligent Systems 6(2), 111–122 (2009)
9. Ensafi, P., Tizhoosh, H.R.: Type-2 Fuzzy Image Enhancement. In: Kamel, M.S., Campilho, A.C. (eds.) ICIAR 2005. LNCS, vol. 3656, pp. 159–166. Springer, Heidelberg (2005)
10. Magoulas, G.D., Prentza, A.: Machine learning in medical applications. In: Paliouras, G., Karkaletsis, V., Spyropoulos, C.D. (eds.) ACAI 1999. LNCS (LNAI), vol. 2049, pp. 300–307. Springer, Heidelberg (2001)
11. Schaefer, G., Hassanien, A.E., Jiang, J.: Computational Intelligence in Medical Imaging Techniques and Applications. CRC Press, Boca Raton (2008)
12. Greco, S., Inuiguchi, M., Slowinski, R.: Fuzzy rough sets and multiple-premise gradual decision rules. International Journal of Approximate Reasoning 41, 179–211 (2006)
13. Xiao, K., Ho, S.H., Hassanien, A.E.: Automatic Unsupervised Segmentation Methods for MRI Based on Modified Fuzzy C-Means. Fundam. Inform. 87(3-4), 465–481 (2008)
14. Lingras, P.: Rough neural networks. In: Proc. of the 6th Int. Conf. on Information Processing and Management of Uncertainty in Knowledge-based Systems, IPMU 1996, Granada, Spain, pp. 1445–1450 (1996)
15. Jain, L.C., Lim, C.P., Nguyen, N.T.: Recent advances in intelligent paradigms fusion and their applications. International Journal of Hybrid Intelligent Systems 5(3), 109–110 (2008)
16. Hirano, S., Sun, X., Tsumoto, S.: Dealing with Multiple Types of Expert Knowledge in Medical Image Segmentation: A Rough Set Style Approach. In: IEEE Inter. Conf. on Fuzzy systems, May 12-17, vol. 2, pp. 884–889 (2002)
17. Peters, J.F.: Near sets. Toward approximation space-based object recognition. In: Yao, J., Lingras, P., Wu, W.-Z., Szczuka, M.S., Cercone, N.J., Ślęzak, D. (eds.) RSKT 2007. LNCS (LNAI), vol. 4481, pp. 22–33. Springer, Heidelberg (2007)
18. Peters, J.F.: Near sets: general theory about nearness of objects. Applied Mathematical Sciences 1(53), 2609–2629 (2007)
19. Peters, J.F.: Tolerance near sets and image correspondence. International Journal of Bio-Inspired Computation 1(4), 239–245 (2009)
20. Peters, J.J.F., Ramanna, S.: Feature selection: Near set approach. In: Raś, Z.W., Tsumoto, S., Zighed, D.A. (eds.) MCD 2007. LNCS (LNAI), vol. 4944, pp. 57–71. Springer, Heidelberg (2008)
21. Peters, J.F., Wasilewski, P.: Foundations of Near Sets. Information Sciences. An International Journal 26, 1–35 (2009)
22. Peters, J.F., Puzio, L.: Anisotropic wavelet-based image nearness measure. Int. J. of Computational Intelligence Systems (in press, 2009)
23. Kerre, E., Nachtegael, M.: Fuzzy techniques in image processing: Techniques and applications. Studies in Fuzziness and Soft Computing, vol. 52. Physica Verlag, Heidelberg (2000)

24. Hu, Q., Xie, Z., Yu, D.: Hybrid attribute reduction based on a novel fuzzy-rough model and information granulation. Pattern Recognition 40(12), 3509–3521 (2007)
25. Martinez, C., Dall Oglio, M., Nesrallah, L., et al.: Predictive value of PSA velocity over early clinical and pathological parameters in patients with localized prostate cancer who undergo radical retropubic prostatectomy. Int. Braz. J. Urol. 30(1), 12–17 (2004)
26. Nelson, D.E., Starzyk, J.A., Ensley, D.D.: Iterated Wavelet Transformation and Signal Discrimination for HRR Radar Target Recognation. IEEE Trans. on Systems, Man, and Cybernetic-part-A, System and Human 33(1), 52–57 (2002)
27. Chandana, S., Mayorga, R.V.: RANFIS: Rough Adaptive Neuro-Fuzzy Inference System. International Journal of Computational Intelligence 3(4), 289–295
28. Pal, S.K.: Rough-Fuzzy Granulation, Rough Entropy and Image Segmentation. In: Wang, G.-Y., Peters, J.F., Skowron, A., Yao, Y. (eds.) RSKT 2006. LNCS (LNAI), vol. 4062, p. 31. Springer, Heidelberg (2006)
29. Scheipers, U., Pesavento, A., Ermert, H., J-Sommerfeld, H., Garcia-Schurmann, M., Kuhne, K., Senge, T., Philippau, S.: Ultrasound Multifeature Tissue Characterization for The Early Detection of Prostate Cancer. In: IEEE Ultrasonics Symposium, pp. 1265–1268 (2001)
30. Starck, J.L., Murtagh, F., Gastaud, R.: New Entropy Measure on the Wavelet Transform and Noise Modeling. IEEE Trans. on Circuits and Systems-II: analog and Digital Signal Processing 45(8), 1118–1124 (1998)
31. Mallat, S.G.: A Theory for Multi-resolution Signal Decomposition: The Wavelet Representation. IEEE Trans. on Pattern Analysis and Machine Intelligence II(7), 674–693 (1989)
32. Shen, J.: A note on wavelets and diffusions. J. Comp. Anal. Appl. 5, 147–159 (2003)
33. Han, S.M., Lee, H.J., Choi, J.Y.: Computer-aided Prostate Cancer Detection using Texture Features and Clinical Features in Ultrasound Image. Journal of Digital Imaging 21(suppl. 1), 121–133 (2008)
34. Zhou, H., Schaefer, G., Shi, C.: Fuzzy C-Means Techniques for Medical Image Segmentation. In: Fuzzy Systems in Bioinformatics and Computational Biology. Studies in Fuzziness and Soft Computing, vol. 242, pp. 257–271. Springer, Heidelberg (2009)
35. Goswami, J., Chan, A.K.: Fundamentals of Wavelets, Theory, Algorithms and Applications. John Wiley & Son, Inc., Chichester (1999)
36. Own, H.S., Hassanien, A.E.: Image Registration Based in Multiresolution Local Contrast Entropy in Wavelet Transform Domain. In: IEEE 14th Inter. Conf. in Digital Signal Processing, Greece, July 1-3, vol. 2, pp. 889–892 (2002)
37. Pawlak, Z.: Rough Sets- Theoretical aspect of Reasoning about Data. Kluwer Academic Publishers, Dordrecht (1991)
38. Perez-Cortes, J.C., Juan, A., Vallada, E.: Textural Analysis of Prostate Cancer In Ttansrectal Ultrasound Images. Biosignal (2002), http://prhlt.iti.es/papers/2002/Perez02.pdf
39. Pawlak, Z.: Rough sets. International J. Comp. Inform. Science 11, 341–356 (1982)
40. Pawlak, Z.: Rough sets. International J. Comp. Inform. Science 11, 341–356 (1982)
41. Pawlak, Z., Grzymala-Busse, J., Slowinski, R., Ziarko, W.: Rough Sets. Communications of the ACM 38(11), 88–95 (1995)
42. Zadeh, L.A.: Fuzzy Sets. Information and Control 8, 338–353 (1965)

Segmentation of Breast Cancer Fine Needle Biopsy Cytological Images Using Fuzzy Clustering

Marek Kowal and Józef Korbicz

University of Zielona Góra, Institute of Control and Computation Engineering,
ul. Podgórna 50, 65-246 Zielona Góra
{M.Kowal,J.Korbicz}@issi.uz.zgora.pl

Abstract. This document presents the stage of research concerning an automatic diagnosis system of breast cancer based on cytological images of FNB (Fine Needle Biopsy). The work concentrates on the image segmentation phase, which is employed to find nucleus in cytological images. The accuracy and correctness of the image segmentation algorithm is a critical factor for successful diagnosis due to the fact that case classification is based on morphometrical features extracted form segmented nucleus. The presented approach to image nucleus segmentation is based on the FCMS (Fuzzy C-Means with Shape function) clustering algorithm. Traditional approaches to image segmentation using clustering algorithms consider clustering pixels in color space in order to recognize objects. The novelty of the presented approach is that the clustering process is conducted in color space but the searched objects must have an arbitrarily defined shape. Simulations and experimental results are included in the work to illustrate the effectiveness of the proposed approach.

Keywords: medical imaging, image segmentation, fuzzy clustering, parameter optimization.

1 Introduction

Breast cancer is the most common cancer among women. The prognosis in breast cancer is strongly dependent on the disease development before any treatment is applied so the chance of recovery is a function of time of the detection of cancer. Modern medicine does not provide one hundred percent reliable, if possible cheap and at the same time non-invasive diagnostic methods for the diagnosis of breast pathology. As a result, in practice the important function acting in breast cancer diagnosis is the so-called triple-test, which is based on the summary of results of three medical examinations with different degrees of sensitivity and it allows to achieving high confidence of diagnosis. The triple-test includes self examination (palpation), mammography or ultrasonography imaging and fine needle biopsy (Underwood, 1987). Fine needle biopsy is collecting nucleus material directly from tumor for microscopic verification. Next, the material (collected cells) is examined using microscope in order to confirm or exclude the presence of cancerous cells. The present approach requires a deep knowledge and experience of the cytologist responsible for diagnosis. In short, some pathologists can diagnose better than others. In order to make the decision independent of the arbitrament factor, morphometric analysis can be applied. Objective analysis of microscopic

J. Koronacki et al. (Eds.): Advances in Machine Learning I, SCI 262, pp. 405–417.
springerlink.com © Springer-Verlag Berlin Heidelberg 2010

images of cells has been a goal of human pathology and cytology since the middle of the 19th century. Early work in this area consisted of simple manual measurements of cell and nuclear size. Along with the development of advanced vision systems and computer science, quantitative cytopathology has become a useful method for the detection of diseases, infections as well as many other disorders (Suri et al., 2002; Hrebień et al., 2008). In this work, we present a method that allows recognizing the malignancy and distinguish cancerous cells from the normal and benign cells. The classification of the tumor is based on morphometric examination of nuclei cells. In contrast to normal and benign nuclei, which are typically uniform in appearance, cancerous nuclei are characterized by irregular morphology that is reflected in several parameters. Morphometric measurements characterizing the shape and size have been mainly used for feature extraction. Features of the nuclei can be extracted from the image after the nuclei are correctly segmented and labeled. This work concentrates mainly on the segmentation phase because this stage of image processing is critical for correct diagnosis and treatment. Taking into account the specificity of the problem, the segmentation task is used to distinguish nuclei from the background and next label all nuclei. One of more often used methods for image segmentation is fuzzy clustering in the form of the Fuzzy C-Means algorithm (FCM). Unfortunately, the algorithm is useless in the case considered because it makes it possible to distinguish the brighter background from darker groups of cell nuclei and it is required to separate each single nuclei from the image in order to measure its morphometric parameters. Finally, the fuzzy c-means with shape algorithm has been adopted to tackle the mentioned problem (Leung et al., 2004; Wang et al., 2007). The original version of the FCMS algorithm was prepared to determine two clusters only (background and demanded objects). The work proposes a modified iterative algorithm in order to overcome the mentioned problem and allows determining all nuclei in the image as separate objects.

The paper is divided into three sections. Section 1 gives an overview of breast cancer diagnosis and fine needle biopsy imaging techniques. Section 2 describes the idea of the FCMS algorithm. Section 3 deals with modifications applied to the original algorithm. Section 4 shows the experimental results obtained using the proposed approach. The last part of the work includes a summary, conclusions and references.

2 FCMS Algorithm

The idea of the segmentation of images using a clustering algorithm usually boils down to a search for natural clusters of pixels represented in color space. With this approach, it is possible to find objects characterized by similar colors, and by applying the fuzzy clustering algorithm the method is more robust against disturbances and image inaccuracy (Babuska, 1998; Rutkowska and Zadeh, 2000; Szczepaniak et al., 2000). Taking into account the fine needle biopsy images of nuclei, the task of segmentation reduces to searching separate objects (nuclei) in order to determine their morphometric parameters. Unfortunately, the application of fuzzy clustering in the form of the FCM algorithm allows us to separate two interesting objects only: the background and all nuclei viewed as one big object. In order to solve the problem, a modified FCM algorithm is used, which was originally developed in the work (Leung et al., 2004; Wang et al., 2004) and used in lip segmentation from the face image.

The FCMS algorithm, similarly to the original FCM algorithm, is based on finding the local minimum of the nonlinear cost function using the Picard iteration through first-order conditions for stationary points (Bezdek, 1981; Babuska, 1998). In the case of the FCMS algorithm, the cost function is defined by the following expression:

$$J(\boldsymbol{U}, \boldsymbol{V}, \boldsymbol{\theta}) = \sum_{x=1}^{X} \sum_{y=1}^{Y} \sum_{k=1}^{C} \left[\mu_{x,y,k}^{m} \left(f_c(\boldsymbol{c}_{x,y}, \boldsymbol{v}_k) + \alpha f_d(x, y, \boldsymbol{\theta}_k)\right)\right], \quad (1)$$

where the matrix $\boldsymbol{U} \in \mathbb{R}^{X \times Y \times C}$ contains the membership degrees of pixels to the defined clusters, $\boldsymbol{V} = [\boldsymbol{v_1}, \boldsymbol{v_2}, \ldots, \boldsymbol{v_C}], \boldsymbol{v_i} \in \mathbb{R}^{C \times q}$ is a matrix which defines the centers of the clusters, $\boldsymbol{\theta}_k$ is a vector of parameters describing the preferred shape of the objects searched in spatial domain, X and Y defines the size of the analyzed image, f_c is a function used to determine the distance between the data points and cluster centers, f_d is a function to evaluate the belong rate of the pixel to the ellipse associated with the tested cluster, $\mu_{x,y,k}$ is the membership of the (x, y)-th pixel in the fuzzy cluster k, $m \in (1, \infty)$ is the fuzziness of the clustering procedure, $\boldsymbol{c}_{x,y}$ is a vector of (x, y)-th pixel parameters which describes its color, $\boldsymbol{v}_k \in \mathbb{R}^q$ is a vector of the coordinates of the k-th cluster center, $\boldsymbol{\theta}_k$ is a vector of ellipse parameters associated with the k-th cluster, α is a weighting parameter. The function f_c defines the metric used to determine the distance between the data points and cluster centers:

$$f_c(\boldsymbol{c}_{x,y}, \boldsymbol{v}_k) = \|\boldsymbol{c}_{x,y} - \boldsymbol{v_k}\|^2 = (\boldsymbol{c}_{x,y} - \boldsymbol{v_k})^T \boldsymbol{A} (\boldsymbol{c}_{x,y} - \boldsymbol{v_k}). \quad (2)$$

The matrix A which occurs in the expression (2) is used to tune the shape and orientation of the clusters in space. In the simplest approach, the matrix $\boldsymbol{A}$ is unitary, thus the distance measure $f_c(\boldsymbol{c}_{x,y}, \boldsymbol{v}_k)$ is an Euclidean norm. In this case, the study metric is defined as a Euclidean distance which measures color dissimilarity in RGB color space. The function f_d incorporates the shape function in the objective function and measures the spatial distance. As the nuclei are more like an ellipse, it was decided to chose the shape function f_d as elliptic function described by the following expression:

$$f_d(x, y, \boldsymbol{\theta}_k) = \left(\frac{\left((x - x_c)\cos\phi + (y - y_c)\sin\phi\right)^2}{w^2} + \right.$$
$$\left. + \frac{\left((x - x_c)\cos\phi - (y - y_c)\sin\phi\right)^2}{h^2} \right)^{n_k}, \quad (3)$$

where $\boldsymbol{\theta}_k = [x_c, y_c, w, h, \phi, n_k]$ denotes the set of parameters that describes the elliptic function, (x_c, y_c) is the center of the ellipse, w and h are respectively the semi-major and semi-minor axis, and $\boldsymbol{\theta_k}$ is inclination angle, exponent n_k ensures a small function values within the ellipse and large vales for the pixels outside the ellipse. Figure 1 shows values of the sample elliptic function f_d for two clusters. Taking into account that each cluster has identical parameters $\boldsymbol{\theta_k}$ but with a different exponent n_k, it is possible to determine two clusters only. The first one represents the background and the second one represents the cell nuclei found in the image.

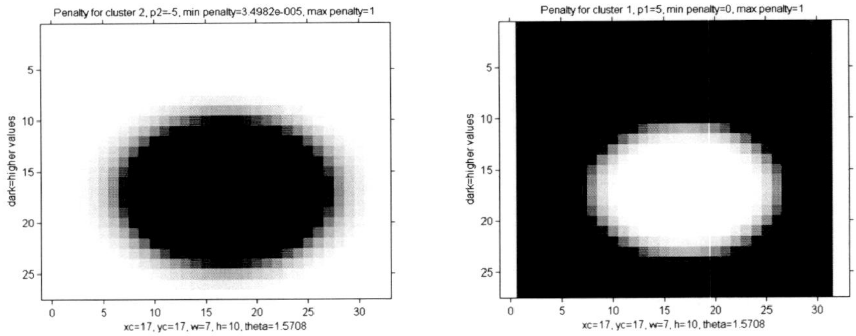

Fig. 1. Sample elliptic function for two clusters

Let us define the vector $\boldsymbol{\theta} = [x_c, y_c, w, h, \phi]$ of parameters which are considered to be tuned and, moreover, are the same for all clusters. Hence the idea of the proposed fuzzy clustering is to minimize the objective function (1):

$$(\boldsymbol{U}^*, \boldsymbol{V}^*, \boldsymbol{\theta}^*) = \arg \min_{(\boldsymbol{U},\boldsymbol{V},\boldsymbol{\theta})} J(\boldsymbol{U}, \boldsymbol{V}, \boldsymbol{\theta}). \tag{4}$$

The minimum of the objective function is calculated by an iterative algorithm using the following expressions for the updating matrices $\boldsymbol{U}$ and $\boldsymbol{V}$:

$$\mu_{x,y,k} = \left[\sum_{j=1}^{C} \left(\frac{f_c(\boldsymbol{c}_{x,y}, \boldsymbol{v}_k) + \alpha f_d(x, y, \theta_k)}{f_c(\boldsymbol{c}_{x,y}, \boldsymbol{v}_j) + \alpha f_d(x, y, \theta_j)} \right)^{\frac{1}{m-1}} \right]^{-1}, \tag{5}$$

$$\boldsymbol{v}_k = \frac{\sum_{x=1}^{X} \sum_{y=1}^{Y} \mu_{x,y,k}^m \boldsymbol{I}_{x,y}}{\sum_{x=1}^{X} \sum_{y=1}^{Y} \mu_{x,y,k}^m}, \tag{6}$$

where $\boldsymbol{I}_{x,y} \in \mathbb{R}^q$ represents a single pixel. The parameters of the ellipse $\boldsymbol{\theta}_k$ are updated in each step of the FCMS algorithm by the gradient descent algorithm using the following general expression:

$$\boldsymbol{\theta}(t+1) = \boldsymbol{\theta}(t) - \gamma \frac{\partial J(\boldsymbol{U}, \boldsymbol{V}, \boldsymbol{\theta})}{\partial \theta}, \tag{7}$$

where

$$\frac{\partial J(\boldsymbol{U}, \boldsymbol{V}, \boldsymbol{\theta})}{\partial \boldsymbol{\theta}} = \alpha \sum_{x=1}^{X} \sum_{y=1}^{Y} \sum_{k=1}^{C} \mu_{x,y,k}^m n_k f_d(x, y, \boldsymbol{\theta}_k)^{n_k - 1} \frac{\partial f_d(x, y, \boldsymbol{\theta}_k)}{\partial \boldsymbol{\theta}}. \tag{8}$$

Equations (5), (6) and (7) are iterated as a Picard iteration to obtain the segmentation of the image. Detailed expressions for updating the parameters of the ellipse can be found in the paper (Leung et al., 2004).

Figure 2 shows a sample image with two cell nuclei and a corresponding objective function calculated for two parameters x_c and y_c (for simplicity, the rest of the ellipse

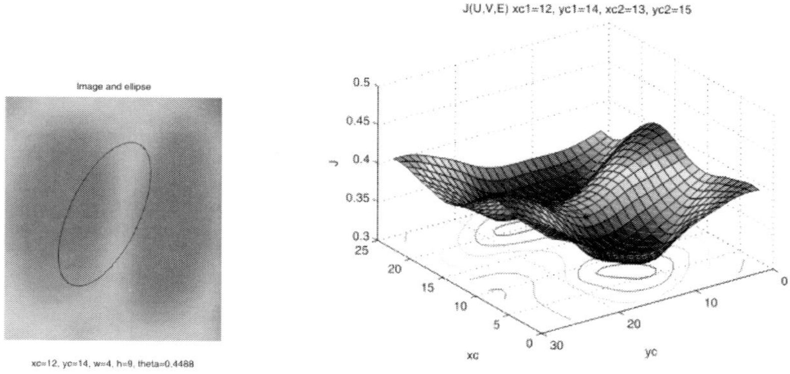

Fig. 2. Original image and the corresponding objective function

parameters are considered to be constant). Two local minima present in Fig. 2 represent two cell nuclei present in the original image. Hence, the gradient descent method can be applied to find the local minima and thus the centers of cell nuclei.

It can be noticed that the objective function is continuous in $(\boldsymbol{U}, \boldsymbol{V}, \boldsymbol{\theta})$ and $\boldsymbol{I}_{x,y}$ is bounded in $\mathbb{R}^q$. Moreover, the second partial derivative of the objective function with respect to $(\boldsymbol{U}, \boldsymbol{V})$ is a positive diagonal matrix for any feasible $\boldsymbol{\theta}$. Following the proof of convergence for FCM in (Bezdek, 1981; Leung et al., 2004), the objective function value of the i-th Picard solution $(\boldsymbol{U}^i, \boldsymbol{V}^i)$ is less than that of $(\boldsymbol{U}^{i-1}, \boldsymbol{V}^{i-1})$:

$$J(\boldsymbol{U}^i, \boldsymbol{V}^{i-1}, \boldsymbol{\theta}^{i-1}) < J(\boldsymbol{U}^{i-1}, \boldsymbol{V}^{i-1}, \boldsymbol{\theta}^{i-1}), \tag{9}$$

and

$$J(\boldsymbol{U}^i, \boldsymbol{V}^i, \theta^{i-1}) < J(\boldsymbol{U}^i, \boldsymbol{V}^{i-1}, \boldsymbol{\theta}^{i-1}). \tag{10}$$

Using the gradient descent method for solving $\boldsymbol{\theta}^i$, the objective function $(\boldsymbol{U}^{i-1}, \boldsymbol{V}^{i-1})$ is decreasing until $\boldsymbol{\theta}^i$ falls to a local minimum or reaches the boundary of the $\boldsymbol{\theta}$-vector space. Taking this into account, the following inequality can be formulated:

$$J(\boldsymbol{U}^i, \boldsymbol{V}^i, \boldsymbol{\theta}^i) < J(\boldsymbol{U}^i, \boldsymbol{V}^i, \boldsymbol{\theta}^{i-1}). \tag{11}$$

Therefore, the Picard iteration is shown to give a local minimum solution. Unfortunately, in the presented approach the update of the parameter vector $\boldsymbol{\theta}$ is computationally very expensive and the solution is extremely dependent upon the initial vales of the calculated parameters. Figure 5 shows the result of fuzzy segmentation applied to find cell nuclei with random initialization of ellipse parameters. It can be observed that the optimization procedure very often leads to the detection of spurious cell nuclei corresponding to the local minimum of the objective function.

The next section tackles the presented problems and finally solves them by applying some modifications to the segmentation procedure presented in this section.

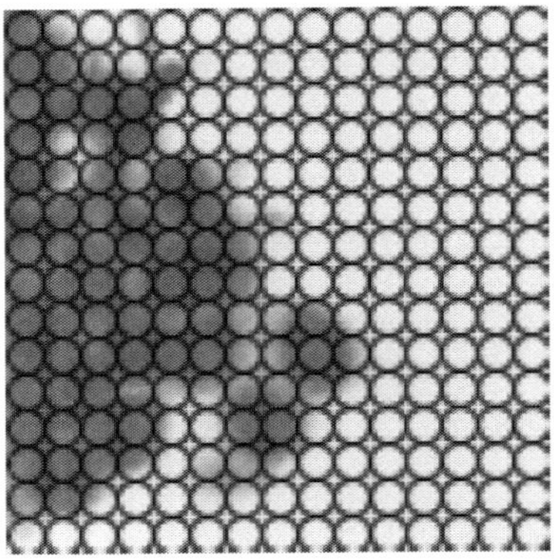

Fig. 3. Determining starting parameters for ellipses

3 Modifications to the FCMS Algorithm

The fuzzy segmentation method presented in the previous section is able to differentiate the objects with different colors, but what is more significant in the case considered it is also designed to differentiate objects with the same color but located in different regions of the image. Such a feature of the algorithm is very useful in the segmentation of cell nuclei. Unfortunately, the original FCMS algorithm (Leung et al., 2004; Wang et al., 2004) has a few drawbacks and limitations when applied to the segmentation of cell nuclei. The main problem arises from the fact that the algorithm is able to find only two clusters, the first one corresponding to the object with an elliptic shape and the second one corresponding to the rest of the image, while it is required to discover as much as possible of cell nuclei in order to make reliable diagnosis based on cell nuclei parameters.

In Section 2 it was mentioned that the FCMS algorithm is computationally very expensive due to gradient descent optimization applied to solve the parameters of the ellipse in each iteration of the whole FCMS procedure. Moreover, the results obtained using such a procedure are very problematic because the objects found are usually not cell nuclei but rather groups of cell nuclei or parts of cell nuclei (Fig. 5).

The mentioned limitations and drawbacks are weakened by changes proposed to the original approach. The first one is to divide the whole process of segmentation into two phases. The first phase is reduced to preprocessing the input image using the FCM algorithm in order to find the color of the cell nuclei and the color of the image background. The second phase is an iterative procedure applied to find cell nuclei in the preprocessed image. The developed modification arises from the observations that computing the parameters of the ellipse in each step of the FCMS algorithm is useless. While the matrices $\boldsymbol{U}$ and $\boldsymbol{V}$ do not represent precisely the proper clusters (cell nuclei and the background), the optimization procedure used to compute the parameters of the ellipse usually leads to finding spurious cell nuclei. Taking into account this fact, it was decided to solve the ellipse parameters optimization problem only once after the FCM algorithm has preprocessed the original image by minimizing (12). So, the modified

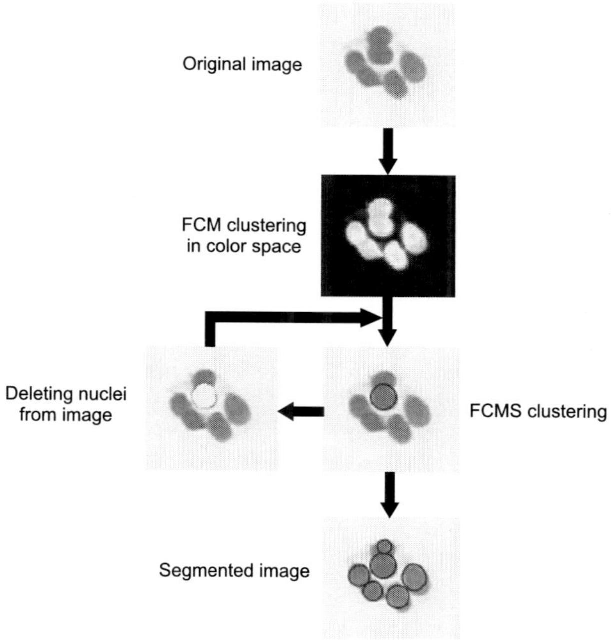

Fig. 4. Segmentation approach using FCM and the gradient descent method

segmentation procedure can be viewed as minimizing the objective function (12) and afterwards minimizing the objective function (13) with respect to $\boldsymbol{\theta}$.

$$J_1(\boldsymbol{U}, \boldsymbol{V}) = \sum_{x=1}^{X} \sum_{y=1}^{Y} \sum_{k=1}^{C} \left[\mu_{x,y,k}^{m} f_c(\boldsymbol{c}_{x,y}, \boldsymbol{v}_k) \right], \tag{12}$$

$$J_2(\boldsymbol{\theta}) = \sum_{x=1}^{X} \sum_{y=1}^{Y} \sum_{k=1}^{C} \left[\mu_{x,y,k}^{m} f_d(x, y, \boldsymbol{\theta}_k) \right]. \tag{13}$$

It was observed during the experiments that the standard FCMS algorithm usually requires about 15 iterations to detect separate cell nuclei and the whole segmentation procedure must be repeated for each cell nuclei present in the image, so the introduced modification dramatically decreased the number of computations required to segment the image.

Unfortunately, the developed method does not solve all of the mentioned problems because, similarly like standard FCMS, it often gets stuck in an inappropriate minimum (Fig. 5). This problem arises from the fact that the parameters of the initial ellipse were generated randomly and the gradient descent method converges to the nearest local minimum. To overcome the problem, it was decided to limit the initial values of ellipses using knowledge about a typical size of the cell nuclei and the multi-start gradient method (Fig. 6). The method allows us to reduce false segmentation of cell nuclei; however, finally the best results was obtained using the Hough like transform to initialize the ellipse. Such an approach proposes systematic generation of a series of ellipses arranged in the form of grids covering the whole image (Fig. 3). Next, the objective function is calculated for each ellipse and the ellipse with the smallest value of the objective function (13) becomes a starting point for the gradient descent tuning procedure. Of course, such a procedure was time consuming, so in order to reduce the complexity, the grids of ellipses was sparse and the parameters w and h were limited by the knowledge about a typical size of the cell nuclei, the parameter ϕ was constant to simplify the exploration. The proposed initialization greatly improved the quality of the segmentation and accelerated the convergence of the gradient method. In order to reduce the number of steps required to reach the minimum by the gradient method even more, an adaptive learning rate was applied to control the level of parameter updates.

The scheme of the final version of the algorithm is presented in Fig. 4. It consists of three steps. In the first one, the original image is preprocessed by the FCM method to find the color of cell nuclei and the background. Next, the parameters of ellipses are initialized using the developed approach. Finally, the gradient method is used to tune the parameters of ellipses in the following iterations. The ellipse found a single algorithm iteration represents a single cell nucleus. In order to facilitate the searching procedure, cell nuclei already detected and described by the corresponding ellipse are erased from the original image by filling in the ellipses using the background color. The stop criteria for the gradient descent algorithm are defined in the form of a minimal value of objective function change and a maximum number of tuning steps. The stop criteria for global algorithm iterations are defined as the maximum number of ellipses found and the minimum objective function value (13). Choosing right values for stop criteria requires some kind of experience from the user. The next section presents some sample results of segmentation using the developed approach.

4 Experimental Results

Testing existing and newly developed algorithms requires to have databases at disposal, on which tests and benchmarks can be realized, especially in the domain of image analysis, where in many problems a domain knowledge needs to be taken into account. In our study we decided to design a new data set that could be applied to a completely automatic process of image analysis. The segmentation of cell nuclei was carried out on cytological material obtained by FNB. Biopsy without aspiration was performed under the control of ultrasonography with a needle of a diameter of 0,5 mm. Smears from the

Table 1. Dataset

Feature	Value
Number of samples (malignant)	50 (25)
Number of images	500
Format of files	BMP
Resolution	704x578
Number of colors	256^3
Dye	h+e

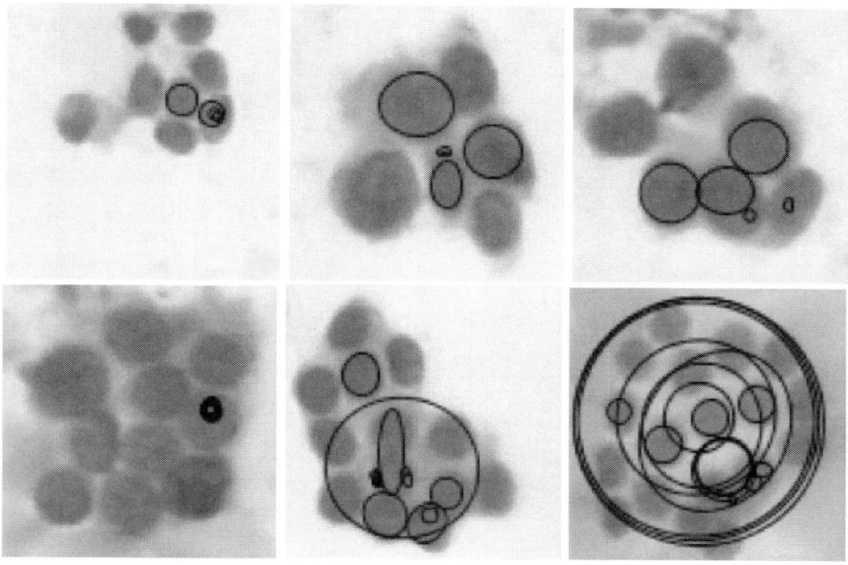

Fig. 5. Segmentation using standard FCMS

material were fixed in spray fixative (Cellfix by Shandon) and dyed with hematoxylin and eosin (h+e). The time between the preparation of smears and their preserving in fixative never exceeded three seconds. The smears were derived from 25 FNB of benign and 25 of malignant lesions collected from 50 patients of the out-patient clinic ONKOMED in Zielona Góra. All cancers were histologically confirmed and all patients with benign disease were either biopsied or followed for a year. The image for digital analysis was generated by a SONY CCD IRIS color video camera mounted atop an AXIOPHOT microscope. The slides were projected into a camera with $10\times$ and $160\times$ objective and a $2,5\times$ ocular. One image was generated for $100\times$ enlargement and nine for $400\times$ enlargement.

Most of the criteria of malignancy are seen in the nuclei of the cells. Therefore, it is essential to isolate the nuclei from the rest of the image. It can be observed that malignant cells in contrast to normal and benign cells are characterized by irregular morphology that is reflected in several morphometric parameters. Morphometric measurements characterizing the shape and size have been mainly used for feature extraction. The extracted features are: size, circularity, perimeter, compactness, lengths of the axis of the ellipse circumscribing the nuclei and the eccentricity of the ellipse circumscribing the nuclei. Details about these features can be found in (Wolberg et al., 2004; Marciniak et al., 2005; Jeleń et al., 2008; Obuchowicz et al., 2008). However, from previous research in the subject it is known that there are big differences in size between benign and malignant cases, and shape factors do not have good discriminative properties (Marciniak et al., 2005). Hence the proposed segmentation approach seems to be very promising in this case due to simplicity in calculating crucial features. In order to test the effectiveness of the proposed segmentation method, a set of 50 manually binary-segmented images was prepared. The result of automatic segmentation was compared

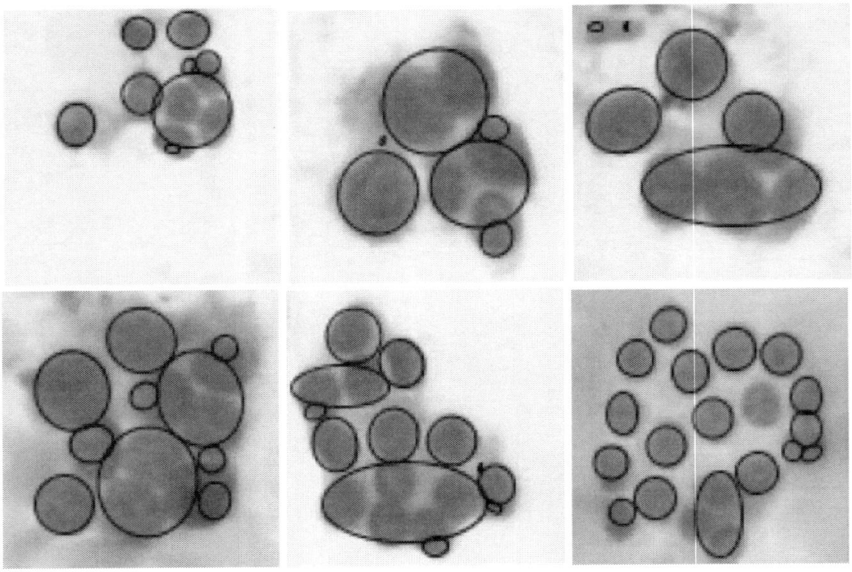

Fig. 6. Segmentation using FCM clustering and gradient descent with random initialization

to that of manual segmentation using the error measure based on the Hamming distance between manually segmented image and automatic segmented image. Since the developed segmentation method produces as a result of segmentation a set of ellipses, it was needed to convert ellipses which represent cells to a binary image by filling in the ellipses. The results of the comparison are summarized in Table 2, where the following error measures were used to compare the segmentation:

$$E_{ave} = \sum_{i=1}^{N} \frac{e_i}{n_i N}, \tag{14}$$

$$E_{min} = \min_{i=1 \dots N} \frac{e_i}{n_i}, \tag{15}$$

$$E_{max} = \max_{i=1 \dots N} \frac{e_i}{n_i}, \tag{16}$$

where e_i is a Hamming distance between the i-th manually segmented binary image and the i-th automatically segmented binary image, n_i is the number of cell nuclei in the i-th image and N is the number of test images. It must be noticed that although FCM and KM clustering achieved better segmentation results with respect to defined criteria than FCMS and two stage FCMS with random initialization, their usage is strongly limited due to the fact that they are not able to determine separated cell nucleus. In order to compare the segmentation results visually, some segmented parts of images are shown in Figs. 5, 6 and 7.

The software to realize the segmentation initially was prepared in the Matlab environment; unfortunately, time consuming computations make this software applicable to

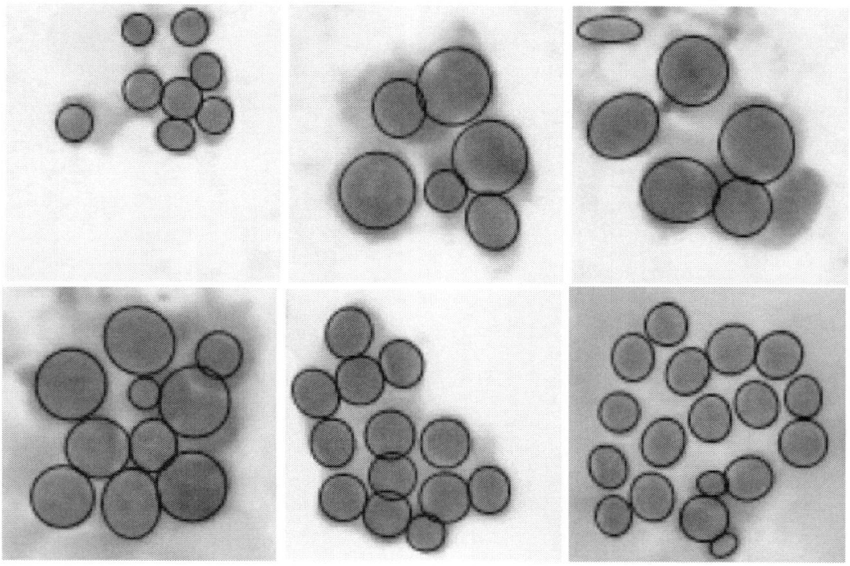

Fig. 7. Segmentation using FCM clustering and gradient descent with grid initialization

Table 2. Segmentation results

Segmentation method	E_{ave}	E_{max}	E_{min}
FCMS	132	25670	29
FCM and gradient descent with random initialization	127	22198	32
FCM and gradient descent with grid initialization	81	16210	14
FCM clustering	102	14129	23
K-mean clustering	125	16509	28

parts of the original images only. Hence the final version of the software was prepared in the Java environment using the Java Advanced Imaging (JAI) library. In order to accelerate the computations, the segmentation algorithm was written as a JAI operator and was prepared to work in a multi-threaded environment. All experiments were done using a PC class machine with a quad core processor.

5 Summary and Conclusions

The paper presents a cell nuclei segmentation method based on fuzzy-c means clustering with a shape function. The results achieved in the experiments seem to be very promising. The main advantage of the method is that the results of the segmentation can be directly used to calculate the morphometric parameters of the cells. Modifications proposed in the paper to the original FCMS algorithm allow decreasing significantly the cost of computations. This achievement is very important due to the fact that nowadays colonoscopes are able to produce images with the resolution of 50000x50000 or even higher. Of course, the presented approach is not able to segment all cases properly so there are a few challenges for the near future. The first one concerns the segmentation of overlapped cell nuclei. The second one regards the adaptation of FCMS clustering to deal with more complex shapes. This will allow calculating parameters of cell nuclei more precisely, and possibly the algorithm can gain more general applications due to the ability to detect objects with different shapes (Gong et al., 2004; Zoller and Varvarigou, 2007; Chatzis and Buhmann, 2008). However, it must be pointed out that the present accuracy of segmentation is so high that the future research will concentrate on the feature extraction and classification stages. The recognition rate in malignancy classification will be a true test for the quality and effectiveness of the developed segmentation method (Zhang, 1996).

References

Babuska, R.: Fuzzy Modelling for Control. Kluwer Academic Publisher, London (1998)

Bezdek, D.: Pattern Recognition with Fuzzy Objective Function Algorithms. Plennum Press, New York (1981)

Chatzis, S.P., Buhmann, T.A.: A fuzzy clustering approach towar hidden markov random field models for enhanced spatially constrained image segmentation. IEEE Trans. on Fuzzy Systems 15(5), 1351–1361 (2008)

Gong, L., Pathak, S.D., Haynor, D.R., Cho, P.S., Kim, Y.: Parametric shape modelling using deformable superellipses for prostate segmentation. IEEE Trans. on Medical Imaging 22(3), 340–349 (2004)

Hrebień, M., Steć, P., Obuchowicz, A., Nieczkowski, T.: Segmentation of breast cancer fine needle biopsy cytological images. Int. J. Appl. Math and Comp. Sci. 18(2), 159–170 (2008)

Jeleń, Ł., Fevens, T., Krzyżak, A.: Classification of breast cancer malignancy using cytological images of fine needle aspiration biopsies. Int. J. Appl. Math and Comp. Sci. 18(1), 75–83 (2008)

Leung, S.-H., Wang, S.-L., Lau, W.-H.: Lip image segmentation using fuzzy clustering incorporating an elliptic shape function. IEEE Trans. on Image Processing 13(1), 51–62 (2004)

Marciniak, A., Obuchowicz, A., Monczak, R., Kołodziński, M.: Cytomorphometry of fine needle biopsy material from the breast cancer. In: Proc. 4th Int. Conf. on Computer Recognition Systems, CORES 2005, pp. 603–609 (2005)

Obuchowicz, A., Hrebień, M., Nieczkowski, T., Marciniak, A.: Computational intelligence techniques in image segmentation for cytopathology. In: Smoliński, T.G., Milanova, M.G., Hassanien, A.-E. (eds.) Computational intelligence in biomedicine and bioinformatics: current trends and applications. Studies in Computational Intelligence, vol. 151, pp. 169–199. Springer, Berlin (2008)

Rutkowska, D., Zadeh, L.: Neuro-fuzzy and soft computing. Special issue of Int. J. Appl. Math. and Comp. Sci. 10(4) (2000)

Suri, J.S., Setarehdan, K., Singh, S. (eds.): Advanced Algorithmic Approaches to Medical Image Segmentation. Springer, London (2002)

Szczepaniak, P., Lisboa, P., Kacprzyk, J. (eds.): Fuzzy Systems in Medicine. Springer, Heidelberg (2000)

Underwood, J.C.E. (ed.): Introduction to Biopsy Interpretation and Surgical Pathology. Springer, London (1987)

Wang, S.-L., Lau, W.-H., Leung, S.-H.: Automatic lip contour extraction from color images. Pattern Recognition 37, 2375–2387 (2004)

Wang, S.-L., Lau, W.-H., Liew, A.W.C., Leung, S.-H.: Robust lip region segmentation for lip images with complex background. Pattern Recognition 40(12), 3481–3491 (2007)

Wolberg, W.-H., Street, W.-N., Mangasarian, O.-L.: Breast cytology diagnosis via digital image analysis. Analytical and Quantitative Cytology and Histology 15, 396–404 (2004)

Zhang, Y.J.: A survey on evaluation methods for image segmentation. Pattern Recognition 29(8), 1335–1346 (1996)

Zoller, T., Varvarigou, T.A.: Robust image segmentation using resampling and shape constraints. IEEE Trans. on Pattern Analysis and Machine Intelligence 29(7), 1147–1164 (2007)

Part V
Machine Learning for Robotics

Automatic Selection of Object Recognition Methods Using Reinforcement Learning

Reinaldo A.C. Bianchi[1], Arnau Ramisa[2], and Ramón López de Mántaras[2]

[1] Centro Universitário da FEI, São Bernardo do Campo, Brazil
rbianchi@fei.edu.br
[2] Artificial Intelligence Research Institute (IIIA-CSIC), Bellaterra, Spain
{aramisa,mantaras}@iiia.csic.es

Abstract. Selecting which algorithms should be used by a mobile robot computer vision system is a decision that is usually made *a priori* by the system developer, based on past experience and intuition, not systematically taking into account information that can be found in the images and in the visual process itself to learn which algorithm should be used, in execution time. This paper presents a method that uses Reinforcement Learning to decide which algorithm should be used to recognize objects seen by a mobile robot in an indoor environment, based on simple attributes extracted on-line from the images, such as mean intensity and intensity deviation. Two state-of-the-art object recognition algorithms can be selected: the constellation method proposed by Lowe together with its interest point detector and descriptor, the Scale-Invariant Feature Transform and Nistér and Stewénius Vocabulary Tree approach. A set of empirical evaluations was conducted using a household mobile robots image database, and results obtained shows that the approach adopted here is very promising.

1 Introduction

Object recognition is a central theme in Computer Vision, with a large variety of applications ranging from medical image analysis to industrial inspection. Unfortunately, there is also a large variety of solutions to the object recognition problem, producing a wider still range of results, depending on the domain of the problem, the sensor hardware, the lightning conditions and the object to be identified. An autonomous mobile robot may employ an object recognition system for a variety of tasks, such as navigation, localization and mapping or interacting with humans. For example, precisely identifying and locating objects in the images means being able to localize themselves better in the real world.

When one thinks about designing an autonomous mobile robot capable of operating in the real world, and the variety of conditions that it will face, one can conclude that choosing, a priori, which object recognition method a robot should have, is not the best design option. In this kind of application, the robot should be able to decide by itself which object recognition method should be used, depending on the current conditions of the world.

In this paper we propose the use of Reinforcement Learning to design a mobile robot that is capable of deciding, during on line world exploration, which method should be

J. Koronacki et al. (Eds.): Advances in Machine Learning I, SCI 262, pp. 421–439.
springerlink.com © Springer-Verlag Berlin Heidelberg 2010

used to identify objects in an indoor environment, aiming also to minimize computing time. To evaluate this idea we implemented a system that is able to choose between two well known object recognition algorithms based on simple attributes extracted on-line from the images, such as mean intensity and intensity deviation. In addition, it is also capable of deciding that an image is not suitable for analysis, and thus discard it.

This paper is organized as follows: Section 2 briefly reviews Reinforcement Learning and its use as a technique to optimize computer vision, image segmentation and object recognition algorithms. Section 3 describes the two object recognition methods used by our robot. Section 4 describe how to use RL to automatic select the object recognition method to be used by the robot and Section 5 describes the domain where this proposal has been evaluated and the results obtained. Finally, Section 6 summarizes some important points learned from this research and outlines future work.

2 Reinforcement Learning and Its Applications in Computer Vision

Reinforcement Learning (Sutton and Barto, 1998) is concerned with the problem of learning from interaction to achieve a goal, for example, an autonomous agent interacting with its environment via perception and action. On each interaction step the agent senses the current state s of the environment, and chooses an action a to perform. The action a alters the state s of the environment, and a scalar reinforcement signal r (a reward or penalty) is provided to the agent to indicate the desirability of the resulting state. In this way, "The RL problem is meant to be a straightforward framing of the problem of learning from interaction to achieve a goal" (Sutton and Barto, 1998).

Formally, the RL problem can be formulated as a discrete time, finite state, finite action Markov Decision Process (MDP) (Mitchell, 1997). Given:

- finite set of states $s \in \mathscr{S}$ that the agent can achieve;
- A finite set of possible actions $a \in \mathscr{A}$ that the agent can perform;
- A state transition function $\mathscr{T} : \mathscr{S} \times \mathscr{A} \rightarrow \Pi(\mathscr{S})$, where $\Pi(\mathscr{S})$ is a probability distribution over $\mathscr{S}$;
- A finite set of bounded reinforcements (payoffs) $\mathscr{R} : \mathscr{S} \times \mathscr{A} \rightarrow \Re$,

the task of a RL agent is to find out a stationary policy of actions $\pi^* : \mathscr{S} \rightarrow \mathscr{A}$ that maps the current state s into an optimal action(s) a to be performed in s, maximizing the expected long term sum of values of the reinforcement signal, from any starting state.

The policy π is some function that tells the agent which actions should be chosen, and is learned through trial-and-error interactions of the agent with its environment. Several algorithms were proposed as a strategy to learn an optimal policy π^* when the model ($\mathscr{T}$ and $\mathscr{R}$) is not known in advance, for example, the Q–learning (Watkins, 1989) and the SARSA (Rummery and Niranjan, 1994) algorithms.

The Q–learning algorithm was proposed by Watkins (1989) as a strategy to learn an optimal policy π^* when the model ($\mathscr{T}$ and $\mathscr{R}$) is not known in advance. Let $Q^*(s,a)$

be the reward received upon performing action a in state s, plus the discounted value of following the optimal policy thereafter:

$$Q^*(s,a) \equiv R(s,a) + \gamma \sum_{s' \in S} T(s,a,s') V^*(s'). \quad (1)$$

The optimal policy π^* is $\pi^* \equiv \arg\max_a Q^*(s,a)$. Rewriting $Q^*(s,a)$ in a recursive form:

$$Q^*(s,a) \equiv R(s,a) + \gamma \sum_{s' \in S} T(s,a,s') \max_{a'} Q^*(s',a'). \quad (2)$$

Let $\hat{Q}$ be the learner's estimate of $Q^*(s,a)$. The Q–learning algorithm iteratively approximates $\hat{Q}$, i.e., the $\hat{Q}$ values will converge with probability 1 to Q^*, provided the system can be modeled as a MDP, the reward function is bounded ($\exists c \in \mathscr{R}; (\forall s,a), |R(s,a)| < c$), and actions are chosen so that every state-action pair is visited an infinite number of times. The Q learning update rule is:

$$\hat{Q}(s,a) \leftarrow \hat{Q}(s,a) + \alpha \left[r + \gamma \max_{a'} \hat{Q}(s',a') - \hat{Q}(s,a) \right], \quad (3)$$

where s is the current state; a is the action performed in s; r is the reward received; s' is the new state; γ is the discount factor ($0 \leq \gamma < 1$); $\alpha = 1/(1 + visits(s,a))$, where $visits(s,a)$ is the total number of times this state-action pair has been visited up to and including the current iteration.

Several researchers have been using RL as a technique to optimize active vision, image segmentation and object recognition algorithms. The area of Computer Vision on which RL was first applied was in Active Vision. Whitehead and Ballard (1991) proposed an adaptive control architecture to integrate active sensory-motor systems with RL based decision systems. Although the work is theoretical and did not make use of real sensors, they were able to describe a system that learns to focus its attention on the relevant aspects of the domain as well as control its behavior, in a simple block manipulation task. Several researchers have been applying RL to active vision since then, for example Minut and Mahadevan (2001) have applied RL for visual attention control, proposing a model of selective attention for visual search tasks, such as deciding where to fixate next in order to reach the region where an object is most likely to be found. Darrell and Pentland (1996a,b) also address visual attention problem: they proposed a gesture recognition system that guides an active camera to foveate salient features based on a reinforcement learning paradigm. An attention module selects targets to foveate based on the goal of successful recognition, learning where to foveate to maximally discriminate a particular gesture. Finally, in Darrell (1998) is shown how a concise representation of active recognition behavior can be derived from hidden-state reinforcement learning techniques.

Paletta and Pinz (2000) have applied RL in an active object recognition system, to learn how to move the camera to informative viewpoints, defining the recognition process as a sequential decision problem with the objective of disambiguating initial object hypotheses. For these authors, "Reinforcement Learning provides then an efficient method to autonomously develop near-optimal decision strategies in terms of sensorimotor mappings" (Paletta et al, 1998). Borotschnig et al (1999) continued in the same line of work, building a system that learns to reposition the camera to capture additional views to improve the image classification result obtained from a single view. More recently, Paletta et al (2005) proposed the use of Q-learning to associate shift of attention actions to cumulative reward with respect to object recognition. In this way, the agent learns sequences of shifts of attentions that lead to scan paths that are highly discriminative with respect to object recognition.

Less work have been done on the use of RL for image segmentation and object recognition. Peng and Bhanu (1998a) used RL to learn, from input images, to adapt the image segmentation parameters of a specific algorithm to the changing environmental conditions, in a closed-loop manner. In this case, the RL creates a mapping from input images to corresponding segmentation parameters. This contrasts with great part of the current computer vision systems whose methodology is open-loop, using image segmentation followed by object recognition algorithms. Peng and Bhanu (1998b) improves the recognition results over time by using the output at the highest level as feedback for the learning system, and has been used to learn the parameters of image segmentation and feature extraction and thereby recognizing 2-D objects, systematically controlling feedback in a multilevel vision system. The same authors presented a general approach to image segmentation and object recognition that can adapt the image segmentation algorithm parameters to the changing environmental conditions, in which segmentation parameters are represented by a team of generalized stochastic learning automata and learned using connectionist reinforcement learning techniques. Results were presented for both indoor and outdoor color images, showing a performance improvement over time for both image segmentation and object recognition using RL (Bhanu and Peng, 2000).

Taylor (2004) also followed this line of research, applying RL algorithms to learn parameters of an existing image segmentation algorithm. Using the Fuzzy ARTMAP artificial neural network, he was able to optimize ten parameters of Wolf and Jolion (2003) algorithm for text detection in still images. The parameters learned by RL were shown to be superior to the parameters previously recommended. Other applications of RL to learn parameters of image segmentation algorithms includes: contrast adaptation (Tizhoosh and Taylor, 2006), finding the appropriate threshold in order to convert an image to a binary one (Yin, 2002; Shokri and Tizhoosh, 2003, 2004, 2008; Sahba et al, 2008) and detection of patterns in satellite images (Hossain et al, 1999).

Finally, Draper et al (1999) modeled the object recognition problem as a Markov Decision Problem, and proposed a theoretically sound method for constructing object recognition strategies by combining CV algorithms to perform segmentation. The authors tested their method in a real system, learning sequences of image processing operators for detecting houses in aerial images.

Reinforcement Learning has been widely used in the Computer Vision field in particular cases, mainly: to optimize the performance of active vision systems; to decide where the focus of attention should be in order to accomplish a certain task; to learn how to move a camera to more informative viewpoints and; to optimize parameters of existing and new computer vision algorithms, such as thresholds, contrast and internal parameters. In these cases, the resulting systems and algorithms have been very successful ones.

RL has also been used for constructing object segmentation and recognition strategies by combining CV algorithms. However, results in this area have not been as good as those of RL applied to active vision or parameter optimization: it usually has been limited to a very specific image domain, such as the one in Draper et al (1999).

The main limitations which arise when using reinforcement learning are, first, that the reward value associated with a situation is usually not directly available, and thus rewards are results of indirect definitions. RL requires that a certain amount of knowledge about the world is available in the form of a training set, which is not the case in many vision tasks. Second, the large space state which difficult convergence of RL algorithms raises performance issues, as the learning phase can take a long time.

3 Two Object Recognition Methods

Two successful general object recognition approaches that have been widely used are the constellation method proposed by Lowe together with its interest point detector and descriptor SIFT (Lowe, 2004) and the Vocabulary Tree algorithm proposed by Nistér and Stewénius (2006).

The first approach is a single view object detection and recognition system which has some interesting characteristics for mobile robots, most significant of which are the ability to detect and recognize objects at the same time in an unsegmented image and the use of an algorithm for approximate fast matching, which reduces the search time by two orders of magnitude for a database of 100,000 keypoints for a 5% loss in the number of correct matches. In this approach, individual descriptors of the features detected in a test image are initially matched to the ones stored in the object database using the Euclidean distance. False matches are rejected if the distance of the first nearest neighbor is not distinctive enough when compared with that of the second. In Figure 1, the matching features between a test and model images can be seen. The presence of some outliers can also be observed.

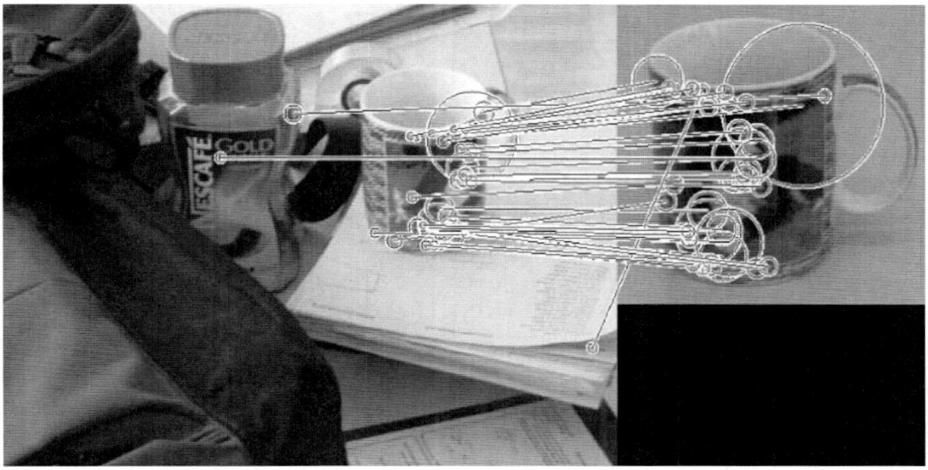

Fig. 1. Matching stage in the Lowe object recognition method

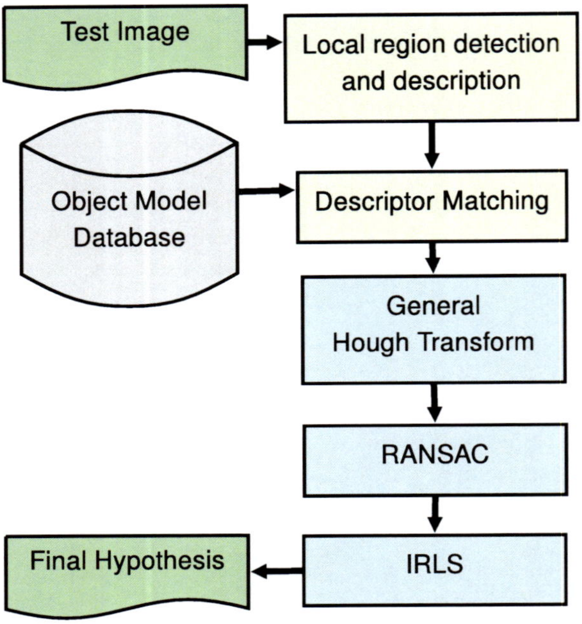

Fig. 2. Diagram for the Lowe algorithm

Once a set of matches is found, the generalized Hough transform is used to cluster each match of every database image depending on its particular transformation (translation, rotation and scale change). Although imprecise, this step generates a number of initial coherent hypotheses and removes a notable portion of the outliers that could potentially confuse more precise but also more sensitive methods. All clusters with at least three matches for a particular image are accepted, and fed to the next stage: a Least Squares is used to improve the estimation of the affine transformation between the model and the test images.

Because of the necessarily broad clusters of the Hough transform, some erroneous matches can still be present and need to be removed. In order to do so, in this work a RANSAC step is added to the Lowe approach. RANSAC labels non-coherent matches as outliers and, additionally, estimates the most probable affine transformation for every hypothesis given its initial set of matches. Hypotheses that lose matches below three are discarded. The hypotheses that remain after the RANSAC step are reasonable outlier-free and a more accurate model fitting algorithm like Iterative Reweighed Least Squares can be used. The diagram of the complete algorithm is presented in Figure 2.

The second approach to object classification that is used in this work is Vocabulary Tree method proposed by Nistér and Stewénius (2006). This approach to object classification comes from the text categorization domain, where the occurrence of certain words in documents is recorded and used to train classifiers that can later recognize the subject of new texts. Recently these approach has been adapted to visual object recognition by different authors (Sivic et al, 2006; Csurka et al, 2004) using local descriptors computed on image features as visual words.

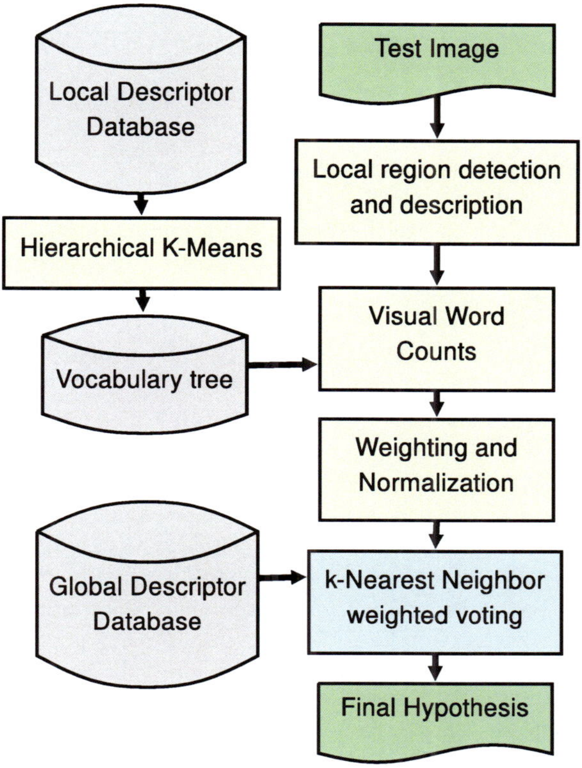

Fig. 3. The Vocabulary Tree algorithm

The visual descriptor space has normally a high dimensionality and, to make it tractable, it is discretized in a codebook. This codebook should be specific enough to distinguish between different descriptor types but general enough to be insensitive to small variations in the local patch. In the Vocabulary Tree approach used in this work, the codebook is a tree built applying hierarchical k-means to a dataset of descriptors. A histogram of descriptor occurrences is built to characterize an image.

Next, a multi-class classifier – the k-NN in this implementation – is trained with the histograms of local descriptor counts. The class of the object in the image is determined as the dominant one in the k nearest neighbors. The diagram of the complete algorithm is presented in Figure 3.

One of the drawbacks of this method is that, in contrast to Lowe's method, is designed to work with pre- segmented images. A straightforward solution to the image segmentation problem is to first apply bilateral filtering to remove texture from the image (Tomasi and Manduchi, 1998; Paris et al, 2007). Next, the Canny edge detector (Canny, 1986) is applied to define the edges in the image and then mathematical morphology operators are applied (one opening followed by and one closing of the image) in order to close the contours that remained open (Haralick et al, 1987). Then, a flood-fill algorithm is used to fill connected areas divided by the edges. A threshold is set to the flood-fill algorithm to preserve strong gradient jumps that have not been detected by the Canny edge detector. Finally, areas which are too small, too large or to thin are removed from the list of segmented regions.

Although both object recognition methods proved their reliability in real world applications (for example, see Ramisa et al (2008), they have their limitations: Lowe's method performs poorly when recognizing sparsely textured objects or objects with repetitive patterns, while the Vocabulary Tree needs an accurate segmentation stage, prior to classification, which can be very time consuming. Furthermore, the method depends on the quality of that segmentation stage to provide good results.

4 Learning to Select Object Recognition Methods

In order to decide which algorithm should be used by the agent, the RL problem was defined as a 2 stage decision problem, with 2 possible actions in each stage: In the first one, the agent must decide if the image contains an object, and thus must be recognized, or if the image does not contain objects, and can be discarded, saving processing time. In the second stage, the agent must decide which object recognition algorithms should be used: Lowe's or the Vocabulary Tree (VT) algorithm (see Figure 4).

To learn how to select the object recognition method appropriate for one image at one stage we propose to use Reinforcement Learning as a classification method. In this approach the space state is defined as a combination attributes extracted from the images plus the possible classification of the image. For example, for the first stage, the state can be defined as a combination of mean image intensity and standard deviation and a value defining if the image is a background and can be discarded or if contain objects.

We also define a new type of action, called "update action". Update actions are not real actions happening in the world, but actions that update the value of a state-action pair $Q(s,a)$ at one state using the value of a neighbor pair. For example, if the space state is composed of image intensity and standard deviation, the $Q(s,a)$ table would be represented as two dimensional matrix containing the possible values of intensity (0 to 255) and standard deviation (0 to 255), and update actions are done between one

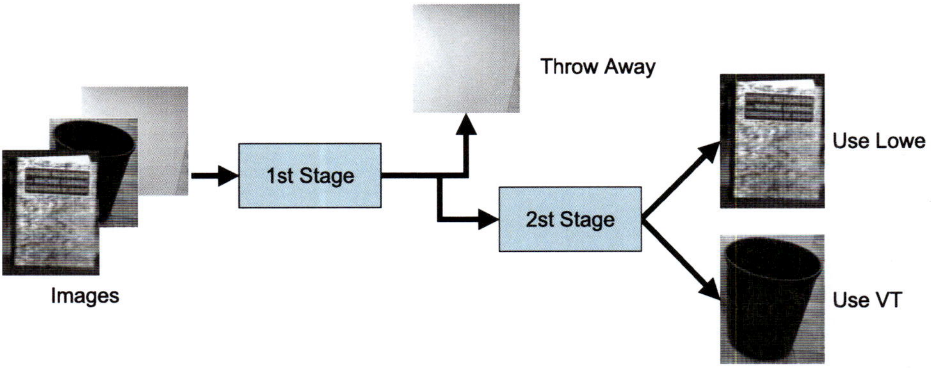

Fig. 4. The two stage decision problem

state and his 4-neighbours located above, below, at left and at right. In a 3-dimensional space state, update actions can be made using 8-neighbours (two in each direction of the space) and so on.

The rewards used during the learning phase are computed using a set of training images. If, during the exploration, the learning agent reaches a state where a training image exists and the state corresponds to the correct classification of the image, the agent receives a reward. Otherwise the reward is zero. For example, in the first stage of the decision, if we have a training image that does not contain an object, with mean intensity value of 50 and standard deviation of 10, a reward is given when the agent moves to the state (mean = 50, std = 10, classification = discard). An interesting propriety of this approach is that the rewards can be pre-computed, creating a reinforcement table that can be used during the learning phase (an example is shown in the next section).

Formally, a MDP can be defined for each stage as:

- The set of update actions $a \in \mathscr{A}$ that the agent can perform, defined as update the Q value using the value of a neighbor.
- The finite set of states $s \in \mathscr{S}$ in this case is the n-dimensional space of values of the attributes extracted from the images plus its classification;
- The state transition function allows updates to be made between any pair of neighbors in the set of states.
- The reinforcements $\mathscr{R} : \mathscr{S} \times \mathscr{A} \to \Re$ are defined using a set of training images.

In this approach, a RL method is used as a classifier, and must have two processing phases: the training phase, where reinforcement learning is performed over a set of pre-classified images, i.e., images to which we know what the best algorithm to use is, and the execution phase, where the results from the learning is used to define which algorithms to apply to other images.

During the training phase, learning an optimal policy to solve the MDP means to learn a mapping from images (or more specifically image attributes) to image classes (or algorithms classes). Although several RL algorithms can be used to do this, the RL algorithm used in this implementation is the Q-learning (Watkins, 1989), because it directly approximates the optimal policy independently of the policy being followed (it is an off-policy method), allowing the state and the action to be executed by the agent to be selected randomly. Using the Q-Learning, at each stage the agent chooses a system state s. Then, it selects an update action to be executed, compute the reward and update the value function.

The learning algorithm used is as follows:

```
Initialize Q(s,a).
Choose a start state s, randomly.
do {
   Choose an action, randomly.
   Execute a, observe s', compute the reward.
   Update the Q value.
   s = s'.
} until the Q values converge.
```

0000000000	1111111110
0000010000	1111111111
2000011000	2011111111
0000001100	2221111111
0200101000	2222111111
0020200110	2222211111
0000010000	2222111111
0200011000	2220111111
2021000000	2221111111

Fig. 5. Example of how the classification works: the reinforcements used (left) and the resulting classification table (right)

To better understand what is happening during the learning phase, we can compare our approach to a robot moving in a two-dimensional grid. Every time the robot finds a "goal" state and receives a reward, the state-action pair where the robot was before reaching the goal state is updated. Every time the robot moves, it iteratively updates the origin state-action pair. By doing this a large number of times, the reward is spread over the Q-table, and a robot will know what to do to reach the goal state (will have learned the optimal policy).

In the learning phase, every time the "robot" reaches a state where there is an image from the training set, it receives a reward, and the state-action pair where the "robot" was before is updated. Every time a new state action pair is randomly chosen, it is iteratively updated. By doing this a large number of times, the reward is spread over the Q-table, and every state-action pair will contain information about what to do with an image with those characteristics (will have learned a mapping from image to actions).

The figure 5 shows an example of the use of RL as an image classification method: in each figure the lines and columns represents two values of the attributes extracted from the images (for example, mean image intensity are lines and standard deviation of the image intensity are the columns), the value represents what is the classification of an image, where zero means that there is not an example with that characteristics, and "1" and "2" are two possible classes.

The figure on the left presents the reinforcement table, created before the training phase using the training images and their classification, which consists of what is the algorithm that was able to classify them. The image on the right shows the results of applying the RL algorithm during the learning phase: a table where the classification was spread over to states where there are no prior examples, and that allows the classification of other images.

To show the applicability of this porposal, experiments and results obtained with this method are presented in the next section.

5 Experiments and Results

To test the algorithm selection method proposed in this work, an image dataset of nine typical household objects (Ramisa et al, 2008) was used. In this dataset, objects are divided in three categories (three objects per category): textured, untextured and textured but with repetitive patterns. Every category consists of three different objects and each object has approximately 20 training images (used for the Vocabulary Tree algorithm). The objects are mugs, books, trashcans, chairs and computer monitors. In Figure 6 one object from each category can be seen.

Also, this image dataset contains test images where the same objects appear. On line, embedded processing means that images acquired by a mobile robot hardly have a resolution greater than one megapixel, and the object to be detected will probably only occupy the image partially. Additionally, movement often implies blurred images. Therefore, the test images include occlusions, illumination changes, blur and other typical nuisances that will be encountered while navigating with a mobile robot. Figure 7 presents sample test images. Finally, this dataset also includes background images, i.e., images that do not contain objects to be recognized, shown in Figure 8.

Several experiments were executed using the typical household objects dataset described. Each experiment consists of a two processing phases: the training of the RL

Fig. 6. Objects from the dataset. First column corresponds to objects with repetitive texture, second to textured objects and third to non-textured objects.

Fig. 7. Test image from the dataset

and the execution phase, where the training quality can be verified. To train the RL, we used 40 test images, from which approximately 160 images containing objects were segmented (using the algorithm in Section 3) and previously classified. Sample of the segmented images are presented in Figure 9. Furthermore, 360 background images, also resulting from the segmentation process, were used. The training was performed according to the algorithm described in the previous section.

To evaluate the result of the learning process statistical validation method called Leave-One-Out was used. Using this method, the RL selection module was trained with all the test images but one, and then the image left out (that can contain several ROI with objects and background) is used to test the result of the learning. This test phase corresponds to the execution phase, which can be used on the real robot during on-line exploration of the environment, and its working diagram is presented in Figure 10.

Six different experiments were conducted, using three different combinations of image attributes as space state descriptors and two different image sizes (the image original size and a 10 by 10 pixels reduced size image). The combinations of image attributes used as space state are: mean and standard deviation of the image intensity (MS); mean

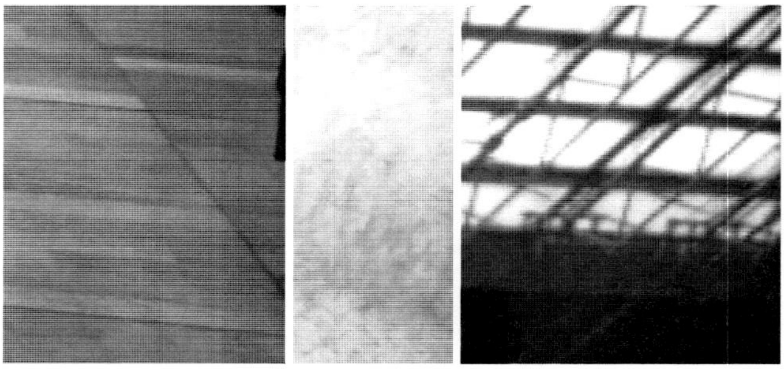

Fig. 8. Background images from the dataset

Fig. 9. Objects segmented from test images of the dataset.

and standard deviation of the image intensity plus entropy of the image (MSE); and mean and standard deviation of the image intensity plus the number of interest points detected by the Difference of Gaussians operator (MSI).

The rewards used during the learning phase were computed using a set of training images. Figure 11 shows part of the reward table built for the first stage of the first experiment (MS). It is a 100 x 130 figure, where the lines represents the possible values of mean intensity of the image, and columns represents 100 possible values in standard deviation of the image intensity (the MS space). There are three kinds of states, marked in the image by: "X", which indicates that in the training set exists an image with this combination of mean and std dev values, "·", which corresponds to points which represents images that does not contain objects (backgrounds) and the rest of the space, which is left without any marking and correspond to points in the MS state where there is no information about it. As it can be seen, this is a sparse image.

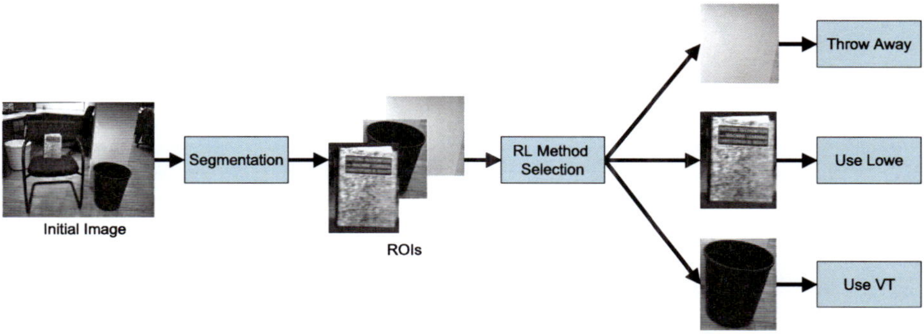

Fig. 10. Execution Phase of the system

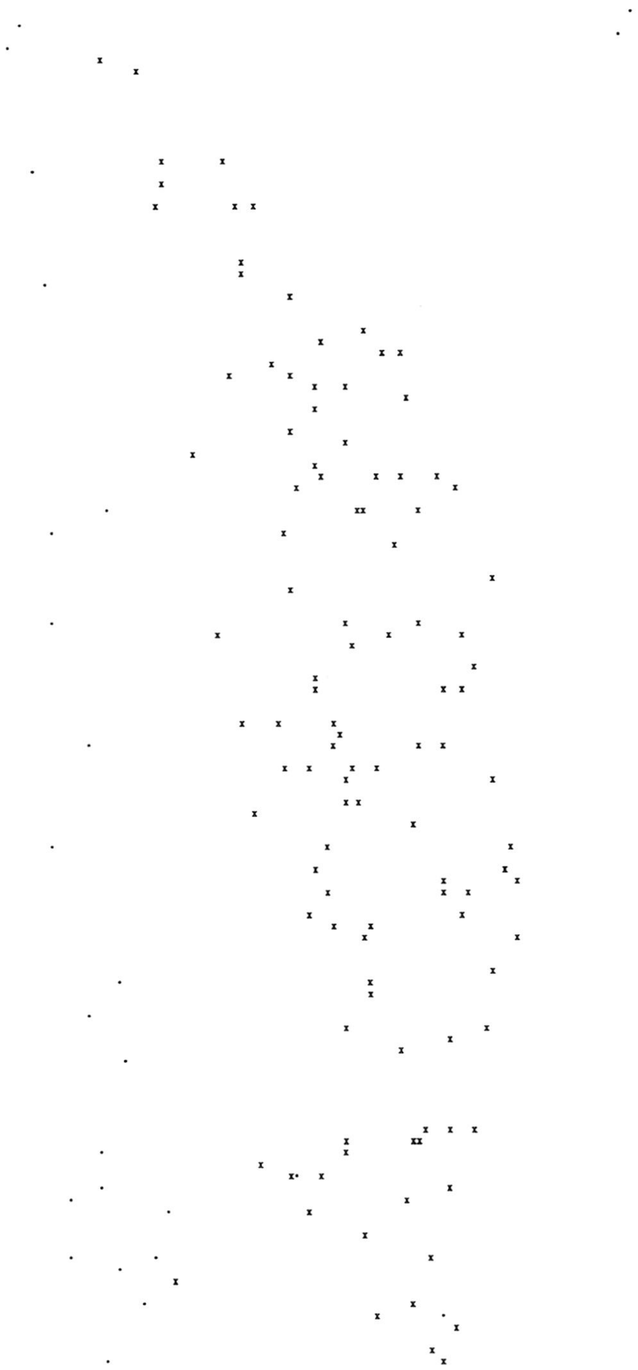

Fig. 11. Image of the reward table built for the first experiment (Mean and Standard deviation of image intensity space)

Fig. 12. Classification table learned in the first experiment

Table 1. Correctly classified images (percentage)

	Full Img			Small Img			Expert
	MS	MSE	MSI	MS	MSE	MSI	
Back	91.9	100.0	98.0	92.6	100.0	98.9	100.0
Lowe	84.5	100.0	44.4	76.0	98.4	38.1	93.2

Table 2. Incorrect classification (percentage)

	Full Img			Small Img			Expert
	MS	MSE	MSI	MS	MSE	MSI	
Back	12.8	1.8	14.2	20.4	2.4	25.3	8.2
Lowe	11.6	1.9	7.9	15.8	1.9	9.9	10.8

During the learning phase (described in Section 4), if the learning agent reaches a state labeled as "X", it receives the value +10 and if it reaches a state labeled as "·", it receives a reward of -10. Otherwise the reward is zero.

Figure 12 shows the results of applying the RL algorithm during the learning phase. As it can be seen, the classification was spread over to states where there are no prior examples, allowing the classification of other images. This table is the one used during the execution phase.

Tables 1 and 2 present the results obtained for the six experiments. The first line of Table 1 shows the percentage times that the agent correctly choose to discard a background image, and the second line shows the percentage of times the agent correctly choose to use the Lowe algorithm, instead of the Vocabulary Tree one. The columns in this table present the results for the six experiments, the first three using the original image and, from the fourth to sixth column, showing the results for the reduced size image. The last column shows the percentage of times a human expert takes the correct action. Table 2 is similar to Table 1, but shows the classification error. The first line shows the percentage of images discarded as background, when they should be analyzed, and line two presents the number of times the Lowe algorithm is chosen, when the correct one is the Vocabulary Tree.

The results shows that the use Reinforcement Learning to decide which algorithm should be used to recognize objects yields good results, for all different combinations of image attributes used space state descriptors, performing better than a human expert in some cases.

These tables also show that the best combination of attributes was mean and standard deviation of the image intensity plus entropy of the image (MSE), which presented very good results for original size images as well as reduced size ones. On the other hand, the use of the number of interest points detected by the Difference of Gaussians operator as space state did not produce good results, failing to choose Lowe's algorithm more than half of the time.

Reinforcement Learning algorithms were implemented in C and Computer Vision algorithms were implemented in C++ using the OpenCV library (Bradski, 2000) and

Matlab. Experiments were executed on a Pentium 4 Computer running Ubuntu Linux and a PowerMac G4 running Mac OS X. The Reinforcement Learning parameters used in the experiments were: the learning rate $\alpha = 0.1$ and the discount factor $\gamma = 0.9$. Values in the Q table were randomly initiated.

6 Conclusion

Reinforcement Learning has been widely used in the Computer Vision field: to optimize the performance of active vision systems; to decide the focus of attention ; to learn how to move a camera; to optimize parameters of existing and new computer vision algorithms, such as thresholds, contrast and internal parameters; and for constructing object segmentation and recognition strategies by combining CV algorithms. Results have been usually very good, but most of the times limited to a very specific image domain.

In this paper we presented a method that uses Reinforcement Learning to decide which algorithm should be used to recognize objects seen by a mobile robot in an indoor environment, based on simple attributes extracted on-line from the images, such as mean intensity and intensity deviation. Another important contribution of this work is a method that allows the use of a Reinforcement Learning algorithm as a Classifier.

The results obtained shows that the use Reinforcement Learning to decide which algorithm should be used to recognize objects yields good results, performing better than a human expert in some cases. To the best of our knowledge, there is no similar approach using automatic selection of algorithms for object recognition.

Future works includes testing other image attributes that can be used as the system's state, for example, the results of applying Gabor Filters in the images, which may give us clue to the existing texture in the image (Fogel and Sagi, 1989; Weldon et al, 1996). The use of other RL algorithms such as SARSA (Rummery and Niranjan, 1994) and Q (λ)(Peng and Williams, 1996) and the study of the use of this technique in others application domains are also other interesting directions of research.

Acknowledgements. This work has been partially funded by the FI grant and the BE grant from the AGAUR, the 2005-SGR-00093 project, supported by the Generalitat de Catalunya, the MID-CBR project grant TIN 2006-15140-C03-01 and FEDER funds. Reinaldo Bianchi is supported by CNPq grant 201591/2007-3.

References

Bhanu, B., Peng, J.: Adaptive integrated image segmentation and object recognition. IEEE Transactions on Systems, Man, and Cybernetics, Part C: Applications and Reviews 30(4), 427–441 (2000)

Borotschnig, H., Paletta, L., Prantl, M., Pinz, A.: Appearance-based active object recognition. Image and Vision Computing 18(9), 715–728 (1999)

Bradski, G.: The OpenCV Library. Dr Dobb's Journal of Software Tools, 120–125 (2000)

Canny, J.: A computational approach to edge detection. IEEE Trans. Pattern Anal. Mach. Intell. 8(6), 679–698 (1986)

Csurka, G., Dance, C., Fan, L., Willamowski, J., Bray, C.: Visual categorization with bags of keypoints. In: Sebe, N., Lew, M., Huang, T.S. (eds.) ECCV/HCI 2004. LNCS, vol. 3058, pp. 1–22. Springer, Heidelberg (2004)

Darrell, T.: Reinforcement learning of active recognition behaviors, Interval research technical report 1997-045. In: Advances in Neural Information Processing Systems, NIPS 1995, vol. 8, pp. 858–864. MIT Press, Cambridge (1995), `http://www.interval.com/papers/1997-045`, portions of this paper previously appeared

Darrell, T., Pentland, A.: Active gesture recognition using learned visual attention. In: Touretzky, D.S., Mozer, M.C., Hasselmo, M.E. (eds.) Advances in Neural Information Processing Systems, vol. 8, pp. 858–864. MIT Press, Cambridge (1996a)

Darrell, T., Pentland, A.: Active gesture recognition using partially observable markov decision processes. In: Proceedings of the 13th International Conference on Pattern Recognition, vol. 3, pp. 984–988 (1996b), doi:10.1109/ICPR.1996.547315

Draper, B.A., Bins, J., Baek, K.: ADORE: Adaptive object recognition. In: Christensen, H.I. (ed.) ICVS 1999. LNCS, vol. 1542, pp. 522–537. Springer, Heidelberg (1998)

Fogel, I., Sagi, D.: Gabor filters as texture discriminator. Biological Cybernetics 61(2), 103–113 (1989), `http://dx.doi.org/10.1007/BF00204594`

Haralick, R.M., Sternberg, S.R., Zhuang, X.: Image analysis using mathematical morphology. IEEE Trans. Pattern Anal. Mach. Intell. 9(4), 532–550 (1987)

Hossain, I., Liu, J., You, J.: Tropical cyclone pattern recognition for intensity and forecasting analysis from satellite imagery. In: IEEE International Conference on Systems, Man, and Cybernetics, IEEE SMC 1999 Conference Proceedings 1999, vol. 6, pp. 851–856 (1999), doi:10.1109/ICSMC.1999.816663

Lowe, D.: Distinctive image features from scale-invariant keypoints. International Journal of Computer Vision 60(2), 91–110 (2004)

Minut, S., Mahadevan, S.: A reinforcement learning model of selective visual attention. In: AGENTS 2001: Proceedings of the fifth international conference on Autonomous agents, pp. 457–464. ACM, New York (2001), `http://doi.acm.org/10.1145/375735.376414`

Mitchell, T.: Machine Learning. McGraw Hill, New York (1997)

Nistér, D., Stewénius, H.: Scalable recognition with a vocabulary tree. In: 2006 IEEE Computer Society Conference on Computer Vision and Pattern Recognition, vol. 2, pp. 2161–2168 (2006)

Paletta, L., Pinz, A.: Active object recognition by view integration and reinforcement learning. Robotics and Autonomous Systems 31(1-2), 71–86 (2000)

Paletta, L., Prantl, M., Pinz, A.: Reinforcement learning of 3-d object recognition from appearance. In: CONALD 1998: Proceedings of the 1998 Conference on Automated Learning and Discovery (1998)

Paletta, L., Fritz, G., Seifert, C.: Q-learning of sequential attention for visual object recognition from informative local descriptors. In: ICML 2005: Proceedings of the 22nd International Conference on Machine Learning, pp. 649–656. ACM, New York (2005), `http://doi.acm.org/10.1145/1102351.1102433`

Paris, S., Kornprobst, P., Tumblin, J., Durand, F.: A gentle introduction to bilateral filtering and its applications. In: SIGGRAPH 2007: ACM SIGGRAPH 2007 courses, p. 1. ACM, New York (2007), `http://doi.acm.org/10.1145/1281500.1281602`

Peng, J., Bhanu, B.: Closed-loop object recognition using reinforcement learning. IEEE Transactions on Pattern Analysis and Machine Intelligence 20(2), 139–154 (1998a)

Peng, J., Bhanu, B.: Delayed reinforcement learning for adaptive image segmentation and feature extraction. IEEE Transactions on Systems, Man, and Cybernetics, Part C: Applications and Reviews 28(3), 482–488 (1998b)

Peng, J., Williams, R.J.: Incremental multi-step q-learning. Mach. Learn. 22(1-3), 283–290 (1996), http://dx.doi.org/10.1007/BF00114731

Ramisa, A., Vasudevan, S., Scaramuzza, D., López de Mántaras, R., Siegwart, R.: A tale of two object recognition methods for mobile robots. In: Gasteratos, A., Vincze, M., Tsotsos, J.K. (eds.) ICVS 2008. LNCS, vol. 5008, pp. 353–362. Springer, Heidelberg (2008)

Rummery, G.A., Niranjan, M.: On-line Q-learning using connectionist systems. Tech. Rep. CUED/F-INFENG/TR 166, Cambridge University Engineering Department (1994)

Sahba, F., Tizhoosh, H.R., Salama, M.M.M.A.: A reinforcement agent for object segmentation in ultrasound images. Expert Syst. Appl. 35(3), 772–780 (2008), http://dx.doi.org/10.1016/j.eswa.2007.07.057

Shokri, M., Tizhoosh, H.R.: Using reinforcement learning for image thresholding. In: IEEE CCECE 2003 Canadian Conference on Electrical and Computer Engineering, vol. 2, pp. 1231–1234 (2003), doi:10.1109/CCECE.2003.1226121

Shokri, M., Tizhoosh, H.R.: Q(lambda)-based image thresholding. In: CRV 2004: Proceedings of the 1st Canadian Conference on Computer and Robot Vision, pp. 504–508. IEEE Computer Society Press, Los Alamitos (2004), http://doi.ieeecomputersociety.org/10.1109/CCCRV.2004.1301490

Shokri, M., Tizhoosh, H.R.: A reinforcement agent for threshold fusion. Appl. Soft Comput. 8(1), 174–181 (2008)

Sivic, J., Schaffalitzky, F., Zisserman, A.: Object Level Grouping for Video Shots. International Journal of Computer Vision 67(2), 189–210 (2006)

Sutton, R., Barto, A.G.: Reinforcement Learning: An Introduction. MIT Press, Cambridge (1998)

Taylor, G.W.: Reinforcement Learning for Parameter Control of Image-Based Applications. MSc Thesis, University of Waterloo, Ontario, Canada (2004)

Tizhoosh, H., Taylor, G.: Reinforced contrast adaptation. International Journal of Image and Graphics 6(3), 377–392 (2006)

Tomasi, C., Manduchi, R.: Bilateral filtering for gray and color images. In: ICCV 1998: Proceedings of the Sixth International Conference on Computer Vision, p. 839. IEEE Computer Society, Washington (1998)

Watkins, C.J.C.H.: Learning from Delayed Rewards. PhD Thesis, University of Cambridge (1989)

Weldon, T.P., Higgins, W.E., Dunn, D.F.: Efficient gabor filter design for texture segmentation. Pattern Recognition 29(12), 2005–2015 (1996), http://dx.doi.org/10.1016/S0031-32039600047-7

Whitehead, S.D., Ballard, D.H.: Learning to perceive and act by trial and error. Mach. Learn. 7(1), 45–83 (1991), http://dx.doi.org/10.1023/A:1022619109594

Wolf, C., Jolion, J.M.: Extraction and recognition of artificial text in multimedia documents. Pattern Anal. Appl. 6(4), 309–326 (2003), http://dx.doi.org/10.1007/s10044-003-0197-7

Yin, P.Y.: Maximum entropy-based optimal threshold selection using deterministic reinforcement learning with controlled randomization. Signal Process. 82(7), 993–1006 (2002), http://dx.doi.org/10.1016/S0165-16840200203-7

Comparison of Machine Learning for Autonomous Robot Discovery

Ivan Bratko

University of Ljubljana, Faculty of Computer and Information Sc.,
Tržaška 25, 1000 Ljubljana, Slovenia
ivan.bratko@fri.uni-lj.si

Abstract. In this paper we consider autonomous robot discovery through experimentation in the robot's environment. We analyse the applicability of machine learning (ML) methods with respect to various levels of robot discovery tasks, from extracting simple laws among the observed variables, to discovering completely new notions that were never explicitly mentioned in the data directly. We first present some illustrative experiments in robot learning in the XPERO European project. Then we formulate criteria for a comparison of learning methods and a systematic list of types of learning or discovery tasks, and discuss the suitability of chosen ML methods for these tasks.

Keywords: Machine learning, robotic discovery, autonomous learning, gaining insights.

1 Introduction

In this contribution we analyse the applicability of machine learning (ML) methods in various autonomous robot discovery tasks. The experimental basis for the comparison comes from the 6th Framework European project XPERO. We first introduce the XPERO project. Then we formulate a list of types of XPERO learning or discovery tasks, a list of applicable ML methods, and a list of criteria for comparison. Finally we discuss the suitability of the methods against the tasks with respect to the chosen criteria.

XPERO is a Sixth Framework European project with full title Learning by Experimentation (www.xpero.org). The scientific goal of XPERO is to investigate the mechanisms of autonomous discovery through experiments in an agent's environment. In XPERO, the experimental domain is the robot's physical world, and the subject of discovery are various quantitative or qualitative laws in this world. Discovery of such laws of (possibly naive) physics would enable the robot to make predictions about the results of its actions, and thus enable the robot to construct plans that would achieve the robot's goals.

In addition to discovering laws that directly enable predictions, XPERO aspires to advance the understanding of the mechanisms that enable *new insights*. In XPERO, an "insight" means something conceptually more general than a law. One possible definition of an insight in the spirit of XPERO is the following: an insight is a new piece of knowledge that makes it possible to simplify the current agent's theory about its

J. Koronacki et al. (Eds.): Advances in Machine Learning I, SCI 262, pp. 441–456.
springerlink.com © Springer-Verlag Berlin Heidelberg 2010

environment. So an insight typically enhances the agent's description language, and thus it should also make further discovery easier because the hypothesis language becomes more powerful and suitable for the domain of application.

Examples of insights are the discoveries of notions like absolute coordinate system, arithmetic operations, notion of gravity, notion of support between objects, etc. An insight would thus ideally be a new concept that makes the current domain theory more flexible and enables more efficient reasoning about the domain. Thus, insights should also make further learning in this domain easier and more effective.

It should be noted that the scientific goals of XPERO are considerably different from the goals of a typical robotics project. Therefore, also the criteria of success in XPERO are different from typical criteria in robotics. In a typical robotics project, the goal may be to improve the robot's performance at carrying out some physical task. To this end, any relevant methods, as powerful as possible, will be applied. In contrast to this, in XPERO we are less interested in improving the robot's performance at some specific task, but in making the robot gaining insights, that is improving the robot's theory and "understanding" of the world. We are interested in finding mechanisms, as generic as possible, that enable the gaining of insights. For such a mechanism to be generic, it has to make only a few rather basic assumptions about the agent's prior knowledge. We are interested in minimizing such "innate knowledge" because we would like to demonstrate how discovery and gaining insights come about from only a minimal set of "first principles". In addition, as our aim is gaining insights, not all machine learning methods are appropriate. Our aim requires that the induced insights can be interpreted and understood, and are not only useful for making predictions. Therefore, new insights have to be represented in *explicit symbolic* form. This requirement largely rules out some otherwise effective machine learning methods, such as neural networks and support vector machines.

2 Experimental Loop, Insights and Evolution of Theories

2.1 Experimental Loop

The discovery process in XPERO takes the form of an "experimental loop" illustrated in Fig.1. The robot starts with some initial knowledge (possibly zero). This is the robot's initial theory of the domain. Then it repeats the steps:

1. Perform experiments and collect observation data
2. Apply a ML method to the data, which results in a new theory
3. Design new experiments aiming at collecting most informative new data
4. Plan the execution of these experiments using the current theory of the domain
5. Repeat the loop at step 1.

When designing next experiments, the robot has to estimate the potential benefits of possible experiments. This can be done by using methods of active learning. Initially, when the robot's knowledge is scarce or zero, it may not be possible to choose next experiments intelligently, so the robot may in such a case just perform some random experiments.

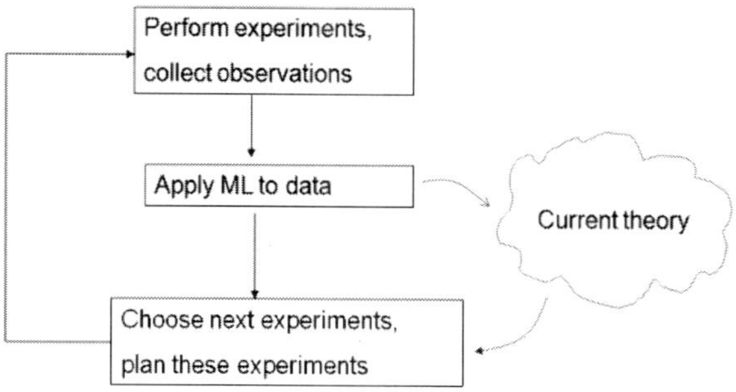

Fig. 1. Experimental loop

A desirable new experiment is defined by measurements to be taken in some situation, e.g. at some robot's position with respect to an object of interest. So each experiment defines a goal for the robot to achieve. To achieve such goals, the robot has to plan sequences of (physical) actions to carry out. This is done in step 4 above. However, there is another difficulty associated with this. When looking for such plans, the robot uses the current theory – nothing better is available at the moment. The current theory is typically imperfect and may thus lead the robot to false or impossible plans. Hopefully, the theory should gradually improve when the experimental loop had been executed several times.

2.2 Insights in Theory Evolution

The exploration process illustrated in Fig. 1 is similar to that of reinforcement learning. However, the approach taken in the present work is different from that normally used in reinforcement learning. One key difference is in the role of insights. The intended effect of insights is illustrated in Fig. 2. The robot's current theory typically evolves as follows. Initially, the robot starts with a small theory, which will become better, and also larger, after first experiments and learning steps. Newly discovered relations and laws of the domain are added to the theory, so the theory keeps growing. Then, when an insight occurs, the insight enables a simplification of the theory, so the theory shrinks. A simplification is possible because an insight gives rise to a better representation language which facilitates a more compact representation of the current knowledge. Then it may start growing again. This process is reminiscent of the evolution of scientific theories where an insight may cause what is sometimes called a "paradigm shift".

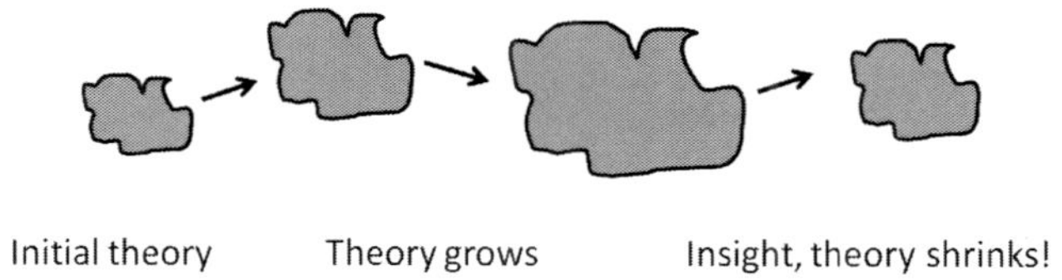

Fig. 2. Evolution of a theory during execution of experimental loop

In this paper we concentrate on the ML box in Fig. 1. For best performance, it is expected that standard techniques of machine learning and data mining will have to be adapted to the specific needs of discovery in XPERO. However, we investigate in this paper what can be achieved, in terms of the above stated robot discovery problem, by straightforward application of standard machine learning techniques to experimental data collected by the robot. In the sequel, we first describe some experiments, and present some illustrative results. Then we specify more systematically a number of generic discovery tasks in XPERO, a repertoire of ML approaches, and analyze these approaches according to their advantages and shortcomings with respect to the tasks.

3 Experimental Domains and Setups

Several setups were used in the experiments with machine learning. In this paper we present results in a simple experimental scenario, called "movability". This setup consisted of a mobile robot moving in the plane with just one additional object, a red block or ball. The robot, equipped with stereo vision capability, could detect the area in the image belonging to the object, robot's distance from the object and the angle between the current orientation of the robot and the angle at which the object was observed. These data were sampled in time at discrete time steps. Also, the robot was aware of its actions, that is moves, expressed as the step distance in the forward direction (i.e. its current orientation), and the "step angle", that is the change of the robot's orientation. Fig. 3 illustrates these parameters.

The experiments according to the setup shown in Fig. 3 were carried out by both a robot simulator (XPERsim, Fig. 4) and various real robots (e.g. Poineer D2X). XPERSim [1] is a robot simulator designed to provide a common simulation system for the XPERO consortium. It enables the replication of experiments, regardless of the physical presence of the robot and speeding up the pace of research. XPERSim is designed to simulate different mobile platforms. One of these platforms is the Khepera II robot from the K-TEAM Switzerland, also used in some of the experiments.

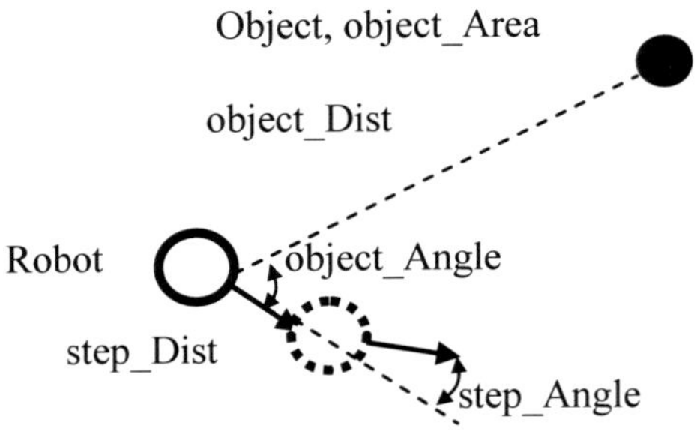

Fig. 3. Movability experiments: relations between the robot and the object, and between the current (solid circle) and the next (dotted circle) robot's positions. The arrows show the robot's orientation, that is the current direction of robot's movement.

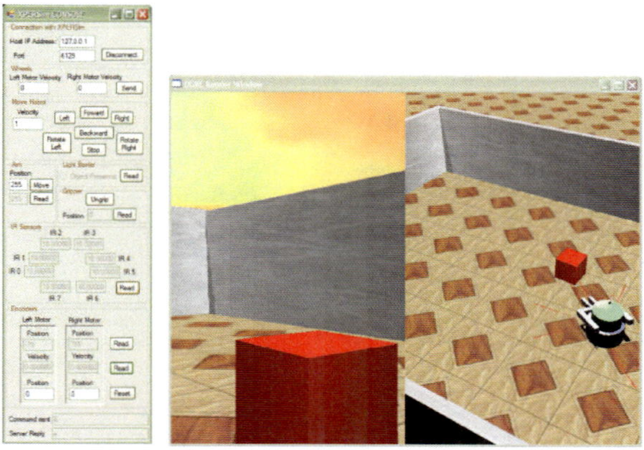

Fig. 4. A snapshot of the XPERsim simulation of Khepera II robot

Pioneer D2X robot was used to run the movability experiments physically [4]. The robot was equipped with odometry, eight sonar sensors and a stereo vision system. The single object used in the experimental setup was a red ball (radius 50mm). Another camera from the top of the scene was used to provide a separate set of data using "world's coordinate system" which was only meant to be used for verification and evaluation of learning methods and not for machine learning. In accordance to the philosophy of using a minimal set of "first principles", these experiments always assume that the robot is not aware of the absolute coordinate system.

The "observation vector" of the robot at some time point consists of the following variables (see Fig. 1):

object_Dist, the distance to the object
object_Angle, the angle of the object w.r.t. current robot's orientation
object_Area, the perceived size of the object as it appears in the image from robot's camera
step_Dist, the distance traveled in the robot's current direction in the last time step (zero, if the robot is turning in place)
step_Angle, the change of angle in the last step (zero if the robot is moving along straight line)

In the learning experiments we learned both "static" and "dynamic" relations. A static relation defines the relation between the variable values at the same time point. An example of a static relation is the function that maps *object_Dist* and *object_Angle* into *object_Area*, where all these variable values are taken at the same time point *t*. The corresponding learning problem is to induce, from given data, a function f for prediction of perceived object area in the camera image from the distance and angle to the object:

object_Area = f(*object_Dist*, *object_Angle*)

In this learning problem, in the usual terminology of ML, *object_Dist* and *object_Angle* are called the attributes, and *object_Area* is called the class.

Dynamic relations are useful to make predictions about variable values at the next time point $t + 1$ (assuming that time points are indexed by integers) from given variable values at time t and robot's actions at time t. An example of learning dynamic relations is to induce a function g that predicts *object_Area* at time $t + 1$:

$$object\_Area(t+1) = g(\ object\_Dist(t), object\_Angle(t), \ step\_Dist(t), step\_Angle(t))$$

4 Machine Learning Techniques Used

Experiments were conducted with a number of different ML techniques chosen so that they constitute a representative sample of ML approaches. For reasons explained in the introduction, our choice of ML techniques was limited to those that induce concept descriptions in explicit symbolic form. The chosen techniques range from the well known ones to the less widely known. The well- known techniques used were: induction of decision trees, regression trees, and induction of if-then rules. The less known methods include the learning of equations (system GoldHorn) and the learning of qualitative models (systems QUIN and Padé). Most of the implementations of these techniques that we used in the experiments are integrated in the Orange ML environment [6]. We briefly describe these techniques and their implementations in the following paragraphs.

Rule learning using CN2. CN2 is a rule induction algorithm that can cope with noisy data [5]. We experimented with the implementation of CN2 in Orange. The output of this algorithm is an ordered set of if-then rules, also known as "decision lists".

Induction of classification and regression trees. In a classification tree (also called decision tree), the internal nodes correspond to selected attributes, and the branches that originate at a node correspond to the values of the node's attribute. The leaves of the tree give the classification that applies to all instances that reach the leaf (i.e. the cases whose attribute values are in agreement with the attribute values along the path from the root of the tree to the corresponding leaf). In the case of noisy data, for example, the classification at a leaf takes the form of a probability distribution over all possible classes. In classification trees, the class is discrete. Regression trees are similar to classification trees, except that the class variable is real-valued. Regression tree learning is more directly applied in our domain where all the variables are continuous. However, classification tree learning is also relevant after discretizing a continuous class variable into intervals. We used the Orange implementation of classification and regression tree induction.

Equation induction with GoldHorn. GoldHorn [8,9] is a machine learning system intended to discover empirical laws, in the form of algebraic equations, that govern the behavior of dynamic systems. Given a behavior of the system, e.g. a sampled execution trace, GoldHorn attempts to find a set of ordinary differential equations that describe the dynamics of the system. GoldHorn does not simply fit the parameters of equations of given forms, but it also constructs *new forms* of equations. The significance of an equation is judged by the degree of fit with the data (correlation coefficient and normalized squared error). It should be noted that, as a result, GoldHorn not only finds differential equations of some predefined forms, but also does *structural* synthesis of new forms of equations.

Learning qualitative models with QUIN. QUIN (Qualitative Induction [14,15]) is a learning program that looks for qualitative patterns in numerical data. QUIN expresses such qualitative patterns by *qualitative trees* that are similar to decision trees, but have monotonic qualitative constraints in the leaves. *Monotonic qualitative constraints (MQCs)* are a kind of monotonicity constraints that are widely used in the field of qualitative reasoning and are a generalization of monotonic function constraint used in QSIM [10]. For example, the MQC $Z = M^{+,-}(X, Y)$ says that Z is monotonically increasing in X and monotonically decreasing in Y. If both X and Y increase, then according to this constraint, Z may increase, decrease or stay unchanged. In such a case, an MQC cannot make an unambiguous prediction of the qualitative change in Z.

Learning qualitative models with Padé. Padé [17] is a set of methods for estimating quantitative or qualitative partial derivatives for points in an attribute space when given a set of learning examples in this space. The name Padé is an acronym for "partial derivative" (and also the name of a famous French mathematician). Once qualitative derivatives are extracted with Padé, one can use arbitrary machine learning algorithms (e.g. C4.5, CN2, Naïve Bayes, SVM etc.) to induce a corresponding qualitative prediction model.

The Orange environment. Most of the above mentioned ML approaches, used in these experiments, are implemented in the Orange ML platform. Orange [6] is a library of C++ core objects and routines that includes a large variety of machine learning and data mining algorithms, plus routines for data input and manipulation. It is a collection of Python-based modules that sit over the core library and offer a versatile environment for developers, researchers and data mining practitioners. Orange also comprises a set of graphical widgets that use methods from the core library and Orange modules and provide a nice user interface. All these make Orange a comprehensive, component-based framework for machine learning and data mining which is constantly growing by adding more and more machine learning methods and algorithms while at the same time improving the existing ones.

Inductive Logic Programming (ILP). ILP is an approach to machine learning that induces logic formulas from learning data. Typically, the learned theory is a set of first-order logic formulas, usually in the form of Prolog clauses. The HYPER program [2] (chapter 19) was used in the experiments. ILP has a particular advantage in that the knowledge known to the agent prior to learning can be stated in a very general form, in logic. This is called background knowledge and is typically specified as a general Prolog program. This is particularly useful in the context of theory evolution because a theory learned at some stage of the theory evolution process can be used as background knowledge in the next stage. The disadvantage of using ILP, on the other hand, is the high computational complexity of ILP.

5 Some Experimental Results

Among many results in learning various static and dynamic relations in the movability scenario with different ML techniques, we here select some examples to highlight some positive and negative points.

In general, the more interesting results were obtained with the learning of qualitative models, both with QUIN and Padé, and with HYPER with respect to discovery of new concepts (effectively extending the description language).

In this review of the results, we are primarily interested in what kinds of descriptions were learned, and whether they offer some meaningful interpretation, or possibly provide insights. We are here less interested in statistical criteria like classification accuracy. Therefore numerical results of this kind are not presented here.

5.1 Results of Learning Qualitative Trees

Here is a qualitative tree, written as a nested if-then rule, for the class variable's *object_Area* qualitative dependence on *object_Dist* and *object_Angle*, induced by QUIN from the XPERSim Khepera traces:

IF *object_Angle* $\leq$ 2.6769
  THEN IF *object_Angle* $\leq$ -13.76
    THEN $M^{+,-}$(*object_Angle, object_Dist*)
    ELSE M^{-}(*object_Dist*)
  ELSE IF *object_Angle* $\leq$ 8.4
    THEN M^{-}(*object_Dist*)
    ELSE $M^{+,-}$(*object_Angle, object_Dist*)

This tree offers a nice qualitative interpretation. When the robot is facing away from the object, the object is not visible, and the area is always zero. The tree above was induced only from examples where the object is visible. The middle two leaves of the tree that apply when the robot is facing the object (angle between -13.76 and 8.4 degrees), say that the area decreases with increasing distance. This is what one would expect, as the perceived object becomes smaller when we move away. The leftmost leaf and the rightmost leaf describe similar decreasing dependence of area on object distance, but something else is also going on there: the disappearing of the object from the camera view when the robot is turning away from the object.

The induced qualitative tree gives a good qualitative explanation of how the area changes with object distance and object angle. Non-symmetry in the angular thresholds in the left and right sub-trees is a blemish due to statistical asymmetry in the data between positive and negative angle. This qualitative tree could be viewed as a first simple discovery. It however falls short of quite being an insight in the particular ambitious sense of XPERO discussed in Section 1.

The learning with Padé also produced sensible results, of similar form as those by QUIN.

5.2 Results of Learning Equations with GoldHorn

The same data was used for learning with GoldHorn. In this case, when static relations were considered only, GoldHorn generated algebraic equations (not differential).

Some equations induced by GoldHorn in this domain, ordered according to the degree of fit between the equation and the date, are:

(1) object_Area * object_Dist = 99.19 + 11.48 object_Area

(2) object_Area^2 * object_Dist = -463.49 + 299.99 object_Area

(3) object_Area^2 = -62.04 + 19.98 object_Area

(4) object_Area * object_Dist^2 = 6232.31 - 6.37 object_Angle^2

(5) object_Area^3 = -470.17 + 19.37 object_Area^2

There are some obvious problems with this approach. The first problem is that GoldHorn induces hundreds of equations and many of them, even those with the highest rankings, are useless or meaningless. Typically, such equations are the result of spurious correlations between the variables and the data, often due to numerical errors. One such spurious and meaningless equation is equation 3 above. The first three equations do not mention object angle and obviously cannot serve as valid models. The third equation and the fifth equation are meaningless. The only (qualitatively) interesting equation is the fourth equation that can be rewritten as:

object_Area = (6232.31 - 6.37 object_Angle^2) / object_Dist^2

Similar to the above qualitative tree, this equation says that the area decreases with object distance. Actually, the equation says more – it correctly states that area decreases with the square of distance. The equation also indicates the decreasing of the area, when a part of the object is disappearing from the camera. However, this is not expressed by a correct form of formula, and is not correct for small object angles, i.e. when the whole object is visible by the camera. As trigonometric functions were not given to GoldHorn as a repertoire of available operators, GoldHorn could not even in principle find a correct functional form of this dependence. Nevertheless, GoldHorn was able to find a form of equation that enabled good fit with the data.

5.3 Gaining Insights through Predicate Invention in ILP

The induced models presented so far (described in more detail in [4]) are concerned with prediction where no new concept was introduced. These may be useful for making predictions, but do not count as "insights" in the more ambitious sense of XPERO described in the Introduction. Therefore in the strict, abstract sense of "gaining insights" in XPERO, more interesting results were obtained with Inductive Logic Programming (ILP) in a somewhat different experimental scenario. Here the robot was pushing blocks in a two-dimensional space. The robot was able to actually move some of the blocks, but some of the blocks could not be moved. After some time, the robot has collected a number of experimental data recorded as ground facts about predicates:

- at(Obj,T,P), meaning object Obj was observed at position P at time T;
- move(Obj,P1,D,P2), meaning command "move Obj from P1 by distance D" resulted in Obj at P2.

Here all positions are two-dimensional vectors. The robot's prior knowledge consisted of the predicates:

- different(X,Y), meaning X and Y are not approximately equal;
- add(X, Y, Z), meaning $Z \approx X + Y$.

These relations are defined as approximations so that they are useful in spite of noise in numerical data.

It should be noted that neither the observations nor the prior knowledge contain the concept of mobility of objects, or any mention of it. There are no examples given of movable and not movable objects. The ILP program HYPER [2] was used on this learning problem to induce a theory of moving in this world (that is learn predicate move/4). The induced theory by HYPER was stated in logic by the following Prolog clauses:

```
move(Obj,Pos,Dist,Pos):-
   not p(Obj).

move(Ob,Pos1,Dist,Pos2):-
   add(Pos1,Dist,Pos2),
   p(Obj).

p(Obj):-
   at(Obj,T1,Pos2),
   at(Obj,T2,Pos2),
   different(Pos1,Pos2).
```

In the clauses above, the variables were renamed for easier understanding. The first clause deals with unmovable objects (after the move command, the position of the object remains unchanged). The second clause handles movable objects. The point of interest is that HYPER invented a new predicate, p(Object) which is true for objects that can be moved. The definition above says that an object is movable if it has been observed at two different positions in time. Thus the robot has come up with a new concept never mentioned in the data or problem definition. The new concept p(Object) enabled the learning system to divide the problem in the two cases. The meaningful name of the newly invented predicate p(Object) would be movable(Object).

In another experiment where many objects were present in the scene, the robot invented another predicate which corresponds to the notion of obstacle. If an object appeared on the trajectory of another, moving object then the stationary object impeded the movement. The ILP learner found it useful to introduce a new predicate whose definition corresponded to such an object – an obstacle. Again, the notion of obstacle was never mentioned in the problem definition, nor in the learning data. The learner just figured it out that such a new notion was useful for explaining the behavior of objects in the robot's world.

The essential learning mechanism in these experiments was predicate invention in ILP. This capability further increases computational complexity, but it enables the gaining of insights in its ambitious "pure form". Details of these experiments are described by Leban et al. [11].

6 Robot Discovery Tasks and Learning Paradigms

Now we formulate a more systematic list of XPERO types of learning and discovery tasks that go beyond the illustrations in the previous section:

(1) ***Learning static functional relations.*** Given a set of observations, all taken at the same time t, find relations between them. This usually consists of finding mappings from a subset of observations into other observations. Examples in the movability scenario were discussed in the previous section such as object area being a function of distance and angle. It should be noted, however, that in this example setting, this relation is only functional for the cases when the angle is determined for all possible orientations of the robot. This may not be the case, for example when the distance is determined through on-board stereo vision (and not by proximity sensors) because then the distance cannot be measured when the robot is facing away from the object.

(2) ***Learning dynamic relations.*** This is about the learning of the dynamics of the robot's world. Dynamic relations include relations between observations, states and actions at different times. They may be expressed in many ways, such as:

- Obs(t+1) = f(Obs(t),Act(t))
  This assumes that current observations Obs(t) are sufficient to determine the consequences Obs(t+1) of any action Act(t).
- State(t+1) = f(State(t),Act(t)), Obs(t) = g(State(t))
  This assumes that the state of the system is given.
- As above, but the state of the system is not known, or is not given as observables, but has to be discovered e.g. as State(t) = f(History(t)) where History(t) = (Obs(t-1), Act(t-1), Obs(t-2), Act(t-2), ...). The learning also needs to discover a suitable length of History. This may give rise to some insights, such as: (a) discovery of the concept of state, and (b) given that the robot assumes the concept of state, it then has to discover its definition in terms of its relation (f above) with observations and actions.

(3) ***Discovering useful notions in the physical world*** of the robot, such as movable(Object) and obstacle(Object, Robot, Intention). The latter means that Object is an obstacle for Robot with respect to the robot's Intentions. Other examples are the notions of containment, force, gravity, friction, support, touch, push, ...

(4) ***Discovering aggregates***: for example stack of blocks, "train" of blocks

(5) ***Discovering notions of cooperation*** of several agents, e.g. two robots jointly pushing a long block

(6) ***Discovering functional concepts***: e.g. a long block used as a tool to push a number of smaller blocks in parallel

Machine learning paradigms considered in this analysis include the following:

(1) ***Attribute-value learning***, such as learning if-then rules, decision trees, regression trees and regression rules.

(2) ***Induction of equations*** from numerical data (programs like Goldhorn and Lagramge [7]).

(3) ***Qualitative learning***, like programs QUIN and Padé. Hypothesis languages in this learning paradigm express qualitative relations in terms of qualitative proportionalities or monotonicity constraints among variables in the system.

(4) ***Q^2 learning*** (also called qualitatively faithful quantitative learning [16]) possibly realised as a combination of qualitative learning with QUIN or Padé, and qualitatively constrained regression e.g. with Qfilter or QCgrid. This is a non-standard approach to attribute-value learning with numerical class variable whereby the induced regression model respects the qualitative properties detected in the learning data.

(5) ***Inductive logic programming*** (ILP) with programs like HYPER, Progol [12] and Aleph [13]. In addition to the learning of logical theories with ILP, we here consider also the learning of dynamic qualitative models [3] of QSIM-type since such models can be naturally represented by logic clauses, as well as simulated by logic programs.

7 Applicability of Learning Methods to Discovery Tasks

The learning paradigms above are roughly ordered according to the power of representation languages, and correspondingly according to increasing computational complexity. So the range of applicability of these paradigms is on the one hand limited by expressiveness of hypothesis languages, and on the other hand by computational complexity. Another limiting factor is the interpretability (understandability) of induced descriptions. We consider this ability as necessary in XPERO with respect to the gaining of insights. For this reason we do not seriously consider the use of some other ML approaches, such as neural networks, support vector machines, and reinforcement learning in its usual form.

Table 1 summarises the suitability of the considered ML approaches for various types of learning/discovery tasks in XPERO. The applicability is assessed in terms of four criteria:

- Expressiveness (E) of the ML method's hypothesis language (representation)
- Predictive accuracy (A) of induced theories; this is here taken as a rough estimate that also takes into account the ability to cope with noise, precision in numerical prediction etc.
- Computational complexity (C)
- Interpretability (I) of induced hypotheses; how easy or difficult is it for the user to figure out the meaning of a hypothesis, and possibly use it for reasoning about the domain.

It should be emphasised that Table 1 only gives some general ideas of the comparative advantages of the potentials of the methods considered, and should not be understood as a definite ranking of the methods that applies to all cases and all situations. Several entries in the table are only tentative, and a more firm assessment would require very substantial further work. In the remainder of this section we comment on the entries in Table 1. The comments are organised according to the ML paradigms.

Table 1. Tentative assessment of applicability of ML approaches to types of task. Criteria: E = expressiveness, A = accuracy, C = complexity, I = interpretability. "+" means above average, "-" means below average, blank means neutral or not applicable.

	Learning static relations	Learning dynamic relations	Discover physical notions	Discover aggreg. notions	Discover agent cooperat.	Discover functional notions
	E A C I	E A C I	E A C I	E A C I	E A C I	E A C I
Att.-value learning	+ +	+ +	-	-	+ +	-
Equation learning	+	+ +	-	-	-	-
Qualitative learning	+ - + +	+ - +			+ +	
Q^2 learning	+ + + +	+ + +			+ + +	
ILP	+ - - +	+ - - +	+ - +	+ - +	+ - +	+ - +

Attribute-value learning. The strength of attribute-value learning comes from the advanced state of development of this ML paradigm, and its good computational efficiency. There are powerful and well-developed techniques for handling noise in the learning data which is important because of measurement errors in sensory data. The accuracy, both numerical and classification accuracy, of these techniques is generally very good. These techniques have been particularly well developed and optimised with respect to accuracy. The limitations of attribute-value learning for the robot discovery tasks comes from the simple hypothesis language used. This language enables good efficiency, and suffices for discovering dependences between a fixed number of predefined variables. But it is too rigid, so it lacks expressive power needed for definition of, and thus for the discovery of, new physical notions (e.g. mobility of objects, obstacles, etc.), aggregate notions, and functional notions.

Induction of equations. Induction of equations is very attractive in the sense that so many basic physics models are also stated by equations. They also are ideally accurate provided that the right equation is found. Of course, in such cases, the interpretability is also good, and they can be used for reasoning symbolically about the domain, for example in control or design applications. Another potential advantage of this approach also lies in the fact that the search space of possible equations can be defined by a grammar that excludes in advance equations that are not possible for the domain in question. The user can state the background knowledge in this way and thus largely improve the search efficiency. Of course, this is only of importance if there is some prior domain-specific knowledge.

On the other hand, all these advantages of equation learning are only possible if the target theory is expressible in terms of the predefined set of operators and functions that the user provides to the equation discovery system. If it is not then the system will only find approximate equations. Typically, there will be many such equations and they may have a misleadingly good fit with the data. This may easily lead to misleading interpretations of the resulting equations which is undesirable, particularly with respect to insights.

Another limitation is that equation discovery only works for numerical models. It is possible to imagine that some physics notions could be discovered this way by looking for repeated sub terms of equations. This would possibly indicate new concepts, such as energy, defined by mere investigation of numerical relations among the variables. The new notion would thus emerge simply as a new formal construct, a convenient new quantity that simplifies the statement of a theory of the domain. Another possibility is the discovery, in the same way, of theorems like the theorem of Pythagoras. It is on the other hard to imagine that equation learning would successfully lead to the discovery of other notions, such as obstacle and mobility of objects, that are not naturally expressed in terms of numerical equations. Also, the discovery of aggregate notions and functional notions seems to be beyond equation learning.

Qualitative learning. Here we consider the learning of QDE models (qualitative differential equation models), and qualitative models as qualitative trees and qualitative if-then rules where a tree's leaves or rule's then-parts are monotonicity constraints between variables. Qualitative abstraction of numerical data should improve the efficiency of learning, and also the interpretability of learned models. Qualitative abstraction aims at removing non-essential numerical details in the data, and only preserving essential, qualitative relations between the observed variables. Interpretability of induced models can also benefit from such abstractions. This may also alleviate the problem of handling noise in quantitative data, and the difficulty of defining detailed numerical dependences that can often simply not be expressed in the available hypothesis language. Typical qualitative representations can be expressed in logic formalisms which may further extend the expressiveness.

An obvious disadvantage of qualitative learning is the absence of numerical accuracy of induced models. If an induced qualitative theory is to be used for controlling the robot numerically then some additional mechanism is needed, for example something like Q^2 learning (see below). A qualitative model can, however, be very useful for the construction of rough, qualitative plans for the robot, possibly using a means-ends planner. A limitation of qualitative abstraction on its own (without extensions to general ILP), does not seem to be sufficient in general for the discovery of physical notions, aggregate notions, and functional notions.

Q^2 learning (qualitatively faithful quantitative learning)**.** Q^2 learning extends qualitative learning with numerical capabilities. The result is a numerical (regression) model that fits well quantitative learning data, under the additional constraint that it also respects the qualitative relations in the domain. Q^2 learning can be implemented in two stages: first, learn a qualitative model from the given data, and then refine the learned qualitative model to a numerical regression model which preserves the induced qualitative relations. This approach should then enjoy the benefits of qualitative learning, and also provide good numerical accuracy.

Inductive logic programming (ILP). The hypothesis language of ILP is typically the language of predicate logic. This is sufficiently expressive to enable, at least in principle, the applicability also to tasks where the other approaches seem to be insufficient. In particular, these tasks are, according to our list of tasks, the discovery of physical, aggregate, and functional notions. For example, a mechanism that is helpful in all these three tasks is that of predicate invention. In the discovery and handling of aggregate notions, such as aggregate objects, the recursion facility is essential. The use of general logic theories as background knowledge allows very natural transition between the learning of theories in increasingly complex worlds.

The general weakness of ILP is ILP's high computational complexity. Another deficiency is lack of quantitative facilities in logic. The latter are typically added to the basic logic formalism of ILP, most elegantly as CLP(R) (constraint logic programming with real numbers), so quantitative aspects should not be a real problem in principle. However, computational complexity which arises from the main strength of ILP, that is the expressiveness of ILP's hypothesis language, is the critical limitation and it seems the situation in this respect is unlikely to change considerably with more efficient algorithms for ILP learning. So, improvements can only be expected from fundamentally enhancing the ways in this approach is applied in concrete situations. For example, one way to dramatically reduce the complexity is through the introduction of hierarchy of learning problems. The problem is solved first at a high level of abstraction, which has low complexity because details are ignored. Such an abstract solution can then hopefully be used to guide the search for solution at a more detailed level of abstraction.

8 Concluding Remarks

In this paper, a list of autonomous robot discovery tasks, and a list of machine learning approaches were considered. Of particular interest to the present work is the notion of "gaining insights". Then important criteria of success were stated (expressiveness, accuracy, computational efficiency, and interpretability). The ML paradigms were compared and assessed in terms of these criteria against the discovery tasks. It should be noted that this assessment is largely tentative due to lack of experimental results at least for some of the tasks and some of the ML approaches. For example, the task of discovering functional notions has scarcely been addressed in past research. Therefore in many aspects of the presented comparison, further work, in particular experimental work, is needed.

Acknowledgements

Research described in this paper was supported by the European Commission, 6th Framework project XPERO, and the Slovenian research agency ARRS, Research program Artificial Intelligence and Intelligent Systems. A number of people contributed to the experimental work described in section 4, including G. Leban and J. Žabkar, and co-authors of paper [4].

References

[1] Awaad, I., Leon, B.: XPERSim: Simulation of the robotic experimenter. XPERO Report on R&D 1 (September 2006)
[2] Bratko, I.: Prolog Programming for Artificial Intelligence, 3rd edn. Addison-Wesley / Pearson, Reading (2001)
[3] Bratko, I., Šuc, D.: Learning qualitative models. AI Magazine 24(4), 107–119 (2003)

[4] Bratko, I., Šuc, D., Awaad, I., Demšar, J., Gemeiner, P., Guid, M., Leon, B., Mestnik, M., Prankl, J., Prassler, E., Vincze, M., Žabkar, J.: Initial experiments in robot discovery in XPERO. In: ICRA 2007 workshop Concept Learning for Embodied Agents, Rome (April 2007)
[5] Clark, P., Niblett, T.: The CN2 induction algorithm. Machine Learning 3, 262–284 (1989)
[6] Demšar, J., Zupan, B.: Orange: Data Mining Fruitful & Fun - From Experimental Machine Learning to Interactive Data Mining (2006), http://www.ailab.si/orange
[7] Džeroski, S., Todorovski, L.: Discovering dynamics. In: Proceedings of the 10th International Conference on Machine Learning, pp. 97–103. Morgan Kaufmann, San Francisco (1993)
[8] Križman, V., Džeroski, S., Kompare, B.: Discovering dynamics from measured data. Electrotechnical Review 62, 191–198 (1995)
[9] Križman, V.: Automatic Discovery of the Structure of Dynamic System Models. PhD thesis, Faculty of Computer and Information Sciences, University of Ljubljana (1998)
[10] Kuipers, B.: Qualitative Reasoning: Modeling and Simulation with Incomplete Knowledge. MIT Press, Cambridge (1994)
[11] Leban, G., Žabkar, J., Bratko, I.: An experiment in robot discovery with ILP. In: Železný, F., Lavrač, N. (eds.) ILP 2008. LNCS (LNAI), vol. 5194, pp. 77–90. Springer, Heidelberg (2008)
[12] Muggleton, S.: Inverse entailment and Progol. New Generation Computing 13, 245–286 (1995)
[13] Srinivasan, A.: The Aleph Manual. Technical Report, Computing Laboratory, Oxford University (2000), http://web.comlab.ox.ac.uk/oucl/research/areas/machlearn/Aleph/
[14] Šuc, D.: Machine Reconstruction of Human Control Strategies, Ph.D. Thesis, Faculty of Computer and Information Sc., University of Ljubljana (2001)
[15] Šuc, D.: Machine Reconstruction of Human Control Strategies. Frontiers Artificial Intelligence Appl., vol. 99. IOS Press, Amsterdam (2003)
[16] Šuc, D., Vladušič, D., Bratko, I.: Qualitatively Faithful Quantitative Prediction. Artificial Intelligence 158(2), 190–219 (2004)
[17] Žabkar, J., Bratko, I., Demšar, J.: Learning qualitative models through partial derivatives by Padé. In: Proc. QR 2007, 21st International Workshop on Qualitative Reasoning, Aberystwyth, Wales, UK, pp. 193–202 (2007)

Multistrategy Learning for Robot Behaviours

Claude Sammut and Tak Fai Yik

ARC Centre of Excellence for Autonomous Systems, School of Computer Science and Engineering, University of New South Wales, Sydney 2052, Australia
claude@cse.unsw.edu.au, tfyik@cse.unsw.edu.au

Abstract. Pure reinforcement learning does not scale well to domains with many degrees of freedom and particularly to continuous domains. In this paper, we introduce a hybrid method in which a symbolic planner constructs an approximate solution to a control problem. Subsequently, a numerical optimisation algorithm is used to refine the qualitative plan into an operational policy. The method is demonstrated on the problem of learning a stable walking gait for a bipedal robot. We use this approach to illustrate the benefits of a multistrategy approach to robot learning.

1 Introduction

In his Inferential Theory of Learning, Ryszard Michalski [16] created a framework to help clarify the relationships between the many approaches to Machine Learning that have been developed. He set his framework in the context of an agent that must adapt its learning strategy, depending on its current task. This is an important point because most research in Machine Leaning is devoted to data analysis, which is usually guided by a human. Therefore, researchers rarely face the problem of a system having to select its own learning behaviour. Moreover, the learning system in an agent is just one component of a larger system that must integrate all aspects of the agent's behaviour. Thus learning does not occur in isolation.

A robot is a classic example of an agent that has many components, all of which must be well integrated to work successfully. Because it operates in the physical world, using sensors and actuators that always have errors, a robot must also cope with uncertainty and unpredictability. All of these characteristics make robotics a rich area for research in learning. In this paper, we examine some aspects of robot learning with particular attention to the role of background knowledge and to a multistrategy learning method that combines symbolic planning and numerical parameter learning. This method enables the robot to acquire motor skills efficiently. Our hypothesis is that a planner can generate constraints for a trial-and-error learner that will dramatically reduce the search space.

In the next section, we introduce the setting for our robot learning experiments and discuss different uses of machine learning in robotics. In section 4, we describe a particular learning problem, namely, a bipedal robot learning to walk. Section 5 introduces the symbolic action model for the robot and the plan generator. Before the plan is passed to the trial-and-error learner, a constraint solver limits the search space of the learner by finding feasible ranges for parameters such as joint angles. Section 6, describes the

J. Koronacki et al. (Eds.): Advances in Machine Learning I, SCI 262, pp. 457–476.
springerlink.com © Springer-Verlag Berlin Heidelberg 2010

learner, which uses a simple hill-climbing technique to search for numerical values that will make the qualitative plan operational. In section 7, we give experimental results. Section 8 contains a discussion of related work. The final section reports our conclusions and future work.

2 Software Components of a Robot

The main distinction between learning for a robot and learning for data mining is that the learning algorithms must work as part of a complex agent with many interacting components required to interact with its environment. Each of these components adds complexity to the robots problem solving and learning. In this section, we list the main components of a robotic system, paying attention to how these components affect learning.

Perception. All sensors are liars. No matter what quality of camera, laser range-finder, etc, is used, any measurement will contain errors. Thus, the robot's decision making and learning must accommodate these errors. In fact learning can be used to build better perceptual models than hand-crafted code. For example, teams engaged in robot soccer competitions frequently learn to classify the colours of markers on the field [28]. Bayesian networks have been used to recognise complex motions [25].

Action. Actuators never do exactly what they are told. If a robot is instructed to drive forward 1 metre, the motors will not move to their precise positions. More importantly, it is not possible to model exactly how the robot will interact with its environment. For example, wheels often slip. This paper will deal with the problem of learning to control actions.

World Model. Except for very simple behaviour-based systems, a robot must build a model of its world so that it can solve problems like navigation and planning. The model not only represents the robots current beliefs about the world, it usually includes estimates about its uncertainty due to measurement and actuation errors. Roboticists often view this as a form of learning in the sense that as new information is acquired, the robot updates its world model and thus adds to its knowledge of the world. Machine Learning researchers may object to this since no generalisation is involved in this "learning". However, most methods are based on some form of Bayesian estimation [14, 30].

Decision Making. In the past, research in knowledge representation and reasoning had assumed that problems due to uncertainty could be isolated to the lower layers of the perceptual system and symbolic representations based, on higher-order logics, could be used to plan and reason about actions. However, it is now clear that uncertainty must be managed. to a greater or lesser extent, throughout the entire system. Thus, we are now seeing increasing use of hybrid systems that incorporate classical planning, backuped up by reactive decision making to handle the unpredictable consequences of the robot's actions [6]. A reactive system may be implemented as a set of situation-action rules, which are very fast to execute. Speed is often critical in a robotics system where dire consequences may result from inaction. However, the flexibility of these rules is limited to the number of situations that can be recognised. Therefore, a planner my be invoked when the robot encounters a novel situation. This is slower to execute but can handle a wider range of situations. Architectures such as SOAR [12] and ICARUS [13] include learning mechanisms that turn plans into reactive behaviours. That is, a form of explanation-based learning [20] is incorporated into the decision making process.

Cooperation and Competition. Robots often work in teams with other robots or with humans. Thus, communication and cooperation become necessary. Problem solving now becomes distributed and the robots may have to operate with varying degrees of autonomy since communications channels are often corrupted or drop out entirely. In teams of heterogeneous robots, agents may communicate to the others their capabilities so that tasks an be allocated to the most appropriate robot. Team coordination can also be used to provide backup when a robot fails. Sometimes, robots have to work in the presence of adversaries, as in robot soccer. In this case, the robots do not know he behaviour of their opponents in advance so they may have behaviours that are, hopefully, robust enough to handle a variety of game play or they may try to adapt to the opponent team. Reinforcement learning has been used to acquire cooperative strategies between team members to avoid opponents [29] and learning opponent models is a major research theme in the RoboCup competition [22].

We can see that each of the above components can benefit from incorporating a learning capability. What kind of learning depends on the task that the component is required to perform. In some cases, training examples can be accumulated so that batch learning of a classifier may be used, for example, collecting cases of opponent game play. For other tasks, trial-and error learning may be more appropriate, for example, learning the best action to take in a particular situation. Where there is uncertainty in perception, Bayesian approaches have been used. This variety in robot learning serves to emphasise Michalski's [17] idea of Task-adaptive Learning.

At present, there is no robot that incorporates learning throughout its architecture. More importantly, there are few systems that have adopted a multistrategy approach, although this would be beneficial. In the following sections, we describe one system that does combine symbolic deliberative reasoning with numerical learning to acquire motion behaviours for a humanoid robot. In our discussion at the end, we reflect on how this approach can be applied to other aspects of robot behaviour.

3 Hybrid Learning

As very young children, we learn many skills by trial-end-error. This kind of learning continues throughout our lives but as we grow older, our ability to observe and communicate with others gives us a new way of learning. We can take advice. We can incorporate this advice into our background knowledge and reason about our actions. For example, when we learn to drive a car, we are not put behind the steering wheel and left to our own devices. If the car has a manual gear shift, the instructor may tell the student that to change gears, the driver has to ease off the accelerator and depress the clutch before moving the gear lever. Of course, the student will not get this right the first time and will probably stall the car, but with some practice, the skill will become

Fig. 1. Cycloid II robot with pressure sensors added to foot pads and tether to PC

automatic. The important point is that without advice, the student has a large number of possible actions to try, most of which are irrelevant to the task. With advice, the search space of possible actions has been reduced enormously so that the number of trials needed to learn the skill is also reduced to a practical level. Our objective in this work is to use background knowledge, provided in a symbolic, qualitative language, to constrain the search through the parameter space of a set of motor actions.

Trial-and-error learning works well as long as the number of state variables and actions is small. Subsequent to early formulations of reinforcement learning [19, 32, 31], many methods have been proposed to deal with the explosion in the search space. These methods include the use of sophisticated value functions, relational reinforcement learning [5] and hierarchical learning [3, 9]. There have been some attempts at creating hybrids of symbolic AI and reinforcement learning [26]. In our approach, we use a classical planner to generate a sequence of qualitative actions. By trial-and-error, we then find numerical values for the action parameters that satisfies the learning goal. Advice is given to the learner in the form of qualitative action models.

We demonstrate the method on the problem of learning a walking gait for a bipedal robot. The robot, a Robotis Cycloid II[1], has 23 degrees of freedom (Figure 1). As demonstrated by highly successful humanoid robots from Honda and Sony, gaits can be constructed by careful modelling and algorithm design. This is a time-consuming process that usually requires very intimate knowledge of the robot. Our approach is to treat the robot dynamics as a "black box", learning the properties needed to make the robot walk. Were we to attempt nave reinforcement learning to generate a gait, the

[1] http://www.robotis.com/zbxe/main

number of trials required would be prohibitive. Instead, a planner constructs a qualitative description of the gait using fairly obvious, common sense knowledge of the main phases in walking. This description is refined using a simple numerical optimisation algorithm. The result is that a sequence of symbolic actions is turned into an operational set of motor commands that respond to feedback from the pressure sensors attached to the robot's feet.

4 The Learning Task

At the outset of this project, our aim was to show that it is possible to do trial-and-error learning on a physical robot without needing so many trials that we would wear out the mechanism or that it would only be possible if an accurate simulation were available. To this end, we targeted a Cycloid II robot, from Robotis, as our problem domain. The robot is unmodified except by the addition of four pressure sensors on the corners of each foot pad.

The Cycloid-II is 415mm tall. It has a total mass of approximately 2.5kg (without batteries) with 23 degrees of freedom. The hips and legs have 12 joints, 8 joints are in the arms with 2 joints in the waist and one in the neck. DX-117 actuators drive all the joints, with an onboard microprocessor to handle motor control and synchronisation. The CM-2 control module is based on the Atmel ATMega128 microprocessor. All planning and learning are performed off-board, on a PC connected to the robot via a serial cable. Force sensing resistors are attached to the four corners of each foot and are connected to the analogue to digital converters on the CM-2.

The objective set for learning is to be able to walk 50cm in a straight line. Since each step moves the robot three or four centimetres, between 12 and 17 steps are need to reach the 50cm target. This is a sufficient number to consider the walk stable and 50cm is short enough that each trial will not take too long. A learning trial is successful if the robot reaches its target. It fails if it falls or the trial lasts longer than a pre-defined time limit. The latter is needed in case the robot takes such small steps that it effectively walks on the spot.

5 Qualitative Representation and Planning

The architecture of the learning system is shown in Figure 2. The first step in constructing a controller for walking is to specify the actions available to the robot. We want to be able to do this in a reasonably intuitive way by the set of actions that the robot can experiment with. We want to do this without having to give exact values of joint angles and motor speeds. Instead, we want to be able to say the equivalent of, "if you bend the need in the positive direction the foot will lift off the ground". We can give quite a lot of advice about walking since it is easy to observe the stages that a human gait goes through (Figure 3).

Actions are described in a STRIPS-style notation [7] which is extended by allowing actions to be parameterised. For example, the action for swaying sideways is shown in Table 1. Before this action is executed, both feet should be on the ground. That is, the forces recorded by the sensors on both feet should be greater than zero. The add list

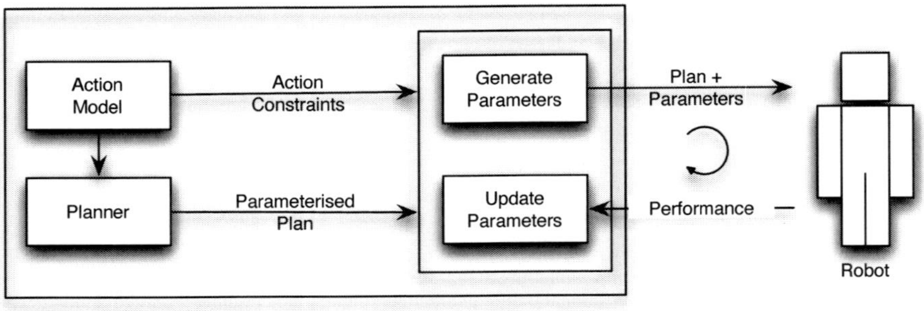

Fig. 2. Architecture of Learning System

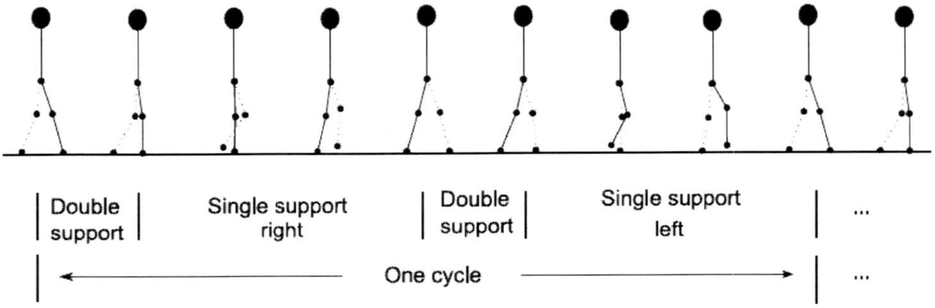

Fig. 3. A human walk cycle (left to right). The left leg is indicated by the dotted line and the right leg by the solid line (from Yik [34]).

for the action specifies that after execution, the hip angle in the lateral plane should be positive, that is, leaning to the left, and the body weight has shifted onto the left leg. For the difference in forces to be significant, we require it to be more than θ. The joint angles are illustrated in Figure 4. For the purposes of the planner, we do not yet need to assign numerical values to parameters like θ and ε. The action specification can be treated as a statement in a form of qualitative physics.

The inequalities in the add list can be treated as constraints on the parameters. After the planner has produced a sequence of actions, these constraints can be collected and passed to a constraint solver. That is, for each action in the sequence, we can also generate bounds on the values of the parameters. Therefore, when the robot begins its trial-and-error search for the parameter values, it can restrict its trials to be within those bounds, thus reducing the search space very substantially.

The "implementation" annotation translates the effects of the action into motor commands. That is, we set the hip joint angles to values specified in the add list. Setting a joint angle is achieved by a simple proportional controller. That is, we do not let the motors snap to their new angles but adjust the speed of rotation proportionally to the difference between the current angle and target angle:

$$\dot{\theta}_i = Gain_i \times (Target_i - \theta_i) \tag{1}$$

Table 1. Specification for "sway sideways" action

$SwaySideways(a) : a \in [Left, Right], b \in [Left, Right], a \neq b$	
Precondition:	*Both feet are on the ground* $ForceSensor(x) > 0 \wedge ForceSensor(y) > 0$
Add:	*Shift body weight onto leg, a* $\theta_{Hip,y} > 0 \wedge ForceSensor(a) > ForceSensor(b) + \varepsilon$
Delete:	*Remove constraints that conflict with the add list*
Implementation:	*Set both hip joints to* $\theta_{Hip,y}$

It is relatively easy to specify the target angle. It is the gain that is more difficult to adjust and therefore is the value we most want to learn. The analogy with learning to drive is useful again. The instructor can tell the student how far to push the clutch but the critical parameter in successfully executing the gear change is the timing, i.e. how fast each action is performed. The same is true for a walking gait. The exact angle of knee or hip bend is less critical than the timing. For example, too rapid a swing can build up momentum and overbalance the robot. Where the constraints specify $\theta > 0$, we will choose a target angle of $90°$, since we know that that is a safe value. If we are willing to sacrifice learning speed, that value of θ can also be left to the trial-end-error learning phase.

5.1 The Planner

The most common use of a planner is to find a sequence of actions that will lead from an initial state to a goal state. Often, the planner employs backwards chaining from the goal, finding actions whose effects include the goal conditions or the preconditions of other actions.

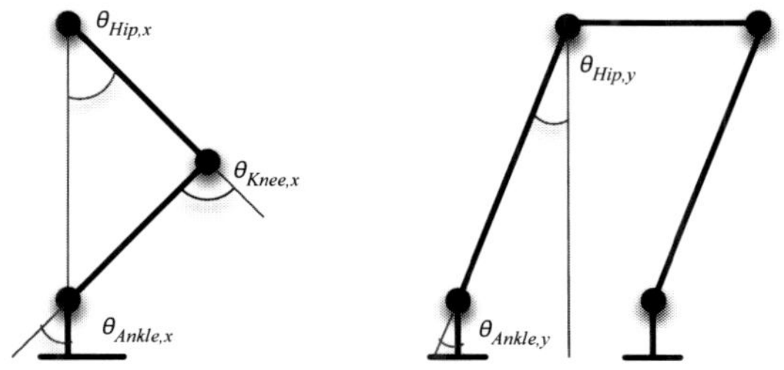

Fig. 4. The left diagram represents the angles that produce motion in the sagittal plane (front-back) and the right diagram represents the angles that produce motion in the lateral plane (left-right)

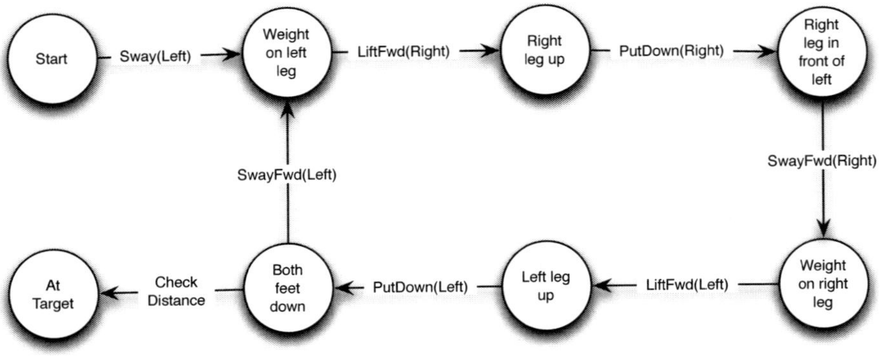

Fig. 5. A looped plan

Creating a plan for walking is somewhat different. In this case, we want to find a sequence of actions that can be repeated so that after each repetition, the distance to the target has been reduced. Figure 5 shows an example of a looping plan. We use a depth-bounded forward search for suitable loops. Because this domain does not have a very large number of possible actions, an exhaustive search is possible. Thus, from the starting state, we perform a depth-first search of all possible action sequences. A loop is detected if a sub-sequence leads to a state that has been visited before. If a possible side-effect of executing this sub-sequence is forward movement, then it becomes a candidate for the learning algorithm. This is somewhat similar to building macros.

Table 2 gives the specification for the "lift leg" action. Note that the precondition of this action is contained in the add list of the "sway sideways" action. Therefore, "lift leg" is a candidate action to follow "sway sideways". Also note that the implementation of "set joint angle" action is qualified. We want the rotation of the hip joint to by faster than that of the knee joint to reduce the chances of the robot over-balancing, therefore, the final implementation annotation places a constraint on the speeds of rotation of the hip and knee joints.

Table 2. Specification for "lift leg" action

$LiftLeg(x) : x \in [Left, Right], y \in [Left, Right], x \neq y$	
Precondition:	*The body weight is on leg, y* $\theta_{Hip,y} > 0 \wedge ForceSensor(y) > ForceSensor(x) + \varepsilon$
Add:	*Lift leg, x* $\theta_{Hip,x} > 0 \wedge \theta_{Knee,x} > 0 \wedge \theta_{Ankle,x} = \theta_{Knee,x} - \theta_{Hip,x} \wedge ForceSensor(x) = 0$
Delete:	*Remove constraints that conflict with the add list*
Implementation:	*Set hip joint to* $\theta_{Hip,x}$ *Set knee joint to* $\theta_{Knee,x}$ *Set ankle joint to* $\theta_{Ankle,x}$ $\dot{\theta}_{Hip,x} > \dot{\theta}_{Knee,x}$

Table 3. Walking plan

Start	→ $Sway(Left)$	→ Weight on left leg
	→ $LiftFwd(Right)$	→ Right leg up
	→ $PutDown(Right)$	→ Right leg in front of left
	→ $ReachedTarget$	→ Terminate
or		
	→ $SwayFwd(Right)$	→ Weight on right leg
	→ $LiftFwd(Left)$	→ Left leg up
	→ $PutDown(Left)$	→ Left leg in front of right
	→ Reached target	→ Terminate
goto start		

Algorithm 1 is a sketch of the loop finding routine. It performs its search until it finds the first feasible loop or fails if it reaches the depth bound. The planner must not only find a loop but one that is likely to move the robot forward. For example, the robot could simply walk on the spot. Since no action in this domain can cause backward motion, we use a simple heuristic that the plan should include at least one action that causes forward motion, for example, $LiftFwd$, which causes a leg to be lifted and swung forward. A larger domain, with a greater variety actions, may require more complex qualitative reasoning to determine if a qualitative plan can be guaranteed to result in an operational policy. However, for our tasks the experimental results show that our simple criterion is effective. Table 3 shows the completed plan generated by algorithm 1.

Algorithm 1. Loop finding

```
  depth ← maxdepth
  plan ← []
  visited ← []
  find_plan(starting state, depth)
  append check for goal to plan

boolean find_plan(state, depth)
  if depth = 0 then
     return false
  else if state ∈ visited then
     if plan contains forward action then
        return true
     else
        return false
  visited ← append state to visited
  for all actions do
     if precondition satisfied by state then
        state ← update state by action effects
        plan ← append action to plan
        if find_plan(state, depth − 1) then
           return true
```

5.2 Constraint Solving

Plan generation results in a sequence of actions required for the robot to walk. It also produces a sequence of qualitative states that contain constraints on parameter values such as joint angles (Table 4). In addition we can define global constraints. These include the minimum and maximum angles achievable by the actuators and their maximum velocities. They also include some constraints that arise from the type of walk we want to achieve. Because this robot does not have toes, it will always walk with its feet horizontal, so we can add constraints to specify that and the fact that the hips move sideways together (Table 5).

We use the ECL^iPS^e [1] constraint logic programing language as our constraint solver. The process begins by applying the global constraints and then applies the constraints from each successive state. Starting with state 1 (Table 4) the robot's first action is sway to shift the weight onto the left leg. The constraints from the resulting state are

Table 4. Sequence of qualitative states produced by the planner

1: Start (standing) $\theta_{Knee,x,Left} > 0 \wedge$ $\theta_{Knee,x,Right} > 0 \wedge$ $\theta_{Hip,x,Left} > 0 \wedge$ $\theta_{Hip,x,Right} > 0 \wedge$ $\theta_{Hip,x,left} = \theta_{Hip,x,Right} \wedge$ $ForceSensor(Left) > 0 \wedge$ $ForceSensor(Right) > 0 \wedge$ $ForceSensor(Left) = ForceSensor(Right)$	5: Weight on right leg $\theta_{Hip,x,Right} > 0 \wedge$ $\theta_{Hip,x,Right} = \theta_{Hip,x,Left} \wedge$ $ForceSensor(Left) > 0 \wedge$ $ForceSensor(Right) > 0 \wedge$ $ForceSensor(Right) > ForceSensor(Left) + \varepsilon$
2: Weight on left leg $\theta_{Knee,x,left} > 0 \wedge$ $\theta_{Knee,x,Right} > 0 \wedge$ $\theta_{Hip,x,left} > 0 \wedge$ $\theta_{Hip,x,Right} > 0 \wedge$ $\theta_{Hip,x,left} = \theta_{Hip,x,Right} \wedge$ $ForceSensor(Left) > 0 \wedge$ $ForceSensor(Right) > 0 \wedge$ $ForceSensor(Left) > ForceSensor(Right) + \varepsilon$	6: Left leg up $\theta_{Knee,x,Right} = 0 \wedge$ $\theta_{Hip,x,Right} = 0 \wedge$ $\theta_{Hip,x,Right} > \theta_{Hip,x,left} \wedge$ $ForceSensor(Left) = 0 \wedge$ $ForceSensor(Right) > 0$
3: Right leg up $\theta_{Knee,x,left} = 0 \wedge$ $\theta_{Hip,x,left} = 0 \wedge$ $\theta_{Hip,x,left} > \theta_{Hip,x,Right} \wedge$ $ForceSensor(Left) > 0 \wedge$ $ForceSensor(Right) = 0$	7: Left leg in front of right $\theta_{Hip,x,Left} > 0 \wedge$ $\theta_{Knee,x,Left} = 0 \wedge$ $\theta_{Hip,x,Right} = 0 \wedge$ $\theta_{Hip,x,Left} > \theta_{Hip,x,Right} \wedge$ $ForceSensor(Left) > 0 \wedge$
4: Right leg in front of left $\theta_{Hip,x,Right} > 0 \wedge$ $\theta_{Knee,x,Right} = 0 \wedge$ $\theta_{Hip,x,Left} = 0 \wedge$ $\theta_{Hip,x,Right} > \theta_{Hip,x,left} \wedge$ $ForceSensor(Left) > 0 \wedge$ $ForceSensor(Right) > 0$	8: Weight on left leg $\theta_{Hip,x,Left} > 0 \wedge$ $\theta_{Hip,x,Left} = \theta_{Hip,x,Right} \wedge$ $ForceSensor(Left) > 0 \wedge$ $ForceSensor(Right) > 0 \wedge$ $ForceSensor(Left) > ForceSensor(Right) + \varepsilon$

Table 5. Global constraints

In the lateral plane:	
$\dot{\theta}_{Hip,x,Left} = \dot{\theta}_{Hip,x,Right}$	Hips move sideways together
$\dot{\theta}_{Ankle,x,Left} = -\dot{\theta}_{Hip,x,Left}$ $\dot{\theta}_{Ankle,x,Right} = -\dot{\theta}_{Hip,x,Right}$	Ankles move relative to hips
In the sagittal plane:	
$\dot{\theta}_{Ankle,x,Left} = \dot{\theta}_{Knee,x,Left} - \dot{\theta}_{Hip,x,Left}$ $\dot{\theta}_{Ankle,x,Right} = \dot{\theta}_{Knee,x,Right} - \dot{\theta}_{Hip,x,Righ}$	Ankles move relative to hips and knees
Other:	
$0 \leq \varepsilon \leq 1$	

applied and then the solver moves to the next state and applies the constraints for lifting the right leg, and so on. Thus, constraints solving is performed sequentially, propagating values from one state to the next, until the final state is reached.

Recall that the critical values to set for the gait are the gains on the proportional controllers associated with each joint angle. The gains are given by equation (1). When ECL^iPS^e is invoked, we are actually solving for the gains, given the constraints. For example, when the left leg is lifted, we want to bend the left hip and left knee joints. From the planner's action and state sequences, we generate input to ECL^iPS^e that includes the following constraints:

$$\dot{\theta}_{Hip,x,Left} = Gain_{Hip,x,Left} \times (\frac{\pi}{2} - \theta_{Hip,x,Left})$$

$$\dot{\theta}_{Knee,x,Left} = Gain_{Knee,x,Left} \times (\frac{\pi}{2} - \theta_{Knee,x,Left})$$

For simplicity, we have chose the target angles for the hip and knee bends to be 90°. Similar constraints are generated for all the angular velocities and their associated gains.

6 Parameter Tuning

After running the constraint solver, we have a parameterised plan in which the parameter values are known within the bounds of the constraints, but we do not yet have precise values for them. The final stage is to determine these values by trial and error. The set of N parameters, where N is domain dependent, forms a continuous, multi-dimensional space. There are thirteen parameters for walking ($N = 13$), including the offset ε, and angular velocity gains for the actions. The parameter values reside within a known volume, specified by the constraints. Our-trial-and-error learner performs a simple hill-climbing search, starting from a random point in the space. Our performance measure is the distance travelled by the robot before it falls over.

To know if one set of parameters is better than another, we must compare their performance by executing the plans on the robot. Because of uncertainty in motor actions, slippage, etc, the same plan will not execute identically each time. Therefore, it is necessary to run several trials to determine the average distance travelled. Since this is time-consuming and potentially damaging to the robot, we must compromise between running many trials to get an accurate mean value versus conserving the hardware by minimising the number of trials. Experimentally, we found that running three trials gave us an adequate average measure for the distance. Therefore, after running three trials (i.e. executing the plan with the parameter values specified by the current point in the search), we compare the distance travelled with the best previously achieved. If the new parameters perform better than the previous best, the search continues from this point. Otherwise, the new parameters are discarded and another point in the search space is chosen. Initially, candidates are generated using a uniform distribution. After a candidate is found that travels a non-zero distance, the candidate generation method changes to use Gaussian random numbers. The new point is found by adding Gaussian noise to each parameter value. If the new point falls within the bounds of the constraints, we attempt another trial, if not, we generate a new set of random numbers. Thus, the algorithm searches for candidates within the neighbourhood of the current candidate until a solution is found or the performance does not improve.This process is summarised in algorithm 2.

Algorithm 2. Parameter tuning by hill climbing search

```
best ← null
best_distance ← 0

repeat
  if best_distance = 0 then
    candidate ← random point in parameter space
  else
    candidate ← add Gaussian noise to best

  if action constraints are not violated then
    execute plan on robot three times
    candidate_distance ← average distance travelled
    if candidate_distance > best_distance then
      best ← candidate
      best_distance ← candidate_distance

until target reached in all three attempts

return best
```

In general, best-first hill climbing algorithms are prone to becoming trapped in local optima. We chose this method for its simplicity. In future work, the algorithm may be replaced by other search methods, however, as we shall see in the next section, the experimental results show that a simple search is generally adequate for this task. This may not be the case in general, therefore more robust search algorithms may need to be investigated.

7 Experimental Results

The main hypothesis of this work is that the number of trials required to learn a gait should be small enough that the above process is practical on a real robot. This hypothesis was confirmed by our experiments. Apart from low-level motor control and sensing, all other programs ran off-board via a tether. That is, the planner, constraint solver and search algorithms executed on a desktop computer and the resulting parameters were downloaded onto the robot. Off-board computation was required mainly because the control processor on the robot is a very simple device. Each trial was started by hand and a harness was attached to the Cycloid II to catch it after it fell. The harness was designed to neither help nor hinder the robot's autonomous control.

Table 6 shows the results from running 15 attempts at learning. Of the 15, 14 succeeded in producing a working gait. The number of trials needed ranged from 8 to 84 with an average of 46. One successful gait is shown in Figure 6 and a learning curve averaged over the 15 experiments is shown in Figure 7. The trial in which no feasible value set could be found suggests that the simple best-first hill climbing search works most of the time but can become trapped in a local maximum. In experiment 6, the robot walked for 16 cm after 50 trials. However, no improvement was observed in the remaining trials, indicating that the search was trapped in a local maximum. Therefore, a mechanism is needed to reduce the chances of this happening.

As discussed earlier, a greedy hill-climbing algorithm can be trapped in a local optimum. This is the reason for the failure in trial 6. Thus, we expect that a different search method would make this approach more robust. Despite the simple algorithm, the results clearly indicate that trial-and-error learning under the constraints of a qualitative plan is practical. It was not possible to provide a baseline comparison against pure

Table 6. Walking plan

Experiment	Number of Trials	Stride Length (cm)	Lift Height (cm)	Walked 50cm	Comment
1	58	6	2	TRUE	
2	59	5	2	TRUE	
3	32	4	2	TRUE	
4	62	7	2	TRUE	
5	28	5	2	TRUE	
6	100	–	–	FALSE	Failed to find a value set
7	23	3	2	TRUE	
8	62	6	2	TRUE	
9	29	4	2	TRUE	
10	84	4	3	TRUE	
11	8	2	1	TRUE	
12	53	4	1	TRUE	
13	17	3	1	TRUE	
14	82	4	3	TRUE	
15	40	6	2	TRUE	

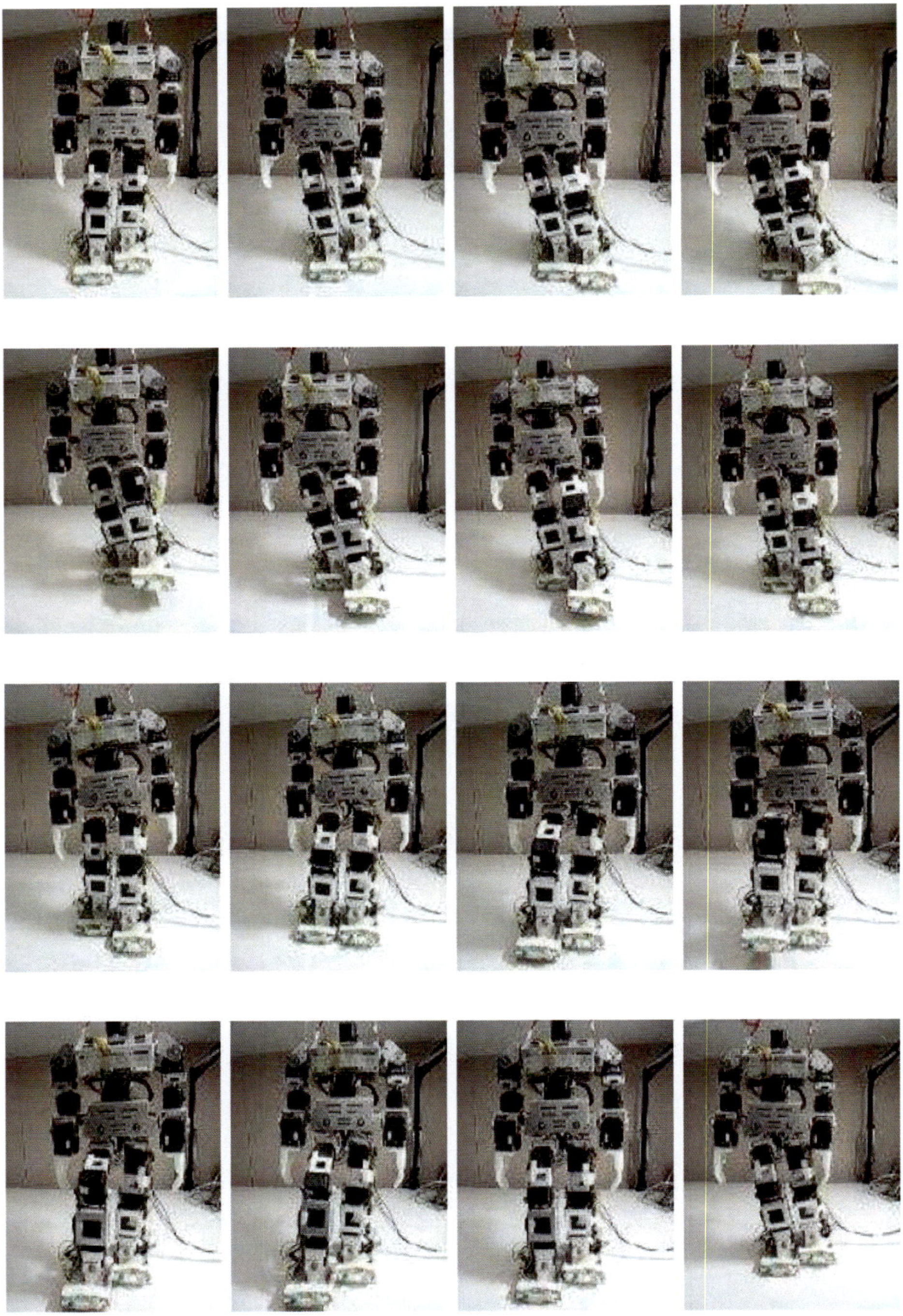

Fig. 6. Front view or a successful gait

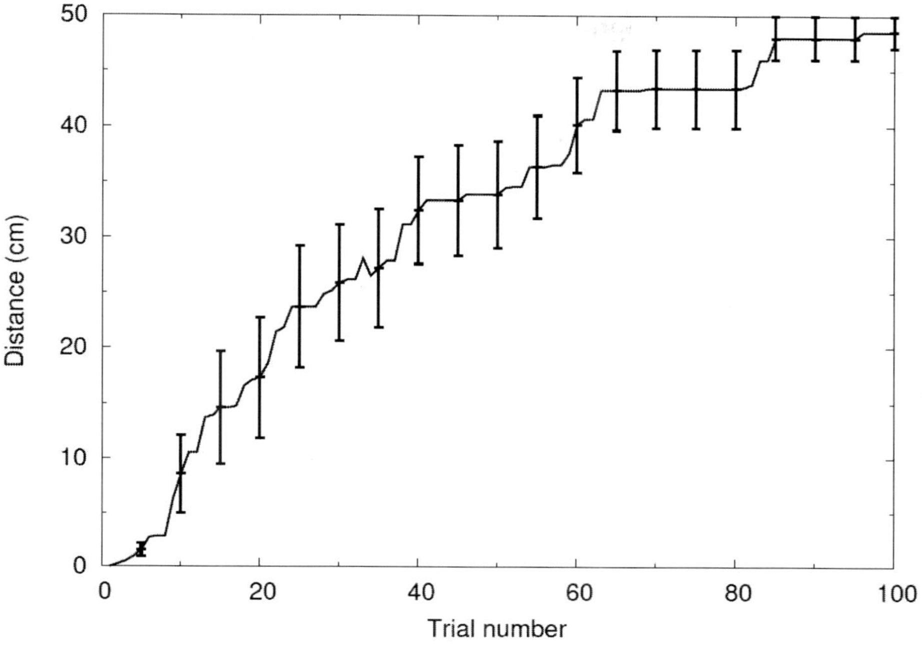

Fig. 7. Average performance of learning trials over 15 runs, walking 50cm

reinforcement learning since such an experiment would have taken so many trials that it would have taken too long and the robot would have worn out before completion.

It is clear from these results that while the method is fairly reliable in producing a working gait, the current set of action models and constraints do not produce anything like optimal results. While the plan specifies the action sequences to be performed, there is a high degree of flexibility in the range of parameters that may be learned. For example, some trials may result in longer strides than others or they may differ in the height the foot is lifted. The present learning system will only find parameters that satisfy the constraints. There is no guarantee of optimality. To produce a gait that is fast, as well as stable, the performance measure should include the time taken to traverse the target distance. An open question is whether it is more effective to try to find an quick and stable walk as a single learning or task or to break the learning task into stages in which a stable walk is acquired first and subsequent learning refines the gait to make it faster.

8 Related Work

The problem of scaling up reinforcement learning has been well studied. In particular, hierarchical reinforcement learning [3, 9], seeks to decompose the learning problem into several layers of abstraction. The more abstract layers generally have smaller search spaces and serve to focus the more specific layers into regions of the search

space that are most likely to contain a solution. In a broad sense, hierarchical reinforcement learning has similar goals to our work. Both use abstractions to constrain search. The main difference between the approaches is that most hierarchical reinforcement learning systems mainly keep each level of abstraction within the reinforcement learning paradigm, whereas we mix symbolic and numerical representations. We believe that qualitative constraints allow more powerful abstractions, which are also comprehensible to
humans.

Ryan [26] introduced a hybrid learning system that incorporated a planner. Like the work presented here, the planner generates a sequence of actions whose implementation is not given. The actions are learned by reinforcement learning. Ryan's method works best in discrete domains. His reinforcement learning could not find parameter values, as is the case with our method.

There has been a substantial amount of work in learning gaits. The majority of this work is conducted in simulation , where the number of trials required to find a solution is not a major consideration. Our goal is to devise a method for learning on a real robot. This may be achieved by performing initial learning in simulation and fine-tuning on the robot. The simulator have be a high-fidelity numerical model or we can consider a planner as performing qualitative simulation. In either case, the related work that we are primarily concerned with are attempts to implement the learned gait on a physical robot. RoboCup is a rich source of data for learning. Hornby et al. [10] used a genetic algorithm to learn a gait for Sony's Aibo robot. Kim and Uther [11] and Kohl and Stone [21] also learned a gait for the Aibo using a numerical optimisation method. This is closely related to our work. However, both of these systems focus entirely on numerical optimisation and are not able to take advantage of background knowledge as our system does.

Wyeth et al. [33] also use an evolutionary algorithm to learn a walk, this time for the GuROO humanoid robot. All learning takes place in simulation, which contains a kinematic model of the robot. Similarly, Zhou *et al* [35] have developed a method using reinforcement learning. The learned gait is tested on a physical robot. In both cases, a detailed model of the robot was known and used to construct a high-fidelity simulator. Our work is intended to learn from the physical robot, without a numerical model.

Ogino has developed several methods for learning bipedal gaits. One method [24] involves parameter tuning of a control system that is already carefully constructed. The result is a much more efficient walk than ours but at the expensive of much more extensive analysis of the robot's dynamics. That is, the system is less flexible in what it is capable of learning. In an application to RoboCup [23] a planner is combined with a reinforcement learner. Planning used dynamic programming. 1500 trials are required to learn tat gait compared with less the 100 for our method. The dynamic programming planner is working at a lower level of abstraction than our planner. We conjecture that the high-level abstraction enables us to cut out a larger part of the search space.

9 Discussion and Future Work

The work presented here is an example of how qualitative explanations can be formulated and given to a robot so that learning problems that might otherwise be intractable can be made practical. The experimental results demonstrate the feasibility of the approach but also point to further development that will make the method more reliable and produce control policies that are more efficient. The main improvements to be studied are:

- incorporating more sophisticated qualitative reasoning to distinguish between plans that are effective versus those that are merely theoretically feasible;
- replacing the best-first hill climbing search by another method less prone to being trapped in local maxima;
- for more complex domains with a larger number of actions, replace the planner with a one that has a more highly directed search.

A further improvement is extending the planner to be able to produce conditional plans. At present, the control policies can adapt variations in the values of the force sensors, i.e., slight differences in the balance of the robot. However, the policies cannot react to any more significant changes such as handling uneven terrain.

In section 2 of this paper, we discussed the various components of a robot system and how learning can be used throughout the whole of the robot architecture. The system presented here only addresses one aspect of robot learning, namely, motor control behaviours. However, the combination of qualitative reasoning and numerical optimisation does give some hints as to how different learning methods could fit into a general framework for learning in a robot.

Robots, like humans, need to use a variety of techniques for reasoning and learning because there is no single method that can do everything well. For example, symbolic representations make it possible to learn complex relational theories about the world. Symbolic representations led themselves to powerful algorithms for generalisation and for planning and problem solving. However, they are not well suited for continuous domains and handling uncertainty is not straightforward. On the other hand, statistical methods are somewhat better at handling uncertainty and continuous domains. They are not so good at representing theories and using them as background knowledge for problem solving or further learning. Statistical relational learning [8] is an attempt to create representations and learning algorithms that unify symbolic and statistical reasoning and learning. However, we can also envisage a heterogeneous system that mixes symbolic and statistical methods in cooperating but distinct modules.

This is the case in the work presented here. The system consists of a planner for symbolic reasoning, a constraint solver for numerical reasoning and a trial-and-error learner for tuning parameters in a continuous problem domain. In general, we envisage an architecture that is organised in hierarchical layers of abstraction (Figure 8). The lower layers provide the rapid response, reactive behaviours needed for robust operation in dynamic environments. They are the immediate connection to the world. The higher layers allow the robot to reflect on its interactions with the world and build theories about the world and how to behave in it. Low-level behaviours may be learned by some form of reinforcement learning but for this to be effective it must be guided by background knowledge about the domain that comes from the higher layers. In our system, we have assumed that the background knowledge has been provided as advice from another skilled agent. Learning action models for planners has already been well studied [27] [2] and future research to incorporate such learning is worth pursuing.

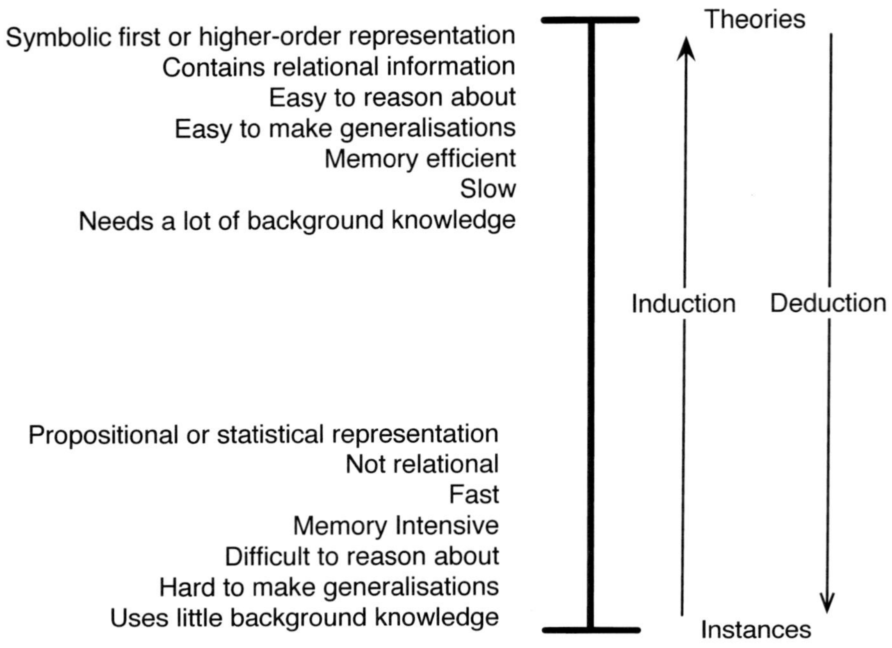

Fig. 8. Knowledge representation, Learning and Reasoning

In conclusion, we believe that multistrategy learning systems, in the spirit of Ryszard Michalski's vision will become increasingly important in the operation of complex agents.

Acknowledgements. Much of the work presented here is based on the PhD dissertation of Tak Fai Yik. This research is supported by the Australian Research Council Centre of Excellence for Autonomous Systems.

A personal note from the first author: I first met Ryszard at a workshop organised by Yves Kodratoff in Bonas, France in 1980. It was my first international workshop as a PhD student and I was very fortunate to meet Ryszard, who agreed to be on my thesis committee and later organised a postdoc at the University of Illinois at Urbana-Champaign, where Ryszard was, at the time. I was also fortunate in meeting Ranan Banerji at the same workshop. He also ended up being on my committee and organised a postdoc with him too! In my interactions with Ryszard, he was always exceedingly generous. Moving to Illinois as young academic from Australia was not an easy thing but Ryszard warmly welcomed me and helped me feel at home. I will always be grateful for the opportunities that he opened up for me.

There is no doubt that Ryszard had a profound impact on our field and was one of the key figures in defining Machine Learning as a legitimate discipline of research. Throughout his career, he tried to understand the foundations of learning systems, producing seminal papers that have influenced several generations of researchers. In later years, Ryszard lamented the fact that Machine Learning seems to be narrowing its field of study of to be just another branch of data analysis. He understood that learning is a fundamental ability of an autonomous system and that learning should be studied in that context. I hope that new researchers entering Machine Learning will continue to read Ryszard's works and take note of his insights.

References

1. Apt, K.R., Wallace, M.: Constraint Logic Programming Using Eclipse. Cambridge University Press, Cambridge (2007)
2. Benson, S., Nilsson, N.J.: Reacting, planning and learning in an autonomous agent. In: Furukawa, K., Michie, D., Muggleton, S. (eds.) Machine Intelligence, vol. 14. Oxford University Press, Oxford (1995)
3. Dietterich, T.G.: The MAXQ method for hierarchical reinforcement learning. In: Proceedings of the Fifteenth International Conference on Machine Learning, pp. 118–126. Morgan Kaufmann, San Francisco (1998)
4. Durrant-whyte, H., Bailey, T.: Simultaneous Localisation and Mapping (SLAM): Part I: The Essential Algorithms. Robotics and Automation Magazine 13, 99–110 (2006)
5. Dzeroski, S., De Raedt, L., Blockeel, H.: Relational reinforcement learning. In: Page, D.L. (ed.) ILP 1998. LNCS (LNAI), vol. 1446. Springer, Heidelberg (1998)
6. Ferrein, A., Lakemeyer, G.: Logic-based robot control in highly dynamic domains. Robotics and Autonomous Systems 56(11), 980–991 (2008)
7. Fikes, R., Nilsson, N.: STRIPS: a new approach to the application of theorem proving to problem solving. Artificial Intelligence 2, 189–208 (1971)
8. Getoor, L., Taskar, B. (eds.): Introduction to Statistical Relational Learning. MIT Press, Cambridge (2007)
9. Hengst, B.: Discovering Hierarchy in Reinforcement Learning with HEXQ. In: Sammut, C. (ed.) Proceedings of the International Conference on Machine Learning, Sydney (2002)
10. Hornby, G.S., Fujita, M., Takamura, S., Yamamoto, T., Hanagata, O.: Evolution of gaits with the sony quadruped robot. In: Proceedings of the Genetic and Evolutionary Computation Conference. Morgan Kaufmann Publishers Inc., San Francisco (1999)
11. Kim, M.S., Uther, W.: Automatic gait optimisation for quadruped robots. In: Australasian Conference on Robotics and Automation, Brisbane (2003)
12. Laird, J., Rosenbloom, P., Newell, A.: Soar: An Architecture for General Intelligence. Artificial Intelligence 33, 1–64 (1987)
13. Langley, P., Choi, D.: A unified cognitive architecture for physical agents. In: Proceedings of the Twenty-First National Conference on Artificial Intelligence. AAAI Press, Boston (2006)
14. Leonard, J.J., Durrant-whyte, H.F.: Simultaneous map building and localization for an autonomous mobilerobot. In: Intelligent Robots and Systems 1991. Intelligence for Mechanical Systems, Proceedings IROS 1991. IEEE/RSJ International Workshop, pp. 1442–1447 (1991)
15. Michalski, R.S.: LEARNING = INFERENCING + MEMORIZING: Basic Concepts of Inferential Theory of Learning and Their Use for Classifying Learning Processes. In: Chipman, S. (ed.) Cognitive Models of Learning (1992)
16. Michalski, R.S.: Inferential Theory of Learning as a Conceptual Basis for Multistrategy Learning. Machine Learning, Special Issue on Multistrategy Learning 11, 111–151 (1993)
17. Michalski, R.S.: Toward a Unified Theory of Learning: Multistrategy Task-adaptive Learning. In: Buchanan, B.G., Wikins, D.C. (eds.) Readings in Knowledge Acquisition and Learning: Automating the Construction and Improvement of Expert Systems. Morgan Kaufmann, San Mateo (1993)

18. Michalski, R.S.: Inferential Theory of Learning: Developing Foundations for Multistrategy Learning. In: Machine Learning: A Multistrategy Approach, vol. IV. Morgan Kaufmann Publishers, San Francisco (1994)
19. Michie, D., Chambers, R.A.: Boxes: An Experiment in Adaptive Control. In: Dale, E., Michie, D. (eds.) Machine Intelligence, vol. 2. Oliver and Boyd, Edinburgh (1968)
20. Mitchell, T.M., Keller, R.M., Kedar-Cabelli, S.T.: Explanation-Based Generalization: A Unifying View. Machine Learning 1(1), 47–80 (1986)
21. Kohl, N., Stone, P.: Policy gradient reinforcement learning for fast quadrupedal locomotion. In: Proceedings of the IEEE International Conference on Robotics and Automation, pp. 2619–2624 (2004)
22. Nardi, D., Riedmiller, M., Sammut, C., Santos-Victor, J. (eds.): RoboCup 2004. LNCS (LNAI), vol. 3276. Springer, Heidelberg (2005)
23. Ogino, M., Katoh, Y., Asada, M., Hosoda, K.: Vision-Based Reinforcement Learning for Humanoid Behavior Generation with Rhythmic Walking Parameters. In: Proceedings of 2003 IEEE/RSJ International Conference on Intelligent Robots and Systems, vol. 2, pp. 1665–1671 (2003)
24. Ogino, M., Hosoda, K., Asada, M.: Learning Energy Efficient Walking with Ballistic Walking. In: 2nd International Symposium on Adaptive Motion of Animals and Machines (2003)
25. Ramos, F.T., Durrant-Whyte, H.F., Upcroft, B.: Learning Articulated Motion Structures with Bayesian Networks. In: 8th International Conference on Information Fusion, Philadelphia (2005)
26. Ryan, M.R.K.: Using Abstract Models of Behaviours to Automatically Generate Reinforcement Learning Hierarchies. In: Sammut, C. (ed.) Proceedings of The 19th International Conference on Machine Learning, Sydney (2002)
27. Sammut, C.A., Hume, D.V.: Observation and Generalisation in a Simulated Robot World. In: Proceedings of the Fourth International Machine Learning Workshop, Los Altos, California (1987)
28. Sammut, C., Hengst, B.: The Evolution of a Robot Soccer Team. In: Jarvis, R.A., Zelinksky, A. (eds.) Robotics Research: The Tenth International Conference, pp. 517–529. Springer, Heidelberg (2003)
29. Stone, P., Sutton, R.S., Kuhlmann, G.: Reinforcement Learning for RoboCup-Soccer Keepaway. Adaptive Behavior 13(3), 165–188 (2005)
30. Thrun, S., Burgard, W., Fox, D.: Probabilistic Robotics. MIT Press, Cambridge (2005)
31. Sutton, R., Barto, A.: Reinforcement Learning: An Introduction. MIT Press, Cambridge (1998)
32. Watkins, C.J.C.H.: Learning with Delayed Rewards. Ph.D. Dissertation, Psychology Department, University of Cambridge, England (1989)
33. Wyeth, G., Kee, D., Yik, T.F.: Evolving a Locus Based Gait for a Humanoid Robot. In: International Conference on Robotics and Intelligent Systems (2003)
34. Yik, T.K.: Locomotion of Bipedal Humanoid Robots: Planning and Learning to Walk. Ph.D. Dissertation, School of Computer Science and Engineering, Universty of New South Wales (2008)
35. Zhou, C., Yue, P.K., Ni, J., Chan, S.-B.: Dynamically stable gait planning for a humanoid robot to climb sloping surface. In: Proceedings of IEEE Conference on Robotics, Automation and Mechatronics, pp. 341–346 (2004)

Part VI
Neural Networks and Other Nature Inspired Approaches

Quo Vadis? Reliable and Practical Rule Extraction from Neural Networks

Joachim Diederich[1], Alan B. Tickle[2], and Shlomo Geva[2]

[1] University of Queensland, Brisbane Q 4072, Australia
[2] Queensland University of Technology, Brisbane Q 4000, Australia

Abstract. Rule extraction from neural network algorithms have been investigated for two decades and there have been significant applications. Despite this level of success, rule extraction from neural network methods are generally not part of data mining tools, and a significant commercial breakthrough may still be some time away. This paper briefly reviews the state-of-the-art and points to some of the obstacles, namely a lack of evaluation techniques in experiments and larger benchmark data sets. A significant new development is the view that rule extraction from neural networks is an interactive process which actively involves the user. This leads to the application of assessment and evaluation techniques from information retrieval which may lead to a range of new methods.

Keywords: rule extraction, neural networks, support vector machines, information retrieval, interaction.

1 Introduction

Rule extraction from neural networks (RENN) aims at obtaining human-comprehensible rules from adaptive, learning based systems in order to explain "how" a decision was made or "why" a certain result was achieved. The field has seen a constant stream of publications over twenty years and there have been very significant applications of the technology (e.g. Baesens et al., 2003). The objective of this chapter is to review some of the recent developments and to point to opportunities for further research.

Andrews et al. (1995) describe the motivation behind rule extraction from neural networks. By way of introduction, these points are briefly reviewed. It is easily possible to find successful applications of RENN meeting the various objectives listed below.

1. User explanation: In symbolic artificial intelligence, the term *`explanation'* refers to an *explicit structure* which can be used for reasoning and learning as well as for the explanation of results to a user. Explanation is required in many application domains to achieve user acceptance, and hence, user explanation remains the primary objective of RENN.
2. Transparency: The provision of 'transparency' of the internal states of an adaptive system a close secondary objective. Transparency means that internal states of the neural network are accessible and can be interpreted unambiguously. Such transparency does allow the exploration of regions in generalization space which may lead to erroneous or sub-optimal decisions. This capability is mandatory if the neural network is to be accepted into *`safety-critical'* problem domains such as air traffic control, the operation of power plants, medical applications *etc*.

J. Koronacki et al. (Eds.): Advances in Machine Learning I, SCI 262, pp. 479–490.
springerlink.com © Springer-Verlag Berlin Heidelberg 2010

3. Software verification: If neural networks are embedded in larger software systems which need to be verified, then clearly this requirement must be extended to adaptive components. Rule extraction algorithms provide mechanisms for either partially or completely *`decompiling'* a neural network. This is about half way to software verification because it allows for a comparison between the extracted rules and the specification.
4. Improving generalization: If a limited or unrepresentative data set has been used in the neural network training process, it is difficult to determine *if and when generalization fails for specific cases* even with evaluation methods such as cross-validation. By expressing learned knowledge as a set of rules, an experienced user can anticipate or predict a *generalization failure* (Andrews et al. 1995).
5. Data exploration: This has been one of the original objectives for rule extraction from neural networks and is essential for data mining and knowledge discovery. As Craven and Shavlik (1994) write *`a (learning) system may discover salient features in the input data whose importance was not previously recognized.'*

1.1 Translucency

The notion of "translucency" describes the degree to which the internal representation of the neural network is accessible to the rule extraction technique (Andrews et al. 1995; Tickle et al. 1998). More broadly, a taxonomy for rule extraction from neural networks has been introduced (the ADT taxonomy in short: Andrews et al. 1995; Tickle et al. 1998) which includes five evaluation criteria: translucency, rule quality, expressive power, portability and algorithmic complexity.

At one end of the translucency spectrum are those rule extraction techniques which view the underlying neural network at the maximum level of granularity i.e. as a set of discrete hidden and output units. Craven and Shavlik (1994) labelled these techniques as "decompositional". The basic strategy of such decompositional techniques is to extract rules at the level of each individual hidden and output unit within the trained neural network.

In contrast to decompositional methods, the strategy of pedagogical approaches is to view the trained neural network at the minimum possible level of granularity i.e. as a single entity or alternatively as a "black box". The focus is on finding rules that map the neural network inputs to outputs (Tickle et al. 1998). In addition to these two main categories, Andrews et al. (1995) also proposed a third category labeled as "eclectic" to accommodate those rule extraction techniques which include elements of both the decompositional and pedagogical approaches.

1.2 Rule Quality

Rule extraction from neural networks adopted criteria for the quality of the extracted rules. The set of criteria for evaluating rule quality includes (Andrews et al. 1995): accuracy, fidelity, consistency and the comprehensibility of the extracted rules. A brief summary is given here while the core issues are discussed below.

A rule set is considered to be accurate if it can correctly classify a set of previously unseen examples from the problem domain (Tickle et al. 1998). Similarly a rule set is considered to display fidelity if it can mimic the behavior of neural network from which it was extracted by capturing all of the information represented in the neural network. An extracted rule set is deemed to be consistent if, under differing training sessions, the neural network generates rule sets which produce the same classifications of unseen examples. Finally, the comprehensibility of a rule set is determined by measuring the size of the rule set (in terms of the number of rules) and the number of antecedents per rule (Tickle et al. 1998).

A rule set extracted from a neural network can have a high level of fidelity if it simulates the generalization behavior of the neural network. Due to the differing *representational bias* (rules in any format are different to hidden units), high fidelity is both an objective and achievement. A rule set may also have a high level of accuracy if it generalizes well; however, it is difficult to achieve high levels of both accuracy and fidelity for the same rule set. This problem has been outlined by Tickle (1998) and will be explored in more detail below.

1.3 Rule Extraction and Function Approximation Networks

An area of significant progress over the last decade has been rule extraction from function approximation neural networks (Setiono et al., 2001, 2002). Neural networks are especially useful for regression since they do not require any prior knowledge about data distribution. Hence, rule extraction algorithms should be available for function approximation networks as well.

The method by Setiono et al. (2001) has been a guide for many approaches that followed: a replacement of the prediction of the trained neural network by a set of multiple linear regression equations. The approach works with a network with a single hidden layer and one linear output unit. The reason for the restriction on the number of hidden layers is to reduce the errors in approximating the hidden unit activation function (the hyperbolic tangent function in this case) by a piece-wise linear function as errors propagate from the input layer to the output layer through the hidden layers (Setiono et al., 2001).

Since many hidden units lead to many rules, Setiono et al. (2001) reduce the number of rules by pruning redundant units. The continuous activation function of each hidden unit is approximated by a 3-piece linear function. Experiments using real-world data sets have been performed by (Setiono et al., 2001) and the results show that the accuracy of the approach is better than those of existing methods based on decision trees and multiple linear regression.

2 Rule Extraction from Neural Networks and Support Vector Machines: The Same Problem?

Most current studies on rule extraction from SVMs focus on decompositional extraction; however, learning-based approaches are also available (Diederich 2008, Barakat and Diederich 2005). The idea is simple: learn what the SVM has learned. For this purpose a data set is divided in two or more parts. The first set is used to train the SVM. The second set does not include targets, the inputs are presented to the SVM and the outputs are obtained from the SVM. Inputs and outputs combined represent a new data set which is used in a second learning episode by use of a machine learning system that produces rules as output.

In the above scenario, pedagogical rule extraction from SVMs is trivial, in particular if the data set is low dimensional. Support vector machines have been designed to process sparse, high dimensional input data. Typical examples are text, speech and image classification. Yet most of the studies available on rule extraction from support vector machines use benchmark data sets that include a limited number of features only. As a result, SVMs are not being used in their core application area and hence rule extraction results are not very meaningful.

If high dimensional data is used, rule extraction from SVMs experiences a problem which can be illustrated by use of an example: text classification. SVMs can achieve good performance with very simple text representation formats such as the "bag-of words" (BOW) technique. BOW uses a document-term matrix such that rows represent documents and columns represent terms (e.g. words).

A disadvantage of the BOW input representation is that after successful classification, it may not be obvious *what* has been learned. In authorship attribution (a classification problem, documents are categorized into classes representing individual authors), an author may have a preference for certain topics and as a result, *an SVM trained on an authorship identification problem may in reality perform topic detection*. In principle, rule extraction techniques are designed to address this very problem: to reveal what has been learned.

While a combination of words constitutes meaning in a natural language, the BOW input representation and hence classification by the SVM *and* the extracted rule set is based on words in isolation. The antecedents in a rule include individual words *completely out of context*. As the set of antecedents includes completely unrelated words, *human or semantic comprehensibility* is low. It can easily be seen that the same problem occurs in image classification as well, if sparse, high dimensional data is used and (micro-) features do not relate to each other, at least not in the eyes of the user.

3 The ADT Taxonomy for Rule Extraction from Neural Networks Reconsidered

The ADT taxonomy has been used widely and this includes rule accuracy and fidelity. These two evaluation criteria, and their relationship, are discussed below.

As indicated previously, a key facet of the ADT taxonomy (Andrews et. al. 1995; Tickle et. al. 1998) is the notion of the "quality" of the rules extracted from the trained ANN. To this end, Andrews et. al. (1995) suggested a set of four general measures of rule quality viz.:

a) rule accuracy (i.e. the extent to which the rule set is able to classify a set of previously unseen examples from the problem domain correctly);
b) rule fidelity (i.e. the extent to which the rule set mimics the behavior of the ANN from which it was extracted);
c) rule consistency (i.e. the extent to which, under differing training sessions, the ANN generates rule sets which produce the same classifications of unseen examples); and
d) rule comprehensibility (e.g. measuring the size of the rule set in terms of the number of rules and the number of antecedents per rule).

For the purposes of this discussion, the intent is to focus on the first two of these measures of rule quality viz. rule accuracy and rule fidelity. Rule fidelity will be discussed first. Essentially, rule fidelity can be viewed as a characteristic feature of the rule extraction from neural networks problem and, as such has attracted far less attention in the broader discussion of the performance of learning algorithms than the more commonly used metric of "accuracy" (and which will be discussed in more detail in a moment). Moreover, and with one important exception, the definition of the term "rule fidelity" as used in the ADT taxonomy is basically consistent with that adopted by later researchers.

For example, Baesens et. al. (2003) defined rule fidelity as the percentage of observations that the extraction algorithm classifies the same way as the ANN. The exception was Diederich (2008a). He introduced a new measure for rule fidelity that he labeled as ROC fidelity. (ROC or Receiver Operating Characteristics analysis is drawn from the area of signal detection theory.) In essence this means that the ANN and the rule set must agree on a minimum number of points on the ROC curve.

Apart from the measurement issue, an aspect of rule fidelity to which Tickle et. al. (1998) drew attention is the possibility of a significant difference emerging between the totality of knowledge embedded within the trained ANN (and the concomitant set of rules that captured this knowledge) and the set of rules that approximated the behavior of the network in classifying the training/ test set (Tickle et. al. 1998). This led the authors to suggest the additional step of validating any rules extracted either by using further test data from the problem domain (and even synthetically-generated test data) or by having the rules scrutinized by a domain expert. This latter idea of having the extracted knowledge/ rules scrutinized by a domain expert was considered to be an important inclusion. Basically it offered a mechanism for identifying and pruning irrelevant knowledge/ rules as well as identifying if the extracted knowledge/ rules afforded new insights into the problem domain.

Garcez et. al. (2001) crystallized these around the concept of "soundness". In their own words, "Soundness is reflected in the very high fidelity achieved in the applications, by assuring that any rule extracted is actually encoded in the network, even if such a rule does not comply with the network's test set. The extraction system is, therefore, bound to produce a set of rules that tries to mimic the network, regardless of the network's performance in the training and test sets."

As previously noted, the concept of rule fidelity is tightly coupled with the rule extraction from neural networks problem. However estimating the accuracy of a set of rules extracted is a subset of the more general problem of measuring and comparing the performance of learning algorithms and, in particular, the performance of supervised classification learning algorithms. As was the case with rule fidelity, the authors of the ADT taxonomy did not prescribe a framework e.g. an experimental design, a set of benchmark data sets, or a statistical testing regime, for determining rule accuracy. Similarly they did not suggest a format for presenting the results e.g. see Little (2008) for a discussion of the widely-used APA format.

However, contemporary with the publication of the ADT methodology, Dietterich (1998) anchored the issue of comparing the performance of supervised classification learning algorithms (and hence the issue of quantitatively estimating "accuracy") firmly in the domain of statistical hypothesis testing. In particular Dietterich (1998) benchmarked five "approximate" statistical tests viz. (1) McNemar's test; (2) a test for the difference of two proportions; (3) the re-sampled *t*-test; (4) the cross-validated *t*-test; and (5) (what the author described as) a new test called the 5x2cv test i.e. 5 replications of 2-fold cross-validation. He focused on standard statistical measures such as Type I error (false-positive) and the "power" of the test, and made a number of important findings as to the efficacy of various tests.

This did not settle the issue. If anything, (and with the benefit of hindsight) the Dietterich (1998) paper arguably seems to mark the starting point of a debate about how to measure the performance of learning algorithms that has continued to the present. For example, whilst accepting the solid statistical foundations on which Dietterich's work was based, Provost et. al. (1998) cast doubts on the utility of classification accuracy as the sole dimension for comparing classifiers. Tellingly, they argued that "comparing accuracies on our benchmark data sets says little, if anything, about classifier performance on real-world tasks" and, in particular, where there are significant differences between costs for Type I (false positive) and Type II (false negative) errors. Provost et. al. (1998) also raised the issue of the selection of benchmark data sets in comparing performance. Their particular concern was that the class distribution in a benchmark data set may differ significantly from the corresponding natural distribution. This, they argued, weakened the credibility of a classifier calibrated using the benchmark data set when applied to the "real world" problem domain.

Instead Provost et. al. (1998) argued the alternative case for using ROC (Receiver Operating Characteristics) analysis. In their view, because ROC curves are "independent of class distributions or error costs" this would enable "stronger and more general conclusions to be made". They then proceeded to outline a method for applying ROC techniques to the problem of comparing machine learning classifiers.

The introduction of ROC as an alternative to accuracy as a measure of performance in machine learning opened a new dimension in the debate. Ling et. al. (2003) then moved the agenda further. They advocated using AUC (i.e. area under the ROC curve) to replace accuracy as a comparative measure for evaluating learning algorithms. Ling et. al. (2003) developed their argument in favor of AUC over accuracy by examining the issue of performance measures from (at least what they saw as) two new dimensions. They labeled these dimensions as "consistency" and "discriminancy". Ling et. al. (2003) were able to show the (statistical) consistency of AUC and accuracy i.e. if, on the basis of AUC measurement, a learning algorithm A is deemed to be better than another learning algorithm B, then a comparison of the performance of A and B on the basis of measuring their accuracy will (consistently) yield the same result. However they also showed that AUC is a more discriminating measure of performance than accuracy, i.e. they demonstrated cases where, on the basis of AUC measurement a difference could be detected between the learning algorithms A and B that was not detectable by using accuracy as the performance measure.

Later, Huang & Ling (2005, 2007) extended this idea further. They suggested a two-level measure (labeled as AUC:acc) that combined both AUC and accuracy. Basically, if the AUC value for the test set is the same when comparing two learning algorithms, then the process is to compare the accuracy to decide which one is better. This, they argued, gives rise to a finer evaluation measure in comparing learning algorithms than using AUC or accuracy alone.

By 2008, the ICML had conducted its third successive workshop devoted to "Evaluation Methods for Machine Learning" without a clear resolution of the underlying issues. At a general level, Demsar (2008) echoed the concerns that had started to appear in social and scientific research domains about the misuse of hypothesis testing in reporting results (Cohen 1994). He also cast further doubt on the utility of hypothesis testing as a basis for comparing machine-learning algorithms and was highly critical of the manner in which results were being presented for publication. Law (2008) went further. She argued that, for a class of problems that included computer vision, natural language processing and music information processing, "accuracy" was not only "unsuitable" but even an "incorrect" measure to evaluate the quality of a learning algorithm in these domains.

Notwithstanding this lack of resolution (and even the mood of pessimism that now seems to pervade the discussion) there is nonetheless a well-defined body of knowledge that would enable researchers in the rule extraction from neural networks area to evaluate their techniques in terms of rule quality in a rigorous and scientific manner (and specifically rule fidelity and rule accuracy – or equivalent surrogates for accuracy such as AUC). In particular, Caruana & Niculescu-Mizil (2006) showed how to conduct such an evaluation using a comprehensive set of eight performance metrics including accuracy, F-score, Lift, ROC Area, average precision, precision/recall break-even point, squared error, and cross-entropy. The extent to which these ideas have extended to the evaluation of techniques within the rule extraction from neural networks community is discussed in the next section.

3.1 Some Comments on the Reporting of Rule Accuracy and Fidelity

The preceding section examined the key "quality" notions of rule accuracy and rule fidelity as they appeared in the original ADT taxonomy and reviewed some of the important new ideas that have influenced subsequent thinking in these areas. Also, as discussed elsewhere in this chapter, the rule extraction from neural networks area has witnessed continuous innovation in the development and application of algorithms and techniques that form the heart of the rule extraction from neural networks process. Hence, it would seem to be instructive to review the advances in rule extraction from neural networks techniques to see how any improvements in performance (and specifically in terms of rule accuracy and rule quality) are actually being measured and reported.

Unfortunately the results of such a review are disappointing. By and large, the vast majority publications seem to follow a similar motif:

1) Synthesize a new technique or enhance an existing technique for rule extraction;
2) Apply the technique either to a data set drawn from a specific problem or to a (usually small) subset of standard problem data sets that, typically, is drawn from the UCI repository (e.g. Iris and Monk);
3) Measure the "accuracy" of the rule extraction technique as the percentage of correctly classified examples in the problem test set;
4) In a similar way, measure the accuracy of the ANN from which the knowledge has been extracted;
5) (Optionally) measure the accuracy obtained from applying a public domain classifier (and notably C4.5 (Quinlan 1993)) against the same data;
6) Perform a straight comparison of the numerical values of the accuracy and highlight any situation where the results for the rule extraction technique are better than either of the other two results.

There are, of course, exceptions. For example Setiono & Thong (2001) and Setionio et. al. (2002) approached the problem in a manner consistent with accepted experimental practice (e.g. Maxwell & Delaney 2004). But, again, these are exceptions. Perhaps, to some extent, this state of affairs could have been anticipated. Arguably, as is the case in the broader machine learning community (Demsar 2008), the focus of effort in the rule extraction from neural networks arena is simply not on experimental design. It is on generating new algorithms – even if they are only slight variations of existing ones. Our review of the literature seems to bear this out. So in essence, one can readily if not reluctantly draw the conclusion that the body of knowledge on how to compare the performance of machine learning algorithms seems to have had little impact on the rule extraction from neural networks research community.

4 Rule Extraction as Interaction

In any real world setting, user explanation is frequently based on an interaction between people (e.g. a teacher and a student) or a machine and a human (e.g. in tutorial or help systems). Cawsey (1993) uses a very simple definition of explanation: "In general an explanation is something which makes some piece of knowledge clear to the hearer. … The explanation is complete when the hearer is satisfied with the reply and understands the piece of knowledge" (Cawsey, 1993, p.1). Hence, explanation is based on a dialogue. As part of this dialogue, explanatory targets may change and may be refined.

The idea of rule extraction as interaction is gaining traction, in particular in the field of data mining. Yamamoto et al. (2008) investigate associative rule mining and note that simply setting minimum support and confidence parameters is not sufficient: "... the process poses many practical challenges to the human miner. First, tuning the input parameter values is difficult, usually requiring a tiresome trial-and-error strategy before reasonable values are attained. Moreover, it is intrinsically difficult to understand the knowledge embedded in the resulting rules, particularly when they involve many items. And, quite often, thousands of rules are extracted and the complete set is hardly treatable by a human" (Yamamoto et al., 2008, p.1212).

Yamamoto et al. (2008) have built a graphical interface to support the interaction between the user and the rule extraction process: "A visualization of the space of k-item sets ... allows the miner to guide rule extraction by exploring the space of item sets. Miners can discard item sets considered not relevant and define clusters of related item sets to perform rule filtering, so that uninteresting rules are removed while preserving item set coverage in the resulting rule set" (Yamamoto et al. 2008, p.1212).

In a similar fashion, it is possible to retrieve more than one explanation (i.e. rule set) from a given neural network. In this case, it is often necessary to select the best explanation. There are two main reasons why it is possible to generate multiple explanations from a neural network: (1) Rule extraction methods include parameters and the selection of certain values results in different rule sets. (2) Rules can be expressed in different ways, e.g. a small rule set with many antecedents can be re-written as a rule set with many, simple rules. While it is generally acknowledged that a small rule set and a limited set of antecedents provides best explanation capability, it is far from obvious that this is always the case. The *information need* of the user has to be taken into consideration and the user may *interact* with the neural network by use of rule extraction.

A user may have a need for multiple explanations because the objective is to explore the generalization space of the neural network. Several rule sets, if considered in turn, may offer best transparency. In another scenario, a user may be interested in the single-best explanation in the form of a few simple rules. In addition, the user may be interested in exploring different parameter sets which lead to different learning results and consequently to different rule sets. It is difficult to consider rule quality criteria without reference to the information need of the user.

The concept of *information need* is central to the discipline of information retrieval. Traditionally in information retrieval, the comparison of systems is performed in two phases – assessment and evaluation. In the assessment phase, in stark contrast with traditional rule extraction evaluation, end-users determine the relevance of pooled results that are returned by a set of retrieval techniques (e.g. Internet search engines). Evaluation of relative performance is then a matter of simple calculations and numerous metrics are used, but all are based on the results of manual assessment. The performance of an information retrieval system is traditionally evaluated by use of "precision" and "recall". Precision is the probability that a document predicted to be genuine truly belongs to this class. Recall is the probability that a genuine document is classified into this class. Another common measurement is the Precision@N which is the precision considering only the first N returned results. It is obviously desirable to have high recall and precision simultaneously, but this is difficult to achieve in information retrieval. A trade-off exists between recall and precision. For instance, a typical user of a search engine would prefer high precision on the first screen of results. A medical practitioner searching a medical document collection may prefer high recall and would be prepared to consider many pages in order to get all relevant material.

The observation that multiple explanations can be extracted from a neural network leads to the application of information retrieval concepts to rule extraction. In the context of rule extraction from neural networks, high precision means that the rule sets extracted are relevant to the user, i.e. match his or her information need. Recall relates to the question "how many of the valid explanations that can be extracted from the neural network are indeed being generated?" Is it possible to generate all possible valid explanations by way of rule extraction that match the information need of the user? In many instances, the number of possible rules can be very large, and a user may not be inclined to consider thousands of rules. Precision@N may be useful, measuring the precision of the first N rules returned by the system.

The ideas outlined above may lead to the application of additional benchmark problems and performance measurements that are commonly used in information retrieval. What is important is that *the view that a single explanation is to be extracted from a neural network is a simplification, and that ultimately end-users must be part of the evaluation*. It may be desirable to extend the notion of "rule quality" to include assessments of multiple explanations in relation to the information need of a user.

5 Rule Extraction from Neural Networks in the Broader Data Mining Area

As may be appreciated from the discussion so far, the rule extraction from neural networks problem is an active and thriving research topic. It has produced and continues to produce new and innovative algorithms and techniques. Also, at least on the basis of the results reported by the individual researchers, rule extraction from neural networks techniques are competitive with other machine learning methods in terms of performance. Importantly, they have the potential to add value in terms of the increased insight and understanding of the problem domain.

However, in spite of this activity and purported success, there is little evidence that rule extraction from neural networks techniques have been accepted into the broader domain of data mining. For example, the seminal paper by Caruana and Niculescu-Mizil (2006) did not include a rule extraction from neural networks technique to complement either the SVM or ANN algorithms they included in their benchmark. (The set of learning algorithms in the benchmark included SVMs, ANNs, Logistic Regression, Naïve Bayes, K-nearest neighbour, Random Forests, Decision Trees, Bagged Trees, and Boosted Stumps). Moreover a review of the eighty data-mining suites listed on the KDD website did not reveal one that appeared to include a rule extraction from neural networks technique as part of its toolbox.

It is possible to speculate as to the reasons for this situation. For example, despite the number of rule extraction from neural networks techniques available, there is no identifiable "off-the-shelf" solution. In most cases the rule extraction from neural networks solution must still be individually crafted for the particular problem domain. Importantly, the effort to do so is non-trivial. Also, to the best of our knowledge, there has been no benchmark of rule extraction from neural networks techniques similar to that conducted by Caruana and Niculescu-Mizil (2006) that guides a potential user as to which technique to apply in a particular problem domain. Moreover, and notwithstanding the published results, the actual benefits of using rule extraction from neural networks as compared with applying a learning technique that produces "rules" directly are not widely appreciated. If rule extraction from neural networks techniques are to gravitate from the periphery to the mainstream of data-mining, then these issues need to be addressed.

What can be done to increase the practical relevance of rule extraction from neural networks? Since the early 90's when rule extraction from neural networks emerged, the kind and magnitude of problems that can be solved with neural networks have changed significantly. Despite the convenience afforded by most of the problems that have been used to test rule extraction in the past, most of these are no longer interesting benchmark problems. The evaluation of rule extraction from neural networks should now be based on large scale problems. Bioinformatics, network security and information retrieval provide fertile grounds for rule extraction. Problem of interest would involve millions of pattern vectors (or documents) and domains where there are many underlying rules to be discovered.

Reported evaluations should include the traditional intrinsic measures of performance. However, if rule extraction from neural networks is to be successful, it is very important to include domain experts in the evaluation by manually assessing ranked lists of rules. The data sets should be made public. In order to facilitate future comparisons, all rules that are validated by domain experts should also be made public.

The size of the data sets, the complexity of the underlying knowledge embedded in the data, and the complexity of the networks that are required to capture it, should inform users on the scalability of rule extraction methods.

References

Andrews, R., Diederich, J., Tickle, A.B.: A Survey and Critique of Techniques For Extracting Rules From Trained Artificial Neural Networks. Knowledge Based Systems 8, 373–389 (1995)

Baesens, B., Setiono, R., Mues, C., Vanthienen, J.: Using neural network rule extraction and decision tables as management science tools for credit-risk evaluation. Management Science 49(3), 312–329 (2003)

Barakat, N., Diederich, J.: Eclectic rule extraction from support vector machines. International Journal of Computational Intelligence 2(1), 59–62 (2005)

Caruana, R., Niculescu-Mizil, A.: An Empirical Comparison of Supervised Learning Algorithms. In: Proceedings of the 23rd International Conference on Machine Learning, Pittsburgh (2006)

Cohen, J.: The earth is round (p <.05). American Psychologist 49, 997–1003 (1994)

Cawsey, A.: Explanation and Interaction. The Computer Generation of Ex-planatory Dialogues. MIT Press, Cambridge (1993)

Craven, M., Shavlik, J.: Using sampling and queries to extract rules from trained neural networks. In: Proceedings of the 11th International Conference on Machine Learning, pp. 37–45 (1994)

Davis, R., Buchanan, B.G., Shortliffe, E.: Production rules as a representation for a knowledge-based consultation program. Artificial Intelligence 8(1), 15–45 (1977)

Demsar, J.: On the Appropriateness of Statistical Tests in Machine Learning. In: The 3rd workshop on Evaluation Methods for Machine Learning In conjunc-tion with ICML 2008, Helsinki, Finland (2008)

d'Avila Garcez, A.S., Broda, K., Gabbay, D.M.: Symbolic knowledge extrac-tion from trained neural networks: A sound approach. Artificial Intelligence 125, 155–207 (2001)

Diederich, J. (ed.): Rule Extraction from Support Vector Machines. Studies in Computational Intelligence, vol. 80. Springer, Heidelberg (2008)

Diederich, J.: Rule Extraction from Support Vector Machines – An Introduction. In: Diederich, J. (ed.) Rule-Extraction from Support Vector Machines. Springer, Heidelberg (2008)

Dietterich, T.G.: Approximate Statistical Tests for Comparing Supervised Classification Learning Algorithms. Neural Computation 10(7), 1895–1923 (1998)

Fung, G., Sandilya, S., Rao, R.: Rule extraction for linear support vector machines. In: Proceedings of the Eleventh ACM SIGKDD International Conference on Knowledge Discovery and Data Mining (2005)

Hayward, R., Nayak, R., Diederich, J.: Using Predicates to Explain Networks. In: ECAI 2000 Workshop: Foundations of Connectionist-Symbolic Integration: Representation, Paradigms and Algorithms, Berlin, Germany (2000)

Huang, J., Ling, C.X.: Using AUC and Accuracy in Evaluating Learning Algorithms. IEEE Transactions on Knowledge and Data Engineering 17(3), 299–310 (2005)

Huang, J., Ling, C.X.: Constructing New and Better Evaluation Measures for Machine Learning. In: The Twentieth International Joint Conference on Artificial Intelligence, pp. 859–864 (2007)

KDD nuggets, `http://www.kdnuggets.com/software/suites.html#N`

Law, E.: The Problem of Accuracy as an Evaluation Criterion. In: The 3rd workshop on Evaluation Methods for Machine Learning, in conjunction with ICML 2008, Helsinki, Finland (2008)

Ling, C.X., Huang, J., Zhang, H.: AUC: A Better Measure than Accuracy in Comparing Learning Algorithms. In: Proceedings 16th Conference of the Canadian Society for Computational Studies of Intelligence, AI 2003, Halifax, Canada, June 11-13. LNCS (LNAI), pp. 329–341. Springer, Heidelberg (2003)

Little, L.: Reporting Results of Common Statistical Tests in APA Format (2008), http://depts.washington.edu/psywc/handouts/pdf/stats.pdf (October 11, 2008)

Maxwell, S., Delaney, H.: Designing experiments and analyzing data: a model comparison perspective, 2nd edn. Lawrence Erlbaum Associates Inc., New Jersey (2004)

Moore, J.D.: A reactive approach to explanation in expert and advice-giving systems. PhD thesis. Los Angeles, University of California (1989)

Moore, J.D., Swartout, W.R.: A Reactive Approach to Explanation. In: IJCAI 1989 International Joint Conference on Artificial Intelligence, pp. 1504–1510 (1989)

Provost, F., Fawcett, T., Kohavi, R.: The Case Against Accuracy Estimation for Comparing Induction Algorithms. In: Proceedings of the Fifteenth International Conference on Machine Learning, pp. 445–453 (1998)

Quinlan, R.: C4.5: Programs for machine learning. Morgan Kaufman, San Mateo (1993)

Setiono, R., Thong, J.Y.L.: An Approach to Generate Rules from Neural Net-works for Regression Problems (2001)

Setiono, R., Leow, W.K., Zurada, J.M.: Extraction of rules from artificial neural networks for nonlinear regression (2002)

Tickle, A., Andrews, R., Golea, M., Diederich, J.: The truth will come to light: directions and challenges in extracting the knowledge embedded within trained artificial neural network. IEEE Transactions on Neural Networks 9(6), 1057–1068 (1998)

Witten, I.H., Frank, E.: Data Mining: Practical Machine Learning Tools and Techniques with Java Implementations. Paperback 162–164 (1999)

Yamamoto, C.H., Oliveira, M.C.F., Rezende, S.O.: Including the user in the knowledge discovery look: interactive itemset-driven rule extraction. In: SAC 2008: Proceedings of the 2008 ACM symposium on Applied computing (March 2008)

Zhang, Y., Su, H.-Y., Jia, T., Chu, J.: Rule extraction from trained support vector machines. In: Ho, T.-B., Cheung, D., Liu, H. (eds.) PAKDD 2005. LNCS (LNAI), vol. 3518, pp. 61–70. Springer, Heidelberg (2005)

Learning and Evolution of Autonomous Adaptive Agents

Vladimir G. Red'ko[1] and Danil V. Prokhorov[2]

[1] Center of Optical Neural Technologies, Scientific Research Institute for System Analysis, Russian Academy of Science,
Vavilova Str., 44/2, Moscow, 119333, Russia
vgredko@gmail.com
[2] Toyota Technical Center, Ann Arbor, MI, USA
dvprokhorov@gmail.com

Abstract. We study a model of evolving populations of self-learning agents and analyze the interaction between learning and evolution. We consider agent-brokers that predict stock price changes and use these predictions for selecting actions. Each agent is equipped with a neural network adaptive critic design for behavioral adaptation. We discuss three cases in which either learning, or evolution, or both, are active in our model. We show that the Baldwin effect can be observed in our model, viz., originally acquired adaptive policy of best agent-brokers becomes inherited over the course of the evolution. Additionally, we analyze influence of neural network structure of adaptive critic design on learning processes.

Keywords: autonomous agents, reinforcement learning, neural networks, adaptive critic designs, evolutionary algorithm.

1 Introduction

The current paper overviews and generalizes our previous work on modeling learning and evolution of autonomous adaptive agents [1, 2]. The control system of the agent ensures autonomous, self-learning agent behavior. The agent control system is based on a neural network adaptive critic design (ACD). The ACD includes two neural networks (NNs), model and critic. The model predicts the state of the environment for the next time step, and the critic is used to select actions on the basis of model predictions. These NNs can be optimized by both learning and evolution. Learning uses reinforcement learning method [3]. Darwinian type of evolution includes mutations of NN synaptic weights and selection of agents that have the best control systems.

The model of autonomous adaptive agents is developed in the framework of the animat approach to artificial intelligence that was proposed in the early 1990s [4, 5]. This research methodology implies understanding intelligence through simulation of artificial animals (*animats*) in progressively more challenging environments. We design the model for simple example of autonomous adaptive agent-brokers that can buy and sell stocks in order to increase their capital. The model is investigated by means of computer simulations.

J. Koronacki et al. (Eds.): Advances in Machine Learning I, SCI 262, pp. 491–500.
springerlink.com © Springer-Verlag Berlin Heidelberg 2010

We analyze specifically the interaction between learning and evolution. It should be noted that the rather non-trivial Baldwin effect [6, 7] can be observed in our model, viz., originally acquired adaptive policy of best agent-brokers becomes inherited over the course of the evolution. In particular, in some simulations acquired features that at initial generations are obtained by means of learning become inherited at fifth generation. This means that NN synaptic weights that are initially adjusted by reinforcement learning are then rediscovered during mutations and selections. Though the evolutionary processes are of Darwinian type, these processes have certain Lamarckian properties.

The paper is organized as follows. Section 2 includes the description of the model. The simulation results are described in Section 3; this Section includes analysis of the interaction between learning and evolution and comparison of two variants of neural network structure of adaptive critic design: with one hidden layer and with two hidden layers. Section 4 concludes the paper.

2 Description of the Model

2.1 Agent Task

Inspired by [8], we implement the adaptive agent-broker which predicts future changes of the stock price and tries to increase its wealth by buying and selling stocks. The agent has its resource distributed into cash and stocks. The sum of these is the net capital of the agent $C(t)$. The agent decision is to modify the variable $u(t)$, which is the fraction of the agent's capital that is currently invested in stocks. The environment is determined by the time series $X(t)$, $t = 1,2,\ldots$, where $X(t)$ is the stock price at the moment t. The goal of the agent is to increase its capital $C(t)$ by changing the value $u(t)$. The capital dynamics is described by

$$\mathrm{C}(t+1) = \mathrm{C}(t)\ \{1 + u(t+1)\ \Delta X(t+1)\ /\ X(t)\}\times[1 - J\ |u(t+1) - u(t)|]\ , \tag{1}$$

where $\Delta X(t+1) = X(t+1) - X(t)$ is the current change of the stock price, and J ($1 > J \geq 0$) is a parameter that takes into account expenses of the agent when buying or selling stocks. The factor in the braces corresponds to the change of the capital as the result of stock price changes. The factor in the square brackets reflects the transaction cost. Following [9], we use the logarithmic scale for the agent resource, i.e., $R(t) = \log C(t)$. The current agent reward $r(t)$ is defined by the expression: $r(t) = R(t+1) - R(t)$:

$$r(t) = \log\ \{1 + u(t+1)\ \Delta X(t+1)\ /\ X(t)\} + \log\ [1 - J\ |u(t+1) - u(t)|]\ . \tag{2}$$

For simplicity we assume that the variable $u(t)$ takes only two values, $u(t) = 0$ (all in cash) or $u(t) = 1$ (all in stock).

2.2 Agent Learning

The agent control system is a simplified ACD. Our adaptive critic scheme consists of two NNs: a model and a critic (see Fig. 1).

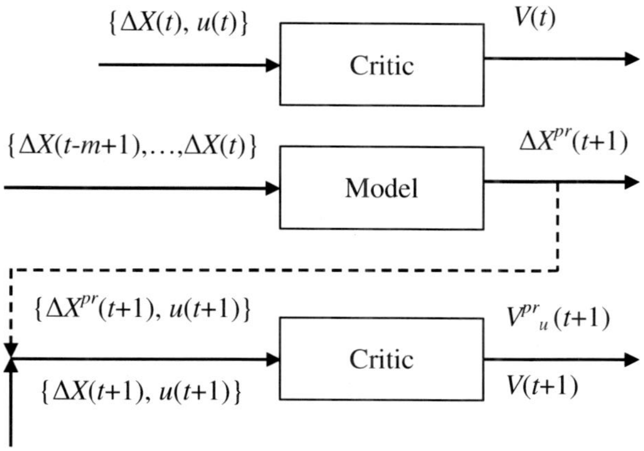

Fig. 1. The scheme of the considered adaptive critic design (ACD). The ACD consists of two neural networks (NNs): a model and a critic. The model predicts changes of the time series. The critic (the same NN is shown in two consecutive moments) forms the state value function for the current state $\mathbf{S}(t) = \{\Delta X(t), u(t)\}$, the next state $\mathbf{S}(t+1) = \{\Delta X(t+1), u(t+1)\}$, and its predictions $\mathbf{S}^{pr}_u(t+1) = \{\Delta X^{pr}(t+1), u(t+1)\}$ for two possible actions, $u(t+1) = 0$ or $u(t+1)= 1$.

The goal of the adaptive critic is to stochastically maximize the utility function $U(t)$ [3]:

$$U(t) = \sum_{j=0}^{\infty} \gamma^j r(t+j), \qquad t = 1,2,\ldots, \tag{3}$$

where $r(t)$ is an instantaneous reward obtained by the agent and γ is the discount factor ($0 < \gamma < 1$). Making the realistic assumption $|\Delta X(t+1)| << X(t)$, we specify that the ACD state $\mathbf{S}(t)$ at moment t depends on two values, $\Delta X(t)$ and $u(t)$: $\mathbf{S}(t) = \{\Delta X(t), u(t)\}$.

The role of the model is to predict changes of the stock time series. The model output $\Delta X^{pr}(t+1)$ is based on m previous values of ΔX: $\Delta X(t-m+1),\ldots,\Delta X(t)$, which are used as the model inputs. The model is implemented as a multilayer perceptron (MLP) with one hidden layer of tanh nodes and linear output. The operation of this neural network is described by the following expressions:

$$\mathbf{x}^{\mathrm{M}} = \{\Delta X(t-m+1),\ldots,\Delta X(t)\},\ y^M_j = \tanh\left(\textstyle\sum_i w^M_{ij} x^M_i\right),\ \Delta X^{pr}(t+1) = \textstyle\sum_j v^M_j y^M_j, \tag{4}$$

where $\mathbf{x}^{\mathrm{M}}$ is the neural network input vector, $\mathbf{y}^{\mathrm{M}}$ is the vector of outputs of hidden layer neurons, w^M_{ij} and v^M_j are neuron synaptic weights.

The critic is intended to estimate the state value function $V(\mathbf{S})$ (estimate of U in (3)) for the current state $\mathbf{S}(t) = \{\Delta X(t), u(t)\}$, the next state $\mathbf{S}(t+1) = \{\Delta X(t+1), u(t+1)\}$, and its predictions $\mathbf{S}^{pr}_u(t+1) = \{\Delta X^{pr}(t+1), u(t+1)\}$ for two possible actions, $u(t+1) = 0$ or $u(t+1) = 1$. The critic is also an MLP of the same structure as the model. The operation of the critic neural network is described by the following expressions:

$$\mathbf{x}^{\mathrm{C}} = \mathbf{S}(t) = \{\Delta X(t), u(t)\},\ \ y^C_j = \tanh\left(\textstyle\sum_i w^C_{ij} x^C_i\right),\ \ V(t) = V(\mathbf{S}(t)) = \textstyle\sum_j v^C_j y^C_j, \tag{5}$$

where $\mathbf{x}^{\mathrm{C}}$ is the neural network input vector, $\mathbf{y}^{\mathrm{C}}$ is the vector of outputs of hidden layer neurons, w^C_{ij} and v^C_j are neuron synaptic weights.

At any moment t, the following operations are performed:

1) The model predicts the next change of the time series $\Delta X(t+1)$.

2) The critic estimates the state value function for the current state $V(t) = V(\mathbf{S}(t))$ and the predicted states for both possible actions $V^{pr}_u(t+1) = V(\mathbf{S}^{pr}_u(t+1))$, where $\mathbf{S}^{pr}_u(t+1) = \{\Delta X^{pr}(t+1), u(t+1)\}$, and $u(t+1) = 0$ or $u(t+1) = 1$.

3) The ε -greedy rule [3] is applied: the action corresponding to the maximum value $V^{pr}_u(t+1)$ is selected with probability $1-\varepsilon$, and an alternative action is selected with probability ε $(0 < \varepsilon << 1)$. (For example, if the action "buy" corresponds to the maximum of $V^{pr}_u(t+1)$, then it is selected with probability $1-\varepsilon$; alternatively, the action "sell" is selected with probability ε.).

4) The selected action is carried out. The transition to the next time moment $t+1$ occurs. The current reward $r(t)$ is calculated in accordance with (2) and received by ACD. The value $\Delta X(t+1)$ is observed and compared with its prediction $\Delta X^{pr}(t+1)$. The NN weights of the model are adjusted to minimize the prediction error by means of the method of error backpropagation [10] and the gradient descent with $\alpha_M > 0$ as the model learning rate.

5) The critic computes $V(t+1)$. The temporal-difference error is calculated:

$$\delta(t) = r(t) + \gamma V(t+1) - V(t) . \tag{6}$$

6) The weights of the critic neural network are adjusted to minimize the temporal-difference error (6) using its backpropagation and the gradient descent with $\alpha_C > 0$ as the critic learning rate.

Note that the error of the model neural network is known, it is equal to the value $E = \Delta X(t+1) - \Delta X^{pr}(t+1)$, so this network is trained to minimize the value E by means of usual error backpropagation method [10]. On the other hand, the critic neural network is learned using rewards and punishments, so this network is trained by means of the temporal difference method that is specially developed in reinforcement learning research [3].

2.3 Evolutionary Algorithm

We consider Darwinian evolution of agent populations. We use a simple version of the genetic algorithm [11,12] without crossovers. Our evolutionary algorithm is described below.

The evolving population consists of n agents. Each agent has a resource that changes in accordance with values of agent rewards: $R(t+1) = R(t) + r(t)$, where $r(t)$ is calculated in (2). Evolution passes through a number of generations, $n_g = 1,2,\ldots$ The duration of each generation n_g is T time steps. At the end of each n_g-th generation, the best agent that maximally increases its resource $R(t)$ during generation time is selected, this best agent gives birth all n agents of the next (n_g+1)-th generation. Any agent has its genome $\mathbf{G}$; the children genomes $\mathbf{G}$ differ from their parent genome by small mutations.

The agent genome $\mathbf{G}$ is equal to the initial synaptic weights of both NNs (model and critic), $\mathbf{G} = \{\mathbf{W}_{M0}, \mathbf{W}_{C0}\}$, these weights are obtained by the agent at the moment of its birth. The evolutionary process is of Darwinian type, so the agent genome $\mathbf{G}$ does not change during the agent life. However, synaptic weights of the NNs $\mathbf{W}_M$ and $\mathbf{W}_C$ are changed during agent life via learning described above.

At the beginning of any generation, the initial resource of each agent is zero, i.e., $R(T(n_g-1)+1) = 0$. The best agent of the generation n_g has the maximum resource $R_{max}(n_g)$ at the end of the generation.

At the beginning of every new (n_g+1)-th generation, we set for each agent $G_i(n_g+1) = G_{best,\ i}(n_g) + \text{rand}_i$, $\mathbf{W}_0(n_g+1) = \mathbf{G}(n_g+1)$, where $\mathbf{G}_{\mathbf{best}}(n_g)$ is selected from the best agent of the previous n_g-th generation, and rand_i is $N(0, P^2_{mut})$, i.e., a normally distributed random number with zero mean and standard deviation P_{mut} (mutation intensity), which is added to each synaptic weight.

Thus, the genome **G** changes only via evolution, whereas the synaptic weights **W** are adjusted only via learning.

3 Results of Simulations

3.1 General Features of Agent Learning and Evolution

The described model was investigated by means of computer simulations. We used two examples of model time series:

1) sinusoid:

$$X(t) = 0.5\ [1 + \sin(2\ t/20)] + 1\ , \tag{7}$$

2) stochastic time series [8]:

$$X(t) = \exp[p(t)/1200]\ ,\ p(t) = p(t-1) + \beta(t-1) + k_1\ \lambda(t)\ ,\ \beta(t) = k_2\ \beta(t-1) + \mu(t)\ , \tag{8}$$

where $\lambda(t)$ and $\mu(t)$ are two random normal processes with zero mean and unit variance, $k_1 = 0.3$, $k_2 = 0.9$.

Some parameters were set to the same values for all simulations. Specifically, we set, $\gamma = 0.9$, $m = 10$, $\alpha_M = \alpha_C = 0.01$, $N_{hM} = N_{hC} = 10$, $\varepsilon = 0.05$, $P_{mut} = 0.1$, where N_{hM} and N_{hC} are numbers of hidden neurons of the model and critic. Other parameters (n, T) were set to different values, depending on the simulation, as specified below. Agent expenses are neglected ($J = 0$) in the main part of simulations.

We analyze the following cases:

1) case L (pure learning); in this case we consider a single agent that learns by means of temporal difference method;

2) case E (pure evolution), i.e., evolving population without learning;

3) case LE, i.e., learning combined with evolution, as described above.

We compare the agent resource values attained during 200 time steps for these three cases of adaptation. For the cases E and LE, we set $T = 200$ (T is generation duration) and record the maximal value of agent resource in a population $R_{max}(n_g)$ at the end of each generation. For the case L (pure learning), we have just one agent whose resource is reset $R(T(n_g-1)+1) = 0$ after the passing of every $T = 200$ time steps for consistency with the cases E and LE. In case L the index n_g is incremented by one after every T time steps, i.e., $R_{max}(n_g) = R(Tn_g)$. The plots R_{max} vs. n_g for the sinusoid (7) are shown in Fig. 2. In order to exclude the decrease of the value $R_{max}(n_g)$ due to the random choice of actions when applying the ε-greedy rule for the cases LE and L, we set $\varepsilon = 0$ after $n_g = 100$ for the case LE and after $n_g = 2000$ for the case L. The results are averaged over 1000 simulations with $n = 10$, $T = 200$.

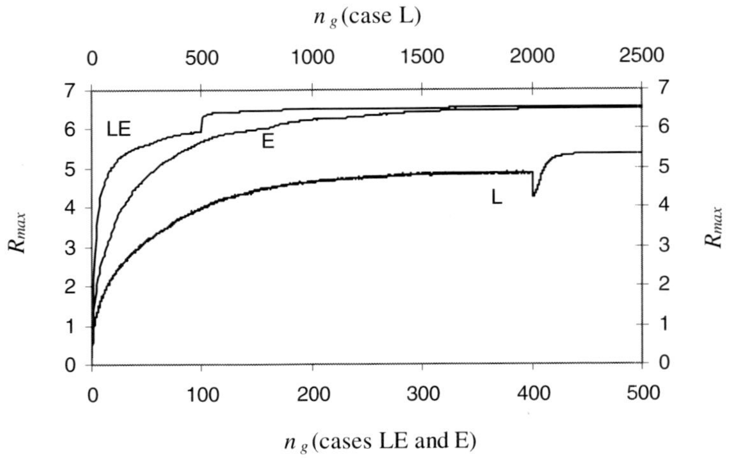

Fig. 2. The plots of $R_{max}(n_g)$ for the sinusoid (7). The curves LE, E and L correspond to the cases of learning combined with evolution, pure evolution and pure learning, respectively. Each point of the plots represents the average over 1000 simulations; $n = 10$, $T = 200$.

We analyzed the agent policy and found that the optimal policy corresponds to the following agent behavior. The agent buys stocks when it predicts that the stock price will rise, and it sells stocks when it predicts a declining stock price. Analysis of agent behavior demonstrates that both pure evolution and learning combined with evolution are able to find the optimal policy. With this policy, the agent attains asymptotic value $R_{max} = 6.5$ (see Fig. 2). For the case L, the asymptotic value of R_{max} is only 5.4. Analysis reveals that the pure learning is able to find only the following satisfactory policy. The agent buys stocks when stock price rises (or falls by a small amount) and sells stocks when stock price falls significantly. So, the agent prefers to keep the capital in stocks in this case.

We described results for the sinusoidal time series (7). It should be noted that similar learning and evolutionary processes are observed for stochastic time series (8).

3.2 Interaction between Learning and Evolution: Baldwin Effect

As shown in Fig. 2 for the case of sinusoid, the pure evolution is able to find the optimal policy in all experiments. However, searching for the optimal policy by means of pure evolution is slower than when combining learning with evolution, as becomes apparent when examining the curves E and LE in Fig. 2. While learning in our model is not optimal by itself, it helps evolution to find better policies.

The role of learning in evolving agent populations can be observed as the Baldwin effect, or the genetic assimilation of initially learned features as a consequence of Darwinian evolution. This effect is found in several experiments, one of which is shown in Fig. 3. We examine how the best agent resource $R_{max}(t)$ changes during the first five generations for the sinusoid time series (7). By the end of each generation, the resource trend is clearly upwards.

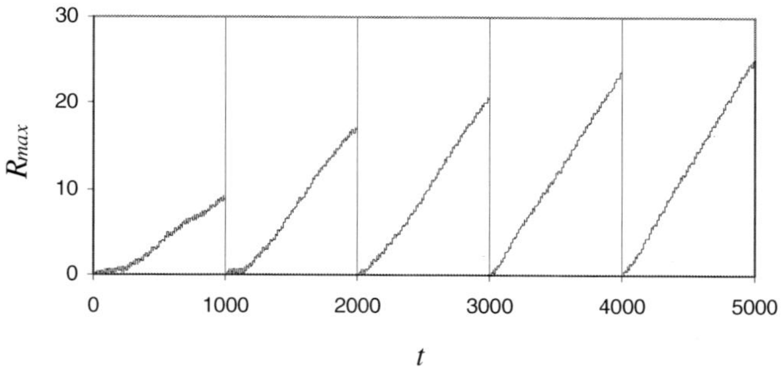

Fig. 3. The plots of the resource $R_{max}(t)$ of the best agent in the population for the first five generations. This is the case of learning combined with evolution on the sinusoid time series; population size n = 10, generation duration T = 1000. The ends of generations are shown by vertical lines. During early generations (generations 1 and 2), there is an obvious delay in the increase of the agent resource. An advantageous policy is found only after some learning period during first 100 to 300 time steps. By the fifth generation, the rapid increase of the resource begins at the start of the generation, demonstrating that the advantageous policy has become inherited.

Fig. 3 shows that during the early generations (generations 1 and 2), any significant increase of the agent resource begins only after a lag of 100 to 300 time steps. The best agent optimizes its policy by learning. Subsequently, the best agents find an advantageous policy faster and faster. By the fifth generation, a newborn agent "knows" a decent policy because it is encoded in its genome **G**, and the learning does not improve the policy significantly. Thus, we can see that the initially learned policy becomes inherited; this corresponds to the Baldwin effect [6, 7]. Actually, simulations demonstrate that though the evolutionary processes are of Darwinian type, these processes have certain Lamarckian properties.

We analyzed different sets of parameters and revealed that the Baldwin effect is observed reliably if generation duration T is greater than 1000. This means that generation duration T should be sufficiently large to ensure significant learning during each generation.

3.3 Influence of Neural Network Structure on Learning Processes

We demonstrated that evolutionary optimization of simple neural networks (NNs) can be more effective than reinforcement learning. In order to analyze the problem in details, we investigate additionally the influence of neural network structure on learning processes in the control system of the agents. Namely, we compare two variants of NNs structures: with one hidden layer (see above) and with two hidden layers. The operation of NNs with one hidden layer is described by equations (4), (5); slightly modified equations can be used to describe the operation of NNs with two hidden layers.

The control system of agents has the same architecture as described above (see Fig. 1); however, the NNs of the model and the critic can have one (first variant) or two (second variant) hidden layers. For both variants we consider three cases of adaptation: L (pure learning), E (pure evolution), and LE, i.e. learning combined with evolution. The main results for first variant are illustrated in Fig. 2.

We analyze the influence of NNs structure on learning (type L of adaptation). The corresponding results are illustrated by Figs. 4 and 5. These figures show the plots of the mean value $R_{max}(n_g)$ along with standard deviation $\sigma(n_g)$ characterizing the distribution of $R_{max}(n_g)$ with respect to different random seeds.

Fig. 4 demonstrates that for the variant 1 (NNs with one hidden layer) the standard deviation $\sigma(n_g)$ decreases and tends to zero meaning that agents find similar policies in all simulations. In contrasts to this, the standard deviation $\sigma(n_g)$ tends to some large asymptotic value for the NNs with two hidden layers (Fig. 5). Analysis reveals that in this variant, leaning is able to find successfully the optimal policy for some set of initial NNs synaptic weights (approximately for 50% of random seeds), while for other initial synaptic weights learning does not find even a satisfactory policy: agents keep their capital in cash all the time and do not increase their resource.

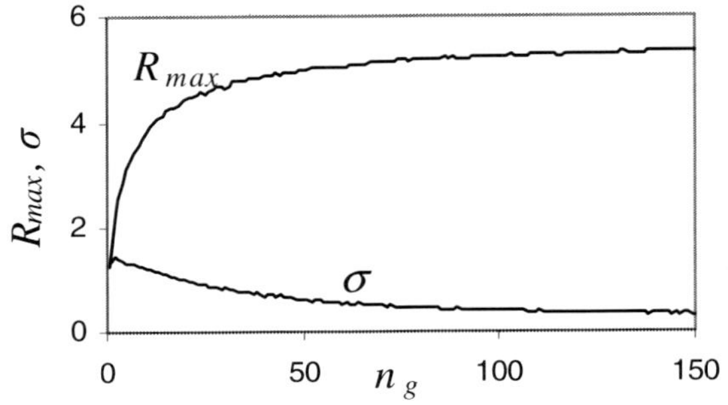

Fig. 4. The plot of $R_{max}(n_g)$ and the standard deviation $\sigma(n_g)$ of $R_{max}(n_g)$ for pure learning (the case L) on the sinusoid time series. Every point of the plot of $R_{max}(n_g)$ represents the average over 1000 simulations; $n = 10$, $T = 200$. The values $\sigma(n_g)$ characterize a distribution of $R_{max}(n_g)$ over different random seeds. NNs have one hidden layer (the variant 1).

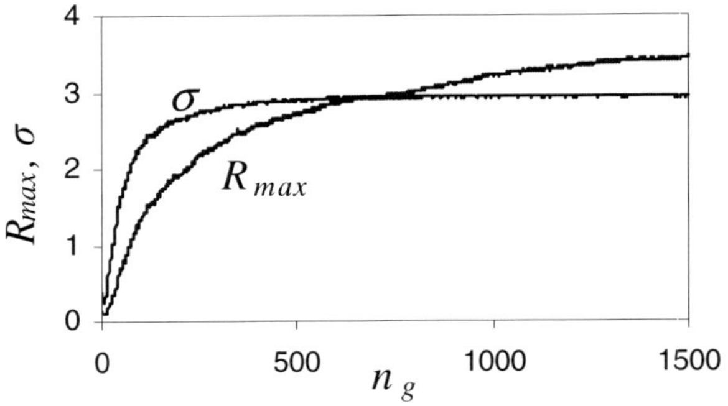

Fig. 5. The plot of $R_{max}(n_g)$ and the standard deviation $\sigma(n_g)$ of $R_{max}(n_g)$ for pure learning (the case L) on the sinusoid time series. Every point of the plot of $R_{max}(n_g)$ represents the average over 1000 simulations; $n = 10$, $T = 200$. The values $\sigma(n_g)$ characterize a distribution of $R_{max}(n_g)$ over different random seeds. NNs have two hidden layers (the variant 2).

Thus, addition of the second hidden layer does result in searching for the optimal policy by means of learning for some random seeds; however, the same type of learning may be very ineffective for other random seeds.

We also analyze evolutionary optimization of agent ACDs. Simulations demonstrate that both evolution (E) and learning combined with evolution (LE) are able to find the optimal policy in control systems with two hidden layer NNs similar to the variant with one hidden layer NNs.

4 Conclusions

We investigated adaptive agents that are able to predict future states, to estimate state value function and to select actions on the basis of predictions and estimations. These adaptive functions can be obtained by means of reinforcement learning in the agent control system. However, our simulations for agents-brokers demonstrate that such control systems are too "intelligent" for the simple agent-broker task. Namely, simple evolution is more efficient in searching for the optimal policy than reinforcement learning. Increase of the number of NNs layers does not improve reinforcement learning essentially. Our simulations demonstrate also that evolution and learning together are more efficient in searching for the optimal agent policy than evolution or learning alone.

We paid special attention to the interaction between learning and evolution. In particular the Baldwin effect is revealed in our model: originally acquired adaptive policy of best agent-brokers becomes inherited during several generations of evolution. This means that certain Lamarckian properties can be observed during Darwinian evolution.

Acknowledgments

The authors thank Dr. Oleg P. Mosalov for his help with computer simulations. This work is partially supported by the Russian Foundation for Basic Research, Grant No 07-01-00180 and the Program of the Presidium of the Russian Academy of Science "Intelligent informational technologies, mathematical modeling, system analysis, and automatics", Project No 2.45.

References

1. Red'ko, V.G., Mosalov, O.P., Prokhorov, D.V.: A model of Baldwin effect in populations of self-learning agents. In: Proceedings of the International Joint Conference on Neural Networks, IJCNN 2005, Montreal, Canada, vol. 3, pp. 1355–1360 (2005)
2. Red'ko, V.G., Mosalov, O.P., Prokhorov, D.V.: A model of evolution and learning. Neural Networks 18, 738–745 (2005)
3. Sutton, R., Barto, A.: Reinforcement Learning: An Introduction. MIT Press, Cambridge (1998)
4. Meyer, J.-A., Wilson, S.W. (eds.): From Animals to Animats: Proceedings of the First International Conference on Simulation of Adaptive Behavior. The MIT Press/Bradford Books, Cambridge (1991)

5. Wilson, S.W.: The animat path to AI. In: [4], pp. 15–21
6. Baldwin, J.M.: A new factor in evolution. American Naturalist 30, 441–451 (1896)
7. Belew, R.K., Mitchell, M. (eds.): Adaptive Individuals in Evolving Populations: Models and Algorithms. Santa Fe Institute Studies in the Sciences of Complexity, vol. 26. Addison-Wesley, Reading (1996)
8. Prokhorov, D., Puskorius, G., Feldkamp, L.: Dynamical neural networks for control. In: Kolen, J., Kremer, S. (eds.) A Field Guide to Dynamical Recurrent Networks, pp. 23–78. IEEE Press, Los Alamitos (2001)
9. Moody, J., Wu, L., Liao, Y., Saffel, M.: Performance function and reinforcement learning for trading systems and portfolios. Journal of Forecasting 17, 441–470 (1998)
10. Rumelhart, D.E., Hinton, G.E., Williams, R.G.: Learning representation by back-propagating error. Nature 323, 533–536 (1986)
11. Holland, J.H.: Adaptation in Natural and Artificial Systems. The University of Michigan Press, Ann Arbor (1975); MIT Press, Boston, MA (1992)
12. Goldberg, D.E.: Genetic Algorithms in Search, Optimization and Machine Learning. Addison-Wesley, Reading (1989)

Learning and Unlearning in Hopfield-Like Neural Network Performing Boolean Factor Analysis

A.A. Frolov[1], D. Húsek[2], I.P. Muraviev[1], and P.Y. Polyakov[3]

[1] Institute of Higher Nervous Activity and Neurophysiology of the Russian Academy of Sciences, Butlerova 5a, 117485 Moscow, Russia
aafrolov@mail.ru, muravevi@mail.ru

[2] Institute of Computer Science Academy of Science of the Czech Republic, Pod Vodárenskou věží 2, 18207 Prague 8, Czech Republic
dusan@cs.cas.cz

[3] Center of Optical Neural Technologies, Scientific Institute for System Analysis, Russian Academy of Sciences, Vavilova 44/2, 119333 Moscow, Russia
pavel.mipt@mail.ru

Abstract. One of the principles used for the transformation of original signal space into a space of lower dimension is a factor analysis based on the assumption that signals are random combinations of latent factors. The goal of the factor analysis is to find factors representation in the signal space (factor loadings) and the contributions of factors into the original signals (factor scores). Recently in [10] we have proposed the general method for Boolean factor analysis based on the Hopfield-like neural network. Due to the Hebbian learning rule the neurons of factor become connected more tightly than other neurons and hence factors can be revealed as attractors of the network dynamics by the random search. The peculiarity of usage the Hopfield-like network for Boolean factor analysis is the appearance of two global spurious attractors. They become dominant and, therefore, prevent successful factors search. To eliminate these attractors we propose a special unlearning procedure. The second unlearning procedure provides the suppression of factors with the largest attraction basins which dominate after suppression of global spurious attractors and prevent the recall of other factors. The origin of the global spurious attractors and the efficiency of the unlearning procedures are investigated in the present paper.

1 Introduction

Factor analysis is one of the most efficient methods to reveal and overcome informational redundancy of high-dimensional signals manifested by the correlation of their component. The main goal of factor analysis methods application is: (1) to reduce the number of variables and (2) to detect structures in the relationships between variables, that is to classify variables. Therefore, factor analysis is applied as a data reduction or structure detection method (the term factor analysis was first introduced by Thurstone[22]). This is a common problem encountered in disciplines such as statistics, data analysis, signal processing, textual data representation and neural network research.

J. Koronacki et al. (Eds.): Advances in Machine Learning I, SCI 262, pp. 501–518.
springerlink.com © Springer-Verlag Berlin Heidelberg 2010

Factor analysis is a procedure which maps original signals into the space of factors. There are many types of factor analyzes. Well-known linear factor analysis implies that each original signal can be presented as

$$\mathbf{X} = \mathbf{FS} = \sum_{l=1}^{L} S_l \mathbf{f}^l \tag{1}$$

where $\mathbf{F}$ is a matrix $N \times L$ of factor loadings (N is the dimensionality of signal space, L is the number of factors) and $\mathbf{S}$ is a vector of factor scores. Each component of $\mathbf{S}$ gives a contribution of the corresponding factor to the original signal. Each vector-column $\mathbf{f}^l$ of loading matrix $\mathbf{F}$ represents the corresponding factor in the signal space. We will call these vectors "factors" below. Mapping the original space to the factor space means that signals are represented by the $\mathbf{S}$ vectors instead of the original $\mathbf{X}$ vectors. There exists many approaches to signal representations in the form (1) based on the supposed model of a signal space. Most popular are: principal component analysis (PCA), when factors are eigenvectors of a covariation matrix; independent component analysis (ICA), when factors are chosen in such a way that factor scores became statistically independent as much as possible; sparse component analysis (SPA), when most factors scores have value "Zero", and nonnegative matrix factorization (NMF), when the elements of X, F and S are nonnegative. Different versions of these approaches are extensively discussed in modern literature (see [11, 13, 17, 18, 19, 20, 23, 25, 26] as examples) and some of them suppose neural network implementation [13, 20].

Boolean factor analysis implies that a composite vector signal has a form of the Boolean sum of weighted binary factors:

$$\mathbf{X} = \bigvee_{l=1}^{L} S_l \mathbf{f}^l. \tag{2}$$

In this case, original signals, factor scores and factor loadings are binary, i.e. they take values 0 or 1. Mapping of the original signal to the factor space means an expression of the signal in terms of the factor scores. We will call l-th factor is mixed in a signal if $S_l = 1$ for this signal.

The methods of Boolean factor analysis are only weekly developed (see, for example [4, 21]) in spite of the fact that the binary form of data presentation is typical for many fields including social science, marketing, zoology, genetics and medicine [16]. The typical example of Boolean factor analysis is so called bars problem [7, 21] which is the benchmark task for the learning of independent objects from complex scenes composed of many objects. In this task each scene is $n \times n$ pixel image containing several of $2n$ possible (one-pixel wide) horizontal and vertical bars (see Fig. 1 as an example). Pixels belonging and not belonging to the bar take value 1 and 0 respectively. Bars are chosen to scenes with equal probabilities. In the point of intersection of vertical and horizontal bars pixel takes the value 1. The Boolean summation of pixels belonging to different bars simulates the occlusion of objects. The goal of the task is to learn all bars as individual objects on the basis of learning set containing M images. In terms of Boolean factorization bars are factors, each image is Boolean superposition of factors and factors scores take values 1 or 0 dependently on whether bars appear in the image. Thus each image can be presented by (2).

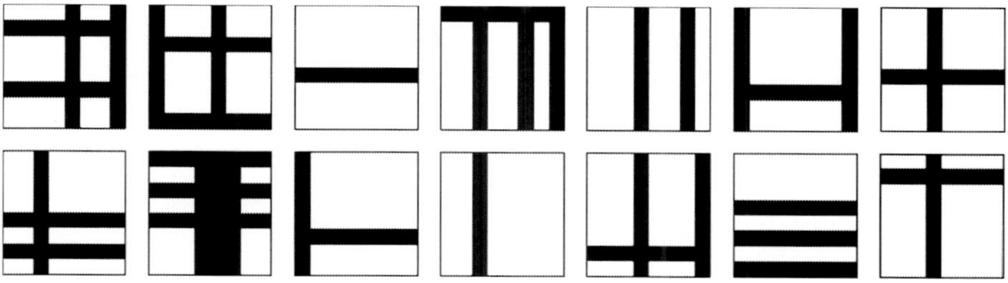

Fig. 1. Example of superpositions of bars in 8×8 pixel images

Boolean factor analysis is also assumed to be a common procedure in brain functioning. According to [3] and [7] environment objects are collections of highly correlated binary properties. For instant, the properties 'furry', 'shorter than a meter', 'has tail', 'moves', barks', etc., are highly correlated, i.e. the combination of these properties is much more frequent than it would be if they were independent. While the components of objects are highly correlated, objects are relatively independent of each other. Objects could therefore be defined as highly correlated sets of binary components that are relatively independent from other such sets. The goal of sensory systems is to detect these correlated components and to represent the objects (that can be recognized as a conjunction of the components) explicitly [1, 2, 7, 24]. In terms of Boolean factor analysis these sets of correlated binary components are factors.

Frolov et al. proposed the general method for Boolean factor analysis [10] based on the Hopfield-like attractor neural network. The method exploits the well known property of a Hopfield network to create attractors of network dynamics by tightly connected neurons. Since neurons representing a factor are activated simultaneously each time when factor appears in the patterns of the learning set and neurons representing different factors are activated simultaneously only by chance, then due to Hebbian learning rule the neurons of factor become connected more tightly than other neurons. Hence the factors can be revealed as attractors of the network dynamics. We have restricted our study to the case of sparsely encoded factors when the number of "Ones" in each factor n is much smaller than the signal dimensionality N. The reason for this is as follows: Boolean superposition of even relatively small numbers of densely encoded factors results in input signals as binary vectors composed almost entirely of entries with value 1. In this extreme case information on factors is lost and therefore a factor analysis of such signals is clearly impossible. We suppose that the number of factors mixed in each signal is rather large (but, nevertheless, much smaller than the total numbers of factors) and therefore factors must be encoded sparsely, i.e. $p = n/N \ll 1$ where n is a number of "Ones" in factor. We call the mean number of factors mixed in each signal the signal "complexity" C. The case of $C = 1$ corresponds to the traditional sparsely encoded Hopfield network [8].

The method includes a special procedure of Hebbian *learning* and two procedures of *unlearning*. During the learning, input signals which are superpositions of some factors are presented to the network. For every signal component there is a certain neuron which takes the state "One" (active) when the corresponding component is "One" and

"Zero" (nonactive) in the opposite case. Due to Hebbian learning, factors constitute *true* attractors of the network dynamics. However, the network dynamics converge to one of the factors only when the initial state falls inside its attraction basin. In other cases it converges to one of the spurious attractors. Since generally there is no *a priori* information on factors the initial network state can only fall into a factor attraction basin by chance. Therefore the ability of a Hopfield-like network to perform factor searches is determined by two aspects: first, by the probability that network activity converges to one of the factors starting from a random state, and second, by the possibility to distinguish between true (factors) and spurious attractors.

The specificity of the proposed network is the existence of two types of spurious attractors: global and local. The main function of the first unlearning procedure is to eliminate from the network dynamics two global spurious attractors which appear when signal complexity C is large. Particularly they are absent for the ordinary Hopfield network which is a particular case of our network with $C = 1$ and therefore they have never been discussed in the literature until [10]. The origin of the global spurious attractors was exemplified in [10] only by computer simulation without special study of the ranges of network parameters where their effect is essential. The main goal of the present paper is to clarify the origin of the global attractors and the performance of the special unlearning procedure analytically.

The second unlearning procedure is dedicated to eliminating factors that are already revealed from the network. Due to this unlearning, factors that dominate in network dynamics and which attract network activity most frequently are gradually deleted from the network memory. Then factors with smaller attraction basins begin to dominate and are in turn deleted until all factors are revealed.

The paper is organized as follows. A method description is given in Sect. 2. Some results of computer simulation illustrating the origin of two global spurious attractors and the performance of the first unlearning procedure which eliminates them from the network dynamics are presented in Sect. 3. Section 4 is dedicated to the second unlearning procedure which eliminates the dominance of factors with the largest Lyapunov functions. Section 5 is a short general discussion.

2 Method Description

We use the fully connected network of the N neurons with binary activity (1 – active, 0 – nonactive). The number of neurons N is equal to the dimensionality of the signal space. Each input signal $\mathbf{X}^m$ ($m = 1...M$, M is the number of signals in the learning set) is presented to the network so that each neuron takes the state equal to the value of the corresponding component of $\mathbf{X}^m$. Signals $\mathbf{X}^m$ are stored in the matrix of synaptic connections $\mathbf{J}'$ according to the modified Hebbian rule:

$$J'_{ij} = \sum_{m=1}^{M} (X_i^m - q^m)(X_j^m - q^m),\ i,j = 1,...,N,\ i{\neq}j,\ J'_{ii} = 0 \tag{3}$$

where bias $q^m = \sum_{i=1}^{N} X_i^m / N$ is the total relative activity of the m-th pattern. As shown below if signals of the learning set are superpositions of some factors in terms of (2) then due to this form of bias, factors become attractors of network dynamics. Unfortunately this learning rule results also in appearance of two global spurious attractors. The unlearning procedure which provides the elimination of the global attractors consists in the subtraction of

$$\mathbf{J}'' = \mathbf{j}''\mathbf{j}''^T/M \tag{4}$$

from $\mathbf{J}'$ where

$$j''_i = \sum_{m=1}^{M} (X_i^m - q^m) = M(q_i - q),\ i = 1..N, \tag{5}$$

$q_i = \sum_{m=1}^{M} X_i^m/M$ is a mean activity of the i-th neuron in the learning set and q is a mean activity of all neurons in the learning set. We denote the resulting matrix of synaptic connections as

$$\mathbf{J} = \mathbf{J}' - \mathbf{J}''. \tag{6}$$

To reveal factors we propose the following two-run recall procedure. Its initialization starts by the presentation of a random initial pattern $\mathbf{X}^{in}$ with k_{in} active neurons. Activity k_{in} which is a parameter of the recall procedure is supposed to be smaller than the activity of any factor. On presentation of $\mathbf{X}^{in}$, network activity $\mathbf{X}$ evolves to some attractor. This evolution is determined by the synchronous discrete time dynamics. At each time step:

$$X_i(t+1) = \Theta(h_i(t) - T(t)),\ i = 1,\cdots,N, \qquad X_i(0) = X_i^{in} \tag{7}$$

where h_i are components of the vector of synaptic excitations

$$\mathbf{h}(t) = \mathbf{J}\mathbf{X}(t), \tag{8}$$

Θ is the step function, and $T(t)$ is an activation threshold.

At each time step of the recall process the threshold $T(t)$ was chosen in such a way that the level of the network activity was kept constant and equal to k_{in}. Thus, on each time step k_{in} "winners" (neurons with the greatest synaptic excitation) were chosen and only they were active on the next time step. As shown in [9], this choice of activation threshold enables the network activity to stabilize in point or cyclic attractors of length two. The fixed level of activity at this stage of the recall process could be ensured by biologically plausible non-linear negative feed-back control accomplished by the inhibitory interneurons. It is worth to note that although the fact of convergence of network synchronous dynamics to point or cyclic attractors of length two was established earlier in [12], it was done for fixed activation threshold T but for fixed network activity it was done first in [9].

When activity stabilizes at the initial level of activity k_{in}, $k_{in} + 1$ neurons with maximal synaptic excitation are chosen for the next iteration step, and network activity evolves to an attractor at the new level of activity $k_{in} + 1$. The level of activity then increases to $k_{in}+2$, and so on, until the number of active neurons reaches the final level k_{fin}. Thus, one trial of the recall procedure contains $k_{fin}-k_{in}$ external steps and several internal steps (usually 2-3) inside each external step to reach an attractor for a given level of activity.

At the end of each external step when the network activity stabilizes at the level of k active neurons a Lyapunov function was calculated by formula:

$$\lambda = \mathbf{X}^T(t+1)\mathbf{J}\mathbf{X}(t)/k, \tag{9}$$

where $\mathbf{X}^T(t+1)$ and $\mathbf{X}(t)$ are two network states in a cyclic attractor (for point attractor $\mathbf{X}^T(t+1) = \mathbf{X}(t)$). In our definition of Lyapunov function its value gives a mean synaptic excitation of neurons belonging to an attractor at the end of each external step. The identification of factors along the trajectories of the network dynamics was based on the analysis of the change of the Lyapunov function and the activation threshold along each trajectory. As shown in [10] Lyapunov function has an apex and activation threshold has a drop at the point $k=n$ where n is a number of active neurons in factor. To match the two properties of the trajectories we subtract the activation threshold from the Lyapunov function calculating $R=\lambda-T$ and its derivative by k. The peak at the curve R' is a reliable index of the factor.

The second unlearning procedure is dedicated to eliminating the revealed factors from the network. When initial states are chosen randomly network activity tends to converge to factors with larger attraction basins. To suppress the dominance of these factors we deleted them from the network memory according to the Hebbian unlearning rule by subtracting ΔJ_{ij} from synaptic connections J_{ij} where

$$\Delta J_{ij} = \eta J[(X_i(t)-r)(X_j(t+1)-r)+(X_i(t+1)-r)(X_j(t)-r),\ j\neq i \tag{10}$$

where $\mathbf{X}(t)$ and $\mathbf{X}(t+1)$ are successive patterns of network activity in the attractors, $r=k/N$ is a relative network activity, $J=\lambda/k$ is a mean weight of synaptic connections between neurons of the factor and η is an unlearning rate. The unlearning ensures that both the point attractors when $X(t+1)=X(t)$ and also the cyclic attractors are suppressed. Due to this unlearning, factors that dominate in network dynamics and which attract network activity most frequently are gradually deleted from the network memory. Then factors with smaller attraction basins begin to dominate and are in turn deleted until all factors are revealed.

3 Elimination of the Global Spurious Attractors

In this section we investigate the nature of the global spurious attractors and the performance of the first unlearning procedure which eliminates them from the network dynamics.

We restrict our analysis to the case when factors are uniformly distributed in B_n^N[1] (i.e. each factor contained exactly n "Ones" and $N-n$ "Zeros"), factors have equal

[1] $B_n^N = \{\mathbf{X}|X_i \in \{0,1\}, \sum_{i=1}^{N} X_i = n\}$.

probabilities to be mixed in patterns of the learning set and exactly C factors are mixed in each pattern. Thus we assume that each signal of the learning set $\mathbf{X}^m$ has the form

$$\mathbf{X}^m = \bigvee_{l=1}^{L} S_l^m \mathbf{f}^l$$

where factor scores S_l^m are uniformly distributed in B_C^L, that is $\sum_{l=1}^{L} S_l^m = C$. Additionally, we assume that $p = n/N \ll 1$, $N \gg 1$ and $L \gg C$.

3.1 Computer Simulation

To clarify the problem we present initially the results obtained on artificial data when patterns in the learning set are superpositions of given factors with known rules of mixing. For the presented example we used the network with $N = 1100$ neurons and $L = 800$ factors. Factors were uniformly distributed in B_{22}^N (i.e., each factor contained exactly $n = 22$ "Ones" and $N - n$ "Zeros"). In each pattern of the learning set, 20 randomly chosen factors were mixed (i.e signal complexity $C = 20$). All factors had equal probabilities to be mixed in patterns. The whole size of the leaning set amounted to $M = 40000$ patterns.

Figure 2 demonstrates several trajectories of network dynamics obtained before (a) and after (b) subtraction of $\mathbf{J}''$ from $\mathbf{J}'$ according to (6). Trajectories started from random states with $k_{in} = 5$ and continued to $k_{fin} = 33$. Only two trajectories with a high Lyapunov function dominate and attract all other trajectories if unlearning procedure is not applied. In this case only a small portion of factors can be revealed. After unlearning the network dynamics drastically changes. The most trajectories with the higher Lyapunov function correspond to factors and a small portion of them with a lower Lyapunov function correspond to local spurious states. We will call these trajectories "true" and "spurious". True trajectories have a specific apex at the point $r = p$ where $r = k/N$

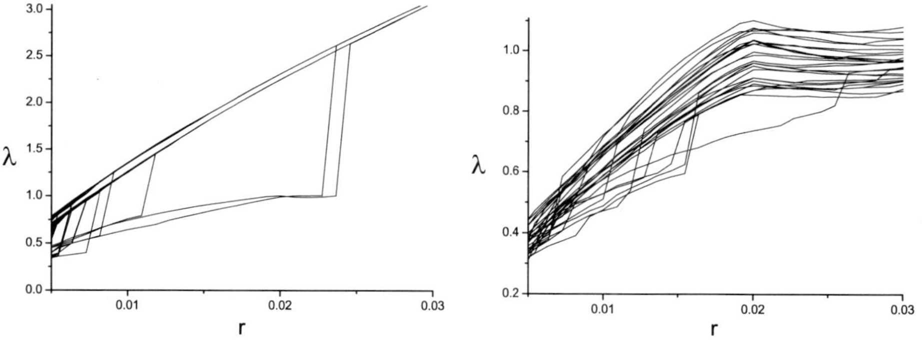

(a) (b)

Fig. 2. Lyapunov function λ dependent on the relative network activity $r = k/N$ before (a) and after (b) subtraction of $\mathbf{J}''$ from $\mathbf{J}'$, $M = 40000$ $N = 1100$, $L = 800$, $C = 20$. Lyapunov function is normalized by its mean value for factors, i.e. at the point $r = p = 0.02$.

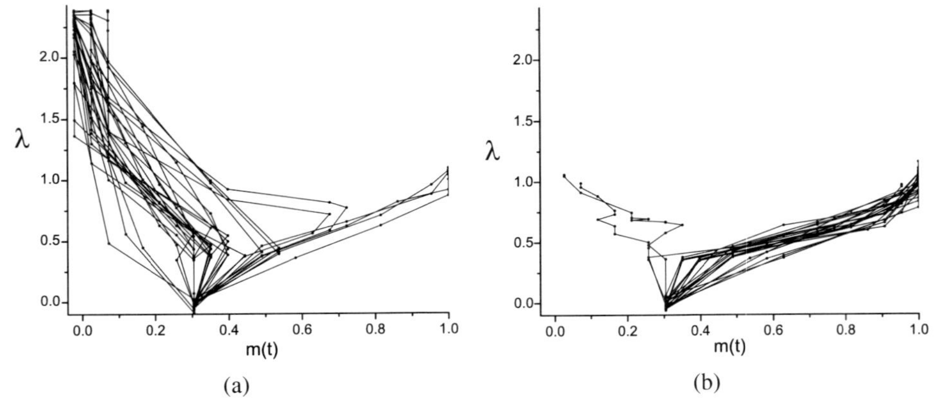

Fig. 3. Trajectories of neurodynamics before (a) and after (b) subtraction of $\mathbf{J}''$ from $\mathbf{J}'$. $M = 40000$, $N = 1100$, $L = 800$, $C = 20$ and $m_{in} = 0.3$. Abscissa is overlap $m(t)$ between the current network activity and the recalled factor, ordinate is Lyapunov function normalized as in Fig. 2.

is the relative network activity and $p = n/N = 0.02$ is the relative activity of factors. When $r < p$, each point at the trajectory corresponds to a fragment of one of the factors and the increase of r occurs due to activation of neurons belonging to the factor that are strongly connected with already active other neurons of the factor. The activation of such neurons results in an increase of the mean synaptic excitation (their Lyapunov function) almost proportional to r. When $r > p$ the increase of r occurs due to activation of some random neurons that are connected with neurons of the factor by weak connections. Hence, the increase of the Lyapunov function for true trajectories sharply slows at the point $r = p$ and tends towards the values of Lyapunov functions for spurious trajectories when r continues to increase.

Another example of the performance of the unlearning procedure is demonstrated in Fig. 3. The trajectories are shown in the plane constituted by axes $[m(t), \lambda(t)]$ where $m(t)$ is the overlap of the current network state $\mathbf{X}(t)$ with the recalled factor $\mathbf{f}^l$, $\lambda(t)$ is the Lyapunov function and t is the time step. The overlap between two patterns of network activity $\mathbf{X}^1$ and $\mathbf{X}^2$ is given by formula

$$m(\mathbf{X}^1, \mathbf{X}^2) = \frac{1}{N\sqrt{p(1-p)q(1-q)}} \sum_{i=1}^{N} (X_i^1 - p)(X_i^2 - q)$$

where p and q are relative activities in patterns $\mathbf{X}^1$ and $\mathbf{X}^2$. According to this formula the overlap between the identical patterns is equal to 1 and the mean overlap between random patterns is equal to 0. We used the same network as in the previous example but trajectories started from the states close to factors ($m_{in} = m(\mathbf{X}^{in}, \mathbf{f}^l) = 0.3$) and having the same activity as factors, i.e. $k = n$ active neurons. If the unlearning procedure is not applied (Fig. 3(a)), only small portion of trajectories converge to factors, i.e. to the final states with $m = 1$. The most trajectories converge to spurious states with m close to 0. If the unlearning procedure is applied (Fig. 3(b)) almost all trajectories converge to factors.

The spurious attractors are two global attractors created by neurons contained in the most and in the least numbers of factors, respectively. To demonstrate this fact we redrew the trajectories in the axes $[m(t), rank(t)]$ (Fig. 4) where $rank(t)$ indicates whether the neurons contained in the most or least numbers of factors contribute to the current network activity. To calculate $rank(t)$, we arranged all neurons in the order of numbers of factors that contained them, so the neurons that were contained in the least number of factors had the least rank and those contained in the most number of factors had the highest rank. The rank of the current activity was calculated as a sum of the ranks of active neurons. The obtained rank was normalized so that the patterns created by the neurons contained in the least number of factors have a rank close to zero, those created by the neurons contained in the most number of factors have a rank close to one and those created by random neurons have a rank close to 0.5. Figure 4(a) demonstrates that the patterns created by the global spurious attractors had ranks close to 0 and 1, while true attractors had ranks around 0.5. As shown in Fig. 4(b) owing to the unlearning procedure most trajectories converge to factors, two global spurious attractors are completely suppressed and trajectories that converge to states far from the factors are attracted by local spurious attractors. These attractors have ranks close to 0.5 and their Lyapunov function value is smaller than that for global spurious attractors (see Fig. 3(b)).

3.2 Analytical Approach

The examples shown in Figs. 2–4 argue into the ability of the unlearning procedure defined by (6) to eliminate two global spurious attractors of network dynamics. In the following we study the range of network parameters where influence of the global attractors are especially important.

Two global spurious attractors dominate when their Lyapunov function exceeds that of true attractors. Thus to find the conditions of their dominance one has to compare Lyapunov functions of true and global spurious attractors. By definition, the Lyapunov function of each attractor can be estimated as a mean synaptic excitation produced in the neurons of attractor by their proper activity. When $r < p$ true attractors are created by fragments of factors. Thus to evaluate the Lyapunov function for true attractors we have to evaluate a synaptic excitation produced in the neurons of factors by activation of their fragments. Along the trajectories of network dynamics those fragments create attractors which neurons are the most tightly connected among the neurons of factors. However, we ignore the specificity of activated fragments and calculate synaptic excitation produced by the fragment as if its neurons were chosen from factor's neurons randomly with equal probabilities. Therefore we obtain slightly underestimated values of the Lyapunov function of true attractors.

Without any loss of generality, we may assume that an activated pattern $\mathbf{X}$ is a fragment of factor $\mathbf{f}^1$. In conformity with [5], we call n and $N-n$ neurons of $\mathbf{f}^1$ with states "One" and "Zero" as "high" and "low" neurons, respectively. According to (3), (4) and (8), the synaptic excitation of i-th neuron can be presented in the form $h_i = \Sigma_i^1 + \Sigma_i^0 - \Sigma_i'$ where

$$\Sigma_i^{\mu} = \sum_{m \in \{m: S_1^m = \mu\}} (X_i^m - q^m) \sum_{j \neq i} (X_j^m - q^m) X_j.$$

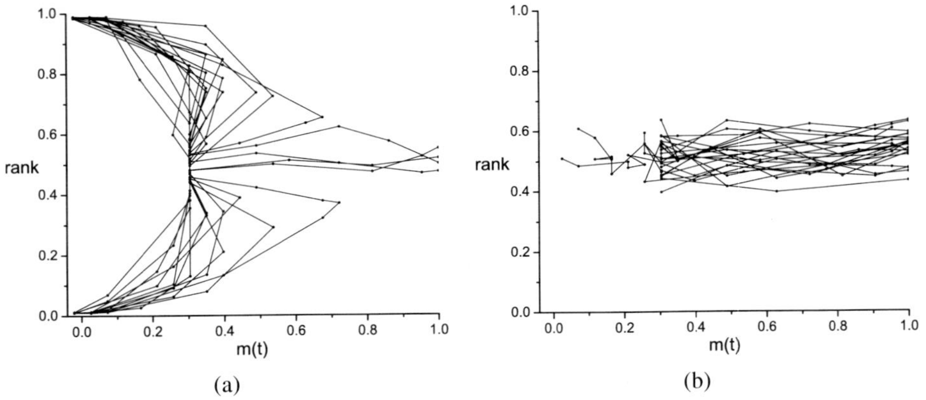

Fig. 4. The same trajectories as in Fig. 3 drawn in the plane $(m(t), rank)$ where *rank* is an index showing contribution of neurons most often or rarely contained in factors set. (a) – before, (b) – after subtraction of $\mathbf{J}''$ from $\mathbf{J}'$.

Two first sums $\Sigma_i^1 + \Sigma_i^0$ is a synaptic excitation of i-th neuron before elimination of the global attractors (i.e. produced by synaptic connections given by $\mathbf{J}'$). The third sum

$$\Sigma_i' = M(q_i - q) \sum_{j \neq i} (q_j - q) X_j$$

is the reduction of the synaptic excitations produced by subtraction of $\mathbf{J}''$ from $\mathbf{J}'$ (see (6)). The sum Σ_i^1 contains M^1 patterns of the training set which include the recalled factor $\mathbf{f}^1$ and Σ_i^0 contains M^0 patterns which do not include it. Since the variance of q^m is of the order $1/N$ (see Appendix 1), one can put

$$q^m = \langle q^m \rangle = q = 1 - (1-p)^C \simeq 1 - \exp(-pC)$$

where $\langle \cdots \rangle$ means averaging over all factor scores and all factors except $\mathbf{f}^1$.

The activated fragment of $\mathbf{f}^1$ contains k high neurons and no low neurons of $\mathbf{f}^1$. For high neurons of $\mathbf{f}^1$

$$\langle \Sigma_i^1 \rangle = \langle M^1 \rangle (1-q)^2 k$$

and for low neurons

$$\langle \Sigma_i^1 \rangle = \langle M^1 \rangle (q' - q)(1-q)k$$

where

$$q' = 1 - (1-p)^{C-1} = 1 - (1-q)/(1-p)$$

is the probability of the neuron to be active in the pattern of the learning set due to the presence of other factors except $\mathbf{f}^1$. Therefore,

$$\langle \Sigma_i^1 \rangle = \langle M^1 \rangle (1-q)^2 (f_i^1 - p)k/(1-p)$$

for both high ($f_i^1 = 1$) and low ($f_i^1 = 0$) neurons.

Since $\mathbf{X}$ is independent of all factors except $\mathbf{f}^1$, then

$$\langle \Sigma_i^0 \rangle = \langle M^0 \rangle k \langle (X_i^m - q^m)(X_j^m - q^m) \rangle.$$

By the definition of q^m, for each pattern of the learning set $\sum_{i=1,N} (X_i^m - q^m) = 0$. Hence

$$\langle (X_i^m - q^m)(X_j^m - q^m) \rangle = -\langle (X_i^m - q^m)^2 \rangle / (N-1) \simeq -q(1-q)/N$$

and $\langle \Sigma_i^0 \rangle = -\langle M^0 \rangle q(1-q)k/N$.

Since $\langle (q_i - q) \rangle$ is evidently equal to 0, then $\langle \Sigma_i' \rangle = 0$ i.e. the subtraction of $\mathbf{J}''$ from $\mathbf{J}'$ does not influence on the mean synaptic excitation of the neurons. Thus

$$\langle h_i \rangle = \langle M^1 \rangle (1-q)^2 (f_i^1 - p)k/(1-p) - \langle M^0 \rangle q(1-q)k/N.$$

Since Lyapunov function for activated fragment of $\mathbf{f}^1$ is the mean synaptic excitation of k active high neurons then its mean Lyapunov function is equal to

$$\lambda_{tr} = [\langle M^1 \rangle (1-q)^2 - \langle M^0 \rangle q(1-q)/N]k.$$

The probability of factor to be mixed in the learning pattern is C/L. Thus $\langle M^1 \rangle = MC/L$ and $\langle M^0 \rangle = M(1 - C/L) \simeq M$ and the first term in the expression for λ_{tr} is of the order MC/L and the second – M/N. Usually $CN/L \gg 1$ and the second term can be ignored (for example for the network used in the computer simulation $CN/L \simeq 30$). Then in the range $k \leq n$ the mean Lyapunov function for true attractor can be estimated as

$$\lambda_{tr} = MCk(1-q)^2/L \tag{11}$$

3.3 Lyapunov Function of Global Spurious Attractors

To estimate a Lyapunov function of global spurious attractors, we take into account that they are created by neurons the most often and the most rare containing in the whole set of factors. We also take into account that $p = n/N \ll 1$ and $r = k/N \ll 1$. The estimation is rather crude but it clarifies the origin of the global spurious attractors.

Let K be the number of factors containing a given neuron. The mean and variance of K are both equal to pL. In Gaussian approximation the number of neurons which belong to $K > K_1$ factors can be estimated as $N\Phi(u_1)$ where $u_1 = (K_1 - pL)/\sqrt{pL}$. To choose rN neurons with the largest K from totally N neurons, one must put $K_1 = pL + u_1\sqrt{pL}$ where u_1 satisfies equation $\Phi(u_1) = r$. On average, each of the chosen neurons belongs to $K_2 = pL + u_2\sqrt{pL}$ patterns where

$$u_2 = \frac{1}{\Phi(u_1)\sqrt{2\pi}} \int_{u_1}^{\infty} u \exp(-u^2/2) du.$$

The probability of one of these neurons to be active during the presentation of a learning pattern is $q' \simeq 1 - \exp(-K_2 C/L) \simeq q + (1-q)Cu_2\sqrt{p/L}$. Then a mean augmentation of synaptic connection between two of these neurons during the presentation of input pattern is $\Delta J = (q' - q)^2 = (1-q)^2 C^2 u_2^2 p/L$ and a mean strength of connection after

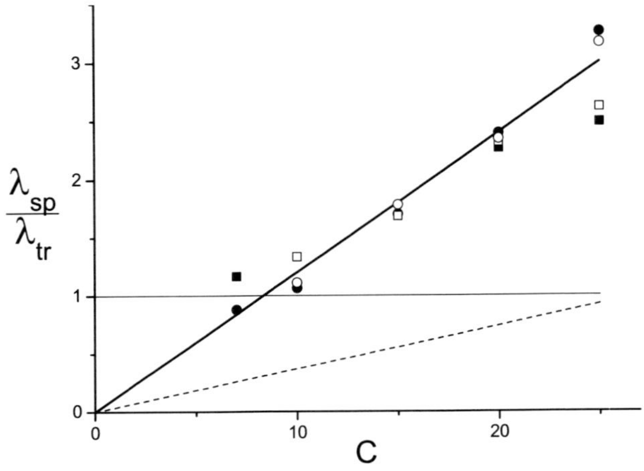

Fig. 5. The ratio of values of Lyapunov function for global spurious and true attractors. Estimations by (13) are shown by a thick solid line for $p = 0.02$ and a dashed line for $p = 0.004$. Points are experimental data averaged over four random sets of factors for $p = 0.02$, $M = 40000$, $N = 1100$: squares – $L = 800$, circles – $L = 1600$, open and full points – attractors created by neurons the most rare and often contained in factors, respectively. The thin horizontal line indicates the equality of values of Lyapunov function for spurious and true attractors. For $p = 0.02$ this line is crossed for $C \approx 10$ indicating the critical value of complexity when spurious attractors become dominate.

presentation of the whole learning set is $J = M\Delta J = M(1-q)^2C^2u_2^2p/L$. Hence the Lyapunov function for this attractor can be estimated as

$$\lambda_{sp} \simeq JrN = MrN(1-q)^2C^2u_2^2p/L.$$

Since $r \ll 1$ then $u_2 \simeq u_1 \simeq [2\ln(1/r)]^{\frac{1}{2}}$. Consequently

$$\lambda_{sp} \simeq 2Mr\ln(1/r)N(1-q)^2C^2p/L \tag{12}$$

Similarly, it is easy to estimate the Lyapunov function for the attractor created by neurons belonging to the smallest number of factors. To do this, it is necessary to replace K_2 in the formula for q' by $K_3 = pL - u_2\sqrt{pL}$ keeping all other equations. This results in the same expression for the Lyapunov function as (12).

According to (11) and (12) the Lyapunov function of true attractors increases proportionally to C while of spurious attractors proportionally to C^2. Their ratio amounts to

$$\lambda_{sp}/\lambda_{tr} \simeq 2Cp\ln(1/r). \tag{13}$$

Figure 5 demonstrates the ratio of Lyapunov functions for spurious and true attractors in dependence on C obtained by computer simulation and by (13) for $r = p = 0.02$. It is shown that (13) gives a rather accurate estimation of this ratio and that the critical complexity, when spurious attractors become dominant, amounts to about $C = 10$ for $p = 0.02$. According to (13), the critical complexity increases when sparseness increases (i.e. p decreases).

Note that formula (12) is valid for the whole range of r while (11) only for $r \leq p$. Thus comparison of values of the Lyapunov function for global spurious and true attractors by (13) is valid only for $r \leq p$.

4 Unlearning of Revealed Factors

After suppression of the global spurious attractors almost all trajectories become true (see Fig. 2(b)). However even in this case the search of all factors could be impossible due to the dominance of some of them. Figure 6 illustrates this statement: the thin line shows the number of found different factors dependent on the number of trials for $M = 50000$, $N = 2000$, $L = 1000$, $C = 20$, $p = 0.02$. Initially the number of new found factors increases proportionally to the number of trials. However, later the search of new factors is locked because the network activity is mainly attracted by the same set of trajectories with the highest Lyapunov function and the proportion of new revealed factors progressively decreases.

This difficulty can be easily overcome by Hebbian unlearning when attractors which appeared during the recall process are deleted from the network memory. The deletion was performed by the anti-Hebb rule (10). Factors, i.e. attractors for which $r = p$, were unlearned by the rate $\eta = 1$. Figure 6 shows the effect of unlearning (thick line). Initially

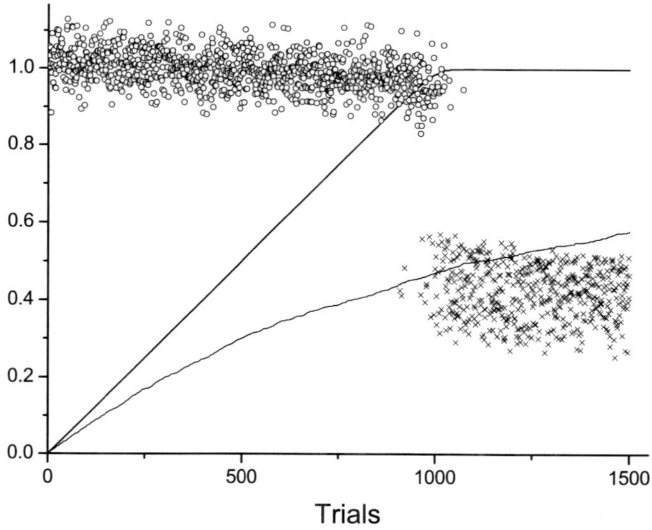

Fig. 6. Lyapunov function (points) normalized by its mean value for factors according to (11) and the portion of found factors (lines) dependent on the number of recall trials for $C = 20$, $N = 2,000$, $L = 1,000$. The values of the Lyapunov function are shown only for the recall with unlearning. In the recall without unlearning, the distribution of experimental points does not depend on the number of trials and corresponds to the initial stage of the recall with unlearning. ∘ – true attractors, × – spurious attractors. The thick and thin lines are the portions of found factors with and without unlearning, respectively.

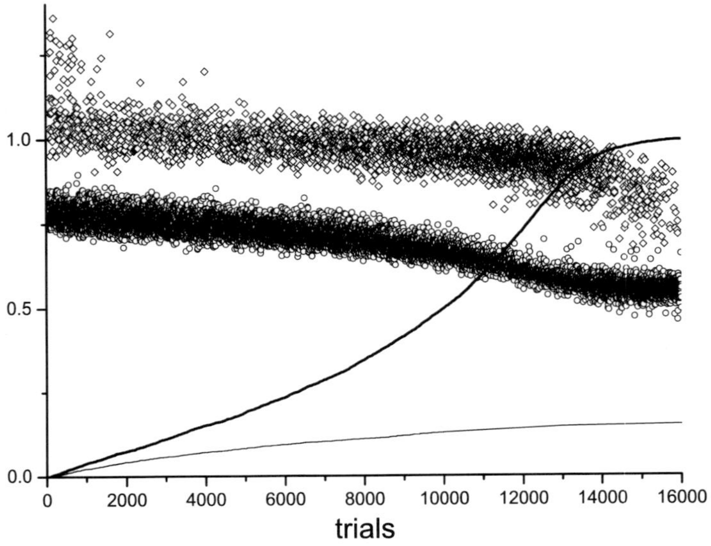

Fig. 7. Relative Lyapunov function (points) and the portion of found factors (lines) dependent on the number of recall trials for $C = 20$, $N = 3000$, $L = 4200$. ◇ – true attractors, ○ – spurious attractors. The thick and thin lines are the portions of found factors with and without unlearning, respectively.

the rate of new factors in the unlearning case is the same as that in the case without unlearning. However, because of unlearning, the rate of the search of new factors does not decrease and to find all 1000 factors only 1000 trials are required, or one trial for one factor.

Circles in Fig. 6 are values of the Lyapunov function at each trajectory at the point of a found factor, i.e. when $k = n$, normalized by its theoretical value given by (11). We show that factors with higher values of the Lyapunov function are found at first, but are then excluded from the network memory and as a result factors with smaller values of the Lyapunov function can be found. When search procedure continue in spite of all existing factors being found and excluded from the memory, next trajectories of network dynamics only correspond to local spurious attractors that are far from factors. The transition in the recall procedure from factors to spurious attractors results in a sharp drop of the Lyapunov function.

The values of parameters used for the network shown in Fig. 6 ensured the absence of local spurious trajectories until all factors were revealed (the probability of spurious trajectory amounts to only 0.004). As shown in [10] the absence of local spurious trajectories is resulted from the relatively small numbers of factors and neurons. To demonstrate the influence of unlearning procedure on factor searches in the case when local spurious trajectories dominate we increase the number of neurons to $N = 3000$ and the number of factors to $L = 4200$. In this case the probability of spurious trajectories amounts to 0.9. Results are shown in Fig. 7. Diamonds and circles are values of the Lyapunov function at true and spurious trajectories, respectively, at the point

$k = n$, normalized by its theoretical value given by (11). Initially the probability of true trajectories in unlearning case is the same as that in case without unlearning and amounts to 0.1. However soon the rate of the search of new factors speeds up. This process is accompanied by the decrease of Lyapunov functions for both true and spurious attractors. As noted above, for true attractors it decreases because of the elimination of factors with high Lyapunov function values. For spurious attractors it decreases because the elimination of factors is equivalent to the network unloading (i.e. the reduction of the number of factors in the network memory), and as shown in [10], the Lyapunov function values for spurious attractors monotonically decreases when the number of learned factors reduces. Since Lyapunov function values decrease faster for spurious attractors, the rate of true attractors increases and at the final stage of the recall process the probability of spurious attractors falls to zero and almost each trial results in the retrieval of a new factor. About 16000 trials are required to reveal all 4200 factors while without unlearning only about 600 factors could be found at all.

5 Discussion

Theoretical analysis and computer simulations revealed that Hopfield-like neural networks with proposed learning and unlearning rules are capable of performing Boolean factor analysis of the signals of a high complexity ($C \gtrsim 10$). This capability is provided by two special unlearning procedures. The first one suppresses two global spurious attractors that become to dominate in the synchronous Hopfield-like network of the common architecture when signal complexity C increases. For example, for $p = 0.02$ the critical complexity of their dominance amounts to $C \simeq 10$. The second unlearning procedure eliminates the dominance of factors with the largest Lyapunov functions (and hence the largest attraction basins). These factors attract neural network activity during the random search and prevent the search of other factors. The second unlearning procedure excludes the found factor from the network memory.

In the whole, our results proved that for a sufficiently large range of parameters our neural network approach can be used efficiently as the method for solution of general statistical Boolean factor analysis task. Our approach is in the main stream of contemporary research of a factor analysis application to data dimension reduction as well as to research of factor analysis methods as such. First, since in our method factor scores are binary (have values 0 or 1), this attempt could be related to the methods of Nonnegative Matrix Factorization (NMF) [26]. Actually there were attempts to apply NMF to Boolean factor analysis [21], however, not very successful. Second, since a signal complexity C is assumed to be much smaller than the total number of factors then the most factor scores values are “Zero” and thus our method could be related to methods of sparse component analysis [11, 17, 18]. Last, since the appearance of different factors in signals is assumed to be statistically independent our method is also related to the methods of independent component analysis [19, 23] and especially to its nonlinear version [14]. Some of these methods utilize neural networks [15, 21] to perform iterative algorithms; however, the learning rules and network dynamics employed are mostly artificially made up to provide the required computations. A major distinction of the current work from the related ones is that natural properties of unmodified

Hebbian learning and dynamics of attractors neural network are implemented in a simple paradigm which has clear neurobiological interpretation.

There are many examples of data in the sciences when the Boolean factor analysis is required [16]. However, Boolean factor analysis with the use of neural network approach seems especially efficient for processing textual data. Since the time of the factor search only slightly increases with the increase of signal space dimensionality, it is most attractive to apply this approach to patterns of very large dimensionality. Texts are good examples of such a kind of signals. The dimensionality of signal space is equal to the number of words in the used dictionary and thus is usually very high. Another example is image processing where the signal dimensionality is also usually high. In image processing our method could be especially efficient because it allows to overcome the problem of objects occlusion that is hardly solved by other methods [21].

In this paper we concentrated on the analysis of origin and method of elimination of two global spurious attractors that appear in the common architecture of the Hopfield-like neural network when applied on Boolean factor analysis. Since these attractors become dominant only when signal complexity C is rather high they have never been observed before. Thus they pose important and completely new phenomenon of attractor neural networks and therefore merit special consideration carried out in the present paper.

The procedure of Hebbian unlearning similar to our second unlearning procedure was previously used only for suppression of *spurious* attractors in the common Hopfield network [6]. It could be related to the fact that previously the problem of the random search of *true* attractors in Hopfield network was not considered at all.

Appendix 1. Estimation of $D\{q^m\}$

By definition, $q^m = (1/N)(\sum_{i=1,N} X_i^m)$. Then

$$D\{q^m\} = \frac{1}{N} D\{X_i^m\} + \frac{N-1}{N} Cov\{X_i^m, X_j^m\}$$

where $D\{X_i^m\} = q(1-q)$. Covariation between X_i^m and X_j^m can be presented in the form $Cov\{X_i^m, X_j^m\} = \langle (1-X_i^m)(1-X_j^m) \rangle - (1-q)^2$. Since factors are statistically independent

$$\langle (1-X_i^m)(1-X_j^m) \rangle = \langle \prod_{l \in \{l: \beta_l^m = 1\}} (1-f_i^l)(1-f_j^l) \rangle = [(1-p)(1-p^*)]^C$$

where $1-p^* = ((1-p)N-1)/(N-1)$ is a probability that the j-th neuron is not active in a given factor under the condition that the i-th neuron is not active in this factor. Thus

$$Cov\{X_i^m, X_j^m\} = (1-q)^2 [(1 - \frac{p}{(1-p)(N-1)})^C - 1] \simeq -\frac{(1-q)^2 pC}{(1-p)N},$$

that means, that $D\{q^m\}$ is of order $1/N$.

References

[1] Barlow, H.: Single units and sesation: a neuron doctrine for perceptual psychology? Perception 1, 371–394 (1972)

[2] Barlow, H.B.: Possible principles underlying the transformations of sensory messages. In: Rosenblith, W.A. (ed.) Sensory communication, pp. 217–234. MIT Press, Cambridge (1961)

[3] Barlow, H.B.: Cerebral cortex as model builder. In: Rose, D., Dodson, V.G. (eds.) Models of the visual cortex, pp. 37–46. Wiley, Chichester (1985)

[4] Belohlavek, R., Vychodil, V.: Formal concepts as optimal factors in Boolean factor analysis: implications and experiments? In: Fifth International Conference on Concept Lattices and Their Applications (2007)

[5] Bucingham, J., Willshaw, D.: On setting unit thresholds in an incompletely connected associative net. Network 4, 441–459 (1993)

[6] Crick, F., Mitchison, G.: The function of dream sleep. Nature 304(5922), 111–114 (1983)

[7] Foldiak, P.: Forming sparse representations by local anti-hebbian learning. Biological Cybernetics 64, 165170 (1990)

[8] Frolov, A.A., Husek, D., Muraviev, I.P.: Informational capacity and recall quality in sparsely encoded hopfield-like neural network: Analytical approaches and computer simulation. Neural Networks 10, 845–855 (1997)

[9] Frolov, A.A., Husek, D., Muraviev, I.P.: Informational efficiency of sparsely encoded hopfield-like autoassociative memory. Optical Memory Neural Networks 12(3), 177–197 (2003)

[10] Frolov, A.A., Husek, D., Muraviev, I.P., Polyakov, P.Y.: Boolean factor analysis by attractor neural network. IEEE Transactions on Neural Networks 18(3), 698–707 (2007)

[11] Georgiev, P., Theis, F., Cichocki, A.: Sparse component analysis blind sourse separation of underdetermined mixters. IEEE Transactions on Neural Networks 16(4), 992–996 (2005)

[12] Goles-Chacc, E., Fogelman-Soulie, F., Pellegrin, D.: Decreasing energy functions as a tool for studying threshold networks. Discrete Mathematics 12, 261–277 (1985)

[13] Jankovic, M.V.: Modulated hebb-oja learning rule - a method for principal subspace analysis. IEEE Transactions on Neural Networks 17(2), 345–356 (2006)

[14] Karhunen, J.: Nonlinear independent component analysis. In: Roberts, S., Everson, R. (eds.) Independent Component Analysis: Principles and Practice, pp. 113–134. Cambridge University Press, Cambridge (2001)

[15] Karhunen, J., Joutsensalo, J.: Representation and separation of signals using nonlinear PCA type learning. Neural Networks 7, 113–127 (1994)

[16] Leeuw, J.D.: Principal component analysis of binary data application to rollcall analysis (2003), http://gifi.stat.ucla.edu

[17] Li, Y., Amari, S., Cichocki, A., Ho, D.C., Xie, S.: Underdetermined blind source separation based on sparse representation. IEEE Trans. Signal Process. 54(2), 423–437 (2006)

[18] Li, Y., Cichocki, A., Amari, S.: Blind estimation of chanal parameters and source components for EEG signals: A sparse factorization approach. IEEE Transactions on Neural Networks 17(2), 419–431 (2006)

[19] Liu, W., Zheng, N.: Non-negative matrix factorization based methods for object recognition. Pattern Recognition Letters 25(8), 893–897 (2004)

[20] Moller, R., Konig, A.: Coupled principal component analysis. IEEE Transactions on Neural Networks 15(1), 214–222 (2006)

[21] Spratling, M.W.: Learning image components for object recognition. Journal of Machine-Learning Reasearch 7, 793–815 (2006)

[22] Thurstone, L.L.: Multiple factor analysis. Psychological Review 38, 406–427 (1931)

[23] Tichavsky, P., Koldovsky, Z., Oja, E.: Performance analysis of the FastICA algorithm and Crame/spl acute/r-rao bounds for linear independent component analysis. IEEE Transactions on Signal Processing, [see also IEEE Transactions on Acoustics, Speech, and Signal Processing] 54(4), 1189–1203 (2006)

[24] Watanabe, S.: Pattern recognition: human and mechanical. Wiley, New York (1985)

[25] Yi, Z., Ye, M., Lv, J.C., Tan, K.K.: Convergence analysis of deterministic discrete time system of oja's PCA learning algorithm. IEEE Transactions on Neural Networks 16(6), 1318–1328 (2005)

[26] Zafeiriou, S., Tefas, A., Bucie, I., Pitas, I.: Exploiting discriminant information in nonnegative matrix factorization with application to frontal face verification. IEEE Transactions on Neural Networks 17(3), 683–695 (2006)

Author Index